CHEMISTRY

IN THE COMMUNITY CHEMCOM

FIFTH EDITION

A PROJECT OF THE AMERICAN CHEMICAL SOCIETY

W. H. FREEMAN AND COMPANY

ChemCom
Fifth Edition Credits

American Chemical Society

Chief Editor: Henry Heikkinen

Project Manager: Angela Powers

Revision Team: Laurie Langdon, Robert Milne, Wendy Naughton

ACS: Sylvia Ware, Michael Tinnesand, Terri Taylor, Helen Herlocker, Jodi Wesemann

ACS Project Editor: Rebecca Strehlow

Technical Reviewer: Conrad Stanitski

Technical Contributers: Janet Cohen, Susan Heikkinen, Jack A. Ladson, COLOR Science Consultancy

Fourth-edition Reviewers: Kirsten Almo, Mark Beehler, Susan Berrend, Martin Besant, Regis Goode, Gary Jackson, Jane Meadows, Walt Shacklett, Terri Taylor

Teacher Contributors: Pat Chriswell, Robert Dayton, Regis Goode, Drew Lanthrum, Joelle Lastica, Steve Long, Cece Schwennsen, Barbara Sitzman, Terri Taylor

W. H. Freeman

Publisher: Susan Finnemore Brennan

Acquisitions Editor/Marketing Manager: Cindi WeissGoldner

Director of Sales and Marketing: Mike Saltzman

Developmental Editor: Rebecca Strehlow

New Media/Supplements Editors: Victoria Anderson/ Amy Shaffer

Photo Editor: Ted Szczepanski

Photo Researchers: Dena Digilio Betz, Julie Tesser

Cover Designer: Blake Logan

Text Design and Layout: Rae Grant Design

Project Editors: Katie Ostler/Denise Kadlubowski, Schawk, Inc.

Illustrations: Network Graphics/Schawk, Inc.

Composition: Schawk, Inc.

Printing and Binding: R.R. Donnelley/Willard

This material is based upon work supported by the National Science Foundation under Grant No. SED-88115424 and Grant No. MDR-8470104. Any opinions, findings, and conclusions or recommendations expressed in this publication are those of the authors and do not necessarily reflect the views of the National Science Foundation. Any mention of trade names does not imply endorsement by the National Science Foundation.

CORE Edition ISBN-13: 978-0-7167-8505-7
(ISBN-10: 0-7167-8505-6)
Student Edition ISBN-13: 978-0-7167-8919-2
(ISBN: 10: 0-7167-8919-1)

CORE Edition Library of Congress Control Number: 2005929254
Student Edition Library of Congress Control Number: 2005938770

Printed in the United States of America
Third printing

ABOUT THE COVER

This is a schematic diagram of a sodium chloride crystal dissolving in water as rendered by Matt Collins of *Scientific American* magazine. For more context, please see Figure 1.39, page 61.

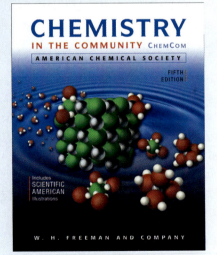

NOTE TO STUDENTS

You are likely taking a chemistry course for the first time. Of course, even if you haven't realized it, you have been immersed in chemistry all of your life. Chemical reactions are involved in foods, fuels, synthetic fabrics, and medicines. Chemistry, which is the study of matter and its changes, is responsible for materials that make up your home, transportation vehicles, what you wear—and, in fact, chemistry is also responsible for sustaining you. As someone wryly observed, you simply can't ignore chemistry, because chemistry won't ignore you.

As society increasingly depends on scientific advances, citizens will be expected to understand scientific phenomena and principles involved in making public-policy decisions. This chemistry course is designed to provide you opportunities to develop and use important chemistry concepts and skills, so you can understand the chemistry behind some issues and problems that may arise within your community.

Each unit in this course introduces a chemistry-related concern related to your life or to your community. You will complete laboratory investigations and other exercises that encourage you to apply your chemistry knowledge and skills to a particular issue or problem. You will seek solutions and weigh consequences of decisions that you and your classmates propose.

We begin with a newspaper article about a water-related emergency in Riverwood, an issue that serves as the theme for the first unit. Can chemistry help solve the problem? Welcome to the study of chemistry and to Riverwood!

IMPORTANT NOTICE

ChemCom is intended for use by high school students in the classroom laboratory under the direct supervision of a qualified chemistry teacher. The investigations described in this book involve substances that may be harmful if they are misused or if the procedures described are not followed. Read cautions carefully, and follow all directions. Do not use or combine any substances or materials not specifically called for in carrying out investigations. Other substances are mentioned for educational purposes only and should not be used by students unless the instructions specifically so indicate.

The materials, safety information, and procedures contained in this book are believed to be reliable. This information and these procedures should serve only as a starting point for good laboratory practices, and they do not purport to specify minimal legal standards or to represent the policy of the American Chemical Society. No warranty, guarantee, or representation is made by the American Chemical Society as to the accuracy or specificity of the information contained herein, and the American Chemical Society assumes no responsibility in connection therewith. The added safety information is intended to provide basic guidelines for safe practices. It cannot be assumed that all necessary warnings and precautionary measures are contained in the document or that other additional information and measures may not be required.

SAFETY AND LABORATORY ACTIVITY

In *ChemCom* you will frequently complete laboratory investigations. While no human activity is completely risk free, if you use common sense, as well as chemical sense, and follow the rules of laboratory safety, you should encounter no problems. Chemical sense is just an extension of common sense. Sensible laboratory conduct won't happen by memorizing a list of rules, any more than a perfect score on a written driver's test ensures an excellent driving record. The true "driver's test" of chemical sense is your actual conduct in the laboratory.

The following safety pointers apply to all laboratory activity. For your personal safety and that of your classmates, make adherence to these guidelines second nature in the laboratory. Your teacher will point out any special safety guidelines that apply to each investigation. Two safety icons appear in your textbook. They appear at the beginning of the laboratory procedure, but apply to the entire investigation.

 When you see the goggle icon you should put on your protective goggles and continue to wear them until you are completely finished in the laboratory.

The caution icon means there are substances or procedures requiring special care. See your teacher for specific information on these cautions.

If you understand the reasons behind them, these safety rules will be easy to remember and to follow. So, for each listed safety guideline:
- Identify a similar rule or precaution that applies in everyday life—for example in cooking, repairing or driving a car, or playing a sport.
- Briefly describe possible harmful consequences if the rule is not followed.

RULES OF LABORATORY CONDUCT

1. Do laboratory work only when your teacher is present. Unauthorized or unsupervised laboratory experimenting is not allowed.

2. Your concern for safety should begin even before the first laboratory investigation. Before starting any laboratory work, always read and think about the details of your laboratory assignment.

3. Know the location and use of all safety equipment in your laboratory. These should include the safety shower, eye wash, first-aid kit, fire extinguisher, fire blanket, exits, emergency warning system, and evacuation routes.

4. Wear a laboratory coat or apron and impact/splash-proof goggles for all laboratory work. Wear closed shoes (rather than sandals or open-toed shoes), preferably constructed of leather or similar water-impervious material, and tie back loose hair. Shorts or short skirts must not be worn.

5. Clear your bench top of all unnecessary material, such as books and clothing, before starting your work.

6. Check chemistry labels twice to ensure that you have the correct substance and the correct solution concentration. Some chemical formulas and names may differ by only a letter or a number.

7. You may be asked to transfer some chemical substances from a supply bottle or jar to your own container. Do not return any excess material to its original container unless authorized by your teacher, as you may contaminate the supply bottle.

8. Avoid unnecessary movement and talk in the laboratory.

9. Never taste any laboratory materials. Do not bring gum, food, or drinks into the laboratory. Do not put fingers, pens, or pencils in your mouth while in the laboratory.

10. If you are instructed to smell something, do so by fanning some of the vapor toward your nose. Do not place your nose near the opening of the container. Your teacher will show you the correct technique.

11. Never look directly down into a test tube; view the contents from the side. Never point the open end of a test tube toward yourself or your neighbor. Never directly heat a test tube in a Bunsen burner flame.

12. Any laboratory accident, however small, should be reported immediately to your teacher.

13. In case of a chemical spill on your skin or clothing, rinse the affected area with plenty of water. If your eyes are affected, rinsing with water must begin immediately and continue for at least 10 to 15 minutes. Professional assistance must be obtained.

14. Minor skin burns should be placed under cold, running water.

15. When discarding or disposing of used materials, carefully follow all provided instructions. Waste chemical substances usually are not permitted in the sewer system.

16. Return equipment, supplies, aprons, and protective goggles to their designated locations.

17. Before leaving the laboratory, make sure that gas lines and water faucets are shut off.

18. Wash your hands before leaving the laboratory.

19. If you are unclear or confused about proper safety procedures, ask your teacher for clarification. If in doubt, ask!

CONTENTS

BRIEF TABLE OF CONTENTS

UNIT 1 WATER: EXPLORING SOLUTIONS 2

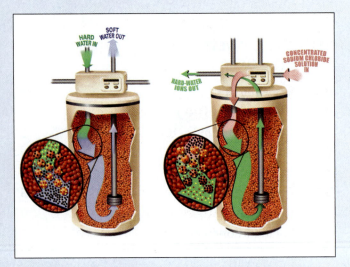

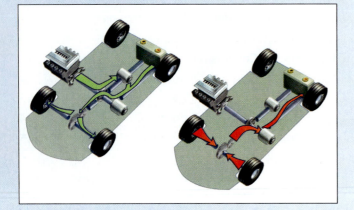

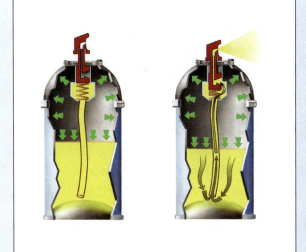

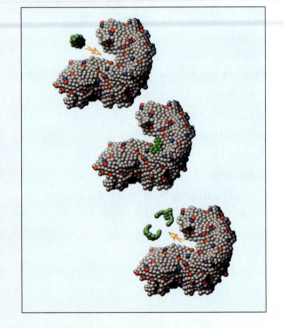

FEATURES

ChemQuandary

Modeling Matter

Particulate-level model of sodium chloride (NaCl) dissolving in water (H₂O). A "snapshot" of the dissolving process, as sodium and chloride ions that make up a salt crystal are surrounded and carried away by water molecules. Emphasizes attractions among electrically-charged ions and polar molecules. **61**

Working knowledge of a home water softener. Illustrates (a) softening hard water for use in the home and (b) recharging the "used" resin so that the process can be repeated. Macroscopic depictions of water flowing through resin-filled tanks are complemented by particulate-level representations of hard and soft ions involved in ion exchange on the resin. **99**

Model of a lithium (Li) atom. A cross-section of an atom illustrating the central nucleus (composed of protons and neutrons) surrounded by electron clouds. Emphasizes differences in volume and mass between an atom's nucleus and its electrons. **121**

Working knowledge of color-change paint. Illustrates a phenomenon whereby objects appear to change colors when viewed at different angles. Depicts light striking the surface of a mica-based pigment at two different viewing angles—one of which makes the paint appear blue, the other green. **201**

Model of a hydrogen (H₂) molecule. A cross-section of a molecule illustrating the bond that forms when two hydrogen atoms share two electrons. Emphasizes attractions between the shared electrons and the protons that make up each atom's nucleus. **223**

Working knowledge of a hybrid gasoline-electric car. Illustrates (a) how both the gasoline engine and electric storage battery contribute to powering the vehicle as it accelerates and (b) how some of that energy is captured and stored in the electric battery when the vehicle's brakes are applied. **292**

Particulate-level model of gases. A macroscopic view of a balloon flying high in the sky combined with particulate-level representations of helium atoms inside the balloon and nitrogen and oxygen molecules outside. Emphasizes the movement and spacing of gas particles. **314**

Working knowledge of an aerosol spray can. A cross-section of an aerosol can before and after the nozzle is pushed down. Shows propellant gases exerting high pressure on the walls of the can and on the liquid product inside the can. Pushing the nozzle opens a passage to the outside, allowing the gas pressure to propel the liquid up the nozzle and out of the can. **397**

Dynamic equilibrium. Depicts water levels in two transparent flasks—one stoppered and one unstoppered. Contrasts relative rates of evaporation and condensation in each flask, as depicted by arrows and particulate-level models of water molecules involved in each process. **435**

Rechargeable batteries. Cross sections of a nickel-metal hydride (NiMH) battery showing direction of electron flow as the battery is charged (in a recharger) and discharged (in an operating flashlight). **459**

Working knowledge of positron emission tomography (PET). Illustrates detection of gamma rays as a PET scan detects a brain tumor in a patient. The illustration zooms in to reveal the cancerous tissue, located near a blood vessel. This is then associated with a particulate-level representation of the production of gamma rays as positrons (emitted from radioisotope tracers within tagged sugar molecules) collide with electrons in cancerous cells. **530**

Nuclear fission and fusion. Particulate-level representations depict (a) smaller nuclei colliding to form a larger nucleus during fusion and (b) fission of a large nucleus into two smaller nuclei due to its collision with a neutron. Highlights energy released as mass is converted into energy during fusion and fission processes. **552**

Enzyme action. A sequence of illustrations that depicts (a) a substrate molecule approaching the active site of an enzyme molecule, (b) the substrate interacting with the enzyme, and (c) two product molecules leaving the enzyme. Emphasizes important enzyme features, including size, shape, specificity, and catalytic behavior. **613**

Working knowledge of emulsifier action. A sequence of illustrations that depicts (a) separate oil and water layers in a glass bowl, (b) addition of an egg yolk, and (c) the mayonnaise that is produced by whisking. Particulate-level representations show how lecithin molecules (contained in the egg yolk), with their nonpolar and polar regions, surround the nonpolar oil molecules and cause them to disperse among polar water molecules—an oil-in-water emulsion is created. **632**

American Chemical Society

Education Division
Sylvia A. Ware, Director

1155 SIXTEENTH STREET, NW
WASHINGTON, DC 20036-4800
Phone 800.227.5558

Dear Reader:

Welcome to *Chemistry in the Community*!

Since the early 1980s, when *Chemistry in the Community (ChemCom)* was first designed and released, *ChemCom* has been used successfully by over an estimated two million students and teachers in many different high school settings. Developed by the American Chemical Society (ACS) with initial funding from the National Science Foundation and several ACS sources, revision of this Fifth Edition has been guided by classroom experiences and advice of *ChemCom* teachers, advances in chemistry, new insights in chemistry teaching and learning, and—most significantly—by chemistry learning expectations highlighted by state and national science education standards.

The five goals of *ChemCom*, as they have been since its creation, are to help students

- develop an understanding of chemistry;
- cultivate problem-solving and critical-thinking skills related to chemistry;
- apply chemistry knowledge to decision-making about scientific and technological issues;
- recognize the importance of chemistry in daily life;
- understand benefits as well as limitations of science and technology.

As in previous editions of this textbook, chemistry concepts are developed on a "need-to-know" basis. That is, each unit of study starts with a significant chemistry-related concern or issue that serves as a motivational framework upon which appropriate chemistry is introduced and developed. Woven into this framework are chemical principles and skills relevant to that particular issue. Each unit is focused upon a community setting—a school, town, region, nation—and also focuses on developing specific chemistry concepts and skills. Each unit ends with a consolidating activity, where the theme is revisited and addressed, applying the chemistry introduced and learned within that unit. Student laboratory investigations, as well as decision-making opportunities and skill-building exercises, help students develop an understanding of chemistry. Although presented within a "need-to-know" context, the content encompasses all key chemical concepts and skills typically found in a first-year high school course.

The following pages will introduce users to various features of this textbook. Great care has been taken to develop the fundamental concepts of chemistry so that student interest, involvement, and learning are encouraged, supported, and cultivated.

Sincerely,

Sylvia A. Ware

Sylvia A. Ware, Director
Education Division
American Chemical Society

PREFACE

Chemistry in the Community (ChemCom) is a high-school general chemistry text. Its seven units cover a full scope of science content and process standards, developing major concepts with laboratory investigations, modeling activities, practice problems, and critical thinking exercises.

ChemCom presents chemical principles on a need-to-know basis. Each unit opens with a real-world community issue, then introduces and develops chemical principles within this context. Each unit concludes with a consolidating activity requiring the application of all the tools and techniques learned throughout the unit.

	UNIT TITLE	REPRESENTATIVE CONCEPTS	SCENARIO
1	Water: Exploring Solutions	Physical and chemical properties, formulas and equations, ions, solutions	Determining the cause of a fish kill
2	Materials: Structure and Uses	Properties, periodicity, atomic structure, the mole	Designing a new coin
3	Petroleum: Breaking and Making Bonds	Bonding, nomenclature, organic chemistry	Marketing an alternative-energy vehicle
4	Air: Chemistry and the Atmosphere	Gas laws, acid–base chemistry, kinetic molecular theory, green chemistry	Evaluating a proposed school bus-idling policy
5	Industry: Applying Chemical Reactions	Oxidation–reduction, kinetics, equilibrium, industrial chemistry, environmental chemistry	Choosing whether to invite an aluminum processing plant or a nitrogen fertilizer plant to the community
6	Atoms: Nuclear Interactions	Atomic structure, nuclear radiation, nuclear reactions, nuclear energy	Informing the public about risks and benefits of nuclear technology
7	Food: Matter and Energy for Life	Biochemistry, energy relationships, organic chemistry	Recommending a new vending-machine policy for the school

UNIT OPENERS

Opener images and questions stimulate thinking. How much do you already know? What are you about to discover?

INVESTIGATING MATTER

Each unit contains laboratory investigations to immerse you in the real world of chemistry. As you learn the processes of science you employ chemical principles, develop analytical skills, and use data and observations to draw conclusions.

Each investigation usually consists of an Introduction, a detailed or student-developed procedure, and Questions. Investigating Matter sections may also include Calculations, Data Analysis, and Post-Lab Activities.

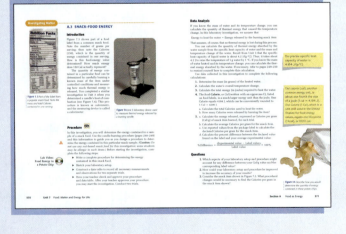

NEW TO THIS EDITION

Scientific American illustrators developed selected artwork for this edition. There are two illustrations for each unit, one to deepen your conceptual understanding and one to share "working knowledge" of a device related to unit concepts. Extended captions complement these teaching visuals.

New photos abound in this edition, and line art has been refined for extra clarity. The text is completely re-designed to be inviting and easy to navigate.

MODELING CHEMISTRY

Modeling Matter activities make abstract chemical ideas easier to grasp. The activities require critiquing and creating visual representations of chemical activity, formulating and revising scientific explanations, developing manipulable models, and using logic and evidence to make decisions.

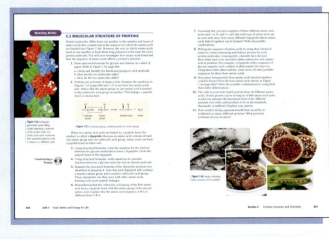

Thousands of people use chemistry every day. **Chemistry at Work** essays introduce you to some of them and show how they apply chemical processes and principles in their careers.

REAL-LIFE SKILLS AND DECISIONS

Making Decisions features give you experience with real-life decision-making strategies—many related to the unit's organizing theme. Each Making Decisions activity asks you to collect and/or analyze data for underlying patterns, and to develop and evaluate answers based on scientific evidence and potential consequences.

ChemQuandary activities stimulate you with puzzling questions.

Developing Skills features focus on problem-solving skills with real-world chemistry situations and practice problems.

Each lettered section in a unit concludes with a **Section Summary** organized around section learning goals. *ChemCom* poses three categories of questions. *Reviewing the Concepts* allows for practice with basic ideas. *Connecting the Concepts* asks you to identify and develop relationships among ideas. *Extending the Concepts* challenges you to seek additional information and apply chemical principles to new situations.

CHEMISTRY IN THE COMMUNITY (*CHEMCOM*) FIFTH EDITION–TEACHER AND STUDENT MEDIA, SUPPLEMENTS, AND SUPPORT

TEACHER RESOURCES

Wraparound Teacher's Edition—this new teacher's edition provides reduced pages from the student text with the teacher's material "wrapped around" the outer edges of these pages. Teacher annotations include answers to problems, suggested activities, and alternative teaching ideas and suggestions, as well as materials and procedures for setting up the laboratory investigations. The teacher's edition offers point-of-use support for both new and experienced teachers, while providing instructional strategies for students with different ability levels and learning styles. In addition, you will find background material, supplemental activities, objectives, and pacing guides for each unit, along with details of how *ChemCom* addresses the National Science Education Standards.

Printed Testbank—the printed testbank contains hundreds of multiple choice and short-answer questions for each unit.

Computerized Testbank—this electronic version of the testbank provides both Windows and Mac formats (on one CD), and is designed to allow teachers to add, edit, and resequence all test items.

Overhead Transparencies—the transparency package is a valuable teaching tool for introducing, clarifying, or reviewing key topics and augmenting instruction with visual learning approaches. Over 100 color transparencies provide teachers with useful graphic organizers of content and figures from the text to prompt class inquiry.

Teacher's Resource Binder—this new ancillary resource includes preliminary skill building activities, classroom handouts, practice problems, lab procedures, and lab worksheets along with reading guides, variable pacing guides, and more! All resources in this binder may be copied for class distribution and multiple uses.

Teacher's Resource CD-ROM—this is designed to help teachers create classroom presentations, enhance instructional time, build websites, and develop other resources, this CD-ROM allows users to search and export all resources listed below by key word:

- All textbook images in JPEG format

- Animations, videos, tutorials, and other interactivities from the student CD-ROM/website

- The entire contents of the Teacher's Resource Binder and other instructor materials in an electronic format.

Professional Development Workshops and Training—ACS-conducted workshops are available throughout the year in various locations. In each workshop, experienced *ChemCom* Teacher Leaders guide participants through the *ChemCom* textbook and instructional philosophy and provide hands-on experience with many laboratory investigations, modeling exercises, and culminating activities. Workshops also introduce participants to many ancillaries new to the fifth edition of *ChemCom*.

STUDENT RESOURCES

CD-ROM—this CD offers a variety of interactive student features, plus a dynamic periodic table, calculator, audio-vocabulary tool, and other useful electronic chemistry learning tools. The interface allows students to search for particular topics, key words, and terms. An auditory glossary facilitates pronunciation of these words.

Website—the *ChemCom* companion website includes all the features and tools offered on the Student CD and more! Access the site at:
www.whfreeman.com/chemcom5e

UNIT 1

Water: Exploring Solutions

WHAT techniques can we use to purify water?

WHAT are the physical properties of water?

WHY do some substances readily dissolve in water and others do not?

HOW does chemistry contribute to effective water treatment?

Fish are dying in Riverwood's Snake River. Why? What are the consequences for the community? Turn the page to find out more about this crisis and the role of water in modern life.

Fish Kill Triggers Riverwood Water Emergency

Severe Water Rationing in Effect

Soon after discovering the fish kill, Riverwood High School students returned to the river to investigate.

BY LORI KATZ
Riverwood News Staff Reporter

Citing possible health hazards, Mayor Edward Cisko announced today that Riverwood will stop withdrawing water from the Snake River and will temporarily shut down the city's water-treatment plant over water-quality concerns provoked by a massive fish kill. Starting at 6 p.m., river water will not be pumped to the plant for at least three days. If the cause of the fish kill has not been determined and corrected by that time, the shutdown will continue indefinitely.

During the plant shutdown, water engineers and chemists from the county sanitation commission and the U.S. Environmental Protection Agency (EPA) will investigate the cause of the major fish kill discovered yesterday. The fish kill extended from the base of Snake River Dam, located upstream from

Riverwood, to the town's water-pumping station.

The initial alarm was sounded when Jane Abelson, 15, and Chad Wong, 16—both students at Riverwood High School—found many dead fish floating in a favorite fishing spot. "We thought maybe someone had poured poison into the reservoir," explained Wong.

Mary Steiner, a Riverwood High School biology teacher, accompanied the students back to the river. "We hiked downstream for almost a mile. Dead fish of all kinds were washed up on the banks and caught in the rocks," Abelson reported.

Ms. Steiner contacted county sanitation commission officers, who immediately collected Snake River water samples for analysis. Chief engineer Hal Cooper reported at last night's emergency meeting that the water samples appeared clear, colorless, and odorless. However, he indi-

cated some concern. "We can't say for certain that the water supply is safe until the cause of the fish kill is determined. It's far better that we take no chances until then," Cooper advised.

Mayor Cisko canceled the community's "Fall Fish-In," which was scheduled to start September 15. No plans to reschedule Riverwood's annual fishing tournament were announced. "The decision was made at last night's emergency town council meeting to start investigating the situation immediately," he said.

After five hours of often-heated debate yesterday, the Riverwood town council finally reached agreement to stop drawing water from the Snake River. Council member Henry McLatchen (also a chamber of commerce member) commented that the decision was highly emotional and unnecessary. He cited financial losses

Dead fish washed up along the banks of the Snake River yesterday afternoon.

see *Fish Kill*, page 5

Fish Kill, from page 1

that motels and restaurants will suffer because of the Fish-In cancellation, as well as potential loss of future tourism dollars due to adverse publicity. However, McLatchen and other council members sharing that view were outvoted by those holding the position that the fish kill, the only one within Riverwood's recorded history, may indicate a public health emergency.

Mayor Cisko assured residents that essential municipal services will not be affected by the crisis. For example, he promised to maintain fire department access to adequate supplies of water to meet firefighting needs.

Arrangements have been made to transport emergency drinking water from Mapleton. The first water shipments by truck are due to arrive in Riverwood by midmorning tomorrow. Distribution points are listed in Section 2 of today's *Riverwood News,* along with guidelines on conserving water during this emergency.

All Riverwood schools will be closed Monday and possibly through Wednesday. No other closings or cancellations have been announced. Local TV and radio will report any schedule changes as they become available.

A public meeting tonight at 8 p.m. at the town hall features Dr. Margaret Brooke, a water expert at State University. She will answer questions concerning water safety and use. Brooke was invited by the county sanitation commission to help clarify the situation for concerned citizens.

Asked how long the water emergency would last, Brooke refused to speculate, saying that she first needed to talk to other scientists conducting the investigation. EPA investigators, in addition to collecting and analyzing water samples, will examine dead fish in an effort to determine what was responsible for the fish kill. Brooke reported that trends or irregularities in water-quality data from Snake River monitoring during the past two years also will play a part in the investigation.

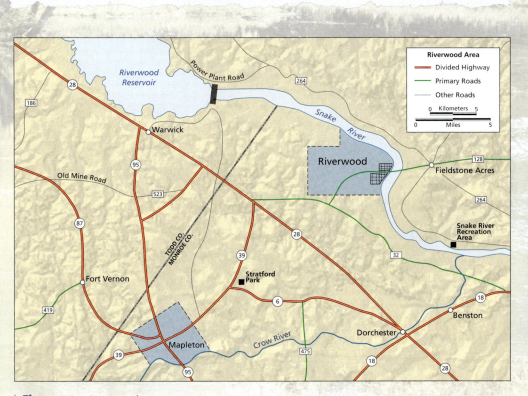

| Figure 1.1 *Riverwood*

Townspeople React to Fish Kill and Riverwood Water Crisis

BY JUAN HERNANDEZ
Riverwood News Staff Reporter

In a series of on-the-street interviews, Riverwood citizens expressed a variety of opinions earlier today about the crisis. "It doesn't bother me," said nine-year-old Jimmy Hendricks. "I'm just going to drink bottled water and canned fruit juice."

"I knew that eventually they'd pollute the river and kill the fish," complained Harmon Lewis, a lifelong resident of Fieldstone Acres, located east of Riverwood. Lewis, who traces his ancestry to original county settlers, still gets his water from a well and will be unaffected by the water crisis. He said that he plans to pump enough well water to supply the children's ward at Community Hospital if the emergency extends more than a few days.

Bob and Ruth Hardy, co-owners of Hardy's Ice Cream Store, expressed annoyance at the inconvenience but were reas-

David Price was excited that school is canceled.

sured by council actions. They were eager to learn the reason for the fish kill and its possible effects on water supplies. The Hardy's daughter Toni, who loves to fish, was worried that late-season fishing would be ruined. Toni and her father won first prize in last year's angling competition.

Riverwood Motel owner Don Harris expressed concern for both the health of town residents and the loss of business due to the tournament cancellation. "I always earn reasonable income from this event and, without the revenue from the Fish-In, I may need a loan to pay bills next spring."

The unexpected school vacation was "great," according to twelve-year-old David Price. Asked why he thought schools would be closed Monday, Price said that all he could think of was that "the drinking fountains won't work."

Elmo Turner, whose residential landscaping has won Garden Club recognition for the past five years, felt reassured on one point. Because of an unusually wet summer, grass watering is unnecessary; lawns are not in danger of drying out due to current water rationing.

The fish kill sparked diverse reactions among Riverwood residents. Don Harris voiced concerns about health and economic impact of the fish kill.

Sources and Uses of Water

Riverwood confronts at least a three-day water shortage. The water emergency has aroused understandable concern among Riverwood citizens, town officials, and business owners. What caused the fish kill? Does the fish kill mean that Riverwood's water supply poses hazards to humans? In the following pages, you will monitor the town's progress in answering these questions as you learn more about water's properties.

A.1 TOWN IN CRISIS

Although Riverwood is imaginary, its problems are not. Residents of many communities have faced these and similar problems. In fact, two water-related challenges confront each of us every day. Can we get enough water to supply our needs? Can we get sufficiently pure water? These two questions serve as major themes of this unit, and their answers require an understanding of water's chemistry and uses.

The notion of water *purity* must be given careful consideration. See Figure 1.2. You will soon learn that the cost of producing a supply of water that is *100%* pure is prohibitively high. Is that level of purity needed—or even desirable? Communities and regulatory agencies are responsible for ensuring the availability of water of sufficiently high quality for its intended uses at reasonable cost. How do they accomplish this task?

Even the apparently simple idea of "water use" presents some fascinating puzzles, as the following *ChemQuandary* illustrates.

Figure 1.2 *This water is obviously clear, but is it pure? What is pure water?*

ChemQuandary 1

WATER, WATER EVERYWHERE

It takes approximately 120 L of water to produce one 1.3-L can of fruit juice. It takes about 450 L of water to place one fried egg on your breakfast plate. Think of possible explanations for these two facts by listing the steps involved in producing and delivering to your home the fruit juice and the egg. Then review each step and consider where water use would occur.

As you just learned, a basic challenge that people have always faced is related to accessing water supplies of sufficiently high quality. In the following activities, you will explore these issues by first evaluating your family's typical water uses and, later, by trying to purify as much of a contaminated-water (foul-water) sample as you can. The challenges for people always remain the same—to obtain purified water at low total cost and to ensure adequate water supplies to meet all necessary uses.

Making Decisions

A.2 USES OF WATER

Keep a diary of water use in your home for three days. On a data table similar to the one shown on the next page, record how often various water-use activities occur. Ask each household member to cooperate and help you.

Check the activities listed on the chart. See Figure 1.3. If family members use water in other ways within the three-day period, add those uses to your diary. Estimate the quantities of water used by each activity.

DATA TABLE

Per Household	Day 1	Day 2	Day 3
Number of persons			
Number of baths			
Number of showers Average duration of a shower (min)			
Number of toilet flushes			
Number of hand-washed loads of dishes			
Number of machine-washed loads of dishes			
Number of washing-machine loads of laundry			
Number of lawn or garden waterings Average duration of a watering (min)			
Number of car washes			
Number of cups of water (estimated) for cooking and drinking			
Number of times water runs in sink Average duration of water running (min)			
Other uses and frequency			

Figure 1.3 *Will you be washing a car in the next three days? If so, record it in your water-use log.*

A.3 FOUL WATER

Introduction

Note: If you have not already done so, carefully read the laboratory safety information on pages iv–v, before beginning this laboratory procedure.

Your objective is to clean up a sample of foul water, producing as much "clean water" as possible, to a point where it could be used for hand-washing. (**Caution:** *Do not test any water samples by drinking or tasting them.*) You will use several different water-purification procedures: oil–water separation, sand filtration, and charcoal adsorption and filtration.

Lab Video: Foul Water

Before starting, read the procedure to learn what you will need to do, note safety precautions, and plan necessary data collecting and observations.

Procedure

1. In your laboratory notebook, prepare a data table similar to the one shown here. Be sure to provide sufficient space to record your entries.

DATA TABLE						
	Volume (mL)	**Color**	**Clarity**	**Odor**	**Presence of Oil**	**Presence of Solids**
Before treatment						
After oil–water separation						
After sand filtration						
After charcoal adsorption and filtration						

2. Using a clean beaker, obtain approximately 100 mL (milliliters) of foul water from your teacher. Measure its volume accurately with a graduated cylinder. See Figure 1.4. Record the actual volume of the water sample in your data table. Leave your sample in the graduated cylinder.

3. Describe in detail the appearance, color, clarity, and odor of your original sample. Record your observations in the "Before treatment" row of your data table.

> The metric prefix *milli-* represents 1/1000 (0.001) of the unit specified. Thus, one milliliter (1 mL) is one-thousandth of a liter, or 0.001 L.

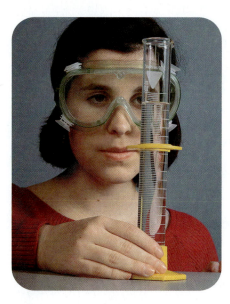

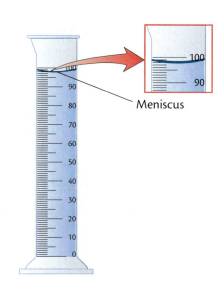

Figure 1.4 *To find the volume of liquid in a graduated cylinder, read the scale at the bottom of the curved part of the liquid (meniscus).*

Meniscus

Oil–Water Separation

As you probably know, if oil and water are mixed and left undisturbed, the oil and water do not noticeably dissolve in each other. Instead, two layers form. Which layer do you think will float on top of the other? Make careful observations in the following procedure to check your answer.

4. Allow your sample to sit in the graduated cylinder for at least one minute.

5. Using a clean, dry Beral pipet, carefully remove as much of the upper liquid layer as possible and place it in a clean, dry test tube.

6. Add several drops of distilled water to the liquid you placed in the test tube. Does the water float on top or sink to the bottom? Is the liquid you removed in Step 5 water? Explain your reasoning, using evidence from your observations to support your answer.

7. Read and record the volume of the liquid sample remaining in the gradated cylinder.

8. Dispose of the liquid in the test tube as directed by your teacher.

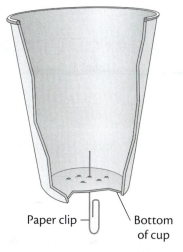

Paper clip — Bottom of cup

Figure 1.5 *Preparing a disposable cup for use in sand filtration.*

Sand Filtration

In **filtration**, solid particles are separated from a liquid by passing the mixture through a material that retains the solid particles and allows the liquid to pass through. The liquid collected after it has been filtered is called the **filtrate**. A sand filter traps and removes solid impurities—at least those particles too large to fit between sand grains—from a liquid.

9. Using a straightened paper clip, poke small holes in the bottom of a disposable cup. See Figure 1.5.

10. Add premoistened gravel and sand layers to the cup as shown in Figure 1.6. (The bottom gravel layer prevents the sand from washing through the holes. The top layer of gravel keeps the sand from churning up when the water sample is poured into the cup.)

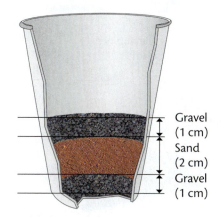

Gravel (1 cm)
Sand (2 cm)
Gravel (1 cm)

Figure 1.6 *Layering gravel and sand for sand filtration.*

11. Gently pour the sample to be filtered into the cup. Catch the filtrate in a beaker as it drains through.

12. Dispose of the used sand and gravel according to your teacher's instructions. (**Caution:** *Do not pour any sand or gravel into the sink!*)

13. Observe the properties of the filtered water sample and measure its volume. Record your results. Save the filtered water sample for the next procedure.

Charcoal Adsorption and Filtration

The pump system in an aquarium often includes a charcoal filter for adsorption of contaminants.

Charcoal **adsorbs,** which means attracts and holds on its surface, many substances that could give water a bad taste, a cloudy appearance, or an odor.

14. Fold a piece of filter paper, as shown in Figure 1.7.

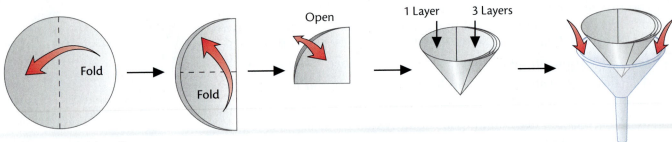

Fold   Fold   Open   1 Layer   3 Layers

| **Figure 1.7** *Folding filter paper.*

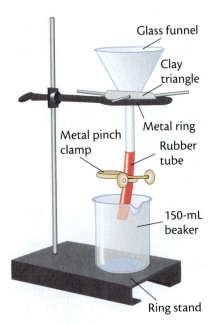

Glass funnel
Clay triangle
Metal pinch clamp
Metal ring
Rubber tube
150-mL beaker
Ring stand

| **Figure 1.8** *Funnel in clay triangle.*

15. Place the folded filter paper in a funnel. Hold the filter paper in position and moisten it slightly so that it rests firmly against the base and sides of the funnel cone.

16. Place the funnel in a clay triangle supported by a ring, as shown in Figure 1.8. Lower the ring so that the rubber tube extends 2 to 3 cm (centimeters) inside a 150-mL beaker.

17. Place no more than one level teaspoon of charcoal in a 125-mL or 250-mL Erlenmeyer flask.

18. Pour the water sample into the flask. Swirl the flask vigorously for several seconds. Then gently pour the liquid through the filter paper. Keep the liquid level below the top of the filter paper; liquid should not flow between the filter paper and the funnel because that might permit unwanted charcoal and other solid matter to seep into the filtrate.

19. If the filtrate is darkened by small charcoal particles, once again filter the liquid through a clean piece of moistened filter paper.

20. When you are satisfied with the appearance and odor of your charcoal-filtered water sample, pour the filtered water sample into a graduated cylinder. Record the final volume and properties of your purified sample.

21. Follow your teacher's suggestions about saving or disposing of your purified sample. Place the used charcoal in the container that is provided for that purpose.

22. Wash your hands thoroughly before leaving the laboratory.

Data Analysis

Record all calculations and answers in your laboratory notebook.

1. What percent of your original foul water sample did you recover as *purified water*? This value is called the **percent recovery.**

2. What volume of liquid (in milliliters) did you lose during the entire purification process?

3. What percent of your original foul-water sample was lost during purification?

To answer the following questions, first collect a list of percent recovery values for water samples from each laboratory group.

4. Construct a **histogram** showing the percent recovery obtained by all laboratory groups in your class. To do so, organize the data into equal subdivisions, such as 90.0–99.9%, 80.0–89.9%, and so forth. Count the number of data points in each subdivision. Then use this number to represent the height of the appropriate bar on your histogram, as illustrated in Figure 1.9. In this sample histogram, for example, you can see that three data points fell between 70.0% and 79.9%.

5. What was the largest percent recovery obtained by a laboratory group in your class? What was the smallest? The difference between the largest and smallest values in a data set is the **range** of those data points. What was the range of percent recovery data in your class?

6. What was the average percent recovery for your class? Compute an **average** value by adding all values together and dividing the sum by the total number of values. The result is also called the **mean** value.

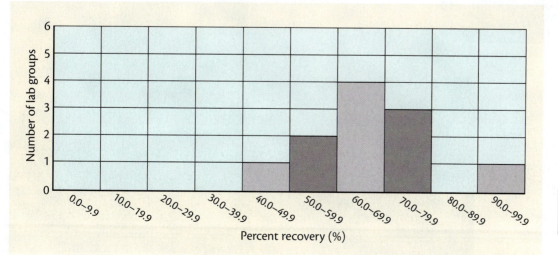

Figure 1.9 *Typical histogram of foul-water recovery data.*

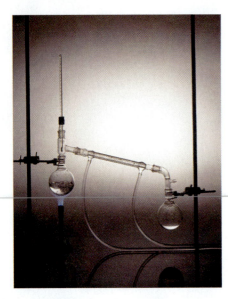

Figure 1.10 *This is a typical laboratory distillation apparatus. It may differ slightly from one that your teacher demonstrates.*

7. The mean is a mathematical expression for the most "typical" or "representative" value for a data set. Another useful expression is the **median** value, or middle value. To find the median for percent-recovery data, list all values in either ascending or descending order. Then find the value in the middle of the list—the point where there are as many data points above as below.

Consider this data set: 1   2   2   4   5   6   7

Three values lower   ↑   Three values higher

Median

If you have an even number of data points, take the average of the two values nearest the middle. What is the median percent recovery of all of your class laboratory results?

Post-Lab Activities

1. Your teacher will demonstrate distillation, which is another technique for water purification. See Figure 1.10.

   a. Write a description of the steps in distillation.
   b. Why did your teacher discard the first portion of distilled liquid?
   c. Why was some liquid purposely left in the distilling flask at the end of the demonstration?

2. Your teacher will organize a test of the **electrical conductivity** of the purified water samples obtained by your class. This test focuses on the presence of dissolved, electrically charged particles in the water. You will also compare the electrical conductivity of distilled water and tap water. (You will learn more about electrically charged particles on pages 37–38.) What do these test results suggest about the purity of your water sample? Support your answer with evidence from your observations.

Figure 1.11 *Above: Particles are suspended in the sample in the beaker on the left. The particles are too small to see, but large enough to reflect light coming from a beam to the left of the beakers. This is called the* Tyndall effect. *Particles in the solution in the beaker on the right are too small to reflect light. Right: The Tyndall effect is also observable in nature.*

3. Your teacher will test the clarity of the various water samples by passing a beam of light through each sample. Observe the results. The differences are due to the presence or absence of the *Tyndall effect.* See Figure 1.11. Referring to your observations, explain what this test suggests about the purity of your water sample.

Questions

1. Is your purified water sample *pure* water? Provide evidence to support your answer.

2. How could you compare the quality of your final water sample with that obtained by other laboratory groups? That is, how should someone judge the success of each laboratory group? Defend your answer using evidence from this investigation.

3. How could you improve the water-purification procedures you followed so that you could recover a higher percent of purified water?

4. a. Estimate the total time you spent purifying your water sample.
 b. In your opinion, did that time investment result in a large enough sample of sufficiently purified water?
 c. It is sometimes said that "time is money."
   i.   If you spent twice as much time purifying your sample, would that extra time investment pay off in higher-quality water?
   ii. If you spent about ten times as much time, would that extra investment pay off? Explain your reasoning.

5. Municipal water-treatment plants do not use distillation to purify water. Why?

A.4 WATER SUPPLY AND DEMAND

Is the United States in danger of running out of water? The answer is both no and yes. The total water available is far more than enough. Each day, some 15 trillion liters (4 trillion gallons) of rain or snow falls in the United States. Only 10% is used by humans. The rest flows back into large bodies of water, evaporates into the air, and falls again as part of a perpetual *water cycle,* or *hydrologic cycle.* So, that is the *no* part of the answer. However, the distribution of rain and snow in the United States does not necessarily correspond to regions of high water use. Figure 1.12 (page 16) summarizes how available water is used in various regions of the country, organized by five major water-use categories.

A U.S. family of four (two adults, two children) uses an average of 1480 liters (390 gallons) of water daily. That approximate volume represents **direct water use,** that which can be directly measured. In addi-

> A trillion gallons is 1 000 000 000 000 gallons, or 10^{12} gallons.

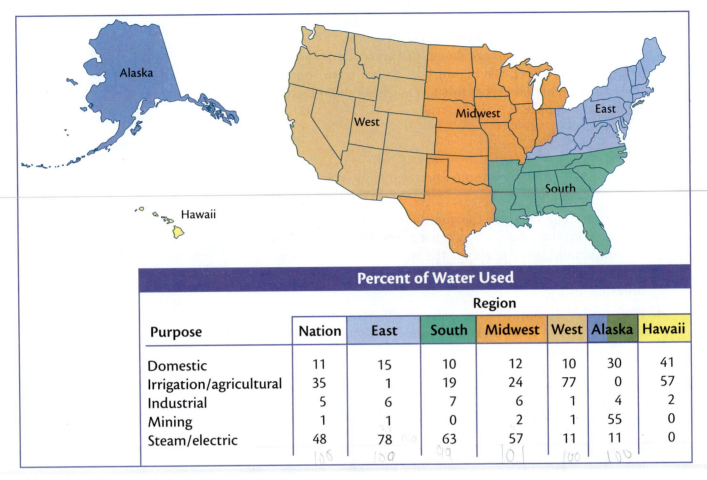

Percent of Water Used

Purpose	Nation	Region East	South	Midwest	West	Alaska	Hawaii
Domestic	11	15	10	12	10	30	41
Irrigation/agricultural	35	1	19	24	77	0	57
Industrial	5	6	7	6	1	4	2
Mining	1	1	0	2	1	55	0
Steam/electric	48	78	63	57	11	11	0

Figure 1.12 *Water use in the United States.*

tion, there is also **indirect water use,** hidden uses of water that you may never have considered. Each time you eat a slice of pizza, some potato chips, or an egg, you are "using" water. Why? Because water was needed to grow and process the various components of each food.

Consider again *ChemQuandary 1* on page 8. At first glance, you probably thought that the volumes of water mentioned were absurdly large. How could so much water be needed to produce one fried egg or one can of fruit juice? These two examples illustrate typical indirect (hidden) uses of water. The chicken that laid the egg needed drinking water. Water was used to grow the chicken's feed. Water was also used for various steps in the process that eventually brought the egg to your home. Even small quantities of water used for these and other purposes quickly add up when billions of eggs are involved!

In a Riverwood newspaper article you read earlier, Jimmy Hendricks was quoted as saying that he would drink bottled water and canned fruit juice until the water supply was turned on again. However, drinking fruit juice from a container involves the use of much *more* water than drinking a glass of tap water. Why? Because the quantity of liquid *in* the container is insignificant when compared with the quantity of water used to *make* the container and possibly to irrigate the fruit trees and process the fruit juice. See Figure 1.13.

Fabricating a metal can, for example, is the primary source of the surprising 120 L of water mentioned on page 8. What examples of hidden water use do you encounter in daily life?

Although people depend on large quantities of water, hardly anyone is aware of how much they actually use. This lack of awareness is understandable because water normally flows freely when taps are turned on—in Riverwood or in your home. Where does all this water come from? Check what you already know about the distribution of this seemingly plentiful resource.

Figure 1.13 *Making canned juice (right) takes large quantities of water. Fruits and other crops require irrigation water, as shown in this aerial photo of a center pivot sprinkler (top). Water is also used to produce steel (bottom) for the juice can.*

Developing Skills

A.5 WATER USE IN THE UNITED STATES

Refer to Figure 1.12 to answer the following questions.

1. For each region in the United States, name the greatest single use of water.

   a. the East   b. the South   c. the Midwest

   d. the West   e. Alaska   f. Hawaii

2. Explain the differences in how water is used in the East and the West. Think about where most people live and where most of the nation's factories and farms are located. What other regional factors help explain the general patterns of water use?

3. List two factors about the weather, economy, or culture that could explain the greatest water use within each of these six U.S. regions.

A.6 WHERE IS THE WORLD'S WATER?

You are probably not surprised to learn that most of Earth's total water (97% of it, in fact) is in the oceans. However, the next largest global water-storage place is not so obvious. Do you know what it is? If you said rivers and lakes, you and many others agree; however, that answer is incorrect. The second largest quantity of water is stored in Earth's ice caps and glaciers. See Figure 1.14. Figure 1.15 shows how the world's supply of water is distributed.

Figure 1.14 *Most of Earth's fresh water is stored in glaciers like this one.*

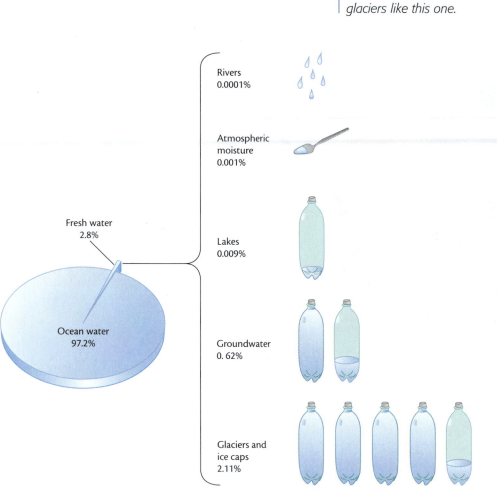

Rivers
0.0001%

Atmospheric moisture
0.001%

Lakes
0.009%

Fresh water
2.8%

Ocean water
97.2%

Groundwater
0. 62%

Glaciers and ice caps
2.11%

| **Figure 1.15** *Distribution of the world's water supply.*

As you know, water can be found in three different physical states at ordinary temperatures. Water vapor in the air is in the **gaseous state.** Water is most easily identified in the **liquid state**—in lakes, rivers, oceans, clouds, and rain. Ice is a common example of water in the **solid state.** What other forms of "solid water" can you identify?

At present, most of the United States is fortunate to have abundant supplies of high-quality water. You turn on the tap, use what you need, and go about your daily routine—giving little thought about how that seemingly unlimited water supply manages to reach you.

In some U.S. regions, water in the gaseous state is experienced as high humidity that contributes to summer discomfort.

City Water

If you live in a city or town, the water pipes in your home are linked to underground water pipes. These pipes bring water downhill from a reservoir or a water tower, usually located near the highest point in town, to all the faucets in the area. Water stored in the tower was previously cleaned and purified at a water-treatment plant. It may have been pumped to the treatment plant from a reservoir, lake, or river. If your home's water supply originated in a river or other body of water, you are using **surface water.** If it originated in a well, you are using **groundwater.** Groundwater must be pumped to the surface.

About one-fifth of the U.S. water supply is held as groundwater.

Rural Water

If you live in a rural area, your home probably has its own water-supply system. A well with a pipe driven deep into an **aquifer** (a water-bearing layer of rock, sand, or gravel) pumps the water to the surface. A small, pressurized tank holds the water before it enters your home's plumbing system.

Not surprisingly, neither groundwater nor surface-water samples are completely pure. When water falls as precipitation (rain, snow, sleet, or hail) and joins a stream or when it seeps far into the soil to become groundwater, it picks up small amounts of dissolved gases, soil, and rock. These dissolved materials are rarely removed from water at the treatment plant or from well water. In the amounts normally found in water, they are harmless. In fact, some minerals found in water (such as iron, zinc, or calcium) are essential in small quantities to human health, or they may actually improve the water's taste.

When the water supply is shut off, as it was in Riverwood, it is usually shut off for a short time. However, suppose a drought lasted several years; or suppose the shortage was perpetual, as it is in some areas of the world. In such circumstances, what uses of water would you give up first? Clearly, using available water for survival purposes would have priority. Nonessential uses would probably be eliminated.

Now it is time to examine your water-use data. Refer to your completed water-use diary (see page 9) to find out how much water you use in your home and for what purposes.

A.7 WATER-USE ANALYSIS

Use Table 1.1 and the data you collected for Making Decisions A.2 to answer the following questions about your household's water use, as well as those of your classmates.

> Table 1.1 indicates that a regular showerhead delivers 19 L each minute. In English units, that is 5 gallons per minute, or 25 gallons of water during a five-minute shower! To visualize that volume of water, think of 47 two-liter beverage bottles or five filled 5-gallon buckets.

1. Calculate the total water volume (in liters) used by your household during the three days.

2. How much water (in liters) did one member of your household use, on average, in one day?

3. Compile the answers to Question 2 for all members of your class by creating a histogram. To review the construction of a histogram, see page 13. (*Hint:* The range in values for each histogram bar must be equal and should be selected so that there are about ten total bars.)

| Table 1.1

WATER REQUIRED FOR TYPICAL ACTIVITIES	
Activity	**Water Volume (L)**
Bathing (per bath)	130
Showering (per minute)	
Regular showerhead	19
Water-efficient showerhead	9
Cooking and drinking (per 10 cups of water)	2
Flushing toilet (per flush)	
Conventional toilet	19
Water-saving toilet	13
Low-flow toilet	6
Watering lawn (per hour)	1130
Washing clothes (per load)	170
Washing dishes (per load)	
By hand (with water running)	114
By hand (washing and rinsing in dishpans)	19
By machine (full cycle)	61
By machine (short cycle)	26
Washing car (running hose)	680
Running water in sink (per minute)	
Conventional faucet	19
Water-saving faucet	9

4. What is the range of the average daily personal water use within your class?

5. Calculate the mean and median values for the class data. Which do you think is more representative of the data set—the mean or median value? That is, which is a better expression of central tendency for these data?

6. Compare your answer from Question 2 with the estimated average volume of water, which is 370 L, used daily by each person in the United States. What reasons can you propose to explain any difference between your value and the national average value?

7. Which is closer to the national average (mean) for daily water use by each person, your answer to Question 2 or the class average in Question 5? What reasons can you give to explain why that value is closer?

Recall that the Riverwood town council arranged to truck water from Mapleton to Riverwood (see Figure 1.16) for three days to meet the needs of Riverwood residents for drinking water and cooking water. The current population of Riverwood is about 19 500.

Figure 1.16 *Emergency workers prepare to distribute bottled water.*

8. On the basis of the water-use data collected and analyzed by your class (pages 8–9 and 20–21), explain how you would estimate the total volume of water that would need to be hauled to Riverwood during the three days.

9. What additional information would help you improve your estimate in Question 8? Why?

10. What assumptions must you make to complete your estimate?

You are now quite aware of the volume of water you use daily. Suppose you had to live with much less water. How would you ration your water for survival and comfort? This is exactly the question that now confronts Riverwood residents.

Figure 1.17 *Faced with water rationing, how essential do you consider these water uses to be? Could water from one activity be reused for the other?*

Making Decisions

Potable water is pure enough for use in drinking and cooking. Grey water is left over from home use (such as showers, sinks, laundry, and dishwashers) and may sometimes be reused for other purposes, especially landscape watering. Black water has contacted toilet wastes and must be chemically treated and disinfected before reuse.

A.8 RIVERWOOD WATER USE

Riverwood authorities have severely rationed home water supplies for three days, while they investigate possible fish-kill causes. The County Sanitation Commission recommends cleaning and rinsing your bathtub, adding a tight stopper, and filling the tub with water. That water will be your family's total water supply for all uses other than drinking and cooking for up to three days. (Recall that water for drinking and cooking will be trucked in from Mapleton.)

Assuming that your household has just one tub of water (150 L, which equals 40 gal) to use and considering the typical water uses (see Figure 1.17) listed here, answer the questions that follow the list.

▶ washing cars, floors, windows, pets
▶ bathing, showering, washing hair, washing hands
▶ washing clothes, dishes
▶ watering indoor plants, outdoor plants, lawn
▶ flushing toilets

1. List three water uses that you could do without.
2. Identify one activity that you could *not* do without.
3. For which tasks could you reduce your water use? How?
4. Impurities added by using water for one particular use may not prevent its reuse for other purposes. For example, you might decide to save hand-washing water and use it later to bathe your dog.

   a. For which activities could you use such impure water?
   b. From which prior uses could this water be taken?

It should now be obvious that clean water is a valuable resource that must not be taken for granted. Unfortunately, water is easily contaminated. In the next section, as Riverwood deals with its water emergency, you will examine some causes of water contamination.

SECTION A SUMMARY
Reviewing the Concepts

Both direct and indirect uses of water must be considered when evaluating water use.

1. Assume that Jimmy Hendricks drank just packaged fruit juice during the water shortage. Does that mean he did not use any water? Explain.

2. List at least three indirect uses of water associated with producing a loaf of bread.

3. The following are three water uses associated with your foul-water laboratory investigation (page 10). Classify each as either a direct or an indirect water use. Explain your answers.

   a. manufacture of the filter paper
   b. premoistening of the sand and gravel
   c. use of water to cool the distillation apparatus

Water can be purified by a variety of techniques.

4. What does it mean to "purify" water?

5. Identify at least three techniques for purifying water.

6. What was removed from your foul-water sample in each step of that investigation?

7. The procedure used in the foul-water laboratory investigation could not convert seawater to water suitable for drinking.

   a. Explain why not.
   b. What additional purification steps would be needed to make seawater suitable for drinking?

The amount of water available in Earth's hydrologic cycle is essentially fixed. The distribution of water is not always sufficient to meet local needs.

8. Has the world's total water changed in the past 100 years? The past 1 million years? Explain.

9. Rank the following locations in order of greatest to least total water abundance on Earth: rivers, oceans, glaciers, water vapor.

10. Consider this quotation: "Water, water, everywhere, nor any drop to drink." Describe a situation in which this would be true.

11. Look at Figure 1.12 on page 16. What percent of water is used in irrigation and agriculture in your U.S. region?

12. Look at Figure 1.15 on page 18. Fresh water makes up 2.8% of Earth's water supply. Calculate the percent of fresh water found in

    a. glaciers and ice caps.
    b. lakes.

Connecting the Concepts

13. Explain why it might be possible that a molecule of water that you drank today was once swallowed by a dinosaur.

14. Consider oil–water separation, sand filtration, charcoal adsorption and filtration, and distillation. Which purification procedure would be *least* practical to purify a city's water supply? Why?

15. A politician campaigning for election guarantees that "every household will have 100% pure water from every tap." Evaluate this promise and predict the likelihood of its success.

16. A carbonated beverage plant has separate water supplies for various uses, such as cleaning, producing the beverage, and employee uses.
    a. How do the water purity standards differ for the various uses?
    b. Why wouldn't it be more efficient to have one water supply for the entire plant?

17. Each person in the United States uses an average of 370 L of water daily. Other sources, however, report that U.S. per capita water use is 4960 L. If both values are correct, explain this apparent discrepancy.

Extending the Concepts

18. You are marooned on a sandy island surrounded by ocean water. A stagnant, murky pond contains the only available water on the island. In your survival kit, you have the following items:

    ▶ one nylon jacket
    ▶ one plastic cup
    ▶ two plastic bags
    ▶ one length of rubber tubing
    ▶ one knife
    ▶ one 1-L bottle of liquid bleach
    ▶ one 5-L glass bottle
    ▶ one bag of salted peanuts

    Describe a plan to produce drinkable water, using only these items.

19. Charcoal-filter materials are available in various sizes—from briquettes to fine powder. List the advantages and the disadvantages of using either large charcoal pieces or small charcoal pieces for filtering.

20. Find out how much charcoal is used in a fish aquarium filter and how fast water flows through the filter. Estimate the volume of water that can be filtered by a kilogram of charcoal. What mass of charcoal would be needed to filter the daily water supply for Riverwood, population 19 500?

21. A group of friends is planning a four-day backpacking hike. Some members of the group favor carrying their own bottled water. Other members of the group wish to carry no water, instead, they will buy bottled water in towns along the trail; still others want to purchase and carry portable filters for use with stream water.
    a. What are advantages and disadvantages of each option?
    b. What additional information would the group need before deciding on a group plan?

Looking at Water and Its Contaminants

Meeting Raises Fish-Kill Concerns

BY CAROL SIMMONS
Riverwood News Staff Reporter

More than 300 concerned citizens, many prepared with questions, attended a Riverwood Town Hall public meeting last evening to hear from scientists investigating the Snake River fish kill.

Dr. Harold Schmidt, a chemist with the Environmental Protection Agency (EPA), expressed regret that the fishing tournament was canceled, but he strongly supported the town council's decision, saying it was the safest course in the long run. He reported that his laboratory is still conducting tests on the river water.

Dr. Margaret Brooke, a State University water specialist, helped interpret information and answered questions. Local physician Dr. Jason Martingdale and Riverwood High School family and consumer sciences teacher Alicia Green joined the speakers for a question-and-answer session.

Brooke confirmed that preliminary water-sample analyses showed no likely cause for the fish kill. She reported that EPA chemists will collect water samples at hourly intervals today to look for any unusual fluctuations in dissolved-oxygen (DO) levels. Fish require an adequate amount of oxygen gas dissolved in water for their survival; unusual DO levels sometimes kill fish.

Concerning possible fish-kill causes, Brooke said that EPA scientists have been unable to identify any microorganisms in the fish that could have been responsible for their death. She concluded that "it must have been something dissolved or temporarily suspended in the water." If dissolved matter were involved, EPA chemists will have to consider the relative amounts of various substances that can dissolve in water, and the effect of water temperature on their solubility. Brooke expressed confidence that further studies would shed more light on the causes of the problem.

Dr. Martingdale reassured citizens that "thus far, no illness reported by either physicians or the hospital can be linked to drinking water." Ms. Green offered water-conservation tips for housekeeping and cooking to make life easier for inconvenienced citizens. The information sheet she distributed is available on the Riverwood city Web site.

Mayor Cisko confirmed that water supplies will again be trucked in from Mapleton today, and he expressed hope that the crisis will last no longer than three days.

Those attending the meeting appeared to accept the emergency situation with good spirits. "I'll never take tap water for granted again," said Trudy Anderson, a Riverwood resident. "I thought scientists would have the answers," puzzled Robert Morgan, head of Morgan Enterprises. "They don't know either! There's certainly more involved in all this than I ever imagined."

As the previous article indicates, water experts attribute the cause of the fish kill to something dissolved or suspended in the Snake River. What might it be? How can the search for the cause be narrowed further? Knowing the properties of water (and of substances that might be found in it) will aid in this task. To understand these properties, you will be introduced to matter at the particulate level. You will also begin to learn the language of chemistry and use it to communicate with your classmates as you investigate the fish kill. And, just like scientists and public policy decision makers, you will gather and use available evidence to make decisions about Riverwood's situation.

Figure 1.18 *Earth as seen from space. Our planet's abundant supply of water makes it unlike any other in our solar system.*

B.1 PHYSICAL PROPERTIES OF WATER

Water is a common substance—so common that we usually take it for granted. We drink it, wash with it, swim in it, and sometimes grumble when it falls from the sky. But are you aware that water is one of the rarest and most unusual substances in the universe? As planetary space probes have gathered data, scientists have learned that the great abundance of water on Earth is unmatched by any planet or moon in our solar system. Earth is usually half-enveloped by water-laden clouds, as you can see in Figure 1.18. In addition, more than 70% of Earth's surface is covered by oceans that have an average depth of more than three kilometers (two miles).

| **Figure 1.19** *What states of water can be observed in this winter scene?*

Water is a form of matter. See Figure 1.19. As you may recall from previous science courses, **matter** is anything that occupies space and has mass. All solids, liquids, and gases are classified as matter. Matter can be distinguished by its characteristic properties. Water has many **physical properties**—properties that can be observed and measured without changing the chemical makeup of the substance. One physical property of a sample of matter is its **density,** which is the mass of material within a given volume. The density of water as a liquid is easy to remember. Because one milliliter (mL) of liquid water at room temperature (25 °C) has a mass of about 1.00 g, the density of this water is 1.00 g/mL. One milliliter of volume is exactly equal to one cubic centimeter (1 cm³), which is pictured in Figure 1.20. Thus, the density of water at 25 °C can also be reported as 1.00 g/cm³.

The **freezing point** of water, which is another physical property, is 0 °C at normal atmospheric pressure. What other physical properties of water can you describe?

Water is the only ordinary liquid found naturally in our environment. Because so many substances dissolve readily in water, quite a few liquids are actually water solutions. Such a water-based solution is called an **aqueous solution.** Even water that seems pure is never entirely so. Surface water contains dissolved minerals as well as other substances. Distilled water used in steam irons and car batteries contains dissolved gases from the atmosphere, as does rainwater.

Pure water is clear, colorless, odorless, and tasteless. The characteristic taste and slight odor of some tap-water samples are caused by substances dissolved in the water. You can confirm this by boiling and then refrigerating a sample of distilled water. When you compare its taste with the taste of chilled tap water, you may notice that *pure* distilled water tastes flat.

Water's physical properties (see Figure 1.21) help to distinguish it from other substances. In the following activity, you will compare the density of water with the density of some other common materials.

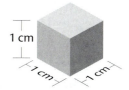

States of Water/ States of Matter

Figure 1.20 *One cubic centimeter (shown actual size). 1 cm³ = 1 mL.*

The density of liquid water actually depends on its temperature—for example, the density of liquid water is 0.992 g/mL at 40 °C and 0.972 g/mL at 80 °C.

Aqueous is based on the Latin root for *water* (aqua).

Figure 1.21 *What properties of water might account for its sheeting action under the swimmer's right arm?*

B.2 DENSITY

You may already be familiar with such physical properties of water as density, boiling point, and melting point. Use your experiences with water and other materials to answer the following questions about density:

> Another approach to characterizing water is to consider how it reacts chemically with other substances. You will learn more about chemical properties of substances in Unit 2.

1. In the foul-water investigation (page 10), you observed that coffee grounds settled to the bottom of the water sample, whereas oil floated on top. Explain this observation in terms of the relative densities of coffee grounds, water, and oil.

2. How does the density of ice compare with that of liquid water? What evidence can you cite to support your answer? (*Hint:* Use your everyday experiences to answer these questions.) What would happen to rivers and lakes (and fish) in colder climates if the relative densities of ice and liquid water were reversed?

3. Suppose you were given a small cube of copper metal. What measurements would you need to make to determine its density? How would you make these measurements in the laboratory?

B.3 MIXTURES AND SOLUTIONS

How can you decide if a water sample is not clean enough to drink? How can dissolved or suspended substances in water be separated and identified? Answers to these questions will be helpful in understanding, and possibly solving, the fish-kill mystery.

When two or more substances combine, and yet the substances retain their individual properties, the result is called a **mixture.** The foul water that you purified earlier is an example of a mixture because it contained coffee grounds, garlic powder, oil, and salt. As you discovered, the components of a mixture can be separated by physical means, such as filtration and adsorption.

When you first examined your foul-water sample, did it look uniform throughout? Most likely, the coffee grounds had settled to the bottom and were not distributed evenly throughout the liquid. The foul water is an example of a **heterogeneous mixture** because its composition is not the same, or uniform, throughout. One type of heterogeneous mixture is called a **suspension** if the solid particles are large enough to settle out or can be separated by using filtration. Water plus coffee grounds and water plus small pepper particles are examples of suspensions.

> A heterogeneous mixture's composition varies from sample to sample.

Figure 1.22 *Milk is an example of a colloid.*

If the particles are smaller than those in a suspension, they may not settle out and, therefore, they may cause water to appear cloudy. Recall what happened when your teacher shined a light through your sample of purified water. The scattering of the light, known as the **Tyndall effect** (see Figure 1.11, page 14), indicated that small, solid particles were still in the water. This type of mixture is called a **colloid.**

A more familiar example of a colloid is whole or low-fat milk (see Figure 1.22), which contains small butterfat particles dispersed in water. These colloidal butterfat particles are not visible to the unaided eye; the mixture appears homogeneous—uniform throughout. Under high magnification, however, individual butterfat globules can be observed suspended in the water. Milk no longer appears homogeneous.

Particles far smaller than colloidal particles also may be present in a mixture. When small amounts of table salt are mixed with water, as in your foul-water sample, the salt dissolves in the water. That is, the salt crystals separate into particles so small that they cannot be seen even at high magnification, nor do the particles exhibit the Tyndall effect when a light beam is passed through the mixture. These particles become uniformly mingled with the particles of water, producing a **homogeneous mixture,** or a mixture that is uniform throughout. All **solutions** are homogeneous mixtures. In a salt solution, the salt is the **solute** (the dissolved substance) and the water is the **solvent** (the dissolving agent). All solutions consist of one or more solutes and a solvent. See Figure 1.23.

Figure 1.23 *Sugar cubes dissolve in tea. In this case, tea is the solvent and sugar is the solute.*

Evidence that something was still dissolved in your purified water sample came from the results of the conductivity test. The positive result (the bulb lit up) indicated that electrically charged particles were dissolved in the mixture. Figure 1.24 contains a summary flowchart of the various types of mixtures, as well as types of substances, which will be discussed next.

Figure 1.24 *Classification of matter*

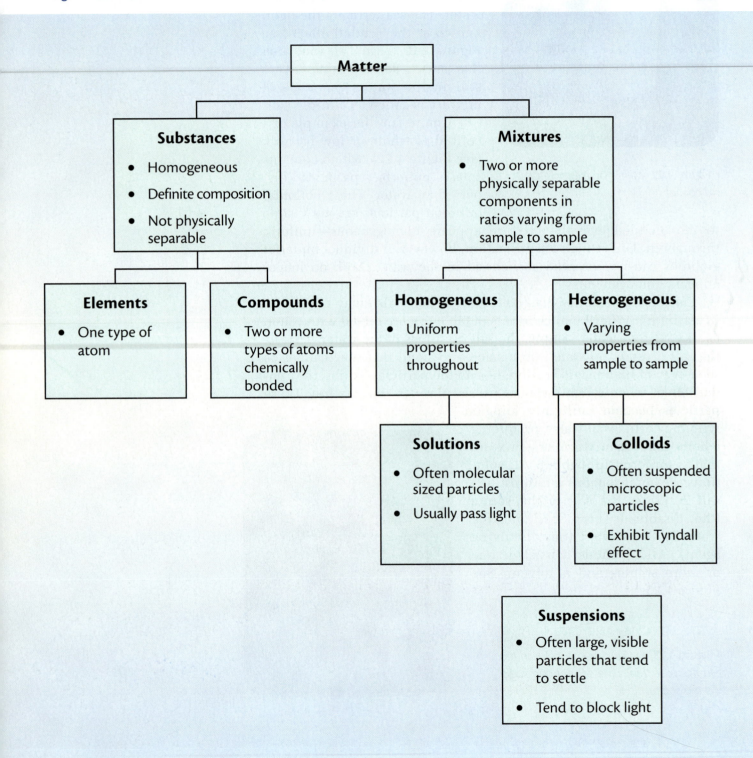

B.4 PARTICULATE VIEW OF WATER

So far in your exploration of water, you have focused on properties that are observable with your unaided senses. Have you wondered why water's freezing point is 0 °C or why certain substances such as salt dissolve in water? To understand why water has its specific properties, you must investigate it at the **particulate level;** that is, at the level of its atoms and molecules.

All matter is composed of atoms. **Atoms** are often called the building blocks of matter. Matter that is made up of only one kind of atom is known as an **element.** For example, oxygen is considered an element because it is composed of only oxygen atoms. Because hydrogen gas contains no atoms other than hydrogen atoms, it, too, is an element. Approximately ninety different elements are found in nature, each having its own type of atom and identifying properties.

What type of matter is a sample of water? Is it an element? A mixture? As you probably know already, water contains atoms of two elements—oxygen and hydrogen. Thus, water is not an element. And, because water's properties are different from those of oxygen and hydrogen, and the oxygen and hydrogen contained in water cannot be readily separated, water cannot be classified as a mixture, either.

Instead, water is an example of a **compound**—a substance that is composed of the atoms of two or more elements linked together chemically in certain fixed proportions. To date, chemists have identified more than 24 million compounds. Compounds and elements are represented by **chemical formulas.** In addition to water (H_2O), some other compounds and formulas with which you may be familiar include table salt (NaCl), ammonia (NH_3), baking soda ($NaHCO_3$), chalk ($CaCO_3$), sugar ($C_{12}H_{22}O_{11}$), and octane (C_8H_{18}).

Each element and compound is considered a **substance** because each has a uniform and definite composition, as well as distinct properties. The smallest unit of a molecular compound that retains the properties of that substance is a **molecule.** Atoms of a molecule are held together by chemical bonds. You can think of chemical bonds as the "glue" that holds atoms of a molecule together. Oxygen is an example of an *element* typically found in molecular form—two oxygen atoms bonded to one another (O_2)—while water is a molecular *compound.* One molecule of water is composed of two hydrogen atoms bonded to one oxygen atom, hence H_2O. An ammonia molecule (NH_3) contains three hydrogen atoms bonded to one nitrogen atom. Figure 1.25 shows representations of some atoms and molecules.

Seeing and Imagining Water

An element cannot be broken down into any simpler substances.

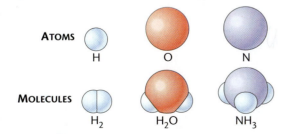

Figure 1.25 *Top row: hydrogen (H), oxygen (O), and nitrogen (N) atoms. Bottom row: hydrogen (H_2), water (H_2O), and ammonia (NH_3) molecules. Models similar to these are used throughout this book to depict atoms and molecules.*

The next activity will give you the opportunity to apply a particulate viewpoint to a variety of common observations.

B.5 PICTURES IN THE MIND

You live in a **macroscopic** world—a world filled with large-scale *(macro)*, readily observed things. As you experience the properties and behavior of bulk materials, you probably give little thought to the particulate world of atoms and molecules. If you wrap leftover cake in aluminum foil, it is unlikely that you think about how the individual aluminum atoms are arranged in the wrapping material. It is also unlikely that you consider what the mixture of molecules making up air looks like as you breathe. And you probably seldom wonder about atomic and molecular behavior when you observe water boiling or an iron nail rusting.

Nevertheless, having a sense of how atoms and molecules might look and behave in substances and mixtures can help you explain everyday phenomena. To develop this sense, it is useful to construct and evaluate **models,** or representations, of atoms and molecules.

Sample Problem: *Draw a model of two gaseous compounds in a homogenous mixture.*

A homogeneous mixture is uniform throughout, so the two compounds should be intermingled and evenly distributed. Also compounds are composed of atoms of two or more different elements linked together by chemical bonds.

Suppose a molecule of one of these compounds contains two different atoms. To represent this molecule, you could draw two differently shaded or labeled circles to denote atoms of the two elements and a line connecting the atoms to indicate a bond, like this:

Now, suppose that the other compound is composed of molecules made up of three atoms, and that two atoms are of the same element. You now need to decide on the order in which the atoms should be connected: the unique atom (Y) could be in the middle, X–Y–X, or on the end, X–X–Y. As long as you draw this imaginary compound the *same way every time,* it does not matter which way you draw it for this activity. However, the way in which atoms are connected in real compounds does, in fact, make a difference; X–Y–X would be quite a different molecule than X–X–Y.

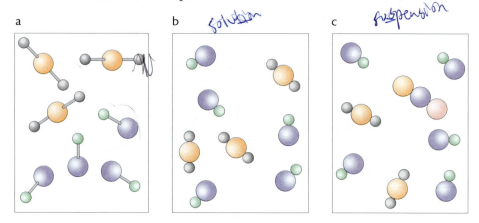

a   b   *solution*   c   *suspension*

Examine models a, b, and c in the illustration on page 32. Which best represents a homogeneous mixture of the two compounds just described? You are correct if you said that model b is the best visual model. The two types of molecules are uniformly mixed, and the atoms are colored to indicate that they represent different elements. In model a, the mixture is not homogeneous because the molecules are not uniformly mixed. Model c contains three different compounds instead of two. Notice that in model a, bonded atoms in each molecule are connected by lines. In models b and c, bonded atoms just touch each other. Both representations are used by chemists; either one is acceptable in this activity.

Now it is your turn to create and evaluate various visual models of matter.

1. Draw a model of a homogeneous mixture composed of three different gaseous elements.

2. What kind of matter does the following model represent? Explain your answer.

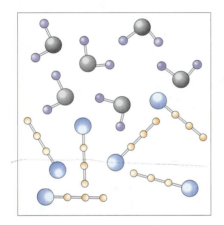

3. Draw a model of a container of each of the following samples of matter. Write a description of key features of each model.

   a. a mixture of gaseous elements X and Z
   b. a two-atom compound of X and Z
   c. a four-atom compound of X and Z
   d. a solution composed of a solvent that is a two-atom compound of L and R, and a solute that is a compound composed of two atoms of element D and one atom of element T

4. Compare each visual model that you created in Question 3 with those of your classmates.

   a. Although the models may look a little different, does each set depict the same type of sample? Comment on any similarities and differences.
   b. Do the differences help or hinder your ability to visualize the type of matter that is depicted? Explain.

5. The element iodine (I) has a greater density in its solid state than in its gaseous state. Draw models that depict and account for this difference at the particulate level. Iodine normally exists as a two-atom molecule.

6. A chemistry student at Riverwood High School was asked to draw a model of a mixture composed of an element and a compound:

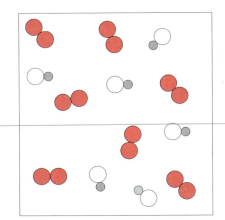

Comment on the usefulness of the student's drawing.

7. You have been interpreting and creating two-dimensional models of three-dimensional molecules.

a. What are some limitations of flat, two-dimensional models?
b. What are some characteristics that good models of matter should have?

As you continue to study chemistry, you will encounter visual models of matter similar to those in this activity. When you see them, think about their usefulness as well as their possible limitations.

B.6 SYMBOLS, FORMULAS, AND EQUATIONS

An international "chemical language" for use in oral and written communication was developed to represent atoms, elements, and compounds. The *letters* in this language's alphabet are **chemical symbols,** which are understood by scientists throughout the world. Each element is assigned a chemical symbol. Only the first letter of the symbol is capitalized; all other letters are lowercase. For example, C is the symbol for the element carbon, and Ca is the symbol for the element calcium. Symbols for some common elements are listed in Table 1.2.

All known elements are organized into the **periodic table of the elements,** which is one of the most useful tools in a chemist's work. A few examples of elements found in the periodic table are shown in Figure 1.26. As you continue your study of chemistry, you will learn more about this table. For now, become familiar with this chart by locating each element listed in Table 1.2 on the periodic table found on page 124.

Words in the language of chemistry are composed of letters (which represent elements) from the periodic table. Each *word* is a **chemical formula,** which represents a different chemical substance.

| Table 1.2

COMMON ELEMENTS	
Name	**Symbol**
Aluminum	Al
Bromine	Br
Calcium	Ca
Carbon	C
Chlorine	Cl
Cobalt	Co
Copper	Cu
Fluorine	F
Gold	Au
Hydrogen	H
Iodine	I
Iron	Fe
Lead	Pb
Magnesium	Mg
Mercury	Hg
Nickel	Ni
Nitrogen	N
Oxygen	O
Phosphorus	P
Potassium	K
Silver	Ag
Sodium	Na
Sulfur	S
Tin	Sn

In the chemical formula of a substance, a chemical symbol represents each element present. A **subscript** (a number written below the normal line of letters) indicates how many atoms of the element just to the left of the subscript are in one unit of the substance.

For example, as you already know, the chemical formula for water is H_2O. The subscript 2 indicates that each water molecule contains two hydrogen atoms. Each water molecule also contains one oxygen atom. However, the subscript 1 is "understood" in the absence of any subscript and is therefore not included in chemical formulas.

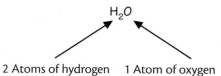

2 Atoms of hydrogen   1 Atom of oxygen

Sample Problem: *The chemical formula for propane, a compound commonly used as a fuel, is C_3H_8. What elements are present in a molecule of propane, and how many atoms of each element are there?*

You are correct if you said each propane molecule consists of *three* carbon atoms and *eight* hydrogen atoms.

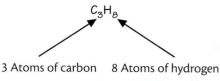

3 Atoms of carbon   8 Atoms of hydrogen

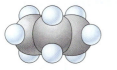

If formulas are words in the language of chemistry, then **chemical equations** can be regarded as chemical *sentences*. Each chemical equation summarizes the details of a particular chemical reaction. **Chemical reactions** entail the breaking and forming of chemical bonds, causing atoms to become rearranged into new substances. These new substances have different properties from those of the original materials.

The expression for the formation of water from its elements

shows that two hydrogen molecules ($2 H_2$) and one oxygen molecule (O_2) react to produce ($\rightarrow$) two molecules of water ($2 H_2O$). The original (starting) substances in a chemical reaction are called **reactants;** their formulas are always written on the left side of the arrow. The new substance or substances formed from the rearrangement of the reactant atoms are called **products;** their formulas are always written on the right side of the arrow. Note that this equation, like all chemical equations, is *balanced*—the total number of each type of atom (four H atoms and two O atoms) is the same for both reactants and products.

| 13 |
| Aluminum |
| **Al** |
| 26.98 |

| 14 |
| Silicon |
| **Si** |
| 28.09 |

| 16 |
| Sulfur |
| **S** |
| 32.07 |

Figure 1.26 *(top) People use aluminum to make a variety of products including foil, cans, and lightweight construction materials. (middle) Silicon has properties that lie between those of metals and nonmetals. It is classified as a metalloid. One of its primary uses is in electronic devices. (bottom) Sulfur is a nonmetal used in products such as fungicides and rubber for automobile tires.*

Perhaps you noticed that in the chemical equation for the formation of water from its elements the reactants hydrogen and oxygen are written with subscripts (H_2 and O_2). Most uncombined elements in chemical equations are represented as single atoms (Cu, Fe, Na, and Mg, for example). A handful of elements are **diatomic molecules;** they exist as two bonded atoms of the same element. Oxygen gas and hydrogen gas are two examples of diatomic molecules. Table 1.3 lists all the common elements that exist as diatomic molecules at normal conditions. It will be helpful for you to remember these elements. Find the diatomic elements in the periodic table. Where are they located?

| Table 1.3

ELEMENTS THAT EXIST AS DIATOMIC MOLECULES

Element	Formula	Element	Formula
Hydrogen	H_2	Chlorine	Cl_2
Nitrogen	N_2	Bromine	Br_2
Oxygen	O_2	Iodine	I_2
Fluorine	F_2		

Developing Skills

B.7 SYMBOLS, FORMULAS, AND EQUATIONS

1. a. Name the element represented by each of these symbols.
    i.   P   v.   Br
    ii. Ni   vi.   K
    iii. Cu   vii. Na
    iv. Co   viii. Fe
    b. Which elements in Question 1a have symbols that correspond to their English names?
    c. Which is more likely to be the same throughout the world—an element's symbol or its name? Explain.

2. For each formula, name the elements and give the number of atoms of each element indicated.

    a. H_2O_2 Hydrogen peroxide   Antiseptic
    b. $CaCl_2$ Calcium chloride   Sidewalk deicer
    c. $NaHCO_3$ Sodium hydrogen carbonate   Baking soda
    d. H_2SO_4 Sulfuric acid   Battery acid

Sample Problem: *Complete an "atom inventory" of this chemical equation:*

$$N_2 \quad + \quad 3\ H_2 \quad \longrightarrow \quad 2\ NH_3$$

2 N atoms   +   6 H atoms $\longrightarrow$ 2 N atoms and 6 H atoms

Note that the total number of atoms of N (nitrogen) and H (hydrogen) does not change during this chemical reaction.

Next, interpret the equation in terms of molecules:

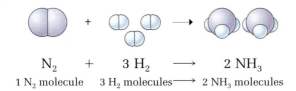

$$N_2 \quad + \quad 3\ H_2 \quad \longrightarrow \quad 2\ NH_3$$

1 N_2 molecule 3 H_2 molecules $\longrightarrow$ 2 NH_3 molecules

One molecule of N_2 reacts with three molecules of H_2 to produce two molecules of the compound NH_3, called ammonia.

Household ammonia is made by dissolving gaseous ammonia in water.

3. The following chemical equation represents the burning of methane, CH_4, to form water and carbon dioxide:

$$CH_4 + 2\ O_2 \longrightarrow CO_2 + 2\ H_2O$$

a. Write a sentence describing the equation in terms of molecules.

b. Identify each molecule as either a compound or an element.

c. Complete an atom inventory for the equation.

d. Provide a visual model (picture in your mind) of the chemical reaction.

Let ⬤ represent CH_4. Let ⬤ represent CO_2.

Use the model of an H_2O molecule depicted in Figure 1.25 (page 31) to draw a molecule of H_2O similar to that for CH_4 and CO_2 shown here.

B.8 THE ELECTRICAL NATURE OF MATTER

Previously, you were introduced to the concept of atoms and molecules. How do the atoms in molecules stick together to form bonds? Are atoms made up of even smaller particles? The answers to these questions require understanding the electrical properties of matter.

You have already experienced the electrical nature of matter, most probably without realizing it. Clothes often display static cling when taken from the dryer. The pieces of fabric stick firmly together and can be separated only with effort. The shock that you sometimes receive after walking across a rug and touching a metal doorknob is another reminder of matter's electrical nature. And if you rub two inflated balloons against your hair, both balloons will attract your hair but repel each other, a phenomenon best seen when the humidity is low. See Figure 1.27.

Figure 1.27 *Rubbing a balloon against your hair results in static electricity. The balloon attracts your hair, even when you hold it away from your head.*

The electrical properties of matter can be summarized as follows:

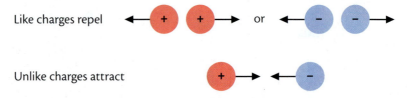

Like charges repel

Unlike charges attract

What are these positive and negative charges? How do they relate to the idea of atoms and molecules? Every electrically neutral (uncharged) atom contains equal numbers of positively charged particles called **protons** and negatively charged particles called **electrons.** For instance, an electrically neutral sodium atom contains 11 protons and 11 electrons. An electrically neutral chlorine atom contains 17 protons and 17 electrons.

In addition to protons and electrons, most atoms contain one or more electrically neutral particles called **neutrons.** Positive–negative attractions between protons in one atom and electrons in another atom provide the attachment that holds atoms together. This "sticky force" (glue) is the chemical bond you read about on page 31.

These ideas will be used in later sections and in future units to explain the properties of substances, the process of dissolving, and chemical bonding. Now you will combine these ideas with your knowledge of atoms, chemical symbols, and chemical names to learn about a class of compounds that generally dissolve to some extent in water. It is possible that one or more of these compounds might be the cause of the Riverwood fish kill.

B.9 IONS AND IONIC COMPOUNDS

Earlier in this unit (page 31), you learned about molecules. Molecules make up one type of compound. Another type of compound is composed of **ions,** which are electrically charged atoms or groups of atoms. Atoms gain or lose electrons to form negative or positive ions, respectively. For instance, a sodium atom can easily lose one electron, resulting in an ion containing 11 protons and 10 electrons. A chlorine atom readily gains one electron. Summing protons and electrons reveals the resulting electrical charge for these two ions:

Sodium ion: 11 protons   (11+ charge)
             + 10 electrons (10− charge)
                  Sodium ion (1+ charge); thus, Na^+

Chloride ion: 17 protons   (17+ charge)
              + 18 electrons (18− charge)
                   Chloride ion (1− charge); thus, Cl^-

Ionic compounds are substances that are composed of positive and negative ions. An ionic compound has no net electrical charge; it is neutral because the positive and negative electrical charges offset each other. The most familiar example of an ionic compound is table salt, sodium chloride ($NaCl$).

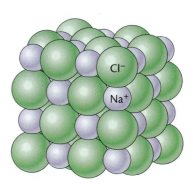

Solid sodium chloride, NaCl(s), consists of equal numbers of positive sodium ions (Na^+) and negative chloride ions (Cl^-) arranged in a three-dimensional network called a *crystal*. In solid ionic compounds, such as table salt, the ions are held together in crystals by attractions among the negative and positive charges. See Figure 1.28. When an ionic compound dissolves in water, its individual ions separate from each other and disperse throughout the water. The designation (aq) following the symbol for an ion, as in Na^+(aq), indicates that the ions are dissolved in water (aqueous) solution.

A negatively charged ion is called an **anion,** and a positively charged ion is called a **cation.** An ion can be a single atom, such as a sodium cation (Na^+) or a chloride anion (Cl^-); or a group of bonded atoms, such as an ammonium cation (NH_4^+) or a nitrate anion (NO_3^-). An ion consisting of a group of bonded atoms is called a **polyatomic** (*many atom*) **ion.** Table 1.4 (page 40) lists the formulas and names of common cations and anions.

The ionic compound calcium chloride, $CaCl_2$, presents a picture similar to NaCl. However, unlike sodium ions, calcium ions (Ca^{2+}) each have a charge of 2+.

You can write formulas for ionic compounds by following two simple rules.

▶ Write the cation first, then write the anion.

▶ The correct formula will contain the fewest positive and negative ions needed to make the total electrical charge zero.

Why are the numbers of chloride ions different in sodium chloride (NaCl) and calcium chloride ($CaCl_2$)? In sodium chloride, the ion charges are 1+ and 1−, thus one cation and one anion result in a total charge of zero. So, the formula for sodium chloride must be NaCl.

When cation and anion charges do not add up to zero, ions of either type must be added until the charges are the same. In calcium chloride, one calcium ion (Ca^{2+}) has a charge of 2+. Each chloride ion (Cl^-) has a charge of 1−; two Cl^- ions are needed to balance a charge of 2+. Thus, two chloride ions (2 Cl^-) are needed for each calcium ion (Ca^{2+}). The subscript 2 written after chlorine's symbol in the formula indicates this. The correct formula for calcium chloride is $CaCl_2$.

Figure 1.28 *You probably use sodium chloride, NaCl, to salt your food (left). A scanning electron micrograph shows the cubic structure of NaCl crystals (center). A space-filling model of NaCl (right) provides information about how the individual chloride ions (Cl^-) and sodium ions (Na^+) are arranged within the salt crystal. What else does this model suggest about sodium and chloride ions?*

| Table 1.4

COMMON IONS

Cations

1+ Charge		2+ Charge		3+ Charge	
Formula	Name	Formula	Name	Formula	Name
H^+	hydrogen	Mg^{2+}	magnesium	Al^{3+}	aluminum
Na^+	sodium	Ca^{2+}	calcium	Fe^{3+}	iron(III)*
K^+	potassium	Ba^{2+}	barium		
Cu^+	copper(I)*	Zn^{2+}	zinc		
Ag^+	silver	Cd^{2+}	cadmium		
NH_4^+	ammonium	Hg^{2+}	mercury(II)*		
Li^+	lithium	Cu^{2+}	copper(II)*		
		Pb^{2+}	lead(II)*		
		Fe^{2+}	iron(II)*		

Anions

1− Charge		2− Charge		3− Charge	
Formula	Name	Formula	Name	Formula	Name
F^-	fluoride	O^{2-}	oxide	PO_4^{3-}	phosphate
Cl^-	chloride	S^{2-}	sulfide	P^{3-}	phosphide
Br^-	bromide	SO_4^{2-}	sulfate		
I^-	iodide	SO_3^{2-}	sulfite		
NO_3^-	nitrate	CO_3^{2-}	carbonate		
NO_2^-	nitrite				
OH^-	hydroxide				
HCO_3^-	hydrogen carbonate (bicarbonate)				
$HOCl^-$	hypochlorite				
SCN^-	thiocyanate				

*Some metal atoms form ions with one charge under certain conditions and another charge under different conditions. To specify the electrical charge for these ions, Roman numerals are used in parentheses after the metal's name.

Formulas for compounds that contain polyatomic ions, such as Na_2CO_3 (sodium carbonate), follow these same rules. However, if more than one polyatomic ion is needed to bring the total charge of the compound to zero, the formula for the polyatomic ion is enclosed in parentheses before the needed subscript is written. For instance, calcium nitrate is composed of Ca^{2+} ions and NO_3^- ions. Its correct formula, written to balance electrical charges, is $Ca(NO_3)_2$. Ammonium sulfate is composed of ammonium (NH_4^+) and sulfate (SO_4^{2-}) ions. *Two* ammonium cations with a total charge of 2+ are needed to match the 2− charge of *one* sulfate anion. Thus, the formula for ammonium sulfate is $(NH_4)_2SO_4$. Note that the entire ammonium ion symbol is enclosed in parentheses; the subscript 2 indicates that *two* cations are needed to complete the formula of ammonium sulfate.

The written name of an ionic compound is composed of two parts. The cation is named first, then the anion. As Table 1.4 suggests, many cations have the same name as their original elements. Anions composed of a single atom, however, have the last few letters of the element's name changed to the suffix *-ide.* For example, the anion formed from fluorine (F) is fluor*ide* (F^-). Thus, KF is named potassium fluor*ide.* The following activity will provide practice in expressing names and writing formulas for ionic compounds, according to the universal language of chemistry.

B.10 IONIC COMPOUNDS

Sample Problem: *What is the composition of potassium chloride, the primary ingredient in table-salt substitutes used by people on low-sodium diets?*

Refer to Table 1.4. The compound described in the sample problem and those described in Statements 1 through 6 should be entered in a table similar to this one:

	Cation	Anion	Formula	Name
Sample	K^+	Cl^-	KCl	potassium chloride
	(Complete this chart for substances 1 through 6.)			

1. $CaSO_4$ is a component of plaster.
2. A substance composed of Ca^{2+} and PO_4^{3-} ions is found in some brands of phosphorus-containing fertilizer.
3. Ammonium nitrate, a rich source of nitrogen, is often used in fertilizer mixtures.
4. $Al_2(SO_4)_3$ is a compound sometimes used to help purify water.
5. Magnesium hydroxide is sometimes called *milk of magnesia* when it is mixed with water.
6. Limestone and marble (see photo) are two common forms of the compound calcium carbonate.

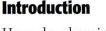
B.11 WATER TESTING

Introduction

How do chemists detect and identify certain substances or ions in water solutions? This investigation will allow you to use a method that chemists, including those investigating the Riverwood fish kill, use to detect the presence of specific ions in water solutions.

The tests that you will conduct are **confirming tests.** That is, a positive test confirms that the ion in question is present. In each confirming test, you will look for a change in solution color or for the appearance of an insoluble material called a **precipitate.** See Figure 1.29. A negative test (no color or precipitate) means one of two things: either the ion is not present, or the ion is present in such low amounts that the chemical test involved is unable to produce a confirming result.

You will conduct **qualitative tests,** which are tests that identify the presence or absence of a particular substance in a sample. In contrast, **quantitative tests** determine the *amount* of a specific substance present in a sample. Most likely, both types of tests would be used to determine the cause of the Snake River fish kill.

You will test for the presence of iron(III) (Fe^{3+}) and calcium (Ca^{2+}) cations, as well as chloride (Cl^-) and sulfate (SO_4^{2-}) anions. These ions often are found in groundwater and tap water. Although you are familiar with the names and symbols for Ca^{2+}, Cl^-, and SO_4^{2-}, the name for Fe^{3+}—iron(III)—may look strange to you. Some elements can form cations with different charges. Iron atoms can lose either two electrons to form Fe^{2+} cations or three electrons to form Fe^{3+} cations. Thus, the name "iron cation" is not helpful enough; it does not distinguish between Fe^{2+} and Fe^{3+}. For this reason, Roman numerals are added to the name to indicate the charge on the ion. Examples of other cations that must include Roman numerals in their names are copper(I), copper(II), cobalt(II), and cobalt(III). The formulas and equations that accompany each of the following tests are provided so that you can see how chemists describe these confirming reactions.

You will complete each confirming test on several different water samples. The first solution will be a distilled-water *blank*—one known not to contain any ions of interest. The second is a **reference solution** (a solution of known composition used as a comparison)—one known to contain the sought ion.

Next, you will test an unknown solution that may contain one or more of the ions. The other samples to test are tap water and the natural water samples that you or your teacher collected. These solutions may or may not contain the ions. To determine whether these

> Technologies are available for detecting ions when their amounts are so small that a confirming test cannot produce a positive result.

Figure 1.29 *A bright yellow precipitate forms when you add a solution of potassiium iodide, KI, to a solution of lead(II) nitrate, $Pb(NO_3)_2$. The precipitate is lead(II) iodide, PbI_2.*

Lab Video:
Water Testing

solutions contain the ions, you will compare the results you obtained with the behavior of your reference samples and with the distilled-water blank.

In your laboratory notebook, prepare four data tables (one for each ion) similar to the one shown. Add rows if you are testing more than one natural-water sample. Note the source of each natural-water sample.

DATA TABLE: *(Name of ion)*		
Solution	**Observations (color, precipitate, etc.)**	**Result (Is ion present?)**
Reference		
Control		
Tap water		
Natural water from ____ (source)		

The following suggestions will help guide your ion analysis:

1. If the ion is present in tap water or natural water, it will probably be present in a smaller amount than in the same volume of reference solution. Therefore, the quantity of precipitate or color produced in the tap-water or natural-water sample will be lower than in the reference solution.

2. When completing an ion test, mix the contents of the well thoroughly, using a toothpick or small glass stirring rod. Do not use the same toothpick or stirring rod in other samples without first rinsing it and wiping it dry.

3. In a confirming test based on color change, so few color-producing ions may be present that it is difficult to determine if the reaction actually took place. Here are two ways to decide whether the expected color is actually present:
   ▶ Place a sheet of white paper behind or under the wellplate to make any color more visible.
   ▶ Place the wellplate on a black or dark surface to make a precipitate more visible.

Before starting, read the procedure to learn what you will need to do, note safety precautions, and plan necessary data collecting and observations.

Procedure 🥽

If the ion of interest is present, a confirming chemical reaction will occur, producing either a colored solution or a precipitate. The chemical equations for the reactions are provided for each ion.

Environmental Cleanup:
It's a Dirty Job...But That's the Point

Wayne Crayton spends his summers touring exotic islands in the Aleutian Islands chain off the coast of Alaska. But it is not just an adventure that he embarks on; it is also his job.

As an Environmental Contaminants Specialist with the U.S. Army Corps of Engineers, Alaska District, Wayne investigates areas that were formerly used as military bases and fueling stations. Wayne and his teammates review and assess the damage (if any) that contaminants have done to key

Wayne works in an office in Anchorage during the rest of the year, analyzing and interpreting data and test results from the field investigations.

areas used by wildlife. Based on their findings, they then develop plans to fix these problems.

As part of investigation planning, the team reviews information to determine what they might find at a given site. For example, historical documents about the site will indicate whether the team members should be looking for petroleum residues or other contaminants. Aerial photography and records from earlier investigations will help to identify specific areas that are potential sites of contamination.

At the site, the team collects soil, sediment, and water samples from the exact location where a contaminant was originally introduced to the environment plus samples from the area over which the contaminant might have spread. The team may also collect small mammals or fish that have been exposed to the contaminants. After collecting the necessary samples, the team members return home quickly because some of the collected samples can degrade or change characteristics soon after collection.

Wayne works in an office in Anchorage during the rest of the year, analyzing and interpreting data and test results from the field investigations. He and his colleagues calculate concentrations of hazardous substances, including organochlorines, polychlorinated biphenyls (PCBs), pesticides, petroleum residues, and trace metals. Then they determine whether any of these substances present a risk to humans or the surrounding ecosystems.

solutions contain the ions, you will compare the results you obtained with the behavior of your reference samples and with the distilled-water blank.

In your laboratory notebook, prepare four data tables (one for each ion) similar to the one shown. Add rows if you are testing more than one natural-water sample. Note the source of each natural-water sample.

Solution	Observations (color, precipitate, etc.)	Result (Is ion present?)
Reference		
Control		
Tap water		
Natural water from (source)		

DATA TABLE: *(Name of ion)*

The following suggestions will help guide your ion analysis:

1. If the ion is present in tap water or natural water, it will probably be present in a smaller amount than in the same volume of reference solution. Therefore, the quantity of precipitate or color produced in the tap-water or natural-water sample will be lower than in the reference solution.

2. When completing an ion test, mix the contents of the well thoroughly, using a toothpick or small glass stirring rod. Do not use the same toothpick or stirring rod in other samples without first rinsing it and wiping it dry.

3. In a confirming test based on color change, so few color-producing ions may be present that it is difficult to determine if the reaction actually took place. Here are two ways to decide whether the expected color is actually present:
   ▶ Place a sheet of white paper behind or under the wellplate to make any color more visible.
   ▶ Place the wellplate on a black or dark surface to make a precipitate more visible.

Before starting, read the procedure to learn what you will need to do, note safety precautions, and plan necessary data collecting and observations.

Procedure

If the ion of interest is present, a confirming chemical reaction will occur, producing either a colored solution or a precipitate. The chemical equations for the reactions are provided for each ion.

Calcium Ion (Ca²⁺) Test

$$Ca^{2+}(aq) \quad + \quad CO_3^{2-}(aq) \quad \longrightarrow \quad CaCO_3(s)$$

Calcium ion   Carbonate ion   Calcium carbonate precipitate

Follow these steps for each sample (distilled-water blank, Ca²⁺ reference, unknown, tap water, and natural water):

1. Place 20 drops of solution into a well of a 24-well wellplate.
2. Add three drops of sodium carbonate (Na_2CO_3) test solution to the well. See Figure 1.30.
3. Record your observations, including the color and whether a precipitate formed.
4. Decide whether Ca²⁺ cations are present and record your results.
5. Repeat for the remaining samples.
6. Discard the contents of the wellplate as directed by your teacher.

Iron(III) Ion (Fe³⁺) Test

$$Fe^{3+}(aq) \quad + \quad SCN^{-}(aq) \quad \longrightarrow \quad [FeSCN]^{2+}(aq)$$

Iron(III) ion   Thiocyanate ion   Iron(III) thiocyanate ion (reddish color)

Follow these steps for each sample (distilled-water blank, Fe³⁺ reference, unknown, tap water, and natural water):

1. Place 20 drops of solution into a well of a 24-well wellplate.
2. Add one or two drops of potassium thiocyanate (KSCN) test reagent to the well.
3. Record your observations, including the color and whether a precipitate formed.
4. Decide whether Fe³⁺ cations are present and record your results.
5. Repeat for the remaining samples.
6. Discard the contents of the wellplate as directed by your teacher.

Chloride Ion (Cl⁻) Test

$$Cl^{-}(aq) \quad + \quad Ag^{+}(aq) \quad \longrightarrow \quad AgCl(s)$$

Chloride ion   Silver ion   Silver chloride precipitate

Follow these steps for each sample (distilled-water blank, Cl⁻ reference, unknown, tap water, and natural water):

1. Place 20 drops of solution into a well of a 24-well wellplate.
2. Add three drops of silver nitrate ($AgNO_3$) test reagent to the well.
3. Record your observations including the color and whether a precipitate formed.
4. Decide whether Cl⁻ anions are present and record your results.
5. Repeat for the remaining samples.
6. Discard the contents of the wellplate as directed by your teacher.

Figure 1.30 *Adding drops to a 24-well wellplate.*

Sulfate Ion (SO_4^{2-}) Test

$$SO_4^{2-}(aq) \quad + \quad Ba^{2+}(aq) \quad \longrightarrow \quad BaSO_4(s)$$

Sulfate ion   Barium ion   Barium sulfate
                                                      precipitate

Follow these steps for each sample (distilled-water blank, SO_4^{2-} reference, unknown, tap water, and natural water):

1. Place 20 drops of solution into a well of a 24-well wellplate.
2. Add three drops of barium chloride ($BaCl_2$) test reagent to the well.
3. Record your observations, including the color and whether a precipitate formed.
4. Decide whether SO_4^{2-} anions are present and record your results.
5. Repeat for the remaining samples.
6. Discard the contents of the wellplate as directed by your teacher.

Questions

1. Why were a reference solution and a blank used in each test?
2. What are some possible problems associated with the use of qualitative tests?
3. These tests cannot absolutely confirm the absence of an ion. Why?
4. How might your observations have changed if you had not cleaned your wells or stirring rods thoroughly after each test?

B.12 PURE AND IMPURE WATER

What is the difference between clean water and pure water?

Now that you have learned about water's properties and about some of the substances that can dissolve in water, you are ready to return to the problem of Riverwood's fish kill. Recall that Riverwood residents had different ideas about the cause of the problem. For example, Harmon Lewis was sure the cause was pollution of the river water.

Families in most U.S. cities and towns receive an abundant supply of clean, but not absolutely pure, water at an extremely low cost. You can check the water cost in your own area; if you use municipal water, your family's water bill will contain the current water cost.

It is useless to insist on absolutely pure water. The cost of processing water to make it absolutely pure would be prohibitively high. Even if cost were not a problem, it would still be impossible to have absolutely pure water. The atmospheric gases nitrogen (N_2), oxygen (O_2), and carbon dioxide (CO_2) will always dissolve in the water to some extent. See Figure 1.31.

Figure 1.31 *Although distilled water (right) is purer than water gushing from this discharge pipe (left), it still contains dissolved gases and thus is not "100%" pure.*

B.13 THE RIVERWOOD WATER MYSTERY

At the beginning of this unit, you read newspaper articles describing the Riverwood fish kill and the reactions of several citizens. Among those interviewed were Harmon Lewis, a longtime resident of Riverwood, and Dr. Margaret Brooke, a water-systems scientist. These two people had very different reactions to the fish kill. Harmon Lewis was certain that human activity—probably some sort of pollution—had caused the fish kill. Dr. Brooke refused to speculate about the cause of the fish kill until she had conducted some tests.

Which of these two opinions comes closer to your own reaction at this point? Complete the following activities to investigate the fish-kill issue further.

1. Reread the fish-kill newspaper reports at the start of Sections A and B (pages 4–6 and page 25).

2. As a group, generate a list of all the fish-kill facts (not opinions) in these articles. Scientists often refer to facts as data. **Data** are objective pieces of information. They do not include interpretation. Figure 1.32 depicts the beginning of a "fact list," and provides some examples to get your group started.

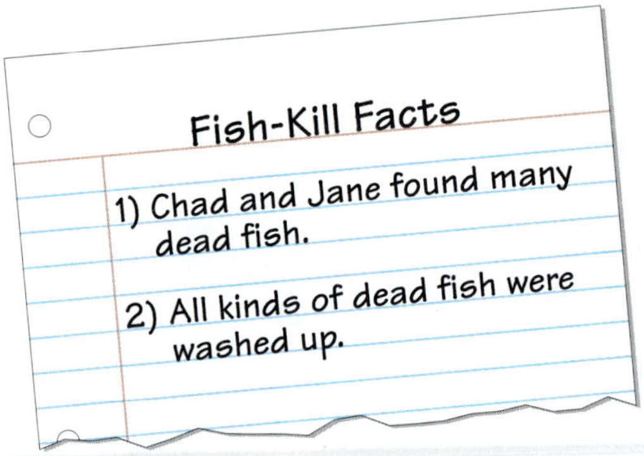

3. As you have read and thought about Riverwood's fish kill, have any questions occurred to you? Perhaps you wish you had more information about the situation. List at least five factual questions you would like to have answered before you decide on the possible causes of the fish kill. Some typical questions might be: Do barges or other commercial boats travel on the Snake River? Were any shipping accidents on the river reported recently?

4. Look over your two lists—one of facts and the other of questions. Although you probably still have unanswered questions about the Riverwood fish kill, perhaps you have some hunches about what did or did not cause it.

   a. Based on the information you now have, list at least two possible fish-kill causes that you consider unlikely. Explain your reasoning.

   b. Can you suggest a probable cause for the fish kill? Be as specific as possible.

Later in this unit, you will have the opportunity to check the reasoning you used in answering these questions.

Fish-Kill Facts

1) Chad and Jane found many dead fish.

2) All kinds of dead fish were washed up.

Figure 1.32 *Some possible Riverwood fish-kill facts.*

B.14 WHAT ARE THE POSSIBILITIES?

The activities that you just completed (gathering data, seeking patterns or regularities among the data, and suggesting possible explanations or reasons to account for the data) are typical of the approaches scientists take in attempting to solve problems. Such scientific methods are a combination of systematic, step-by-step procedures and logic, as well as occasional hunches and guesses.

A fundamental, yet difficult, part of a scientist's work is knowing what questions to ask. You have listed some questions that might be posed concerning the cause of the fish kill. Such questions help focus a scientist's thinking. Often a large problem can be reduced to several smaller problems or questions, each of which is easier to manage and solve.

The number of possible causes for the fish kill is large. Scientists investigating this problem must find ways to eliminate some causes and focus on the more promising causes. They try to disprove all but one cause or they produce conclusive proof in support of a specific cause.

As you recall, water specialist Dr. Brooke studied possible causes of the Snake River fish kill. She concluded that if the actual cause were water-related, it would have to be the result of something dissolved or suspended in the water.

In Section C, you will examine several categories of water-soluble substances and consider how they may be implicated in the fish kill. The mystery of the Riverwood fish kill will be confronted at last!

Environmental Cleanup:
It's a Dirty Job...But That's the Point

Wayne Crayton spends his summers touring exotic islands in the Aleutian Islands chain off the coast of Alaska. But it is not just an adventure that he embarks on; it is also his job.

As an Environmental Contaminants Specialist with the U.S. Army Corps of Engineers, Alaska District, Wayne investigates areas that were formerly used as military bases and fueling stations. Wayne and his teammates review and assess the damage (if any) that contaminants have done to key

Wayne works in an office in Anchorage during the rest of the year, analyzing and interpreting data and test results from the field investigations.

areas used by wildlife. Based on their findings, they then develop plans to fix these problems.

As part of investigation planning, the team reviews information to determine what they might find at a given site. For example, historical documents about the site will indicate whether the team members should be looking for petroleum residues or other contaminants. Aerial photography and records from earlier investigations will help to identify specific areas that are potential sites of contamination.

At the site, the team collects soil, sediment, and water samples from the exact location where a contaminant was originally introduced to the environment plus samples from the area over which the contaminant might have spread. The team may also collect small mammals or fish that have been exposed to the contaminants. After collecting the necessary samples, the team members return home quickly because some of the collected samples can degrade or change characteristics soon after collection.

Wayne works in an office in Anchorage during the rest of the year, analyzing and interpreting data and test results from the field investigations. He and his colleagues calculate concentrations of hazardous substances, including organochlorines, polychlorinated biphenyls (PCBs), pesticides, petroleum residues, and trace metals. Then they determine whether any of these substances present a risk to humans or the surrounding ecosystems.

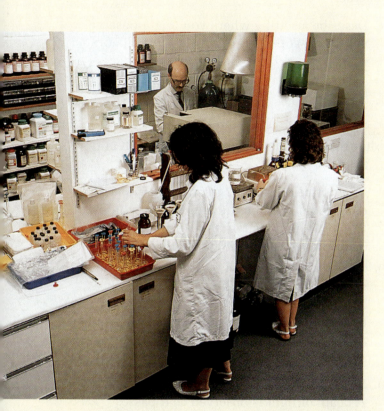

Solving Scientific Problems . . .

This icon indicates an opportunity to consult resources on the World Wide Web. See your teacher for further instructions.

Scientists often solve problems in unique ways—ways that are different from the methods used in other areas of academic research.

- Outline the problem-solving steps that Environmental Contaminants Specialists use in planning their investigations.

- Compare the steps used by these scientists with the steps that you have used in completing laboratory investigations.

- Conduct a World Wide Web search for any U.S. Army Corps of Engineers or U.S. Environmental Protection Agency investigations or projects that might be underway in your community.

Using these results, the team recommends procedures for removing and treating contaminated soil and other material. In some situations, they might decide that the best solution is to do nothing; the cleanup itself could destroy wetlands, disturb endangered wildlife, or have other negative effects on the environment. The U.S. Army Corps of Engineers uses the team's recommendations to direct the work of the contractor performing the actual cleanup.

SECTION B SUMMARY
Reviewing the Concepts

Substances can be distinguished, in part, by their physical properties.

1. What is a physical property?
2. Identify three physical properties of water.
3. How does the density of solid water compare to the density of liquid water?

4. Describe a setting where you might observe water as a solid, a liquid, and a gas all at the same time.

Physically combining two or more substances produces a mixture. Mixtures can be either heterogeneous or homogeneous.

5. How are heterogeneous and homogeneous mixtures different?
6. When gasoline and water are poured into the same container, they form two distinct layers. What do you need to know to predict which liquid will be on top?
7. Identify each of the following materials as a solution, a suspension, or a colloid.
   a. a medicine accompanied by instructions to "shake before using"
   b. Italian salad dressing
   c. mayonnaise
   d. a cola soft drink
   e. an oil-based paint
   f. milk

8. You notice beams of light passing into a darkened room through the blinds on a window. Does this demonstrate that air in the room is a solution, a suspension, or a colloid? Explain.
9. Sketch a visual model on the molecular level, similar to those you drew in the Modeling Matter activity (page 32), that represents each of the following types of mixtures. Label and explain the features of each sketch.
   a. a solution   b. a suspension
10. Suppose you have a clear, red liquid mixture. A beam of light is observed as it passes through the mixture. Over a period of time, no particles settle to the bottom of the container. Classify this mixture as a solution, a colloid, or a suspension, and provide evidence to justify your choice.

An element is composed of only one type of atom; compounds consist of two or more types of atoms. Both elements and compounds are considered substances.

11. Define the term *substance* and give two examples.
12. Classify each of the following substances as an element or a compound.
    a. CO   c. HCl   e. $NaHCO_3$ g. I_2
    b. Co   d. Mg   f. NO
13. Draw a model of a molecular compound made up of two atoms of A and one atom of D.

14. Look at these models:

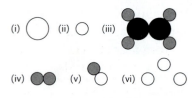

(i)   (ii)   (iii)
(iv)   (v)   (vi)

a. Which models represent elements?
b. Which models represent compounds?

A chemical formula indicates the composition of a substance. A chemical equation represents a reaction of one or more substances to form one or more new substances.

15. What two pieces of information does a chemical formula provide?

16. Name the elements and list the number of each atom indicated in the following substances:

a. phosphoric acid, H_3PO_4 (used in soft drinks)
b. sodium hydroxide, NaOH (found in some drain cleaners)
c. sulfur dioxide, SO_2 (an air pollutant)

17. Represent each chemical equation by drawing the molecules and their component atoms. Use circles of different sizes, colors, or shading for atoms of each element. Let ●●● represent a hydrogen peroxide molecule, H_2O_2.

a. $H_2(g) + Cl_2(g) \longrightarrow 2\ HCl(g)$
b. $2\ H_2O_2(aq) \longrightarrow 2\ H_2O(l) + O_2(g)$

c. Using complete sentences, write word equations for the chemical equations given in Questions 17a and 17b. Include the numbers of molecules involved.

18. Write chemical equations that represent the following word equations:

a. Baking soda ($NaHCO_3$) reacts with hydrochloric acid (HCl) to produce sodium chloride, water, and carbon dioxide.
b. During respiration, one molecule of glucose, $C_6H_{12}O_6$, combines with six molecules of oxygen gas to produce six molecules of carbon dioxide and six molecules of water.

An atom is composed of smaller particles (protons, neutrons, and electrons), each possessing a characteristic mass and electrical charge. An electrically neutral atom has equal numbers of protons and electrons.

19. For each of the following elements, identify the number of protons or electrons needed for an electrically neutral atom.

a. carbon: 6 protons   ___ electrons
b. aluminum: ___ protons   13 electrons
c. lead: 82 protons   ___ electrons
d. chlorine: ___ protons   17 electrons

20. Decide whether each of the following atoms is electrically neutral.

a. sulfur: 16 protons   18 electrons
b. iron: 26 protons   24 electrons
c. silver: 47 protons   47 electrons
d. iodine: 53 protons   54 electrons

Ionic compounds are composed of positively and negatively charged ions (atoms that have lost or gained electrons), combined to give the compound no net electrical charge.

21. Classify each of the following as an electrically neutral atom, an anion, or a cation.

a. O^{2-} c. Cl   e. Hg^{2+}
b. Li   d. Ag^+

22. For each particle in Question 21, indicate whether the electrical charge or lack of electrical charge was from a neutral atom gaining electrons, losing electrons, or neither.

23. Write the symbol and show the electrical charge (if any) on the following atoms or ions:

a. hydrogen with 1 proton and 1 electron
b. sodium with 11 protons and 10 electrons
c. chlorine with 17 protons and 18 electrons
d. aluminum with 13 protons and 10 electrons

24. Write the name and formula for the ionic compound that can be formed from these cations and anions:

a. K^+ and I^- d. Ba^{2+} and OH^-
b. Ca^{2+} and S^{2-} e. NH_4^+ and PO_4^{3-}
c. Fe^{3+} and Br^- f. Al^{3+} and O^{2-}

25. Describe differences between qualitative and quantitative tests.

26. What is a confirming test?

27. In the water-testing investigation (pages 42–45), what was the purpose of

  a. the reference solution?
  b. the distilled-water blank?

28. Using the procedure outlined in the water-testing investigation, a student tests a sample of groundwater for iron and observes no color change. Should the student conclude that no iron is present? Explain your answer.

Connecting the Concepts

29. Given an unknown mixture,

  a. what steps would you follow to classify it as a solution, a suspension, or a colloid?
  b. describe how each step would help you to distinguish among the three types of mixtures.

30. Explain the possible risks in failing to follow the direction "shake before using" on the label of a medicine bottle.

31. Why is it useful for element symbols to have international acceptance?

32. Draw a model of a solution in which water is the solvent and oxygen gas (O_2) is the solute.

33. Is it possible for water to be 100% "chemical free?" Explain.

Extending the Concepts

34. Compare the physical properties of water (H_2O) with the physical properties of the elements from which it is composed.

35. Some elements in Table 1.2 (page 34) have symbols that are not based on their modern names (such as K for potassium). Look up their historical names and explain the origin of their symbols.

36. The symbols of elements are accepted and used by chemists in all nations, regardless of the country's official language. However, the name of an element often depends on language. For example, the element N is *nitrogen* in English, but *azote* in French. The element H is *hydrogen* in English, but *Wasserstoff* in German. Investigate the names of some common elements in a foreign language of your choice. What are the meanings or origins of the foreign element names that you find? How do those meanings or origins compare with those for the corresponding English element names?

37. Investigate and report on why 100% pure water would be unsuitable for long-term human consumption—even if taste were not a consideration.

38. Using an encyclopedia or an Internet search, compare the maximum and minimum temperatures naturally found on the surfaces of Earth, the Moon, and Venus. The large amount of water on Earth is one of the factors limiting the natural temperature range on the planet. Suggest ways that water accomplishes this. As a start, find out what *heat of fusion, specific heat capacity,* and *heat of vaporization* mean.

39. Look up the normal freezing point, boiling point, heat of fusion, and heat of vaporization of ammonia (NH_3). If a planet's life forms were made up mostly of ammonia rather than water, what special survival problems might those life forms face? What temperature range would an ammonia-based planet need to support those life forms?

Investigating the Cause of the Fish Kill

The challenge facing investigators of the Riverwood fish kill is to decide what was present in the Snake River water that caused this crisis. In this section, you will learn about water as a solvent, how water solutions behave and are described, and the types of substances that dissolve in water. This background will ensure that you have the knowledge and skills needed to evaluate Snake River water data and, finally, to determine the cause of the fish kill.

C.1 SOLUBILITY OF SOLIDS IN WATER

What Is Solubility?

Could something that dissolved in the Snake River have caused the fish kill? As you already know, a variety of substances can dissolve in natural waters. To decide whether any of these possible substances could harm fish, it will be helpful to know how water solutions form and are described. For example, how much of a certain solid substance will dissolve in a given amount of water?

Imagine preparing a water solution of potassium nitrate, KNO_3. As you stir the water, the solid white crystals dissolve. The resulting solution remains colorless and clear. In this solution, water is the solvent, and potassium nitrate is the solute.

What will happen if you add a second scoopful of potassium nitrate crystals and stir? These crystals also may dissolve. However, if you continue adding potassium nitrate without adding more water, eventually some potassium nitrate crystals will remain undissolved—as a solid—on the bottom of the container, no matter how long you vigorously stir. At this point, the solution is said to be **saturated.** The maximum quantity of a substance that will dissolve in a certain quantity of water (for example, 100 g) at a specified temperature is called its **solubility** in water. In this example, the solubility of potassium nitrate might be expressed as "grams potassium nitrate per 100 g water" at a specified temperature.

From everyday experience, you probably know that both the size of the solute crystals and the vigor and duration of stirring affect how long it takes for a sample of solute to dissolve at a given temperature. But, with enough time and stirring, will even more potassium nitrate dissolve in water? It turns out that the solubility of a substance in water is a characteristic of the substance and cannot be changed by any extent of stirring over time.

> In a solution, the solvent is the dissolving agent, and the solute is the dissolved substance.

Temperature and Solubility

So what *does* affect the actual quantity of solute that dissolves in a given amount of solvent? As you can see from Figure 1.33, the mass of solid solute that will dissolve in 100 g water varies as the water temperature changes from 0 °C to 100 °C. The graphical representation of this relationship is called the solute's **solubility curve.**

Each point on the solubility curve indicates a solution in which the solvent contains as much dissolved solute as it normally can at that temperature. Such a solution is called a *saturated solution.* Thus, each point along the solubility curve indicates the conditions of a saturated solution. Look at the curve for potassium nitrate (KNO_3) in Figure 1.33. Locate the intersection of the potassium nitrate curve with the vertical line representing 40 °C. Follow the horizontal line to the left and read the value. According to the solubility curve, the solubility of potassium nitrate in 40 °C water is 60 g KNO_3 per 100 g water.

Sample Problem 1: *At 50 °C, how much potassium nitrate will dissolve in 100 g water to form a saturated solution?*

This value from Figure 1.33—80 g KNO_3 per 100 g water—represents the solubility of potassium nitrate in 50 °C water. By contrast, the solubility of potassium nitrate in cooler 20 °C water is only about 30 g KNO_3 per 100 g water. (Ensure that you can read this 30-g value from the Figure 1.33 graph.)

Note that the solubility graph for sodium chloride (NaCl) is nearly a horizontal line. What does this mean about the solubility of sodium chloride as temperature changes? Compare the curve for sodium chloride with the curve for potassium nitrate (KNO_3), which rises steeply as temperature increases. You should be able to conclude that for some solutes, such as potassium nitrate (KNO_3), solubility in water is greatly affected by temperature, whereas for others, such as sodium chloride (NaCl), the change is only slight.

Sample Problem 2: *Consider a solution containing 80 g potassium nitrate in 100 g water at 60 °C. Locate this point on the graph. Locate where this temperature falls in relation to the solubility curve. What does this information tell you about the level of saturation of the solution?*

Because each point on the solubility curve represents a saturated solution, any point on a graph *below* a solubility curve must represent an unsaturated solution. An **unsaturated solution** is a solution that contains less dissolved solute than the amount that the solvent can normally hold at that temperature.

Sample Problem 3: *What would happen if you cooled this solution to 40 °C?*

Follow the line representing 80 g to the left. You should expect that some solid KNO_3 crystals would form and fall to the bottom of the beaker. In fact, this is likely to occur. Sometimes, however, you can

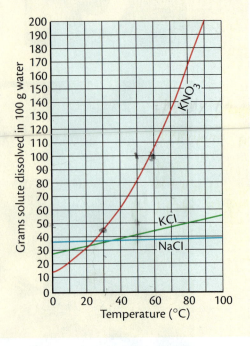

Figure 1.33 *Relationship between solute solubility in water and temperature.*

Figure 1.34 *As the water in Emerald Pool (Yellowstone National Park) cools and evaporates at the edges, it becomes supersaturated and precipitates begin to form.*

cool a saturated solution without causing any solid crystals to form, producing an unstable solution that contains more solute than could usually be dissolved at that temperature. This type of solution is called a **supersaturated solution.** (Note that this new point lies *above* the solubility curve for potassium nitrate.) Merely agitating a supersaturated solution or adding a "seed" crystal to the solution often causes the *extra* solute to appear as solid crystals and settle to the bottom of the beaker, or precipitate. The remaining liquid then contains the amount of solute that represents a stable, saturated solution at that temperature.

One example of crystallization from a supersaturated solution may be familiar to you—making rock candy. Here is how it is done: A water solution is first supersaturated with sugar. When seed crystals are added to the supersaturated solution, they cause excess dissolved sugar to crystallize from the solution onto a string hanging into the liquid.

Mineral deposits around a hot spring are another example of crystallization from a supersaturated solution. See Figure 1.34. Water emerging from a hot spring is saturated with dissolved minerals. As the solution cools, it becomes supersaturated. The rocks over which the solution flows act as seed crystals, causing the formation of more mineral deposits.

Developing Skills

C.2 SOLUBILITY AND SOLUBILITY CURVES

Sample Problem 1: *What is the solubility of potassium nitrate at 70 °C?*
The answer is found from the solubility curve for potassium nitrate given in Figure 1.33. The solubility of potassium nitrate in 70 °C water is about 134 g KNO_3 per 100 g water.

Sample Problem 2: *At what temperature will the solubility of potassium nitrate be 25 g per 100 g water?*

Think of the space between 20 g and 30 g on the y-axis in Figure 1.33 as divided into two equal parts, then follow an imaginary horizontal line at "25 g/100 g" to its intersection with the curve. Follow a vertical line down to the x-axis. Because the line falls halfway between 10 °C and 20 °C, the desired temperature must be about 15 °C.

As you have seen, the solubility curve is quite useful when you are working with 100 g water. For example, the solubility curve indicates that 60 g potassium nitrate will dissolve in 100 g water at 40 °C. But what happens when you work with other quantities of water?

Sample Problem 3: *How much potassium nitrate will dissolve in 150 g water at 40 °C?*

You can reason the answer in the following way. The quantity of solvent (water) has increased from 100 g to 150 g—1.5 times as much solvent. That means that 1.5 times as much solute can be dissolved. Thus: 1.5×60 g $KNO_3 = 90$ g KNO_3.

The answer can also be expressed as a simple proportion, which produces the same answer once the proportion is solved for x:

$$\frac{60 \text{ g } KNO_3}{100 \text{ g } H_2O} = \frac{x \text{ g } KNO_3}{150 \text{ g } H_2O}$$

$$150 \text{ g } H_2O \times 60 \text{ g } KNO_3 = 100 \text{ g } H_2O \times x \text{ g } KNO_3$$

$$x \text{ g } KNO_3 = 150 \text{ g } \cancel{H_2O} \times \frac{60 \text{ g } KNO_3}{100 \text{ g } \cancel{H_2O}} = 90 \text{ g } KNO_3$$

Refer to Figure 1.33 when answering the following questions.

1. a. What mass (in grams) of potassium nitrate (KNO_3) will dissolve in 100 g water at 60 °C?
   b. What mass (in grams) of potassium chloride (KCl) will dissolve in 100 g water at the same (60 °C) temperature?

2. a. You dissolve 25 g potassium nitrate in 100 g water at 30 °C, producing an unsaturated solution. How much more potassium nitrate (in grams) must be added to form a saturated solution at 30 °C?
   b. What is the minimum mass (in grams) of 30 °C water needed to dissolve 25 g potassium nitrate?

3. a. A supersaturated solution of potassium nitrate is formed by adding 150 g KNO_3 to 100 g water, heating until the solute completely dissolves, and then cooling the solution to 55 °C. If the solution is agitated, how much potassium nitrate will precipitate?
   b. How much 55 °C water would have to be added (to the original 100 g water) to just dissolve all of the KNO_3?

C.3 CONSTRUCTING A SOLUBILITY CURVE

Introduction

In this investigation, you will collect experimental data to construct a solubility curve for succinic acid ($C_4H_6O_4$), which is a molecular compound. Before you proceed, think about how your knowledge of solubility can help you gather data to construct a solubility curve.

▶ How can you use the properties of a saturated solution?

▶ What solvent temperatures can you investigate?

▶ How many times should you repeat the procedure to be sure of your data?

Discuss these questions with your partner or laboratory group. Your teacher will then discuss with the class how you will gather data and will demonstrate safe use of the equipment and materials involved.

Before starting, read the following safety and procedure sections to learn what you will need to do, note safety precautions, and plan necessary data collecting and observations. For guidance, refer to the sample data table.

DATA TABLE	
Trial	**Crystal Height (mm)**
1	
2	
...	
Average	

SAMPLE FOR REFERENCE ONLY

Safety

Keep the following precautions in mind while completing this laboratory investigation.

▶ Succinic acid is slightly toxic if ingested by mouth. Wash your hands thoroughly at the end of the investigation.

▶ Avoid spilling any succinic acid crystals or solution on the hot plate; its products are acrid smoke and irritating fumes.

▶ Never stir a liquid with a thermometer. Use a stirring rod.

▶ Use insulated tongs or gloves to remove a hot beaker from a hot plate. Hot glass burns bare fingers!

▶ Dispose of all wastes as directed by your teacher.

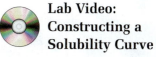
Lab Video:
Constructing a
Solubility Curve

Procedure 🥽

1. To prepare a warm-water bath, add approximately 300 mL water to a 400-mL beaker. Heat the beaker, with stirring, either to 45 °C, 55 °C, or 65 °C, as agreed upon in your prelab class discussion. Ensure that each student team sharing your hot plate is investigating the same temperature. Using gloves or beaker tongs, carefully remove the beaker when it reaches the desired temperature. (*Note:* Do not allow the water-bath temperature to rise more than 5 °C above the temperature you have chosen.) Return the beaker to the hot plate as needed to maintain the appropriate water-bath temperature.

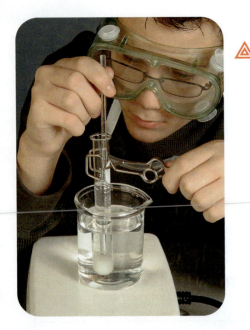

Figure 1.35 *Stirring a test tube in a water bath.*

Figure 1.36 *Decanting a solution from a test tube.*

A millimeter is 1/10 of a centimeter.

Figure 1.37 *Measuring crystal height.*

2. Place between 4 g and 5 g succinic acid in each of two test tubes. (**Caution:** *Be careful not to spill any succinic acid crystals or solution. If you do, clean up and dispose of the succinic acid as directed by your teacher.*) Add 20 mL distilled water to each test tube.

3. Place each test tube in the warm-water bath and take turns stirring the succinic acid solution with a glass stirring rod every 30 s for 7 min (Figure 1.35). Each minute, place the thermometer in the test tube and monitor the temperature of the succinic acid solution, ensuring that it is within 2 °C of the temperature that you have chosen.

4. At the end of seven minutes, carefully decant the clear liquid from each test tube into a separate, empty test tube, as demonstrated by your teacher and as shown in Figure 1.36.

5. Carefully pour the hot water from the beaker into the sink.

6. Prepare an ice bath by filling a new 400-mL beaker with water and ice.

7. Place the two test tubes containing the clear liquid in the ice bath for two minutes. Stir the liquid in each test tube gently once or twice. Remove the test tubes from the ice water. Allow the test tubes to sit at room temperature for 5 min. Observe each test tube carefully during that time. Record your observations.

8. Tap the side of each test tube and swirl the liquid once or twice to cause the crystals to settle evenly on the bottom of the test tube.

9. Measure the height of crystals collected in millimeters (mm). See Figure 1.37. Have your partner or laboratory group measure the same crystal sample height and compare your results. Report the average crystal height for your two test tubes to your teacher.

10. Rinse the succinic acid crystals from the test tubes into a collection beaker designated by your teacher. Make sure that your laboratory area is clean.

11. Wash your hands thoroughly before leaving the laboratory.

Data Analysis

1. Find the mean crystal height obtained by your entire class for each temperature reported.

2. Plot the mean crystal height in millimeters (*y*-axis) versus the water temperature in degrees Celsius (*x*-axis).

Turn to page 13 to review how to calculate a mean.

Questions

1. Why is it useful to collect data from more than one trial at a particular temperature?

2. How did you make use of the properties of a saturated solution at different temperatures?

3. Did all the succinic acid that originally dissolved in the water crystallize out of the solution? Provide evidence to support your answer.

4. Given pooled class data, did you have enough data points to make a reliable solubility curve for succinic acid? Would the curve be good enough to make useful predictions about succinic acid solubility at temperatures you have not yet investigated? Explain your answers.

5. What procedures in this investigation could lead to errors? How would each error affect your data?

6. Using your knowledge of solubility, propose a different procedure for gathering data to construct a solubility curve.

C.4 DISSOLVING IONIC COMPOUNDS

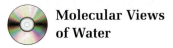
Molecular Views of Water

You have just investigated a compound dissolving in water. What you observed is called a *macroscopic* phenomenon. However, chemistry is primarily concerned with what happens at the *particulate* level—atomic and molecular phenomena that cannot be easily observed. You have seen that temperature, agitation, and time all contribute to dissolving a solid material. But how do the atoms and molecules of solute and solvent interact to make this happen?

Experiments indicate that although the entire water molecule is electrically neutral, the electrons are not evenly distributed throughout its structure. A **polar molecule** has an uneven distribution of electrical charge, which means that each molecule has a partial positive region at one end and a partial negative region at the other end.

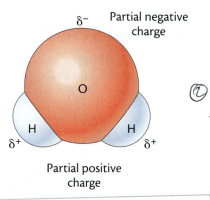

Figure 1.38 *Polarity of a water molecule. The δ⁺ and δ⁻ indicate partial electrical charges.*

δ⁻ Partial negative charge

O

H   H

δ⁺   δ⁺

Partial positive charge

Evidence also verifies that a water molecule has a bent or V-shape, as illustrated in Figure 1.38, rather than a linear, sticklike shape as in H–O–H. The *oxygen end* is an electrically negative region that has a greater concentration of electrons (shown as δ⁻) compared with the two *hydrogen ends,* which are electrically positive (shown as δ⁺). The Greek symbol δ (delta) means "partial"—thus, partial positive and partial negative electrical charges are indicated. Because these charges are equal and opposite, the molecule as a whole is electrically neutral.

Polar water molecules are attracted to other polar substances and to substances composed of electrically charged particles. These electrical attractions make it possible for water to dissolve a great variety of substances. Many solid substances, especially ionic compounds, are crystalline. In ionic crystals, positively charged cations are surrounded by negatively charged anions, with the anions likewise surrounded by cations. The crystal is held together by attractive forces between the cations and the anions. The substance will dissolve only if its ions are so strongly attracted to water molecules that the water molecules can "tug" the ions away from the crystal.

Water molecules are attracted to ions located on the surface of an ionic solid, as shown by the models in Figure 1.39. The water molecule's negative (oxygen) end is attracted to the crystal's cations. The positive (hydrogen) ends of other water molecules are attracted to the anions of the crystal. When the attractive forces between the water molecules and the surface ions are strong enough, the bonds between the crystal and its surface ions become strained, and the ions may be pulled away from the crystal. Figure 1.39 uses models of water molecules and solute ions to illustrate the results of water tugging on solute ions. The detached ions become surrounded by water molecules, producing an ionic solution.

Modeling Matter: Attraction Between Particles

Using the description and illustrations of this process, can you decide which factors influence whether an ionic solid will dissolve in water? Because dissolving entails competition among three types of attractions—those between solvent and solute particles, between solvent particles themselves, and between particles within the solute crystals—properties of both solute and solvent determine whether two substances will form a solution. Water is highly polar; therefore, it is effective at dissolving charged or ionic substances. However, if positive–negative attractions between cations and anions within the crystal are sufficiently strong, a particular ionic compound may be only slightly soluble in water.

The water solubility of some ionic compounds is extremely low, indeed. For example, at room temperature, solid lead(II) sulfide (PbS) has a solubility of only about 10^{-14} g (0.000 000 000 000 01 g) per liter of water solution or 0.000 01 parts per billion (ppb).

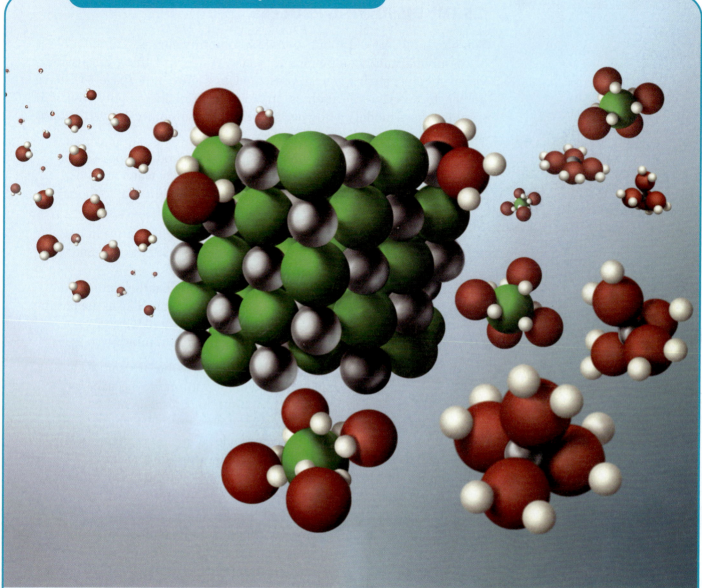

Figure 1.39 *Particulate-level model of sodium chloride (NaCl) dissolving in water (H₂O).* Sodium chloride crystals (common table salt) consist of positively-charged sodium ions (Na⁺, gray spheres) and negatively-charged chloride ions (Cl⁻, green spheres). Every water molecule is polar; the oxygen (red) side has partial negative electrical charge and the hydrogen (white) side has partial positive electrical charge. Thus, as water molecules approach the sodium chloride crystal, hydrogen attracts chloride ions and oxygen attracts sodium ions. Several water molecules surround each ion and carry it away from the crystal.

As you study this two-dimensional representation, try to visualize it in three dimensions and put the dissolving process "in motion." Think about these ions and molecules interacting at different times, such as immediately after the crystal drops into the water, after the crystal seems to disappear completely, or if the water has fully evaporated.

C.5 THE DISSOLVING PROCESS

You have now learned about solubility, solubility curves, and the process of dissolving ionic compounds in water. As part of these discussions, visual models, such as those presented in Figure 1.39 (page 61), have helped to describe the process of dissolving. In this activity, you will combine these models with your knowledge of solubility curves to create new models of ions dissolved in water.

Figure 1.40 *Solubility curve for potassium chloride.*

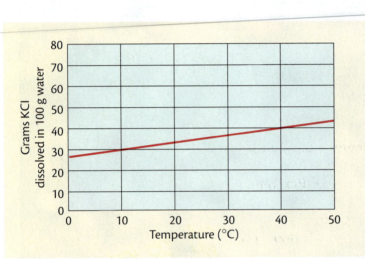

1. Suppose you dissolved 40 g potassium chloride (KCl) in 100 g water at 50 °C. You then let the solution cool to room temperature, about 25 °C.

   a. What changes would you see in the beaker as the solution cooled? See Figure 1.40.
   b. Draw models of what the contents in the beaker would look like at the particulate level at 50 °C, 40 °C, and 25 °C. Keep these points in mind:

   You should consider whether the sample at each temperature is *saturated* with excess precipitate present, *unsaturated,* or *supersaturated,* and draw the model accordingly. The solubility curve in Figure 1.40 and the particles depicted in Figure 1.39 will be helpful. It is impossible to draw all the ions and molecules in this sample. The ions and molecules you draw will represent a simpler version of what happens.

Dissolving Ionic Compounds

Modeling Matter: Ionic Solutions

2. An unsaturated solution will become more concentrated if you add more solute. Decreasing the total volume of water in the solution (such as by evaporation) also increases the solution's concentration. Consider a solution made by dissolving 20 g KCl in 100 g water at 40 °C.

   a. Draw a model of this solution.
   b. Suppose that while the solution was kept at 40 °C, one-fourth of the water evaporated.
      i. Draw a model of this final solution and describe how it differs from your model of the original solution.
      ii. How much water must evaporate at this temperature to create a saturated solution?

3. A solution may be diluted (made less concentrated) by adding water.

   a. Draw a model of a solution containing 10.0 g KCl in 100 g water at 25 °C.
   b. Suppose you diluted this solution by adding another 100 g water with stirring. Draw a model of this 25 °C new solution.
   c. Compare your models in Questions 3a and 3b. What key feature is different in the two models? Why?

C.6 SOLUTION CONCENTRATION

The general terms *saturated* and *unsaturated* are not always adequate to describe the properties of solutions that contain different amounts of solute. A more precise description of the quantity of solute in a solution is needed—a quantitative expression of concentration.

Solution **concentration** refers to how much solute is dissolved in a specific quantity of solvent or solution. You have already worked with one type of solution concentration expression: The water-solubility curves in Figure 1.33 (page 54) reported solution concentrations as the mass of a substance dissolved in a particular mass of water.

Another way to express concentration is with **percent** values. For example, dissolving 5 g table salt in 95 g water produces 100 g solution with a 5% salt concentration (by mass). "Percent" means parts solute per hundred total parts (solute plus solvent). So, a 5% salt solution could also be reported as a five parts per hundred salt solution (5 pph salt). However, percent is much more commonly used.

For solutions containing considerably smaller quantities of solute (as are found in many environmental water samples, including those from the Snake River), concentration units of **parts per million (ppm)** are sometimes useful. As you might expect, for very low concentrations, **parts per billion (ppb)** is often used. For example, the maximum concentration of nitrate allowed in drinking water is much too low to be conveniently written as a percent (it is 0.0010%); instead, it is written as 10 ppm.

Sample Problem: *What is the concentration of a 1% salt solution expressed in ppm?*

Because 1% of 1 million is 10 000, a 1% salt solution is 10 000 ppm (parts per million).

Although you might not realize it, the notion of concentration is part of daily life. For example, preparing beverages from premixed concentrate, adding antifreeze to water in an automobile radiator, and mixing pesticide or fertilizer solutions all require the use of solution concentrations. The following activity will help you review solution concentration, as well as gain experience with the chemist's use of this concept.

> Just as percent means "for every hundred," per million means "for every million." To express a concentration using parts per million (ppm), divide the mass of solute by the mass of solvent and multiply by 1 000 000.

> Nitrates are anions commonly found in natural waters as a result of fertilizer runoff. The EPA has established a maximum concentration limit of nitrates in drinking water at 10 ppm to prevent undesirable health effects resulting from elevated nitrate levels in the body.

Developing Skills

C.7 DESCRIBING SOLUTION CONCENTRATIONS

Sample Problem 1: *A common intravenous saline solution used in medical practice contains 4.5 g NaCl dissolved in 495.5 g sterilized distilled water. What is the concentration of this solution, expressed as percent by mass?*

Because this solution contains 4.5 g NaCl and 495.5 g water, its total mass is 500.0 g. Expressing a solution's concentration in *percent by mass* involves taking the mass of solute and dividing it by the mass of solution, then multiplying by 100. In the sample problem, that means dividing 4.5 g by 500.0 g and then multiplying by 100 to get 0.90%.

$$\text{Percent by mass} = \frac{4.5 \text{ g}}{500.0 \text{ g}} \times 100\% = 0.90\%$$

Another way to look at it is to remember that *percent* means "per hundred." How many grams of NaCl do you have for each 100 g solution? Because there are 500 g solution, we are dealing with 5 times as much material as in 100 g of solution. If there are 4.5 g NaCl in 500.0 g solution, there must be 1/5 that much NaCl in 1/5 of the solution—1/5 of 4.5 g NaCl is 0.90 g NaCl. That means there are 0.90 g NaCl per 100 g solution, or that the solution is 0.90% NaCl by mass, which is the same answer we arrived at previously.

Sucrose, $C_{12}H_{22}O_{11}$, is ordinary table sugar.

Sample Problem 2: *One teaspoon of sucrose, with a mass of 10.0 g, is dissolved in 240.0 g water. What is the concentration of the solution, expressed as grams sucrose per 100 g solution? As percent sucrose by mass?*

This solution contains 10.0 g sucrose and 240.0 g water; therefore, its total mass is 250.0 g. Because 100.0 g solution would contain 2/5 as much solution, it would also contain 2/5 as much solute, or 4.0 g sucrose. Thus, 100.0 g solution contains 4.0 g sucrose, a concentration of 4.0 g sucrose per 100.0 g solution. To express this concentration in the form of a percent by mass,

$$\text{Percent by mass} = \frac{4.0 \text{ g}}{100.0 \text{ g}} \times 100\% = 4.0\%$$

Now answer the following questions.

1. One teaspoon of sucrose is dissolved in a cup of water. Identify
   a. the solute.   b. the solvent.

2. What is the concentration of each of these solutions expressed as percent sucrose by mass?

   a. 17 g sucrose dissolved in 183 g water
   b. 30.0 g sucrose dissolved in 300.0 g water

3. a. What is the concentration of each of these solutions expressed as ppm?
   i.   0.0020 g iron(III) ions dissolved in 500.0 g water
   ii. 0.25 g calcium ions dissolved in 850.0 g water
   b. Expressed in ppm, what is the concentration of each solution in Question 2?

4. A saturated solution of potassium chloride is prepared by dissolving 45 g KCl in 100.0 g water at 60 °C.

   a. What is the concentration of this solution?
   b. What would be the new concentration if 155 g water were added?

5. How would you prepare a *saturated solution* of potassium nitrate (KNO_3) at 25 °C?

Now that you know how solutions of ionic compounds are formed and described, it is time to consider some possible causes of the Riverwood fish kill. The following information will suggest some possible "culprits"—substances that can dissolve or be suspended in water and harm living things. These substances include heavy metals, acids, bases, and other substances—even dissolved oxygen gas. Although many of these substances are naturally found in water sources, their concentration levels can positively or negatively affect aquatic life.

C.8 INAPPROPRIATE HEAVY-METAL ION CONCENTRATIONS?

Many metal ions, such as iron(II) (Fe^{2+}), potassium (K^+), calcium (Ca^{2+}), and magnesium (Mg^{2+}), are essential to the health of humans and other organisms. Humans obtain these ions primarily from foods, but the ions may also be present in drinking water.

Not all metal ions that dissolve in water are beneficial, however. Some **heavy-metal ions,** called "heavy metals" because their atoms have greater masses than those of essential metallic elements, are harmful to humans and other organisms. Among heavy-metal ions of greatest concern in water are cations of lead (Pb^{2+}) and mercury (Hg^{2+}). Lead and mercury are particularly likely to cause harm because they are widely used and dispersed in the environment. Heavy-metal ions are toxic because they bind to proteins in biological systems (such as in your body), preventing proteins from performing their normal tasks. As you might expect, because proteins play many roles in the body, heavy-metal poisoning effects are severe. They include damage to the nervous system, brain, kidneys, and liver, which can even lead to death.

Unfortunately, heavy-metal ions are not removed as waste as they move through the food chain. They become concentrated within the bodies of fish and shellfish, even when their abundance in the surrounding water is only a few parts per million. Such aquatic creatures then become hazardous for humans and other animals to consume.

In very low concentrations, heavy-metal ions are hard to detect in water, and they are even more difficult and costly to remove. So, how can heavy-metal poisoning be prevented? An effective way is to prevent heavy-metal ions from entering water systems in the first place. This prevention can be accomplished by producing and using alternate materials that do not contain these ions and, thus, are not harmful to health or the environment. Such practices, which prevent pollution by eliminating the production and use of hazardous substances, are examples of **Green Chemistry.** These practices are applicable to heavy metals and to many other types of pollution.

Figure 1.41 *Lead is a soft metal, as demonstrated when a lead weight is squeezed (Left and center). Paints containing lead compounds were used in houses until the government banned such use in 1978 (right).*

Lead Ions (Pb²⁺)

Lead is probably the heavy metal most familiar to you. Its symbol, Pb, is based on the element's original Latin name *plumbum;* it is also the basis of the word *plumber.* Water pipes in ancient Rome were commonly made of lead.

Lead and lead compounds (see Figure 1.41) have been, and in some cases still are, used in pottery, automobile electrical storage batteries, solder, cooking vessels, pesticides, and paints. One compound of lead and oxygen, red lead (Pb_3O_4), is the primary ingredient in paint that protects bridges and other steel structures from corrosion.

Although lead water pipes were used in the United States in the early 1800s, they were replaced by iron pipes after people discovered that water transported through lead pipes could cause lead poisoning. In modern homes, copper or plastic water pipes are now used to prevent any contact between household water and lead.

Until the 1970s, the molecular compound *tetraethyl lead,* $Pb(C_2H_5)_4$, was added to gasoline to produce a better-burning automobile fuel. Unfortunately, the lead entered the atmosphere through automobile exhaust as lead oxide. Although the phaseout of leaded gasoline reduced lead emissions, lead contamination remains in soil surrounding heavily traveled roads. In some homes built before 1978 and not since repainted, the flaking of old leaded paint is another source of lead poisoning, particularly among children who may eat the flaking paint (see Figure 1.41).

Mercury Ions (Hg²⁺)

Mercury is the only metallic element that is a liquid at room temperature (see Figure 1.42). In fact, its symbol comes from the Latin *hydrargyrum,* meaning quicksilver or liquid silver.

> Some urban home construction included lead pipes through the 1930s; lead solder was even used, at times, until the early 1970s, posing health challenges within some homes.

> Gasoline containing tetraethyl lead was known as "leaded gasoline."

Mercury has several uses, some due specifically to its liquid state. It is an excellent electrical conductor; therefore, it is used in "silent" light switches. People also use it in medical and weather thermometers, thermostats, mercury-vapor street lamps, fluorescent light bulbs, and some paints. Elemental mercury can be absorbed directly through the skin, and its vapor is quite hazardous to health. At room temperature, some mercury vapor will always be present if liquid mercury is exposed to air; therefore any mercury spill should be reported and carefully cleaned up.

Because mercury compounds are toxic, they are useful in eliminating bacteria, fungi, and agricultural pests when used in antiseptics, fungicides, and pesticides. In the 18th and 19th centuries, mercury(II) nitrate, $Hg(NO_3)_2$, was used in making felt hats, popular at that time (see Figure 1.42). After unintentionally absorbing this compound through their skin for several years, hatmakers often suffered from mercury poisoning. Their symptoms included numbness, staggered walk, tunnel vision, and brain damage, thus giving rise to the expression "mad as a hatter."

Figure 1.42 *Like other metals, mercury is very shiny. Unlike other metals, mercury is a liquid at room temperature (left). During the 18th and 19th centuries, people used mercury compounds to make felt hats (below).*

The sudden release of a large amount of heavy-metal ions might cause a fish kill, depending on the particular metal ion involved, its concentration, the species of fish present, and other factors. Was such a release responsible for the Riverwood fish kill? As you read about other possible causes of the fish kill, keep in mind questions that are relevant to all potential culprits, such as:

▶ Is there a source of this material along the Snake River within 15 km of Riverwood?

▶ What concentration of this solute would be toxic to various species of fish?

C.9 INAPPROPRIATE pH LEVELS?

You have probably heard the term *pH* used before, perhaps in connection with acid rain or hair shampoo. What is pH, and could it possibly help account for the fish kill in Riverwood?

The **pH scale** is a convenient way to measure and report the acidic, basic, or chemically neutral character of a solution. Nearly all pH values are in the range from 0 to 14, although some extremely acidic or basic solutions may be outside this range. At room temperature, any pH value less than 7.0 indicate an acidic condition; the lower the pH, the more acidic the solution. Solutions with pH values greater than 7.0 are basic; the higher the pH, the more basic the solution. Basic solutions are also called **alkaline** solutions.

Quantitatively, a change of one pH unit indicates a *tenfold* difference in acidity or alkalinity. For example, lemon juice, with a pH of about 2, is about 10 times as acidic as some soft drinks, which have a pH of about 3.

Acids and bases, some examples of which are listed in Table 1.5 and shown in Figures 1.43a and 1.43b, exhibit certain characteristic properties. For example, the vegetable dye *litmus* turns blue in basic solution and red in acidic solution. Both acidic and basic solutions conduct electricity. Each type of solution has a distinctive taste and a distinctive feel on your skin. (*Caution: Never test these sensory properties in the laboratory.*) In addition, concentrated acids and bases are able to react chemically with many other substances. You are probably familiar with the ability of acids and bases to corrode materials, which is a type of chemical behavior.

Most **acids** are made up of molecules including one or more hydrogen atoms that can be released rather easily in water solution. These *acidic* hydrogen atoms are usually written first in the formula for an acid. See Table 1.5.

> Vinegar is an acid you probably have tasted; that common kitchen ingredient is considered a dilute solution of acetic acid.

Figure 1.43a *Vinegar contains acetic acid, and some soft drinks contain both carbonic acid and phosphoric acid. Many fruits and vegetables also contain acids.*

NAMES, FORMULAS, AND COMMON USES OF SOME ACIDS AND BASES

Name	Formula	Common Uses
Acids		
Acetic acid	$HC_2H_3O_2$	In vinegar (typically a 5% solution of acetic acid)
Carbonic acid	H_2CO_3	In carbonated soft drinks
Hydrochloric acid	HCl	Used in removing scale buildup from boilers and for cleaning materials
Nitric acid	HNO_3	Used in making fertilizers, dyes, and explosives
Phosphoric acid	H_3PO_4	Added to some soft drinks to give a tart flavor; also used in making fertilizers and detergents
Sulfuric acid	H_2SO_4	Largest-volume substance produced by the chemical industry; used in automobile battery fluid
Bases		
Calcium hydroxide	$Ca(OH)_2$	Present in mortar, plaster, and cement; used in paper pulping and removing hair from animal hides
Magnesium hydroxide	$Mg(OH)_2$	Active ingredient in milk of magnesia
Potassium hydroxide	KOH	Used in making some liquid soaps
Sodium hydroxide	$NaOH$	A major industrial product; active ingredient in some drain and oven cleaners; used to convert animal fats into soap

| **Table 1.5**

Many **bases** are ionic substances that include hydroxide ions (OH^-). Sodium hydroxide, $NaOH$, and barium hydroxide, $Ba(OH)_2$, are two examples. Some bases, such as ammonia (NH_3) and baking soda (sodium hydrogen carbonate, $NaHCO_3$), contain no OH^- ions, but they still produce basic solutions because they react with water to generate OH^- ions, as illustrated by the following equation:

$$NH_3(g) \; + \; H_2O(l) \longrightarrow NH_4^+(aq) \; + \; OH^-(aq)$$

| **Figure 1.43b** *Milk of magnesia and oven cleaner contain bases. Use Table 1.5 to identify the bases contained in these and other common household items.*

Purifying Water Supplies by Municipal Pre-Use Treatment

BY RITA HIDALGO
Riverwood News Staff Reporter

Today, many rivers—such as the Snake River in Riverwood—are both a source of municipal water and a place to release wastewater (sewage). Therefore, the water is cleaned twice, both before and after the water is used. *Pre-use purification,* often called *water treatment,* takes place at a municipal filtration and treatment plant. It is the focus of this article.

A typical water-treatment process begins when intake water flows through metal screens that prevent fish, sticks, beverage containers, and other large objects from entering the water-treatment plant.

Plant operators may add chlorine, which is a powerful disinfecting agent, during the treatment process to kill disease-causing organisms. This step is known as *prechlorination.* Operators add crystals of alum—aluminum sulfate, $Al_2(SO_4)_3$—and slaked lime—calcium hydroxide, $Ca(OH)_2$—to remove suspended particles, such as colloidal clay, from the water. (Suspended particles can give water an unpleasant, murky appearance.) The alum and slaked lime react to form aluminum hydroxide, $Al(OH)_3$, which is a sticky, jellylike material that

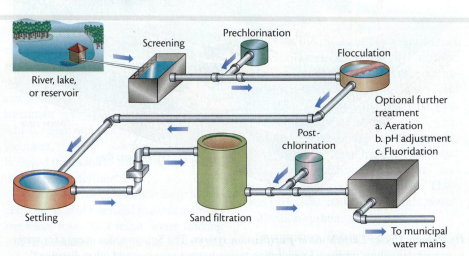

Municipal Treatment: A human-made water-purification system. Surface water is commonly cleaned at a municipal water-treatment plant before being distributed to homes and businesses. Various steps in the cleaning process remove suspended materials, kill disease-causing organisms, and may remove odors or adjust pH levels.

see *Municipal Treatment,* page 6

Name	Formula	Common Uses
Acids		
Acetic acid	$HC_2H_3O_2$	In vinegar (typically a 5% solution of acetic acid)
Carbonic acid	H_2CO_3	In carbonated soft drinks
Hydrochloric acid	HCl	Used in removing scale buildup from boilers and for cleaning materials
Nitric acid	HNO_3	Used in making fertilizers, dyes, and explosives
Phosphoric acid	H_3PO_4	Added to some soft drinks to give a tart flavor; also used in making fertilizers and detergents
Sulfuric acid	H_2SO_4	Largest-volume substance produced by the chemical industry; used in automobile battery fluid
Bases		
Calcium hydroxide	$Ca(OH)_2$	Present in mortar, plaster, and cement; used in paper pulping and removing hair from animal hides
Magnesium hydroxide	$Mg(OH)_2$	Active ingredient in milk of magnesia
Potassium hydroxide	KOH	Used in making some liquid soaps
Sodium hydroxide	NaOH	A major industrial product; active ingredient in some drain and oven cleaners; used to convert animal fats into soap

| **Table 1.5**

Many **bases** are ionic substances that include hydroxide ions (OH^-). Sodium hydroxide, NaOH, and barium hydroxide, $Ba(OH)_2$, are two examples. Some bases, such as ammonia (NH_3) and baking soda (sodium hydrogen carbonate, $NaHCO_3$), contain no OH^- ions, but they still produce basic solutions because they react with water to generate OH^- ions, as illustrated by the following equation:

$$NH_3(g) \ + \ H_2O(l) \ \longrightarrow \ NH_4^+(aq) \ + \ OH^-(aq)$$

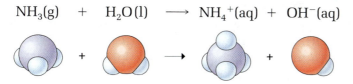

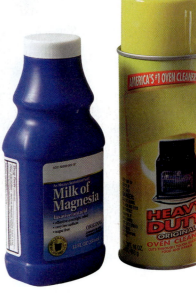

| **Figure 1.43b** *Milk of magnesia and oven cleaner contain bases. Use Table 1.5 to identify the bases contained in these and other common household items.*

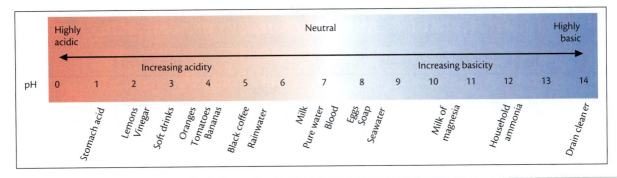

	Highly acidic							Neutral							Highly basic

Increasing acidity → ← Increasing basicity

pH: 0 1 2 3 4 5 6 7 8 9 10 11 12 13 14

Stomach acid — Lemons, Vinegar — Soft drinks — Oranges, Tomatoes, Bananas — Black coffee — Rainwater — Milk, Pure water, Blood — Eggs, Soap, Seawater — Milk of magnesia — Household ammonia — Drain cleaner

Figure 1.44 *The pH values of some common materials.*

Expert freshwater anglers sometimes plan to fish in water with pH between 6.5 and 8.2.

What about substances that display neither acidic nor basic characteristics? Chemists classify these substances as chemically *neutral*. Water, sodium chloride (NaCl), and table sugar (sucrose, $C_{12}H_{22}O_{11}$) are all examples of chemically neutral compounds.

At 25 °C, a pH of 7.0 indicates a chemically neutral solution. The pH values of some common materials are shown in Figure 1.44.

As you can see in Figure 1.44, rainwater is naturally slightly acidic. This is because the atmosphere contains some substances—carbon dioxide (CO_2) for one—that produce acidic solutions when dissolved in water. Both acidic and basic solutions have effects on living organisms—effects that depend on their pH (Figure 1.45). When the pH of rivers, lakes, and streams is too low (meaning high acidity), fish-egg development is impaired, thus hampering the ability of fish to reproduce. Bodies of water with low (acidic) pH values also tend to increase the concentrations of metal ions in natural waters by leaching metal ions from surrounding soil. These metal ions can include aluminum ions (Al^{3+}), which are toxic to fish when present in sufficiently high concentration. High pH (basic contamination) is a problem for living organisms primarily because alkaline solutions are able to dissolve organic materials, including skin and scales.

The EPA requires that drinking water be within the pH range of 6.5 to 8.5. However, most fish can tolerate a slightly wider pH range, from about 5.0 to 9.0, in lake or river water.

On a normal day, the pH of Snake River water in Riverwood ranges between 7.0 and 8.0, nearly optimal for freshwater fishing. Could the pH have changed abruptly, killing the fish? If so, was acidic or basic contamination responsible for the Riverwood crisis?

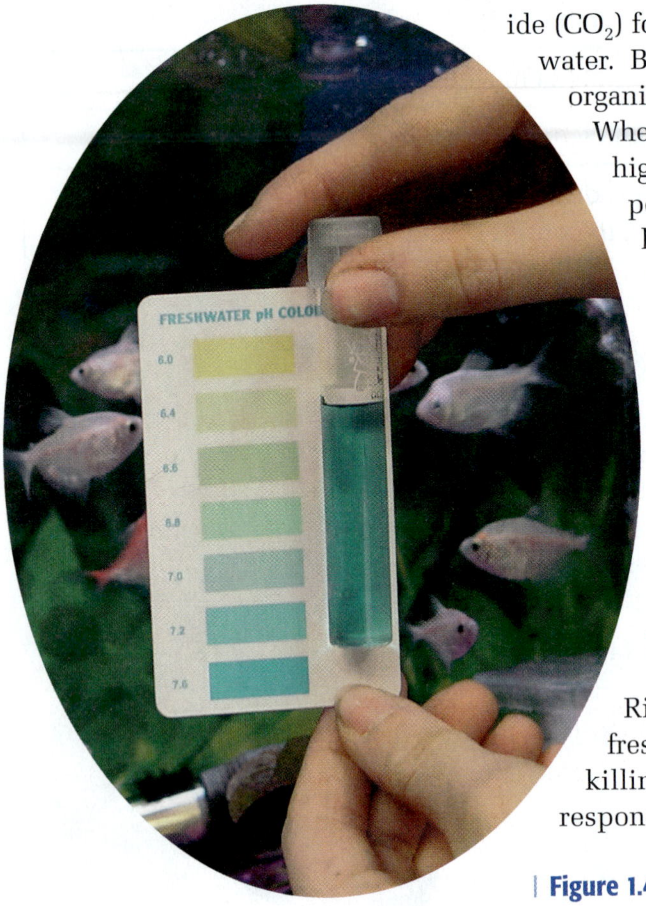

Figure 1.45 *Testing the pH of an aquarium.*

C.10 INAPPROPRIATE MOLECULAR SUBSTANCE CONCENTRATIONS?

Until now, substances considered suspects in the Riverwood fish kill have been ionic compounds that dissolve in water, releasing ions. Do other types of substances dissolve in water and possibly present a hazard to aquatic life? Some substances, such as sugar and ethanol, dissolve in water, but not as ions. These substances belong to a category known as **molecular substances** because they are composed of molecules.

Unlike ionic substances, which are crystalline solids at normal conditions, molecular substances are found as solids, liquids, or gases at room temperature. Some molecular substances, such as oxygen (O_2) and carbon dioxide (CO_2), have little attraction among their molecules and are, thus, gases at normal conditions. Molecular substances such as ethanol (ethyl alcohol, C_2H_5OH) and water (H_2O) have larger between-molecule attractions, which cause these "stickier" molecules to form liquids at normal conditions. Other molecular substances with even greater between-molecule attractions—succinic acid ($C_4H_6O_4$), for example—are solids at normal conditions. Stronger attractive forces hold these molecules together more tightly, in effect determining in which state the substance can be found at normal conditions.

Ions and ionic compounds were introduced on page 38.

You investigated the solubility behavior of succinic acid earlier. See page 57.

| Oxygen gas | Carbon dioxide | Water | Succinic acid | Ethanol |
| (O_2) | (CO_2) | (H_2O) | ($C_4H_6O_4$) | (C_2H_5OH) |

What determines the solubility of a molecular substance in water? The attraction of a substance's molecules for each other compared to their attraction for water molecules plays a major part. But what causes these attractions? The distribution of electrical charge within molecules has a great deal to do with it.

Most molecular compounds contain atoms of nonmetallic elements. As you learned earlier, these atoms are linked together by the attraction of one atom's positively charged nucleus for another atom's negatively charged electrons. If differences in electron attraction between atoms are large enough, electrons move from one atom to another, forming ions. This is what happens in a reaction between a metallic atom and a nonmetallic atom as an ionic compound is formed. The ability of an element's atoms to attract shared electrons when bonding within a compound is known as the element's **electronegativity.** In molecular substances, these differences in electron attraction, or electronegativity, are not large enough to cause ions to form, but they may cause the electrons to be unevenly distributed among the atoms.

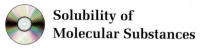

Solubility of Molecular Substances

Figure 1.46 *A painter must use paint thinner—a nonpolar solvent—to clean oil-based paint (also nonpolar) from this paintbrush. Water, a polar solvent, will not dissolve the paint. This is an example of "like dissolves like."*

You already know that the oxygen end of a water molecule is electrically negative compared with its positive hydrogen end. This is because oxygen is more electronegative than is hydrogen; thus, electrons are partially pulled away from the two hydrogen atoms toward the oxygen atom. The result is a *polar molecule.* Water molecules are polar, and water is a common example of a polar solvent. Such charge separation (and resulting molecular polarity) is found in many molecules whose atoms have sufficiently different electronegativities.

Polar molecules tend to dissolve readily in polar solvents, such as water. For example, water is a good solvent for sugar or ethanol, which are both composed of polar molecules. On the other hand, nonpolar liquids are good solvents for nonpolar molecules. Nonpolar cleaning fluids are used to dry-clean clothes because they readily dissolve nonpolar body oils found in fabric of used clothing. By contrast, nonpolar molecules (such as those of oil and gasoline) do not dissolve well in polar solvents (such as water or ethanol).

This pattern of solubility behavior—polar substances dissolve in polar solvents, nonpolar substances dissolve in nonpolar solvents—is often summarized in the generalization "like dissolves like." This rule also accounts for why nonpolar liquids are usually ineffective in dissolving ionic and polar molecular substances. See Figure 1.46.

Were dissolved molecular substances present in the Snake River water where the fish died? Most likely *yes,* at least in small amounts. Were they responsible for the fish kill? That depends on which molecular substances were present and at what concentrations. And that, in turn, depends in part on how each solute interacts with water's polar molecules.

In the following investigation, you will compare the solubility behavior of some typical molecular and ionic substances.

Investigating Matter

C.11 SOLVENTS

Introduction

The *Riverwood News* reported that Dr. Brooke believes a substance dissolved in the Snake River is one likely fish-kill cause. She based her judgment on her chemical knowledge and experience. Dr. Brooke also has a general idea about which contaminating substances she can initially rule out: those that cannot dissolve appreciably in water. Such background knowledge helps Dr. Brooke (and other chemists) reduce the number of water tests required in the laboratory.

To investigate possible solutes in the Snake River, we need to be sure we know what the terms *soluble* and *insoluble* mean. Is anything truly *insoluble* in water? It is likely that at least a few molecules or ions of any substance will dissolve in water. Thus, the term *insoluble* actually refers to substances that are only very, very slightly soluble in water. An earlier margin note (page 60) revealed that lead(II) sulfide

Various molecular substances may normally be present at very low concentrations—so low that we observe no harm to living things.

(PbS), which is technically considered insoluble, actually has a very low solubility of about 10^{-14} g (0.000 000 000 000 01 g) per liter of water at room temperature. Also, chalk is considered insoluble in water, even though 1.53 mg calcium carbonate ($CaCO_3$), which is the main ingredient in chalk, can dissolve in 100 g water at 25 °C.

> 1.53 mg = 0.001 53 g

In this investigation, you will first address the solubilities of various molecular and ionic solutes in water. These solubility data, along with toxicity data, will help you rule out some solutes as probable causes of the fish kill. You will then test other solvents and examine solubility data for any general patterns that may be apparent.

Your teacher will tell you which solutes to investigate. List them in your laboratory notebook.

You will plan your own procedure and construct your own data table for recording your data and observations. See Figure 1.47. There are no absolute rules for constructing data tables. However, scientists follow basic guidelines when organizing their laboratory data and observations. For instance, your data table should be well-organized and clear. It must be comprehensible to someone who has not completed the investigation. Keep this in mind as you plan your procedure and data collection.

Figure 1.47 *Planning a procedure for investigating solubility.*

Part I: Designing a Procedure for Investigating Solubility in Water

Your teacher will direct you to discuss with either the entire class or your laboratory partner a procedure for testing the room-temperature solubilities of substances listed in your laboratory notebook. (If you have conducted solubility tests before, it might be useful to recall how you completed them.) With your partner, design a step-by-step procedure to determine whether each solute is soluble (S), slightly soluble (SS), or insoluble (I) in room-temperature water.

The following questions will help you design your procedure.

1. What particular observations will allow you to judge how well each solute dissolves in water? That is, how will you decide whether to classify a given solute as soluble, slightly soluble, or insoluble in water?

2. Which variables will you need to control? Why?

3. How should the solute and solvent be mixed—all at once or a little at a time? Why?

In designing your procedure, keep these concerns in mind.

> In this investigation, *insoluble* (I) means that solubility is so low that you could observe no solute dissolving.

▶ Avoid any direct contact of your skin with any solutes.

▶ Follow your teacher's directions for waste disposal.

When you and your partner have agreed upon a written procedure and an accompanying data table, ask your teacher to check and approve your written plans.

Part II: Investigating Solubility in Water

Follow your teacher-approved procedure to investigate the water solubility of each specified solute. Record all data in your data table.

Part III: Investigating Solubility in Ethanol and Lamp Oil

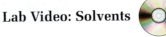

Lab Video: Solvents

The task of determining what may have caused the fish kill can be simplified somewhat by focusing on substances that will dissolve appreciably in water. However, in dealing with other solubility-based problems, chemists sometimes find it helpful to use solvents other than water. Ethanol and lamp oil serve that role in this investigation.

You and your partner will investigate the solubility in ethanol and lamp oil of some or all solutes from Part II.

In addition, test the solubility of water in ethanol and in lamp oil. Gaining experience with three liquid solvents—lamp oil, ethanol, and water—will allow you to recognize some general patterns about solubility behavior.

Can you use the same procedure you designed for Part II? If not, what parts of that procedure should be revised? In considering your Part III procedure, keep the questions listed in Part I in mind. Write down the procedure and construct the data table or tables that you plan to use for this investigation. (*Note:* Before starting this task, be sure to read the next paragraph completely.)

Have your written procedure and data table or tables approved by your teacher. Before you start the laboratory work, test your interpretation of the Part II results by predicting what you think will be observed about solubility in each case. Include these predictions in your data table for Part III. Then collect and record data for both solvents.

Wash your hands thoroughly before leaving the laboratory.

Questions

Part II

1. According to your data, which of the solutes tested are *least* likely to be dissolved in the Snake River? Explain, and support your explanation with evidence from your completed investigation.

2. Compare your data with those of the rest of the class. Are there any differences? If so, how can you explain those differences?

Part III

3. a. How does the behavior of ethanol as a solvent for sodium chloride compare with that of water?
   b. How does ethanol's behavior compare with that of lamp oil?

4. a. Were any of your solubility observations unexpected?

   b. If so, explain what you expected, why you expected it, and how your expectations compare with what you actually observed.

5. Based on your data, what general pattern of solubility behavior can you summarize and describe?

6. Predict the solubility behavior of each solid solute you tested in

   a. hexane, a liquid that is essentially insoluble in water.

   b. ethylene glycol, a liquid that is very soluble in ethanol.

7. a. Given that water is a polar solvent and lamp oil is a nonpolar solvent, classify each molecular solute tested as polar or nonpolar.

   b. How did you decide?

8. How useful is the rule "like dissolves like" for predicting solubility? Explain your answer on the basis of your results.

9. In Part II, water was the solvent, but in Part III, it was a solute.

   a. How can it be both?

   b. How can you decide whether a substance is a solute or a solvent?

702-705

EW 11 8-9 odds

Chinese → english

> The familiar saying "oil and water do not mix" has a sound chemical basis!

You now know about the solubility of some solids and liquids in water. As you read further to learn about the behavior of gases in natural waters, consider the possibility that the presence or absence of a dissolved gas was responsible for the Snake River catastrophe.

C.12 INAPPROPRIATE DISSOLVED OXYGEN LEVELS?

You have noted that the solubility of ionic and molecular solids in water usually increases when the water temperature is increased. Does this phenomenon extend to gases dissolved in solution? To build an understanding of the dissolving behavior of gases, look at Figure 1.48, which shows the solubility curve for oxygen gas plotted as milligrams oxygen gas dissolved per 1000 g water.

What is the solubility of oxygen gas in 20 °C water? In 40 °C water? As you can see, increasing the water temperature causes the gas to be less soluble! Note also the magnitude of the values for oxygen solubility. Compare these values with those for solid solutes, as shown in Figure 1.33 (page 54). At 20 °C, about 30 g potassium nitrate will dissolve in 100 g water. In contrast, only about 9 mg (0.009 g) oxygen gas will dissolve in 10 times more water (1000 g) at this temperature. It should be clear that most gases are far less soluble in water than are many ionic solids.

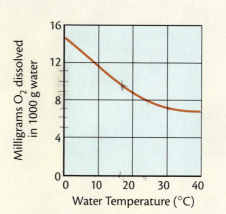

Figure 1.48 Solubility curve for O_2 gas in water in contact with air.

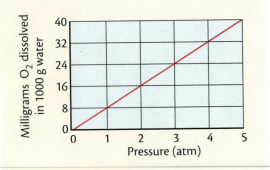

Figure 1.49 *Relationship between solubility and pressure of O_2 at 25 °C.*

The modernized metric unit for pressure is the pascal (Pa): 1 atmosphere (atm) = 101 325 Pa.

When you investigated the solubility of molecular and ionic solids, you found that solubility depended on two factors—temperature and the nature of the solvent. The solubility of a gas depends on these two factors as well. But it also depends on gas pressure. Referring to Figure 1.49, note what happens to oxygen's solubility as the pressure of oxygen gas above it is increased. Does more or less gas dissolve in the same amount of water? What happens if the O_2 gas pressure is doubled? For example, look at the solubility of O_2 at one atmosphere and at two atmospheres of gas pressure. Also consider the shape of the graph line. What type of relationship does a linear graph line indicate?

As you have deduced by now, gas solubility in water is directly proportional to the pressure of that gaseous substance on the liquid. You see one effect of this relationship every time you open a can or bottle of carbonated soft drink. As the carbon dioxide (CO_2) gas pressure on the liquid is reduced by opening the container to the air, some dissolved CO_2 gas escapes from the liquid in a rush of bubbles.

Because not very much $CO_2(g)$ is present in air, carbon dioxide gas from a high-pressure tank must be forced into the carbonated beverage just before the beverage container is sealed. This increases the amount of carbon dioxide that dissolves in the beverage. After saturating the beverage with carbon dioxide under a very high pressure of CO_2, pressure is quickly reduced when the can or bottle is opened, and the beverage solution now becomes supersaturated with CO_2. As a result, CO_2 gas quickly begins to leave the solution (hence, you observe fizzing) and continues to leave the solution (see Figure 1.50) until the solution is again saturated with CO_2, now in relation to the lower CO_2 pressure in the atmosphere above the liquid. In other words, dissolved CO_2 escapes from the liquid until the gas reaches its lower solubility at this lower pressure. When fizzing stops, you might describe the beverage as having gone flat. Actually, excess dissolved CO_2 has simply escaped into the air; the resulting "flat" beverage is still saturated with CO_2 at the new conditions.

Figure 1.50 *Carbon dioxide in a carbonated soft drink.*

C.13 TEMPERATURE, DISSOLVED OXYGEN, AND LIFE

On the basis of what you know about the effect of temperature on solubility, you may wonder if the temperature of the Snake River had something to do with the fish kill. As you just learned, water temperature affects how much oxygen gas can dissolve in the water. Various forms of aquatic life, including the many fish species, have different requirements for the concentration of dis-

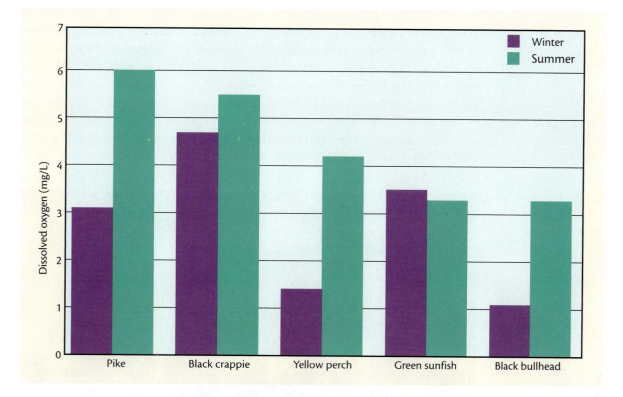

solved oxygen needed to survive. Figure 1.51 provides this information for selected fish species.

How, then, does a change in the temperature of natural waters affect fish internally? Fish are *cold-blooded* animals; their body temperatures rise and fall with the surrounding water temperature. If the water temperature rises, the body temperatures of fish also rise. This increase in body temperature, in turn, increases fish metabolism, a complex series of interrelated chemical reactions that keep fish alive. As these internal reaction rates speed up, the fish eat more, swim more, and require more dissolved oxygen. With rising temperatures, the rate of metabolism also increases for other aquatic organisms, such as aerobic bacteria, that compete with fish for dissolved oxygen.

As you can see, an increase in water temperature affects fish by decreasing the amount of dissolved oxygen in the water and by increasing the oxygen consumption of fish. A long stretch of hot summer days sometimes results in large fish kills, where hundreds of fish literally suffocate because of insufficient dissolved oxygen. Table 1.6 summarizes the maximum water temperatures at which selected fish species can survive.

Figure 1.51 *Dissolved-oxygen requirements of various fish.*

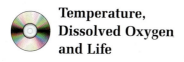 **Temperature, Dissolved Oxygen and Life**

| Table 1.6

MAXIMUM WATER TEMPERATURE TOLERANCE FOR SOME FISH SPECIES (24-HOUR EXPOSURE)		
	Maximum Temperature	
	°C	**°F**
Trout (brook, brown, rainbow)	24	75
Channel catfish	35	95
Lake herring (cisco)	25	77
Largemouth bass	34	93
Northern pike	30	86

Sometimes hot summer days are not to blame. Often, high lake or river temperatures can be traced to human activity. Many industries, such as electrical power generation, depend on natural bodies of water to cool heat-producing processes. Cool water is drawn from lakes or rivers into an industrial or power-generating plant, and devices called *heat exchangers* transfer thermal energy (heat energy) from the processing area to the cooling water. The heated water is then released back into the lakes or rivers, either immediately or after the water has partly cooled, as shown in Figure 1.52.

| Figure 1.52 *Power plant cooling water is often returned to lakes or streams.*

There is a lower limit on the amount of dissolved oxygen needed for fish to survive. Is there an upper limit? Can too much dissolved oxygen be a problem for fish? In nature, both oxygen and nitrogen gas are present in the air at all times. When oxygen gas dissolves, so does nitrogen gas. This fact turns out to be significant when considering the upper limit for dissolved gases. When the total amount of dissolved gases—primarily oxygen and nitrogen—reaches between 110% and 124% of saturation (a state of supersaturation), a condition called *gas-bubble trauma* may develop in fish.

This situation is dangerous because the supersaturated solution causes gas bubbles to form in the blood and tissues of fish. Oxygen-gas bubbles can be partially utilized by fish during metabolism, but nitrogen-gas bubbles block capillaries within the fish. This blockage results in the death of the fish within hours or days. Gas-bubble trauma can be diagnosed by noting gas bubbles in the gills of dead fish if they are dissected promptly after death. Supersaturation of water with

Figure 1.53 *Froth at the base of a dam can trap large quantities of air. This can result in water that is supersaturated with dissolved oxygen and nitrogen gas.*

nitrogen and oxygen gases can occur at the base of a dam or a hydro-electric project, as the released water forms "froth," trapping large quantities of air, as shown in Figure 1.53.

So, back to the original question: *What caused the Riverwood fish kill?* You have considered several possible causes. How will you decide which was the actual cause? You will start by examining water-related measurements collected by scientists and engineers on the Snake River. From these data and what you have learned, you will decide the actual cause of the fish kill.

Making Decisions

C.14 DETERMINING THE CAUSE OF THE FISH KILL

Snake River watershed data have been collected and monitored since the early 1900s. Although some measurements and methods have changed during that time, excellent data have been gathered, particularly in recent years (see Figure 1.54, page 80). Data are available for the following factors:

- ▶ water temperature and dissolved oxygen
- ▶ rainfall
- ▶ water flow
- ▶ dissolved molecular substances
- ▶ heavy metals
- ▶ pH
- ▶ nitrate and phosphate levels
- ▶ organic carbon

Your teacher will assign you to a group to study some of the data just listed. Each group will complete data-analysis procedures for its assigned data. After the groups have finished their work, they will share their analyses and the class will draw conclusions about factors possibly related to the fish kill.

The following background information will help you complete the analysis of data assigned to you by your teacher.

Data Analysis

Interpreting graphs of environmental data requires a slightly different approach from the one you used to interpret a solubility curve. Rather than seeking a predictable relationship, you will be looking for regularities or patterns among the values. Any major irregularity in the data may suggest a problem related to the water factor being evaluated. The following suggestions will help you prepare and interpret such graphs.

▶ Choose your scale so that the graph is large enough to fill most of the available space on the graph paper.

▶ Assign each regularly spaced division on the graph paper a convenient, constant value. The graph-paper line interval value should be easily divided by eye, such as 1, 2, 5, or 10; rather than awkward values, such as 6, 7, 9, or 14.

Figure 1.54 *Researchers monitor water quality by checking for methane emissions (left) and measuring dissolved oxygen (right). Similar procedures provided the Snake River data.*

▶ An axis scale does not have to start at *zero,* particularly if the plotted values cluster in a narrow range not near zero.

▶ Label each axis with the quantity and unit being plotted.

▶ Plot each point. Then draw a small symbol around each point, like this: ⊙. If you plot more than one set of data on the same graph, distinguish each by using a different color or small geometric shape to enclose the points, such as ⊡, ▽, or △.

▶ Give your graph a title that will readily convey its meaning and purpose to readers.

▶ If you use technology—such as a graphing calculator or computer software—to prepare your graphs, ensure that you follow the guidelines just given. Different devices and software have different ways to process data. Choose the appropriate type of graph (scatter plot and bar graph, for example) for your data.

Next, follow the steps listed below when preparing and plotting your graphs, and answer the questions that follow.

1. Prepare a graph for each of your group's Snake River data sets. Label the *x*-axis (independent axis) with the consecutive months indicated in the data. Label the *y*-axis (dependent axis) with the water factor measured, accompanied by its units.

2. Plot each data point, and connect the consecutive points with straight lines.

3. Is any pattern apparent in your group's plotted data?

4. Can you offer possible explanations for any pattern or irregularities that you detect?

5. What arguments might someone make to challenge your explanation?

6. How might fish be adversely affected by the pattern or patterns that your group has identified?

7. Do you think the data analyzed by your group might help account for the Snake River fish kill? Why? How?

8. Prepare to share your group's data-analysis findings in a class discussion. During the class discussion, take notes on key findings reported by each data-analysis group. Also note and record significant points raised in the data-analysis discussions.

Your class will reassemble several times during your study of Section D to discuss and consider implications of the water-analysis data you have just processed. In particular, guided by the patterns and irregularities found in your analysis of Snake River data, your class will seek an explanation or scenario that accounts for the observed data and for the resulting Riverwood fish kill. Good luck!

SECTION C SUMMARY
Reviewing the Concepts

The solubility of a substance in water can be expressed as the quantity of that substance that will dissolve in a certain quantity of water at a specified temperature.

1. Explain why three teaspoons of sugar will completely dissolve in a serving of hot tea, but will not dissolve in an equally sized serving of iced tea.

2. What is the maximum mass of potassium chloride (KCl) that will dissolve in 100.0 g water at 70 °C?

3. If the solubility of sugar (sucrose) in water is 2.0 g/mL at room temperature, what is the maximum mass of sugar that will dissolve in
   a. 100.0 mL water?
   b. 355 mL (12 oz) water?
   c. 946 mL (1 qt) water?

4. Rank the substances in Figure 1.33 (page 54) from most soluble to least soluble at
   a. 20 °C.   b. 80 °C.

Solutions can be classified as unsaturated, saturated, or supersaturated. In quantitative terms, the concentration of a solution expresses the relative quantities of solute and solvent in a particular solution.

5. Distinguish between the terms *saturated* and *unsaturated.*

6. Using the graph on page 54, answer these questions about the solubility of potassium nitrate, KNO_3:
   a. What maximum mass of KNO_3 can dissolve in 100 g water if the water temperature is 20 °C?
   b. At 30 °C, 55 g KNO_3 is dissolved in 100 g water. Is this solution saturated, unsaturated, or supersaturated?
   c. A saturated solution of KNO_3 is formed in 100.0 g water at 75 °C. If some solute precipitates as the saturated solution cools to 40 °C, what mass (in grams) of solid KNO_3 should form?

7. You are given a solution of KNO_3 of unknown concentration. What will happen when you add a crystal of KNO_3, if the solution is
   a. unsaturated?   b. saturated?
   c. supersaturated?

8. A 35-g sample of ethanol is dissolved in 115 g water. What is the percent concentration of the ethanol, expressed as percent ethanol by mass?

9. Calculate the masses of water and sugar in a 55.0-g sugar solution that is labeled 20.0% sugar by mass.

10. The EPA maximum standard for lead in drinking water is 0.015 mg/L. Express this value as parts per million (ppm).

The polarity of water helps explain its ability to dissolve many ionic solids.

11. What makes a water molecule polar?

12. Draw a model that shows how molecules in liquid water generally arrange themselves relative to one another.

13. Which region of a polar water molecule will be attracted to a
   a. K^+ ion?   b. Br^- ion?

Heavy-metal ions can be useful resources, but some are toxic if introduced into biological systems, even in small amounts.

14. Why are heavy metals called *heavy?*

15. List three symptoms of heavy-metal poisoning.

16. List two possible sources of human exposure to
    a. lead.   b. mercury.

Solutions can be characterized as acidic, basic, or chemically neutral on the basis of their observed properties.

17. What ion is found in many bases?

18. What element is found in most acids?

19. Classify each sample as acidic, basic, or chemically neutral:
    a. seawater (pH = 8.6)
    b. drain cleaner (pH = 13.0)
    c. vinegar (pH = 2.7)
    d. pure water (pH = 7.0)

20. Using Figure 1.44 on page 70, decide which is more acidic:
    a. a soft drink or a tomato
    b. black coffee or pure water
    c. milk of magnesia or household ammonia

21. How many times more acidic is a solution at pH 2.0 than a solution at pH 4.0?

22. List three negative effects of inappropriate pH levels on aquatic organisms.

The solubility of a molecular substance in water depends on the relative strength of solute–water attractive forces, compared to competing solute–solute and water–water attractive forces.

23. Distinguish between polar and nonpolar molecules.

24. Would you select ethanol, water, or lamp oil to dissolve a nonpolar molecular substance? Explain.

25. Why does table salt (NaCl) dissolve in water but not in cooking oil?

26. Explain the phrase "like dissolves like."

27. Explain why you cannot satisfactorily clean greasy dishes with just plain water.

The solubility of a gaseous substance in water depends on the water temperature and the external pressure of the gas.

28. With increasing water temperature, how does the solubility of oxygen change?

29. How many milligrams of O_2 will dissolve in 1000.0 g water at 2.5 atm? (*Hint:* Use Figure 1.49 on page 76.)

30. As scuba divers descend, the pressure increases on the gases they breathe. How does the increasing pressure affect the amount of gas dissolved in their blood?

31. Given your knowledge of gas solubility, explain why a bottle of warm cola produces more fizz when opened than does a bottle of cold cola.

32. Using the graph on page 77, determine
    a. which type of fish requires more dissolved oxygen in water in the winter than in the summer.
    b. how much more dissolved oxygen in water is required by yellow perch in the summer than in the winter.

Connecting the Concepts

33. Many mechanics prefer to use waterless hand cleaners to clean their greasy hands. Explain
   a. what kind of materials are likely to be found in these cleaners.
   b. why using these cleaners is more effective than washing with water.

34. From each of these pairs, select the water source more likely to contain the higher concentration of dissolved oxygen. Give a reason for each choice.
   a. a river with rapids or a calm lake
   b. a lake in spring or the same lake in summer
   c. a lake containing only black crappie fish or a lake containing only black bullhead fish

35. Fluorine has the highest electronegativity of any element. Fluorine and hydrogen form a polar bond. Which atom in HF would you expect to have a partial positive charge? Explain.

36. Suppose there are two identical fish tanks with identical conditions except for temperature. Tank A is maintained at a temperature 5 °C higher than is Tank B. Which tank could support the greater number of similar fish? Explain.

37. Using Figure 1.48 on page 75,
   a. determine the solubility of oxygen at 20 °C.
   b. express the answer you obtained in ppm.

Extending the Concepts

38. Read the label on a container of baking soda. Compare it with the label on a container of baking powder. Which one contains an acid ingredient? Suggest a reason for including the acid in the mixture.

39. Describe how changes in solubility due to temperature could be used to separate two solid, water-soluble substances.

40. The continued health of an aquarium depends on the balance of the solubilities of several substances. Investigate how this balance is maintained in a freshwater aquarium.

41. The pH of rainwater is approximately 5.5. Rainwater flows into the ocean. The pH of ocean water, however, is approximately 8.7. Investigate the reasons for this difference in pH.

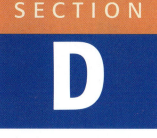

Water Purification and Treatment

Editorial

Attendance Urged at Special Council Meeting

A special town council meeting Wednesday will address three questions: (1) Who is responsible for the fish kill? (2) Who should pay the expense of trucking water to Riverwood during the three-day water shutoff? and (3) Who should pay for any losses resulting from the cancellation of the fishing tournament? Answers to these questions have financial consequences for all Riverwood taxpayers.

Those testifying at next week's public meeting include representatives of industry and agriculture, scientists who completed the river-water analyses, and consulting scientists and engineers who have investigated the cause of the fish kill. Chamber of commerce members, representing Riverwood store owners; representatives from the county sanitation commission; and officials of the Riverwood Taxpayer Association also will make presentations at the meeting.

We urge you to attend and participate in this meeting. Many questions remain. Was the fish kill an act of nature or was it due to human error? Was there evidence of negligence? Should the town's business community be compensated, at least in part, for financial losses from the fish kill? If so, how should they be compensated and by whom? Who should pay for the drinking water brought to Riverwood? Can this entire situation be prevented in the future? If so, at what expense? Who will pay for it?

The *Riverwood News* will set aside part of its Letters to the Editor column in the coming days for your comments on these questions and other matters related to the recent water crisis. For useful background information on water quality and treatment, we have prepared a special feature in today's paper.

Purifying Water Through the Hydrologic Cycle

The Hydrologic Cycle

BY RITA HIDALGO
Riverwood News Staff Reporter

Riverwood residents want to know how clean their water is. It should not take a crisis to focus attention on their concerns.

This article provides details on how natural water-purification systems work. A companion article in today's *News* examines municipal water-treatment methods—procedures that mimic, in part, water-purification processes at work in nature's water cycle.

Until the late 1800s, Americans obtained water from local ponds, wells, and rainwater holding tanks. Wastewater and even human wastes were discarded into holes, dry wells, or leaching cesspools (pits lined with broken stone). Some wastewater was simply dumped on the ground.

By 1880, about one-quarter of U.S. urban households had flush toilets, and municipalities were constructing sewer systems. However, as recently as 1909, sewer wastes were often released without treatment into lakes and streams, from which water supplies were drawn at other locations. Many community leaders believed that natural waters would purify themselves indefinitely.

Waterborne diseases increased as the concentration of intestinal bacteria in drinking water rose. As a result, water filtering and chlorination soon began. However, municipal sewage—the combined waterborne wastes of a community—remained generally untreated. Today, sewage treatment is part of every U.S. municipality's water-processing procedures.

Nature's water cycle, the hydrologic cycle, includes water-purification steps that address many potential threats to water quality. Thermal energy from the Sun causes water to evaporate from oceans and other water sources. Dissolved heavy metals, minerals, or molecular substances do not evaporate and, thus, are left behind.

This natural process accomplishes many of the same results as distillation. Water vapor rises, condenses into tiny droplets in clouds, and—depending on the temperature—eventually falls as rain or snow. Raindrops and snowflakes are nature's purest form of water, containing only dissolved atmospheric gases. However, human activities release a number of gases into the air, making today's rain less pure than it used to be.

When raindrops strike soil, the rainwater collects impurities. Organic substances deposited by living creatures become suspended or dissolved in the rainwater. A few centimeters below the soil surface, bacteria feed on these substances, converting them into carbon dioxide, water, and other simple compounds. Such bacteria thus help repurify the water.

As water seeps farther into the ground, it usually passes through gravel, sand, and even rock. Waterborne bacteria and suspended matter are filtered out. Thus, three processes make up nature's water-purification system.

- Evaporation, followed by condensation, removes nearly all dissolved substances.
- Bacterial action converts dissolved organic contaminants into a few simple compounds.
- Filtration through sand and gravel removes nearly all suspended matter.

> Environmentalists sometimes remark that "we are all downstream from someone else."

see *Purifying Water*, page 5

Purifiying Water, **from page 3**

Given appropriate conditions, people could depend solely on nature to purify their water. *Pure* rainwater is the best natural supply of clean water. If water seeping through the ground encountered enough bacteria for a long enough period of time, all organic contaminants could be removed. Flowing through sufficient sand and gravel would remove suspended matter from the water. However, nature's system only works well if it is not overloaded.

If slightly acidic groundwater (pH less than 7.0) passes through rocks that contain slightly soluble compounds, such as magnesium and calcium minerals, chemical reactions with these minerals may add substances to the water rather than remove substances in the water. In this case, the water may contain an increased concentration of dissolved minerals.

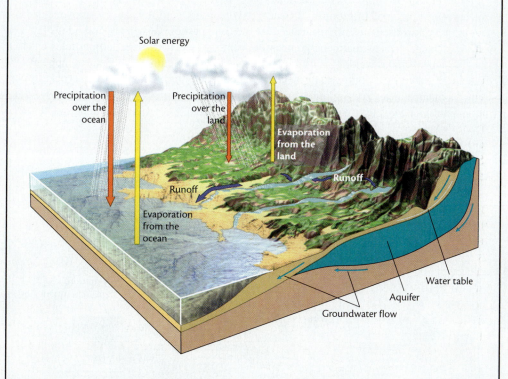

Hydrologic Cycle: Earth's water-purification system. The Sun provides energy for water to evaporate (yellow arrows). Evaporation leaves behind minerals and other dissolved substances. Water vapor condenses and falls as precipitation (red arrows), which runs off the land (blue arrows) to join surface water sources (lakes, streams, and rivers) or groundwater sources beneath Earth's surface. Both surface water and groundwater are sources of municipal and agricultural water. Any water—when returned to the surface—can evaporate and continue flowing through the hydrologic cycle. Throughout the cycle, evaporation, bacterial action, and filtration purify water.

Purifying Water Supplies by Municipal Pre-Use Treatment

BY RITA HIDALGO
Riverwood News Staff Reporter

Today, many rivers—such as the Snake River in Riverwood—are both a source of municipal water and a place to release wastewater (sewage). Therefore, the water is cleaned twice, both before and after the water is used. *Pre-use purification,* often called *water treatment,* takes place at a municipal filtration and treatment plant. It is the focus of this article.

A typical water-treatment process begins when intake water flows through metal screens that prevent fish, sticks, beverage containers, and other large objects from entering the water-treatment plant.

Plant operators may add chlorine, which is a powerful disinfecting agent, during the treatment process to kill disease-causing organisms. This step is known as *prechlorination.* Operators add crystals of alum—aluminum sulfate, $Al_2(SO_4)_3$—and slaked lime—calcium hydroxide, $Ca(OH)_2$—to remove suspended particles, such as colloidal clay, from the water. (Suspended particles can give water an unpleasant, murky appearance.) The alum and slaked lime react to form aluminum hydroxide, $Al(OH)_3$, which is a sticky, jellylike material that

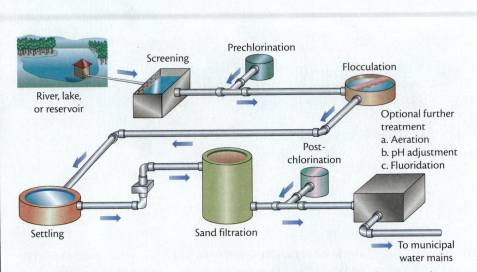

Municipal Treatment: A human-made water-purification system. Surface water is commonly cleaned at a municipal water-treatment plant before being distributed to homes and businesses. Various steps in the cleaning process remove suspended materials, kill disease-causing organisms, and may remove odors or adjust pH levels.

see *Municipal Treatment,* page 6

Municipal Treatment, from page 3

traps and removes the suspended particles, a process called *flocculation.* Operators then allow the aluminum hydroxide (holding trapped particles from the water) and other solids remaining in the water to settle to the tank bottom. They remove any suspended materials that do not settle out by filtering the water through sand.

In the *post-chlorination* step, operators adjust the chlorine concentration in the water to ensure that a low, but sufficient, concentration of chlorine remains in the water, thereby protecting the water from bacterial infestation.

One or more additional steps may take place, depending on community procedures. Some plants spray water into the air to remove odors and improve its taste, a process known as *aeration.*

Water may sometimes be acidic enough to slowly dissolve metallic water pipes. This process not only shortens pipe life, but it may also cause copper (Cu^{2+}), as well as cadmium (Cd^{2+}) and other undesirable ions, to enter the home water supply. A plant may add lime—calcium oxide (CaO), a basic substance—to neutralize such acidic water, thus raising its pH to a proper level.

As much as about 1 ppm of fluoride ion (F^-) may be added to the treated water in a process known as *fluoridation.* Even at this low concentration, fluoride ions can reduce tooth decay.

D.3 WATER PURIFICATION

Refer to the two water-treatment articles by Rita Hidalgo (pages 86 and 88) to answer the following questions.

1. Compare natural water-treatment steps to the treatment steps in municipal wastewater systems.

   a. What are key similarities?
   b. What are key differences?

2. After reading the two water-treatment articles, a Riverwood resident wrote a letter to the *Riverwood News* proposing that the town's water-treatment plant be shut down. The reader pointed out that this would save taxpayers considerable money because "natural water treatment can meet our needs just as well." Do you support the reader's proposal? Explain your answer.

Chlorine is probably the best known and most common substance used for water treatment; it is found not only in community water supplies but also in swimming pools. The following article provides background on the role of chlorine in water treatment.

Chlorine in Public Water Supplies

BY **RITA HIDALGO**
Riverwood News Staff Reporter

The single most common cause of human illness in the world is unhealthful water supplies. Without a doubt, adding chlorine to public water supplies has helped save countless lives by controlling water-borne diseases. In water, chlorine kills disease-producing microorganisms.

In most municipal water-treatment systems, chlorination, which is the addition of chlorine to the water supply to kill harmful organisms, often takes place in several different ways:

- Chlorine gas, Cl_2, is bubbled into the water. This substance is not very soluble in water. Chlorine does react with water, however, to produce a water-soluble, chlorine-containing compound.

- A water solution of sodium hypochlorite, $NaOCl$, which is the active ingredient in household bleach, is added to the water.

- Calcium hypochlorite, $Ca(OCl)_2$, is dissolved in the water. Available as both a powder and small pellets, calcium hypochlorite is often used in swimming pools. It is also a component of some solid household products sold as bleaching powder.

Regardless of how chlorination takes place, chemists believe that chlorine's most active form in water is hypochlorous acid ($HOCl$). This substance forms whenever chlorine, sodium hypochlorite, or calcium hypochlorite dissolves in water.

Unfortunately, a potential problem is associated with adding chlorine to municipal water. Under some conditions, chlorine in water can react with organic compounds produced by decomposing animal and plant matter to form substances that, if in sufficiently high concentrations, can be harmful to human health.

One group of such substances is known as the *trihalomethanes* (THMs). A common THM is chloroform ($CHCl_3$), a *carcinogen,* which is a substance that is known to cause cancer.

Because of concern about the possible health risks associated with THMs, the Environmental Protection Agency has placed a current limit of 80 parts per billion (ppb) on total THM concentration in municipal water-supply systems.

Possible risks associated with THMs must be balanced, of course, against the benefits of chlorinated water.

Flocculator–clarifier units at a municipal water-treatment facility.

D.5 CHLORINATION AND THMs

Operators of municipal water-treatment plants have several options for eliminating possible THM health risks highlighted in the newspaper article that you just read. However, each method has its disadvantages.

▶ They can pass treatment-plant water through an activated charcoal filter. Activated charcoal can remove most organic compounds from water, including THMs. *Disadvantage:* Charcoal filters are expensive to install and operate. Disposal also poses a problem because it is hard to clean used filters of contamination. Filters must be replaced relatively often.

▶ They can completely eliminate chlorine and use ozone (O_3) or ultraviolet light to disinfect the water. *Disadvantage:* Neither ozone nor ultraviolet light protects the water once it leaves the treatment plant. Treated water can be infected by bacteria—through faulty water pipes, for example. In addition, ozone can pose toxic hazards if not handled and used properly.

▶ They can eliminate prechlorination. Chlorine would be added only once, after filtering the water and removing much of the organic material. *Disadvantage:* The chlorine added in post-chlorination can still promote formation of THMs, even if to a lesser extent than with prechlorination. Additionally, a decrease in chlorine concentration might allow bacterial growth in the water.

Making Decisions

D.6 BOTTLED WATER VERSUS TAP WATER

Water that has been chlorinated sometimes has a characteristically bad taste. In general, when people do not like the taste of tap water or they think that the available water is unsafe to drink or they do not have access to other sources of fresh water, or just for convenience, they may buy bottled water (see Figure 1.55). This bottled water may come from a natural source, such as a mountain spring, or it may be processed at the bottling plant.

Is this water any better for you than tap water that has been processed by a municipal water treatment plant or that comes directly from ground or surface water?

Figure 1.55 *Which is better, bottled water or tap water? Why?*

Could it possibly be harmful? What determines water quality? How can the risks and benefits of drinking water from various sources be assessed?

As usual, answering challenging questions requires gathering reliable data and information, weighing alternatives, and making informed decisions. Working with a partner, answer the following questions:

1. In your view, what two or three factors should you consider in deciding whether to drink tap water or bottled water?

2. For each factor listed in your answer to Question 1, what factual information would be needed to establish the advantages and disadvantages of drinking bottled water rather than tap water?

Investigating Matter

D.7 WATER SOFTENING

Lab Video:
Water Softening

Introduction

The goal of municipal water-treatment facilities is to provide water that is safe to drink, it is not to produce completely pure water. One aspect—water hardness—is not usually addressed at the municipal-treatment level.

Water containing an excess of dissolved calcium (Ca^{2+}), magnesium (Mg^{2+}), or iron(III) (Fe^{3+}) ions is known as **hard water.** Hard water does not form a soapy lather easily. River water usually contains low concentrations of hard-water ions. However, as groundwater flows over limestone, chalk, and other minerals that contain calcium, magnesium, and iron, it often gains higher concentrations of these ions, thus producing hard water. See Table 1.7 and Figure 1.56.

In this investigation, you will explore several ways of softening water by comparing three possible water treatments and determining

The formula $CaSO_4 \cdot 2H_2O$ indicates there are two molecules of water for every unit of $CaSO_4$ contained in this solid crystal.

| Table 1.7

SOME MINERALS CONTRIBUTING TO WATER HARDNESS		
Mineral	**Chemical Composition**	**Chemical Formula**
Limestone or chalk	Calcium carbonate	$CaCO_3$
Magnesite	Magnesium carbonate	$MgCO_3$
Gypsum	Calcium sulfate	$CaSO_4 \cdot 2H_2O$
Dolomite	Calcium carbonate and magnesium carbonate combination	$CaCO_3 \cdot MgCO_3$

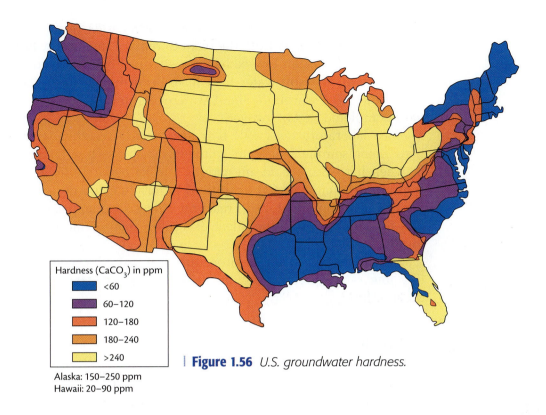

Hardness (CaCO$_3$) in ppm

■	<60
■	60–120
■	120–180
■	180–240
■	>240

| **Figure 1.56** *U.S. groundwater hardness.*

Alaska: 150–250 ppm
Hawaii: 20–90 ppm

whether they remove calcium ions from a hard-water sample: (1) sand filtration, (2) treatment with Calgon, and (3) treatment with an ion-exchange resin.

Calgon, which contains sodium hexametaphosphate (Na$_6$P$_6$O$_{18}$), and similar commercial products remove hard-water cations, including calcium ions, by causing them to become part of larger, soluble anions. For example:

$$2\ \text{Ca}^{2+}(aq) \quad + \quad (\text{P}_6\text{O}_{18})^{6-}(aq) \quad \longrightarrow \quad [\text{Ca}_2(\text{P}_6\text{O}_{18})]^{2-}(aq)$$

| Calcium ion from hard water | Hexametaphosphate ion from Na$_6$P$_6$O$_{18}$ | Calcium hexametaphosphate anion |

Calcium hexametaphosphate is a water-soluble anion that binds calcium ions and prevents them from forming insoluble compounds. Calgon also contains sodium carbonate (Na$_2$CO$_3$), which softens water by removing hard-water cations as precipitates, such as calcium carbonate (CaCO$_3$). The equation for this reaction follows. Solid calcium carbonate particles are washed away with the rinse water.

$$\text{Ca}^{2+}(aq) \quad + \quad \text{CO}_3{}^{2-}(aq) \quad \longrightarrow \quad \text{CaCO}_3(s)$$

| Calcium ion from hard water | Carbonate ion from sodium carbonate | Calcium carbonate precipitate |

Another water-softening method relies on **ion exchange.** Hard water is passed through an ion-exchange resin, such as that found in home water-softeners. The resin consists of millions of tiny, insoluble, porous beads capable of attracting and binding cations.

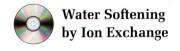

Water Softening by Ion Exchange

Hard-water ions are retained on the ion-exchange resin, while soft-water ions are released from the resin to replace the hard-water ions in the solution. Thus, an ion-exchange resin must be recharged occasionally when all the soft-water ions have been displaced by hard-water ions. You will learn more about water-softening procedures after you have completed this investigation.

Two laboratory tests will help you decide whether your original hard-water sample has been softened. The first test is the reaction between hard-water calcium cations and carbonate anions (added as sodium carbonate, Na_2CO_3, solution) to form a calcium carbonate precipitate. The equation for this reaction is described on page 93. The second test is to observe the effect of adding soap to the water sample, forming a lather.

Before starting, read the procedure to learn what you will need to do, note safety precautions, and plan necessary data collecting and observations.

Procedure

1. In your laboratory notebook, prepare a data table similar to the one that follows.

DATA TABLE

Test	Filter Paper		Filter Paper and Sand		Filter Paper and Calgon		Filter Paper and Ion-Exchange Resin	
	Distilled Water	Hard Water	Distilled Water	Hard Water	Distilled Water	Hard Water	Distilled Water	Hard Water
Reaction with sodium carbonate (Na_2CO_3)								
Degree of cloudiness (turbidity) with Ivory soap								
Height of suds (mm)								

2. Prepare the lab setup as shown in Figure 1.57. Lower the tip of the funnel stem into a test tube supported in a test-tube rack.

3. Fold a piece of filter paper; insert it in the funnel.

4. You will repeat Steps 5 through 7 several times. This first time, your funnel should contain only the filter paper; it serves as the control.

5. Pour about 10 mL of distilled water into the funnel. Do not pour any water over the top of the filter paper or between the filter paper and the funnel wall. Collect the filtrate in the test tube.

6. Divide the filtrate into two equal volumes of approximately 5 mL, using two clean test tubes. Add 10 drops of sodium carbonate (Na_2CO_3) solution to one of the test tubes. Does a precipitate form? Record your observations. A cloudy precipitate indicates that the Ca^{2+} ion (a hard-water cation) was not removed. Clean the test tube thoroughly with tap water and rinse with distilled water.

7. Take the second test tube containing 5 mL of filtrate and add one drop of Ivory liquid hand soap (*not* liquid detergent). Label the test tube and set it aside for Steps 12 through 15.

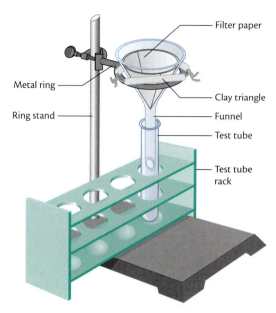

Figure 1.57 *Filtration setup.*

8. Repeat Steps 5 through 7, using 10 mL of hard water instead of distilled water.

9. Install a new piece of filter paper and fill it one-third full of sand. Repeat Steps 5 through 7 by pouring distilled water through the sand; then repeat Steps 5 through 7, pouring hard water through the sand. Dispose of the filter paper as directed by your teacher.

10. Install a new piece of filter paper and fill it one-third full of Calgon. Repeat Steps 5 through 7 for distilled water and hard water. Dispose of the solids held in your filter paper as directed by your teacher. (*Note:* The Calgon-treated filtrate may appear blue due to other softener additives. They will not cause a problem. However, if the filtrate appears cloudy, which means that some Calgon powder passed through the filter paper, use a new piece of filter paper and filter the test-tube liquid again.)

11. Install a new piece of filter paper and fill it one-third full of ion-exchange resin. Repeat Steps 5 through 7 for each type of water tested. Dispose of the solids held in your filter paper as directed by your teacher.

12. Take each test tube from Step 7 and stir gently. Wipe the stirring rod before inserting it into another test tube.

13. Compare the cloudiness of the soap solutions. Record your observations. The greater the cloudiness, the greater the quantity of soap that dispersed. The quantity of dispersed soap determines the cleaning effectiveness of the solution.

14. Stopper each test tube and then shake vigorously, as demonstrated by your teacher. The more suds that form, the softer the water. Measure the height of suds in each test tube and record your observations.

15. Clean the test tubes thoroughly with tap water and rinse with distilled water.

16. Wash your hands thoroughly before leaving the laboratory.

Questions

1. Which was the most effective water-softening method? Support your answer with evidence from observations and data you gathered. Suggest why this method worked best.

2. What relationship can you describe between the amount of hard-water ion (Ca^{2+}) remaining in the filtrate and the dispersion (cloudiness) of Ivory liquid hand soap?

3. What effect does this relationship have on the cleansing action of the soap?

D.8 WATER AND WATER SOFTENING

Hard Water and Soap Scum

Hard water causes some common household problems. It interferes with the cleaning action of soap. As you observed, when soap mixes with soft water, it disperses to form a cloudy solution topped with a sudsy layer. In hard water, however, soap reacts with hard-water ions to form insoluble compounds (precipitates). These compounds appear as solid flakes or a sticky scum. The precipitated soap is no longer available for cleaning. Worse yet, soap curd can deposit on clothes, skin, and hair. The structural formula for this substance, the product of the reaction of soap with calcium ions, is shown in Figure 1.58.

If hydrogen carbonate (bicarbonate, HCO_3^-) ions are present in hard water, boiling the water causes solid calcium carbonate ($CaCO_3$) to form. The reaction removes undesirable calcium ions and, thus, softens the water. However, the solid calcium carbonate can produce rocklike scale inside tea kettles, household hot-water heaters, and even power-plant boilers. This scale (the same compound found in marble and limestone) acts as thermal insulation, partly blocking heat flow to the water. More time and energy are required to heat the water. Such

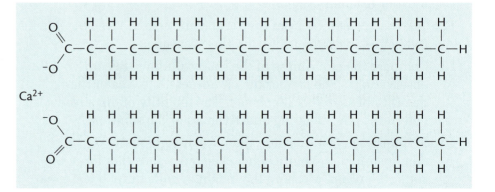

Figure 1.58 *Structural formula of a typical soap scum. This substance is calcium stearate.*

Figure 1.59 *In the 1960s and early 1970s, detergent molecules not decomposed by microorganisms caused some waterways to fill with sudsy foam, such as shown here. The development of biodegradable detergent molecules put an end to this unusual kind of water pollution.*

deposits can also form in home water pipes. In older homes with this problem, water flow can be greatly reduced.

Fortunately, it is possible to soften hard water by removing some dissolved calcium, magnesium, or iron(III) ions. Adding sodium carbonate to hard water (as you did in the preceding investigation) was an early method of softening water. Sodium carbonate (Na_2CO_3), known as *washing soda,* was commonly added to laundry water along with the clothes and soap. Hard-water ions, precipitated as calcium carbonate ($CaCO_3$) and magnesium carbonate ($MgCO_3$), were washed away in the rinse water.

Water-softening substances in common use today include borax, trisodium phosphate, and sodium hexametaphosphate (as in Calgon). As you learned, Calgon does not tie up hard-water ions as a precipitate, but rather as a new, soluble ion that does not react with soap.

Most cleaning products sold today contain *detergents* rather than soap. Synthetic **detergents** act like soap, but they do not form insoluble compounds with hard-water ions. Unfortunately, many early detergents were not easily decomposed by bacteria in the environment; that is, they were not biodegradable. At times, "mountains" of foamy suds were observed in natural waterways. See Figure 1.59. These early detergents also contained phosphate ions (PO_4^{3-}) that encouraged extensive algae growth, choking lakes and streams. Because most of today's detergents are biodegradable and phosphate free, they do not cause these problems.

If you live in a hard-water region, your home plumbing may include a **water softener** (see Figure 1.60). Hard water flows through a tank containing an ion-exchange resin similar to the one that you used in the water-softening investigation. Initially, the resin is filled with sodium cations (Na^+). Calcium and magnesium cations in the hard water are attracted to the resin and become attached to it. At the same time, sodium cations leave the resin and dissolve in the water. Thus, undesirable hard-water ions are exchanged for sodium ions, which do not react to form soap curd or water-pipe scale.

As you might imagine, the resin eventually fills with hard-water ions and must be regenerated. Concentrated salt water (containing sodium ions and chloride ions) flows through the resin, replacing the hard-water ions held on the resin with sodium ions. Released hard-water ions wash down the drain with excess salt water. Because this process takes several hours, it is usually completed at night. After the resin has been regenerated, the softener is again ready to exchange ions with incoming hard water.

Water softeners are most often installed in individual homes. Other water treatment is done at a municipal level, both in Riverwood and in other cities.

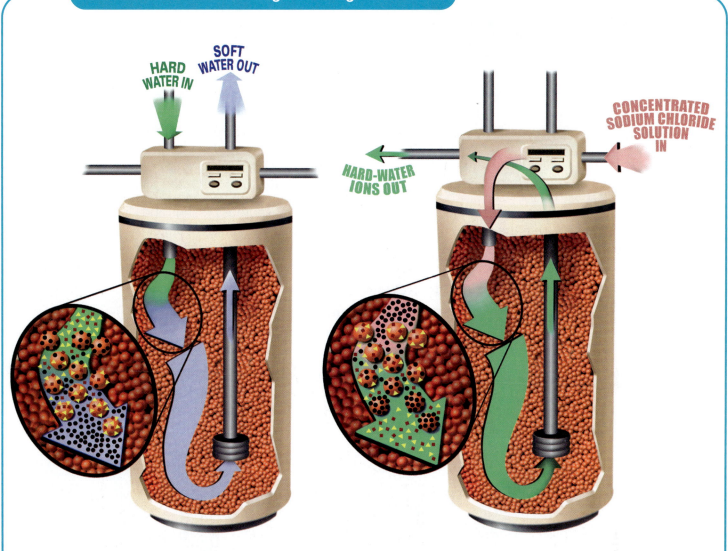

Figure 1.60 *Home Water Softener.* *Hard water, containing excess calcium (Ca^{2+}), magnesium (Mg^{2+}), or iron(III) (Fe^{3+}) cations, prevents soap from lathering easily and can cause scale to build up inside pipes. Thus, where hard water poses a problem, many people install home water softeners. Water softeners contain ion-exchange resin, depicted here as orange beads. Left: Initially, sodium ions (Na^+, depicted as black dots in the magnified portion) are attached to the resin. As hard water flows into the tank, hard-water cations such as Mg^{2+} and Ca^{2+} (yellow triangles and red squares) become attached to the resin, releasing Na^+ ions into the water. This softened water flows through pipes to faucets, water heaters, and washing machines. Right: Eventually, all Na^+ ions are replaced by hard-water ions; no further ion exchange can occur. Then, the resin is "recharged" with Na^+ ions (black dots) by passing concentrated sodium chloride solution through the system. This displaces hard-water ions (yellow triangles and red squares) from the resin; they flow from the water softener into a drain. Thus, regenerated resin—with Na^+ ions again attached—is ready to soften hard water again.*

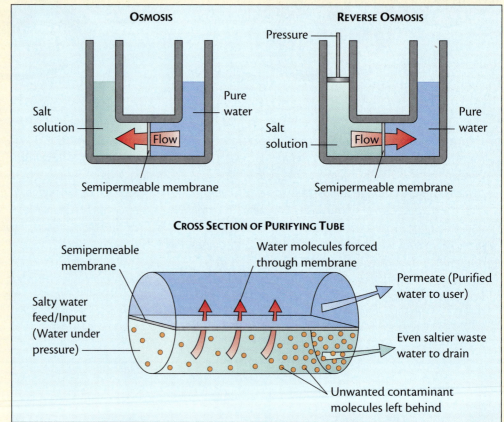

CHEMISTRY at Work

Purifying Water Means More Than Going with the Flow

When you drink from a water fountain, do you ever wonder where the water comes from? In some parts of the country, drinking water is provided by people such as Phil Noe.

Phil is the Production Manager at Island Water Association (IWA), which provides water for Florida's Sanibel and Captiva islands. "We have a limited supply of fresh water," says Phil, "so we built a plant that lets us use water from our aquifers." Aquifers are underground layers of

"We have a limited supply of fresh water," says Phil, "so we built a plant that lets us use water from our aquifers."

permeable (porous) sand and limestone that contain large quantities of water.

After pumping the brackish, undrinkable water from the Suwanee Aquifer to IWA's processing plant, Phil and his coworkers remove most of the salt and minerals, producing water that is purer than many mountain streams through a process known as *reverse osmosis*. The accompanying illustration is a comparison of osmosis with reverse osmosis. The level of

OSMOSIS

Salt solution — Pure water
Flow
Semipermeable membrane

REVERSE OSMOSIS

Pressure
Salt solution — Pure water
Flow
Semipermeable membrane

CROSS SECTION OF PURIFYING TUBE

Semipermeable membrane
Water molecules forced through membrane
Salty water feed/Input (Water under pressure)
Permeate (Purified water to user)
Even saltier waste water to drain
Unwanted contaminant molecules left behind

Top Left: *Osmosis occurs naturally when water in a dilute solution passes through the semipermeable membrane into a concentrated solution.*
Top Right: *In reverse osmosis, pressure must be applied to a concentrated solution to force water through the semipermeable membrane, producing purified water.*

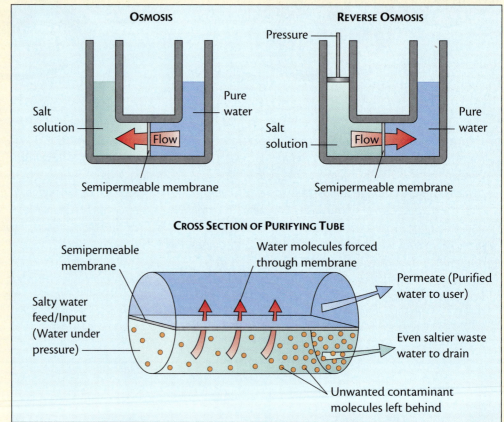

After filtering to remove the larger solids, the water passes through equipment where reverse osmosis is used to produce purified water.

Saving up for a windy day.

Aquatic wildlife is unaffected by the salt water left over from the reverse osmosis process.

pressure applied in reverse osmosis must exceed the osmotic pressure, which tends to move the water from a region of higher vapor pressure to one of lower vapor pressure. Because osmotic pressure depends only on the concentration of the salt solution, it is known as a *colligative property.*

Aquifers are located at depths ranging between 240 and 310 m (700–900 ft). Water pumped from these aquifers contains an average of 3000 ppm total dissolved solids (TDS). Feed water is pumped through long stretches of pipes that contain salt-filtering membranes. In the process, the purified water, called the *permeate,* separates from the salts and other dissolved solids. Eventually, the water is purified to levels as low as 100 ppm TDS.

Because Florida is subject to hurricanes and other tropical storms, Sanibel Island's water system maintains 15 million gallons of purified water in storage. As a result, island customers can manage without the plant in operation during brief periods of severe weather with no more inconvenience than several weeks without lawn sprinklers.

Web Search . . .

- Search the World Wide Web for other communities that rely on reverse osmosis for all or part of their water supply.

- Most communities use rivers, lakes, or wells to supply their water needs. Search the Web for information about municipal water-supply systems that use alternative water purification processes.

- What other kinds of water purification systems are in use in the United States?

- Use the Web to gather information about the cost of water for household use in various parts of the United States. Is there a connection between the cost of water and the location of the community? For example, is water more expensive in the desert than it is near the ocean?

SECTION D SUMMARY
Reviewing the Concepts

Water can be purified through the actions of the hydrologic cycle or through municipal treatment.

1. Make a diagram of the hydrologic cycle and label all processes.

2. List three major processes that occur in natural water purification and, for each, identify the contaminants that the process removes.

3. How are the properties of aluminum hydroxide related to the process of flocculation?

4. Why is calcium oxide (CaO) sometimes added in the final steps of municipal water treatment?

5. Fluoride, an ingredient in many types of toothpaste, is sometimes added to municipal water supplies in the last stage of water treatment. How much fluoride is added and what is its purpose?

Chlorination is commonly used to treat and purify water for human consumption.

6. What are advantages of chlorinated drinking water compared to untreated water?

7. Is there a disadvantage to using chlorination in water treatment? Explain.

8. Water from a clear mountain stream may require chlorination to make it safe for drinking. Explain.

9. List two alternatives to the use of chlorination in municipal water treatment.

Hard water can be softened by removing hard-water ions as a precipitate or as large soluble ions.

10. What are two problems associated with the use of hard water?

11. Identify three common hard-water ions.

12. When a sample of well water was mixed with a few drops of sodium carbonate solution, a precipitate formed. What does the formation of a precipitate indicate about the water sample?

13. Hard water often tastes better than distilled water. Explain.

14. Which water source in a given locality would probably have harder water, a well or a river? Explain.

15. Sketch two molecular-level representations of an ion-exchange resin bead—one bead before and one bead after treatment of hard water.

Connecting the Concepts

16. A simple test of water hardness is to add soap to the water sample and shake. Explain how you can measure the quantity of soap suds to assess the relative hardness of the water.

17. Explain why hard water can decrease the efficiency of a boiler in a steam-generated, electric power plant.

18. Explain what would happen to Earth's hydrologic cycle if water evaporation suddenly stopped.

19. One unique characteristic of water is that it is present in all three physical states (solid, liquid, and gas) in the range of temperatures found on Earth. How would the hydrologic cycle be different if this were not true?

20. Why does the EPA limit the concentration of THMs to 80 ppb instead of requiring their total elimination from municipal water supplies?

21. Compare how the various processes used in the foul-water investigation (page 10) are similar to steps in the natural purification of water.

22. Some physicians recommend consuming about 2 L of water daily. Municipal water supplies may contain up to 1 ppm fluoride. Assume that you drink 2 L of water per day. At 1 ppm fluoride, how many grams of fluoride ion would you consume in

a. one day?   b. one week?   c. one year?

Extending the Concepts

23. How much water would you need to drink to get your minimum daily requirement of calcium from water that contains 300 ppm Ca^{2+}?

24. Explain why we find hard-water stains in old sinks around hot-water faucets more often than around cold-water faucets.

25. Research the sources and production of a brand of bottled water. Report on its origin and identify the substances that are removed and added before the water is bottled and sold.

26. Discuss the health effects associated with sodium-based ion-exchange resins used in home water softening systems.

27. Compare activated charcoal with reverse osmosis in home water filtration systems.

28. Describe and evaluate the practicality of two desalination processes for making seawater suitable for drinking.

FISH KILL—FINDING THE SOLUTION

Fish Kill Cause Identified

Meeting Tonight

BY ORLANI O'BRIEN
Riverwood News Staff Reporter

Mayor Edward Cisko announced at a news conference earlier today that the cause of the fish kill in the Snake River has been determined. The details of the accident cause will be released at a town council meeting tonight. As of today, levels of all dissolved materials in the river are normal and the water should be considered safe.

Accompanying Mayor Cisko was Dr. Harold Schmidt of the Environmental Protection Agency. Schmidt dissected fish taken from the river within a few hours of their death. He also directed the team that analyzed accumulated river-water data in efforts that led to determining the cause of the fish kill. Schmidt gave assurances that the town's water supply is "fully safe to drink."

Mayor Cisko refused to elaborate on the reasons for the accident. However, he invited the public to the special council meeting at 8 p.m. tonight at the town hall. The council will discuss events that caused the fish kill and how costs associated with the three-day water shutoff will be paid. Several area groups, as well as invited experts, plan to make presentations at tonight's meeting.

Mayor Cisko announced that the fish-kill mystery has been solved.

TOWN COUNCIL MEETING

Your teacher will assign you to one of the following Riverwood groups that will participate in the special town council meeting (similar to Figure 1.61), and will give you some background information about your group's viewpoints on the impact of the fish kill:

▶ Agricultural cooperative representatives
▶ Consulting engineers
▶ Consulting scientists

- ▶ County sanitation commission members
- ▶ Mining company representatives
- ▶ Power company officials
- ▶ Riverwood Chamber of Commerce members
- ▶ Riverwood Taxpayer Association members
- ▶ Riverwood Town Council members

Meeting Rules and Penalties for Rule Violations

1. The presentation order is decided by council members and announced at the start of the meeting.
2. Each group will have a specified time for its presentation. Time cards will notify each speaker of time remaining.
3. If a member of another group interrupts a presentation, the offending group will be penalized thirty seconds for each interruption, to a maximum of two minutes.

Figure 1.61 *A town meeting in session in New Jersey. Open meetings such as this encourage citizen involvement in discussions of local issues and in related decision-making challenges.*

If the group has already made its presentation, it will forfeit its rebuttal time.

LOOKING BACK AND LOOKING AHEAD

The Riverwood water mystery is solved! In the end, scientific data and analysis provided the answer. Now, human ingenuity will provide strategies to prevent the recurrence of such a crisis. In the course of solving the problem, the citizens of Riverwood learned about the water that they take for granted—abundant, clean water flowing steadily from their taps—and gained a greater appreciation for it.

Although Riverwood and its citizens exist only on the pages of this textbook, their water-quality crisis could be very real. The chemistry-related facts, principles, and procedures that clarified their problem and its solution have applications in your own home and community.

Although the fish-kill mystery has been solved, your exploration of chemistry has only just begun. Many issues related to chemistry in the community remain; water and its chemistry are only one part of a larger story.

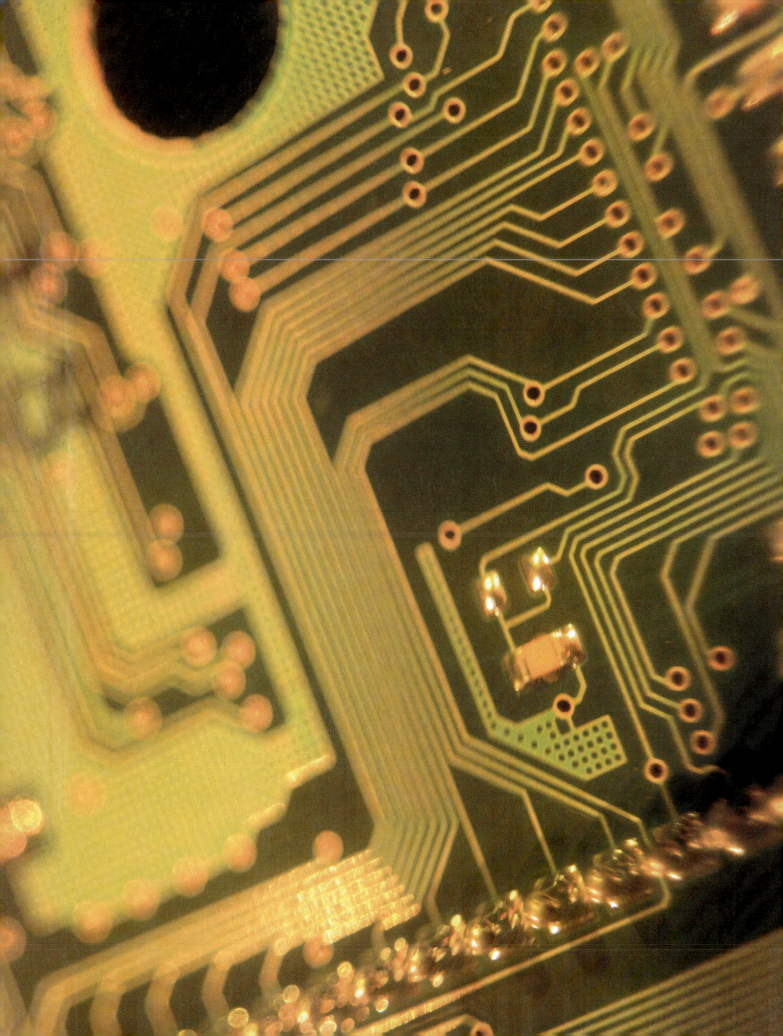

UNIT 2

Materials: Structure and Uses

HOW can chemists explain the chemical and physical properties of matter?

SECTION A
Why We Use What We Do (page 110)

WHERE do we find mineral resources and how are they processed?

SECTION B
Earth's Mineral Resources (page 134)

WHAT information do chemical equations convey about matter and its changes?

SECTION C
Conserving Matter (page 153)

HOW can chemists modify matter to make it more useful?

SECTION D
Materials: Designing for Desired Properties (page 184)

Your congresswoman has invited you and your classmates to submit a design for a new half-dollar coin. What will be the coin's composition? What makes a material best suited for its intended use? Turn the page to learn the answers— perhaps you can become the design winner.

The Honorable María Gonzales
United States House of Representatives

MEMORANDUM

TO: District 12 High School Principals and Chemistry Teachers

FROM: The Hon. Maria Gonzales
        U.S. House of Representatives

SUBJECT: Coin-Design Competition

As you may know, I plan to introduce a bill in the House of Representatives authorizing the production of a new half-dollar coin by the U.S. Mint. The purpose of this memorandum is to inform you about a contest my office is sponsoring for high school chemistry students in your congressional district. The competition involves proposing a possible design for the new coin. The winning design will be used as an example in House Committee deliberations and on the floor of the House as I seek support from colleagues for authorization of the new half-dollar coin.

Each high school within your congressional district may submit one complete design. The student or team of students who creates the winning design will be honored, along with the rest of their chemistry class, at an open house at my local office. In addition, the winning student or team will travel with my staff to Washington, D.C., for the introduction of my bill and for a public presentation of the suggested coin design.

A complete coin-design proposal must include the following information:
- Full name(s) and address(es) of the designer(s)
- Chemistry teacher's name and course title
- School name and phone number
- Coin diameter, thickness, and mass
- Detailed drawing or actual model of the coin, enlarged 5 times for clarity
- Specifications for the composition of the coin's material
- Plans for obtaining or creating the materials used in the coin
- Two-page rationale for key decisions made in the coin's design

All completed proposals will be due in my office within six weeks of receipt of this memorandum.

MG/hs

Encouraged by the success of the recent quarter coin series featuring each state, Congressional Representative Maria Gonzales plans to introduce a bill authorizing the U.S. Mint to produce a new half-dollar coin. To solicit unique coin-design ideas from her community, she is sponsoring a contest for high-school chemistry students in her district to propose coin designs.

Representative Gonzales realizes that her congressional colleagues will request information about the proposed coin. Thus, as her memo suggests, every aspect of the half-dollar coin—from its appearance to its size and composition—is to be included in the design. Because only one design can be submitted from each school, your class will decide which team's coin proposal will be submitted for this competition.

As you consider your coin's design, you will learn about Earth's mineral resources and how nations use them. You will learn why certain materials are used for particular new products, such as coins, and how those materials are obtained from available resources. Throughout this unit, keep in mind how such chemical knowledge can help guide your design of the new coin.

Why We Use What We Do

Every produced object, old or new, is composed of materials selected for their specific properties. What makes a particular material best for a particular use? You can begin to answer this question by exploring some properties of materials.

A.1 PROPERTIES MAKE THE DIFFERENCE

Properties of Metals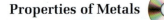

In this unit, you will consider the design of something that you use every day—money in the form of coins. Throughout history, people have used many different items as money: beads, stones, printed paper, and precious metals, to name a few. See Figure 2.1. What characteristics make a material suitable or useful to produce coins? How important is appearance or cost? What other characteristics or properties can you suggest?

As you already know, every substance has characteristic properties that distinguish it from other substances. These characteristic properties include **physical properties** such as color, density, and odor—properties that can be determined without altering the chemical makeup of the material. Physical properties and the ability (or inability) of a material to undergo physical changes, such as melting, boiling, and bending, often influence the use of that material. In a **physical change,** the material remains the same, although its form appears to have changed.

When a substance changes into one or more *new* substances, it has undergone a **chemical change.** A substance's **chemical properties,** which

Figure 2.1 *What properties should you consider when you design a coin?*

relate to any kind of chemical changes it undergoes, often determine the substance's usefulness. Consider the common chemical change of iron rusting. The tendency of a metal to form an oxide, such as when iron rusts, is the chemical property that accounts for this chemical change. You can often detect a chemical change by observing one or more indications of a change, such as the formation of a gas or solid, a permanent color change, or a temperature change, which indicates that thermal energy has been absorbed or given off. Figure 2.2 illustrates some physical and chemical changes of copper.

In the following activity, you will classify some characteristics of common materials as either physical or chemical properties.

A.2 PHYSICAL AND CHEMICAL PROPERTIES

Developing Skills

Sample Problem 1: *Consider this statement: Copper compounds are often blue in color. Does the statement describe a physical or chemical property?*

To answer this, first ask yourself a question: Was the substance chemically changed as its color was observed? If the answer is *no,* then the statement describes a physical property; if the answer is *yes,* then the statement describes a chemical property. You can observe this property—color—without changing the chemical makeup of a copper compound. Color is a characteristic physical property of many compounds.

Sample Problem 2: *Consider this statement: Oxygen gas supports the burning of wood. Does the statement refer to a physical or chemical property of oxygen gas?*

If you apply the same key question—is there a change in the identity of the wood and the oxygen?—you will arrive at the correct answer. As you have no doubt noted, the ash left over from a campfire looks nothing like the original wood that was burned. In fact, the burning—or **combustion**—of wood involves chemical reactions between wood and oxygen that change both reactants. The reaction products of ash, carbon dioxide, and water vapor are very different from wood and oxygen. Thus, the statement refers to a chemical property of oxygen (as well as of wood).

Wood burning

Now it's your turn. Classify each of the following statements as describing either a physical property or a chemical property. (*Hint:* Decide whether the chemical identity of the material does or does not change when the property is observed.)

1. Pure metals have a high **luster** (are shiny and reflect light).

2. The surfaces of some metals become dull when exposed to air.

3. Nitrogen gas, which is a relatively nonreactive element at room temperature, can form nitrogen oxides at the high temperatures of an operating automobile engine.

4. Milk turns sour if left too long at room temperature.

5. Diamonds are hard enough to be used as a coating for drill bits.

6. Metals are typically **ductile** (can be drawn into wires).

7. Leavened bread dough increases in volume if it is allowed to rise before baking.

8. Unreactive argon gas, rather than air, is used to fill many light bulbs to prevent the metal filament wire inside the bulb from being destroyed through oxidation.

9. Generally, metals are better conductors of heat and electricity than are nonmetals.

Metallic luster

Bread rising

A.3 PROPERTIES MATTER: DESIGNING THE PENNY

As you might imagine, you must weigh many considerations when selecting materials for a specific use. A material with properties well suited to a purpose may be either unavailable in sufficient quantity or too expensive. Alternatively, a material may have undesirable physical or chemical properties that can limit its use. In these and other situations, you can often find another material with most of the sought-after properties and use it instead.

The cost of a material is a concern when manufacturing coins and printed currency. Just imagine what would happen if the declared value of a coin were less than the cost of its component metals. How would this affect the production and circulation of the coin? This situation nearly occurred in the United States about one-quarter century ago. In the early 1980s, copper became too expensive to be used as the primary metal in pennies. In other words, the cost of the copper composing a penny was becoming just about as great as the face value of the penny. Zinc, another metallic element, was chosen to replace most of the copper in all post-1982 pennies. Zinc is about as hard as copper, and it has a density (7.13 g/cm³) that is close to the density of copper metal (8.96 g/cm³). Zinc is also readily available, and it is less expensive than copper.

Unfortunately, zinc is also more chemically reactive than copper. During World War II, copper metal was in short supply. To conserve that resource, zinc-plated steel pennies—known to coin collectors as "white cents" or "steel cents"—were created in 1943. The new pennies quickly corroded. As you can see in Figure 2.3, these pennies also looked considerably different from traditional copper pennies. The production of zinc-plated pennies ended within a year.

The problems associated with using zinc in pennies were solved in the early 1980s. In the new design, the properties of copper were used where they were most needed—on the coin's surface—and the properties of zinc were used where they were useful, within the coin's body. All post-1982 pennies are 97.5% zinc. They are composed of a zinc core surrounded by a thin layer of copper metal, which is added to increase the coin's durability and maintain its familiar appearance. Figure 2.4 shows a cross-section of a post-1982 penny.

Every substance has its own set of physical and chemical properties. However, with millions of substances available, how can anyone identify the *best* substance or material to meet a given need? Fortunately, all substances are made of a relatively small number of building blocks—the atoms of the different chemical elements. Knowing the similarities and differences among atoms of elements and among combinations of those atoms can greatly simplify the challenge of matching a substance to appropriate uses.

Figure 2.3 *Zinc–copper and copper pennies (top); "new" and corroded zinc-plated steel pennies (bottom).*

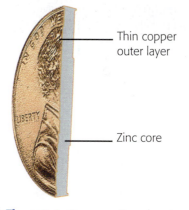

Thin copper outer layer

Zinc core

Figure 2.4 *Cross-section showing the structure and composition of a post-1982 penny.*

COMMON ELEMENTS AND THEIR SYMBOLS

Name	Symbol
Aluminum	Al
Antimony	Sb
Argon	Ar
Barium	Ba
Beryllium	Be
Bismuth	Bi
Boron	B
Bromine	Br
Cadmium	Cd
Calcium	Ca
Carbon	C
Cesium	Cs
Chlorine	Cl
Chromium	Cr
Cobalt	Co
Copper	Cu
Fluorine	F
Gold	Au
Helium	He
Hydrogen	H
Iodine	I
Iron	Fe
Krypton	Kr
Lead	Pb
Lithium	Li
Magnesium	Mg
Manganese	Mn
Mercury	Hg
Neon	Ne
Nickel	Ni
Nitrogen	N
Oxygen	O
Phosphorus	P
Platinum	Pt
Potassium	K
Radon	Rn
Silicon	Si
Silver	Ag
Sodium	Na
Sulfur	S
Tin	Sn
Tungsten	W
Uranium	U
Zinc	Zn

| Table 2.1

A.4 THE CHEMICAL ELEMENTS

You learned in Unit 1 that all matter is composed of atoms. One element differs from another because its atoms have properties that differ from those of other elements. See Figure 2.5. More than 100 chemical elements are now known. Table 2.1 lists some common elements and their symbols. An alphabetical list of all elements (names and symbols) can be found on pages 128–129.

Elements can be classified in several ways, according to the similarities and the differences in their properties. Two major classes of elements are *metals* and *nonmetals.* **Metals** include such elements as iron (Fe), tin (Sn), zinc (Zn), and copper (Cu). Carbon (C) and oxygen (O) are examples of **nonmetals.** Everyday experience has given you some knowledge of metallic and nonmetallic properties. The upcoming investigation will let you further explore common properties of metals and nonmetals.

Several elements, called **metalloids,** have properties that are intermediate to those of metals and nonmetals. That is, metalloids exhibit both metallic and nonmetallic properties. Examples of metalloids include silicon (Si) and germanium (Ge), which are commonly used in the computer industry.

What properties of matter can we use to distinguish metals, nonmetals, and metalloids? The next investigation will help you find out.

Figure 2.5 *Each of these elements (clockwise from top right: sulfur, antimony, iodine, phosphorus, copper, and bismuth) is composed of chemically identical atoms. Which elements appear metallic? Which appear nonmetallic?*

A.5 METAL OR NONMETAL?

Introduction

In this investigation, you will explore several properties of seven elements and then decide whether each element is a metal, a nonmetal, or a metalloid. You will examine the color, luster, and form of each element, and you will also attempt to crush each sample with a hammer. In addition, you or your teacher (as a demonstration) will test the substance's ability to conduct electricity. Finally, you will determine the reactivity of each element with two solutions: hydrochloric acid, $HCl(aq)$, and copper(II) chloride, $CuCl_2(aq)$.

Before starting, read the procedure to learn what you will need to do, note safety precautions, and plan necessary data collecting and observations. In your data table, create six columns: one column will be used to list the elements tested, and the other five columns will be used to record each result for appearance, conductivity, crushing, reactivity with copper(II) chloride, and reactivity with acid.

Lab Video: Metal or Nonmetal

The symbol (aq) means that the substance is dissolved in water; thus, it indicates an aqueous solution.

Procedure

1. Construct a data table appropriate for recording the data you will collect in this investigation.

2. *Appearance:* Observe and record the appearance of each element, including physical properties such as color, luster, and form. You can record the form as nonmetallic (like table salt, NaCl, or baking soda, $NaHCO_3$) or metallic (like iron, Fe).

3. *Conductivity:* If an electrical conductivity apparatus is available, use it to test each sample. (***Caution:*** *Avoid touching the bare electrode tips; some may deliver an uncomfortable electric shock.*) Touch both electrodes to the element sample, but do not allow the electrodes to touch each other. See Figure 2.6. If the bulb lights, even dimly, electricity is flowing through the sample. Such a material is called a **conductor.** If the bulb fails to light, the material is a **nonconductor.**

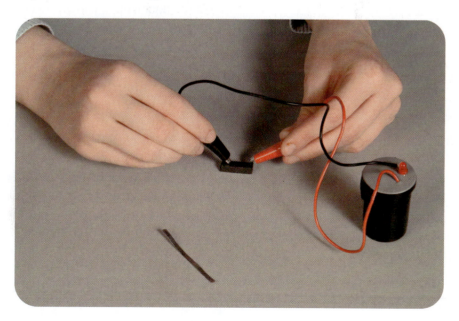

Figure 2.6 *Testing a sample for electrical conductivity.*

Figure 2.7 *Testing a sample for malleability.*

4. *Crushing:* Gently tap each element sample with a hammer as shown in Figure 2.7. Based on the results, decide whether the sample is **malleable,** which means it flattens without shattering when struck, or **brittle,** which means it shatters into pieces.

5. *Reactivity with copper(II) chloride.*

   a. Label seven wells of a clean wellplate *A* through *G.*

   b. Place a sample of each element in its well. The ribbon or solid wire samples provided by your teacher will be less than 1 cm in length. Other samples should be between 0.2 g and 0.4 g. You can estimate that mass as being no larger than the size of a match head.

   c. Add 15 to 20 drops of 0.1 M copper(II) chloride ($CuCl_2$) to each sample.

   d. Observe each system for three to five minutes—changes may be slow. A change in a sample's appearance may indicate a chemical reaction. Decide which elements reacted with the copper(II) chloride and which did not. Record these results.

   e. Discard the wellplate contents as instructed by your teacher.

6. *Reactivity with acid.*

   a. Repeat Steps 5a and 5b.

    b. Add 15 to 20 drops of 0.5 M HCl to each well that contains a sample. (***Caution:*** *0.5 M hydrochloric acid (HCl) can chemically attack skin if allowed to remain in contact for a long time. If any hydrochloric acid accidentally spills on you, ask a classmate to notify your teacher immediately. Wash the affected area immediately with tap water and continue rinsing for several minutes.*)

   c. Observe and record each result. The formation of gas bubbles may indicate that a chemical reaction has occurred. Decide which elements reacted with the hydrochloric acid and which did not. Record these results.

   d. Discard the wellplate contents as instructed by your teacher.

7. Wash your hands thoroughly before leaving the laboratory.

Questions

1. Classify each property tested in this investigation as either a physical property or a chemical property.

2. Sort the seven coded elements into two groups based on similarities in their physical and chemical properties.

3. Which element or elements could fit into either group? Why?

4. Using the following information, classify each tested element as a metal, a nonmetal, or a metalloid:

    ▶ Metals have a luster, are malleable (can be hammered into sheets), and conduct electricity.

    ▶ Many metals react with acids; many metals also react with copper(II) chloride solution.

    ▶ Nonmetals are usually dull in appearance, are brittle, and do not conduct electricity.

    ▶ Metalloids have some properties of both metals and nonmetals.

You have been introduced to one classification scheme for elements: metals, nonmetals, and metalloids. However, the quantity of detailed information about all the elements is enormous. When you are choosing or designing materials for specific uses, the more information you have about the elements (including the similarities and the differences among them), the better your decisions will be. Where is knowledge about the elements conveniently organized? You have already been introduced to one answer—the *periodic table.* Now you will explore its origins and gain a greater understanding of the chemical information that the periodic table conveys.

A.6 THE PERIODIC TABLE

By the mid-1800s, chemists had identified about 60 elements. Five of these elements were nonmetals that are gases at room temperature: hydrogen (H), oxygen (O), nitrogen (N), fluorine (F), and chlorine (Cl). Two liquid elements were also known, the metal mercury (Hg) and the nonmetal bromine (Br). The rest of the known elements were solids with widely differing properties.

To organize information about the known elements, several scientists tried to place elements with similar properties near one another in a chart. Such an arrangement is called a *periodic table.*

Dimitri Mendeleev, a Russian chemist, published a periodic table in 1869. We use such a table today. In some respects, the periodic table has a pattern that resembles a monthly calendar, in which weeks repeat on a regular (periodic) seven-day cycle. Figure 2.8 shows a stamp honoring Mendeleev, and Figure 2.9 shows a moon-phase chart, another type of periodic behavior.

Figure 2.8 *Dimitri Mendeleev (1834–1907) studied trends in the physical and chemical properties of elements. This stamp's left field includes 19th century data on aluminum, gallium, and indium. What particular information does the stamp provide about these elements?*

The periodic tables of the 1800s were organized according to two characteristics of elements. First, chemists knew that atoms of different elements have different masses. For example, hydrogen atoms have the lowest mass, oxygen atoms are about 16 times more massive than hydrogen atoms, and sulfur atoms are about twice as massive as oxygen atoms (making them about 32 times more massive than hydrogen atoms). Based on such comparisons, an *atomic weight* was assigned to each element in Mendeleev's periodic table. This atomic weight then became one of the two criteria for arranging elements.

The other criterion for organizing elements was their respective "combining capacity" with other elements, such as chlorine and oxygen. Atoms of various elements differ in the way that they combine with atoms of another element. For example, one atom of potassium (K) or cesium (Cs) combines with only one atom of chlorine (Cl) to produce the compound KCl or CsCl. Such one-to-one compounds can be represented as ECl, where E stands for the element combining with chlorine.

One atom of magnesium (Mg) or of strontium (Sr) combines with *two* atoms of chlorine to produce the compound $MgCl_2$ or $SrCl_2$, which can be represented in general terms as ECl$_2$. Atoms of other elements may combine with three or four chlorine atoms to produce compounds with the general formula ECl$_3$ or ECl$_4$.

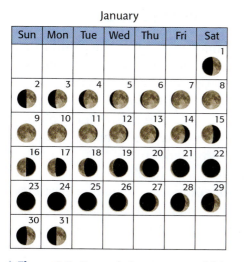

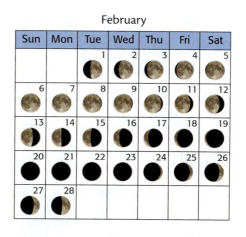

Figure 2.9 *Several phenomena exhibit periodicity—the moon's phases, the seasons, and the calendar. Can you think of other common periodic examples?*

In the first periodic table, elements with similar chemical properties were placed in the same vertical column. Horizontal arrangements were based on the increasing atomic weights of the elements. In the activity that follows, you will develop a classification scheme for some elements in much the same way that Mendeleev did.

See the modern periodic table, page 124.

Making Decisions

A.7 GROUPING THE ELEMENTS

You will receive a set of 20 element data cards. Each card lists some properties of a particular element.

1. Arrange the cards in order of increasing atomic weight.

2. Try sorting the cards into several different groups. Each group should include elements with similar properties. You might need to try several methods of grouping according to properties before you find one that makes sense to you.

3. Examine the cards within each group for any patterns. Arrange the cards within each group in some logical sequence. Again, trial and error may be a useful method for accomplishing this task.

4. Observe how particular element properties vary from group to group.

5. Arrange all the card groups into some logical sequence.

6. Select the most reasonable and useful patterns within and among the card groups. Then tape the cards onto a sheet of paper to preserve your pattern for later classroom discussion.

A.8 THE PATTERN OF ATOMIC NUMBERS

Creators of early periodic tables were unable to explain the similarities in properties found among neighboring elements. For example, we know now that all elements in the leftmost column of the periodic table are very reactive metals. All elements listed in the rightmost column are unreactive (noble) gases. See Figure 2.10. The reason for such patterns, which were discovered more than 50 years after Mendeleev's work, serves as the basis for the modern periodic table.

As you will recall from Unit 1, all atoms are composed of smaller particles, including equal numbers of positively charged protons and negatively charged electrons. The number of protons in an atom, called the **atomic number,** distinguishes atoms of different elements. For example, each sodium atom (and only atoms of sodium) contains 11 protons; the atomic number of sodium is 11. Each carbon atom contains 6 protons. If the number of protons in an atom is 9, it is a fluorine atom; if 12, it is a magnesium atom. Thus, the atomic number identifies each atom as a particular element.

Figure 2.10 *When excited by an external energy source, neon's electrons produce brilliant orange-red light emissions. Other excited gases or inner coatings on the tube allow the emission of light of many distinctive colors.*

In the modern periodic table of the elements, elements are placed in sequence according to their increasing atomic number (number of *protons*). However, because electrically neutral (uncharged) atoms contain equal numbers of protons and electrons, the periodic table is also sequenced by the number of *electrons* contained in neutral atoms of each element.

Early periodic tables, much like the one you just constructed, used atomic weights to organize the elements. Although this method produces reasonable results for elements with relatively small atomic weights, it does not work well for more massive atoms. The reason for this is the existence of another small particle that also contributes to the atomic weight, the electrically uncharged *neutron*. The total mass of an atom is largely determined by the combined mass of protons and neutrons in its nucleus. The **nucleus** is a concentrated region of positive charge (due to protons) in the center of an atom. See Figure 2.11.

> Mass number and atomic number are not the same.

The total number of protons and neutrons in the nucleus of an atom is called the **mass number.** Electrons make up the rest of an atom; however, because each electron is about 1/2000 the mass of a proton or neutron, the total mass of electrons does not contribute significantly to the mass of an atom.

While all atoms of a particular element have the same number of protons, the number of neutrons can differ from atom to atom of an element. For example, carbon atoms always contain 6 protons, but they

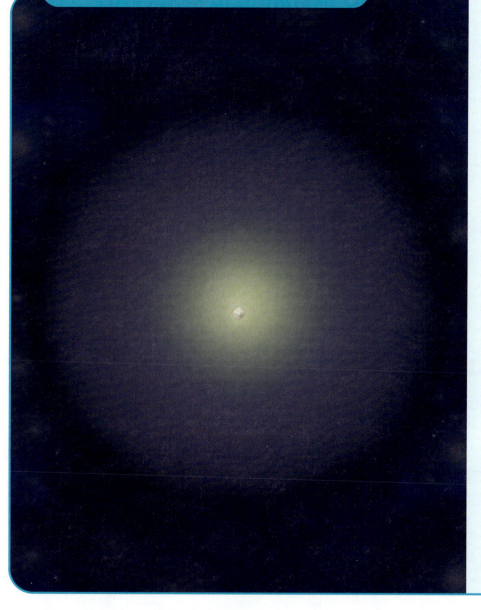

Figure 2.11 *Model of a lithium (Li) atom. Each electrically neutral lithium atom contains three protons, three electrons, and either three or four neutrons. In this model, the protons and three neutrons reside in the Li nucleus (center). Electrons are not depicted as individual particles; they occupy "electron clouds" around the nucleus. Two electrons (yellow cloud) remain close to the nucleus; the third electron (fainter cloud) is relatively further away. Electron clouds occupy most of an atom's volume; the nucleus accounts for nearly all the atom's mass.*

Note that an atom does not have a distinctly defined "outer edge," but, instead, has a rather fuzzy outer region.

may contain 6, 7, or 8 neutrons. Thus, individual carbon atoms can have mass numbers of 12, 13, or 14. For example, 6 protons + 7 neutrons = mass number 13. Atoms with the same number of protons but different numbers of neutrons, such as these carbon atoms, are called **isotopes.** In other words, isotopes are atoms of the same element with different mass numbers.

Is there a connection between the atomic numbers used to organize the modern periodic table and the properties of elements used by nineteenth-century chemists to create their periodic tables? If there is, what is that connection? Continue reading to explore the relationship between atomic numbers and the properties of elements.

Isotopes are the major reason for fractional atomic weights found on the periodic table. For example, the atomic weight of chlorine is listed as 35.45. These values represent the average mass of all of the atoms of an element.

A.9 PERIODIC VARIATION IN PROPERTIES

Your teacher will assist you in identifying the atomic numbers of the twenty elements you considered earlier in this unit. Use these atomic numbers and information about each element's properties to prepare the two graphs described below. Look for patterns between atomic numbers and element properties as you construct the plots.

Graph 1: Trends in a Chemical Property

> Follow the graphing guidelines you learned in Unit 1 (pages 80–81).

1. On a sheet of graph paper, draw a set of axes, and title the graph "Trends in a Chemical Property."

2. Label the x-axis "Atomic Number of E." What is the range of possible numbers that you will need to display on your graph? Scale your x-axis with these values in mind.

3. Label the y-axis "Oxygen Atoms per Atom of E." What is the range of possible numbers that you will have to display on your graph? Scale your y-axis with these values in mind.

4. Construct a bar graph, as demonstrated in Figure 2.12, by plotting the oxide data from the element cards. For example, if no oxide forms, the height of the bar will be 0 because oxygen atoms do not form a compound with atoms of E. If E_2O (1 oxygen atom for 2 E atoms) forms, the height of the bar is 0.5, which is the number of oxygen atoms for each E atom in the compound. Similarly, the heights of the graph bars for other oxides are 1 for EO, 1.5 for E_2O_3, 2 for EO_2, and 2.5 for E_2O_5. Do you understand why?

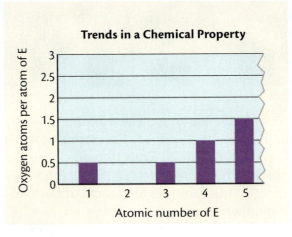

| **Figure 2.12** *Sample bar graph for oxide data.*

5. Label each bar with the actual symbol of the element E involved in that compound.

Graph 2: Trends in a Physical Property

6. On a separate sheet of graph paper, draw a set of axes, and title the graph "Trends in a Physical Property."

7. Label the *x*-axis "Atomic Number" and number it from 1 to 20. What range of possible numbers will you have to display on your graph? Scale your *x*-axis with these values in mind.

8. Label the *y*-axis "Boiling Point (K)." What range of possible numbers will you have to display on your graph? Scale your y-axis with these values in mind. Use as much of the graph paper as possible to plot these kelvin temperatures. (*Hint:* See the margin note to learn how to convert temperature values from degrees Celsius to kelvins.)

9. Construct a bar graph as in Step 4, this time using the boiling point data from the element cards, as shown in Figure 2.13. (*Note:* Do not include data for the element with atomic number 6. The boiling point of this element (carbon) would be quite far off the graph paper.)

10. Label each bar with the actual symbol of the element it represents.

After your graphs are completed, answer the following questions.

1. Does either bar graph reveal a repeating, or cyclic, pattern? (*Hint:* Focus on elements represented by very large or very small values.) Describe any patterns you observe.

2. Are these graphs consistent with patterns found in your earlier grouping of the elements? Explain.

3. Based on these two bar graphs, why is the chemist's organization of elements called a *periodic* table? (*Hint:* Look up the meaning of *periodic* in the dictionary.)

4. Where are elements with the highest *oxide numbers* located on the periodic table?

5. Where are elements with the highest *boiling points* located on the periodic table?

6. Explain any trends you noted in your answers to Questions 4 and 5.

7. Predict which element should have the lowest boiling point: selenium (Se), bromine (Br), or krypton (Kr). Use evidence from your graphs to explain how you decided.

8. Using your graphs, predict the pattern in boiling points and oxide numbers for the next 5 to 8 elements, starting with gallium (Ga).

Temperature in kelvins (K) is related to temperature in degrees Celsius (°C) by $K = 273.16 + °C$.

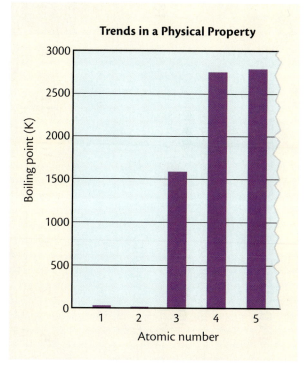

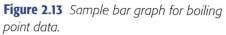

Figure 2.13 *Sample bar graph for boiling point data.*

PERIODIC TABLE OF THE ELEMENTS

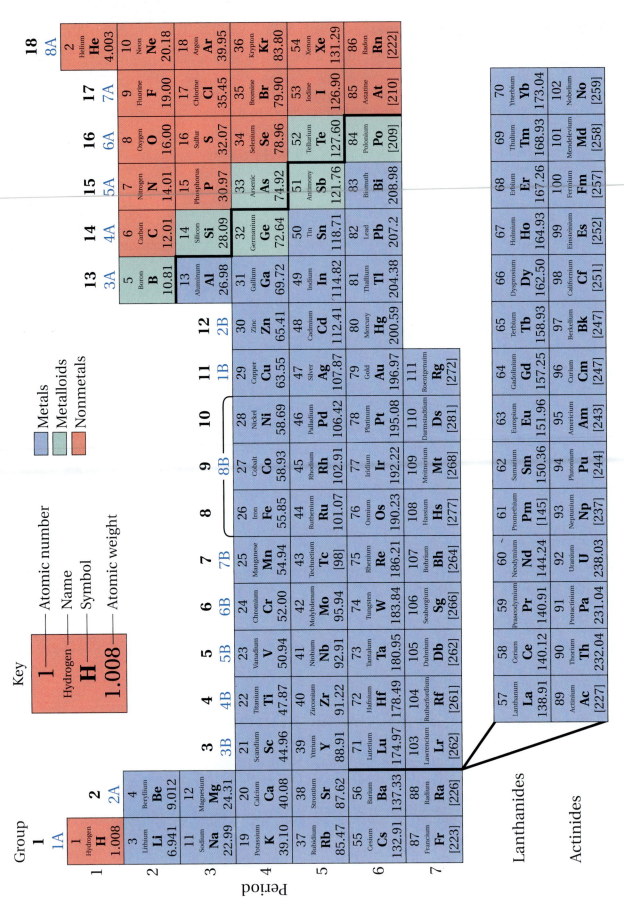

Key

1	—— Atomic number
Hydrogen	—— Name
H	—— Symbol
1.008	—— Atomic weight

| Metals |
| Metalloids |
| Nonmetals |

A.10 ORGANIZATION OF THE PERIODIC TABLE

When the first 20 elements are listed in order of increasing atomic numbers and grouped according to similar properties, they form horizontal rows called **periods**. This **periodic relationship** among elements is summarized in the modern periodic table shown on page 124. To become more familiar with the periodic table, locate the 20 elements you grouped earlier. How do their relative positions compare with those shown on your chart?

Each vertical column in the periodic table contains elements with similar properties. Each of these columns is called a **group** or **family** of elements. For example, the **alkali metal family** consists of the six elements (starting with lithium) in the first column at the left of the table. See Figure 2.14. Each element (E) in this family is a highly reactive metal that forms an ECl chloride and an E_2O oxide. By contrast, at the right side of the table, the **noble gas family** consists of very unreactive (or even chemically inert) elements; only xenon (Xe) and krypton (Kr) are known to form any compounds under normal conditions. The group containing fluorine, chlorine, and bromine—in the column just to the left of noble gases—is called the **halogen family**. Halogens readily form 1− ions; as anions, they are a component of many ionic compounds.

The arrangement of elements in the periodic table provides an orderly summary of the key characteristics of each element. By knowing the major properties of a certain chemical family, you can predict some properties of any element in that family. This knowledge can be very helpful when evaluating elements for possible uses. All elements are also listed alphabetically in Table 2.2 (pages 128–129).

Figure 2.14 *Sodium is an alkali metal. All alkali metals have similar physical and chemical properties, so they are placed in the same column (family) in the periodic table.*

Like sodium chloride (NaCl), all chlorides and oxides of alkali metal elements are ionic compounds.

Developing Skills

A.11 PREDICTING PROPERTIES

Some element properties can be estimated by averaging the respective properties of the elements located just above and just below an element on the periodic table. That is how Mendeleev predicted the properties of several elements unknown in his time. He was so convinced these elements existed that he purposely left gaps for them in his periodic table, along with a prediction of some of their properties. When those elements were discovered shortly thereafter, they fit exactly as expected. Mendeleev's fame rests largely on the accuracy of these predictions.

For example, germanium (Ge) was not known when Mendeleev proposed his periodic table. However, in 1871 he predicted the existence of germanium, calling it *ekasilicon.* Not only did Mendeleev predict germanium's existence, he also accurately predicted many of its chemical and physical properties, based on known properties of other elements in its same family.

Eka- means "standing next in order."

Sample Problem: *Given that the density of silicon (Si) is 2.3 g/cm³ and the density of tin (Sn) is 7.3 g/cm³, estimate the density of germanium (Ge).*

These three elements are in the same group in the periodic table. Germanium is below silicon and above tin. (You can verify this by locating these elements in the periodic table.) Thus, the predicted density of germanium can be estimated by averaging the densities of silicon and tin, arriving at a calculated value of 4.8 g/cm³.

When germanium was discovered in 1886, its density was found to be 5.3 g/cm³, which is within about 10% of the earlier estimated density. The periodic table helped guide Mendeleev (and now you, too) to make a useful prediction.

Formulas for chemical compounds can also be predicted from relationships established in the periodic table. For example, carbon and oxygen form carbon dioxide (CO_2). What formula would be predicted for a compound of carbon and sulfur? The periodic table indicates that sulfur (S) and oxygen (O) are in the same family. Knowing that carbon and oxygen form CO_2, a logical—and quite correct—prediction would be CS_2 (carbon disulfide). Now it's your turn:

1. The element krypton (Kr) was not known in Mendeleev's time. Given that the boiling point of argon (Ar) is −186 °C and of xenon (Xe) is −112 °C, estimate the boiling point of krypton.

2. a. Estimate the melting point of rubidium (Rb). The melting points of potassium (K) and cesium (Cs) are 337 K and 302 K, respectively.

   b. Do you expect the melting point of sodium (Na) to be higher or lower than that of rubidium (Rb)? Explain. On what evidence did you base your answer?

3. Mendeleev knew that silicon tetrachloride ($SiCl_4$) existed. Using his periodic table, he correctly predicted the existence of *ekasilicon,* an element just below silicon in the periodic table. Predict the formula for the compound formed by Mendeleev's *ekasilicon* and chlorine.

4. Here are formulas for several known compounds: NaI, $MgCl_2$, CaO, Al_2O_3, and CCl_4. Using that information, predict the formula for a compound formed from:
   a. C and F
   b. Al and S
   c. K and Cl
   d. Ca and Br
   e. Sr and O

A.12 WHAT DETERMINES PROPERTIES?

While any atom is uniquely identified by the number of protons contained in its nucleus, the physical and particularly the chemical properties of that atom (how it interacts with other atoms) are governed largely by the number and arrangement of the atom's electrons.

A major difference between the atoms of metals and nonmetals is that metal atoms lose electrons much more easily than do nonmetal atoms. Under suitable conditions, one or more outer electrons may be removed from a metal atom. This results in metallic elements forming positive ions (cations) because the number of positively charged protons remains unchanged while the total number of negatively charged electrons has been decreased.

Some physical properties of metals depend on attractions among their atoms. For example, stronger attractions among atoms of a metal result in higher melting points. The melting point of magnesium is 650 °C, whereas that of sodium is 98 °C. Thus, attractions among the atoms in magnesium metal must be stronger than those in sodium metal.

Chemical and physical properties of nonmetals and compounds are also explained by the makeup of their atoms, ions, or molecules and by attractions among these particles. As you learned in Unit 1, the abnormally high melting and boiling points of water are due to the strong attractions among water molecules.

Understanding the properties of atoms is the key to predicting, and even manipulating, the behavior of materials. Combined with a bit of imagination, this information allows chemists to find new uses for materials and to create new chemical compounds to meet specific needs.

> Chemists synthesize several thousand new compounds each year.

A.13 IT'S ONLY MONEY

Based on what you have learned so far, you can start to make some decisions about your coin design. A good first step is to specify some characteristics that are *necessary* or *desirable* in the material you will use. (For example, a high-melting-point material is *required;* after all, who would want their coins to melt in hot sunlight? However, a metallic luster is only *desirable.*) Apply your knowledge of existing coins, as well as what you have learned about the properties of elements, to answer the following questions:

1. What physical properties must the coin material have?

2. What other physical properties are desirable?

3. What chemical properties are required of the coin's material?

4. What other chemical properties are desirable?

5. Which would make the best primary material for the new coin: a metal, a nonmetal, or a metalloid? Explain.

6. What factors or desirable coin characteristics did you consider when you answered Question 5?

Save your answers to these questions; they will help guide your coin-design work later in this unit.

| Table 2.2

CHART OF THE ELEMENTS

Element	Symbol	Atomic Number	Atomic Weight	Element	Symbol	Atomic Number	Atomic Weight
Actinium	Ac	89	[227]	Erbium	Er	68	167.26
Aluminum	Al	13	26.98	Europium	Eu	63	151.96
Americium	Am	95	[243]	Fermium	Fm	100	[257]
Antimony	Sb	51	121.76	Fluorine	F	9	19.00
Argon	Ar	18	39.95	Francium	Fr	87	[223]
Arsenic	As	33	74.92	Gadolinium	Gd	64	157.25
Astatine	At	85	[210]	Gallium	Ga	31	69.72
Barium	Ba	56	137.33	Germanium	Ge	32	72.64
Berkelium	Bk	97	[247]	Gold	Au	79	196.97
Beryllium	Be	4	9.012	Hafnium	Hf	72	178.49
Bismuth	Bi	83	208.98	Hassium	Hs	108	[277]
Bohrium	Bh	107	[264]	Helium	He	2	4.003
Boron	B	5	10.81	Holmium	Ho	67	164.93
Bromine	Br	35	79.90	Hydrogen	H	1	1.008
Cadmium	Cd	48	112.41	Indium	In	49	114.82
Calcium	Ca	20	40.08	Iodine	I	53	126.90
Californium	Cf	98	[251]	Iridium	Ir	77	192.22
Carbon	C	6	12.01	Iron	Fe	26	55.85
Cerium	Ce	58	140.12	Krypton	Kr	36	83.80
Cesium	Cs	55	132.91	Lanthanum	La	57	138.91
Chlorine	Cl	17	35.45	Lawrencium	Lr	103	[262]
Chromium	Cr	24	52.00	Lead	Pb	82	207.2
Cobalt	Co	27	58.93	Lithium	Li	3	6.941
Copper	Cu	29	63.55	Lutetium	Lu	71	174.97
Curium	Cm	96	[247]	Magnesium	Mg	12	24.31
Darmstadtium	Ds	110	[281]	Manganese	Mn	25	54.94
Dubnium	Db	105	[262]	Meitnerium	Mt	109	[268]
Dysprosium	Dy	66	162.50	Mendelevium	Md	101	[258]
Einsteinium	Es	99	[252]	Mercury	Hg	80	200.59

Table 2.2

CHART OF THE ELEMENTS (CONTINUED)

Element	Symbol	Atomic Number	Atomic Weight	Element	Symbol	Atomic Number	Atomic Weight
Molybdenum	Mo	42	95.94	Samarium	Sm	62	150.36
Neodymium	Nd	60	144.24	Scandium	Sc	21	44.96
Neon	Ne	10	20.18	Seaborgium	Sg	106	[266]
Neptunium	Np	93	[237]	Selenium	Se	34	78.96
Nickel	Ni	28	58.69	Silicon	Si	14	28.09
Niobium	Nb	41	92.91	Silver	Ag	47	107.87
Nitrogen	N	7	14.01	Sodium	Na	11	22.99
Nobelium	No	102	[259]	Strontium	Sr	38	87.62
Osmium	Os	76	190.23	Sulfur	S	16	32.07
Oxygen	O	8	16.00	Tantalum	Ta	73	180.95
Palladium	Pd	46	106.42	Technetium	Tc	43	[98]
Phosphorus	P	15	30.97	Tellurium	Te	52	127.60
Platinum	Pt	78	195.08	Terbium	Tb	65	158.93
Plutonium	Pu	94	[244]	Thallium	Tl	81	204.38
Polonium	Po	84	[209]	Thorium	Th	90	232.04
Potassium	K	19	39.10	Thulium	Tm	69	168.93
Praseodymium	Pr	59	140.91	Tin	Sn	50	118.71
Promethium	Pm	61	[145]	Titanium	Ti	22	47.87
Protactinium	Pa	91	231.04	Tungsten	W	74	183.84
Radium	Ra	88	[226]	Uranium	U	92	238.03
Radon	Rn	86	[222]	Vanadium	V	23	50.94
Rhenium	Re	75	186.21	Xenon	Xe	54	131.29
Rhodium	Rh	45	102.91	Ytterbium	Yb	70	173.04
Roentgenium	Rg	111	[272]	Yttrium	Y	39	88.91
Rubidium	Rb	37	85.47	Zinc	Zn	30	65.41
Ruthenium	Ru	44	101.07	Zirconium	Zr	40	91.22
Rutherfordium	Rf	104	[261]				

Note: A value in square brackets is the mass number of the isotope with the longest half-life.

SECTION A SUMMARY
Reviewing the Concepts

The physical properties of a substance can be determined without altering the substance's chemical makeup; physical changes alter a substance's physical properties. Chemical properties describe how a substance reacts chemically through its transformation into one or more different substances.

1. Classify each of the following as a chemical or a physical property:

   a. Copper has a reddish brown color.
   b. Propane burns readily.
   c. CO_2 gas extinguishes a candle flame.
   d. Honey pours more slowly than does water.

2. Classify each of the following as a chemical or a physical property:

   a. Metal wire can be bent.
   b. Ice floats in water.
   c. Paper is flammable.
   d. Sugar is soluble in water.

3. Classify each of the following as a chemical or a physical change:

   a. A candle burns.
   b. An opened carbonated beverage fizzes.
   c. Hair curls as a result of a "perm."
   d. As shoes wear out, holes appear in the soles.

4. Classify each of the following as a chemical or a physical change:

   a. A cut apple left out in the air turns brown.
   b. Flashlight batteries lose their "charge" after extended use.
   c. Dry cleaning removes oils from clothing.
   d. Italian salad dressing separates over time.

5. For each of your answers in Question 4, give evidence for your classification as a chemical or physical change.

6. a. List the steps involved in making chocolate chip cookies from scratch.
   b. Classify each step in Question 6a as involving either a chemical change or a physical change.

Elements can be classified as metals, nonmetals, or metalloids according to their physical and chemical properties.

7. Classify each property as characteristic of metals or nonmetals:

   a. shiny in appearance
   b. does not react with acids
   c. shatters easily
   d. electrically conductive

8. Classify each of these elements as a metal, a nonmetal, or a metalloid:

   a. tungsten   c. krypton
   b. antimony   d. sodium

9. List the names and symbols of two elements that are metalloids.

10. What would you expect to happen if you tapped a sample of each of the following elements with a hammer?

    a. iodine   c. phosphorus
    b. zirconium   d. nickel

11. List two properties that make nonmetals unsuitable for electric wiring.

12. List three properties that make metals suitable for use in coins.

Elements are arranged in the periodic table based on their properties. Elements with similar chemical properties are placed in the same columns. Physical properties vary in predictable patterns across rows and down columns.

13. Give another term for each of these features of the periodic table:

    a. row   b. column

14. Give the names and symbols of two elements other than lithium in the alkali metal family.

15. Consider the noble gas family:

    a. Where are noble gases located on the periodic table?

    b. Name one physical property that noble gases share.

    c. Name one chemical property that noble gases share.

16. Given a periodic table and the formulas $BeCl_2$ and AlN, predict the formula for a compound containing

    a. Mg and F   b. Ga and P

17. The melting points of sodium (Na) and rubidium (Rb) are 98 °C and 39 °C, respectively. Estimate the melting point of potassium (K).

18. Would you expect the boiling point of chlorine to be higher or lower than that of iodine? Explain.

The mass number of an atom is the sum of its number of protons and neutrons. The number of protons in an atom (the atomic number) of a given element distinguishes it from atoms of all other elements.

19. Copy and complete the following table for each electrically neutral atom.

Element Symbol	Number of Protons	Number of Neutrons	Number of Electrons
a	6	6	6
b	6	7	6
Ca	*c*	21	*d*
e	*f*	117	78
U	*g*	146	*h*

20. Using Figure 2.11 (page 121) as a model, illustrate the number of protons, neutrons, and electrons in an atom of

    a. beryllium.   b. nitrogen.   c. neon.

21. A student is asked to explain the formation of a lead(II) ion (Pb^{2+}) from an electrically neutral lead atom (Pb). The student says that a lead atom must have gained two protons to make the ion. How would you correct this student's mistaken explanation?

The mass of an atom depends largely on the number of protons and neutrons contained within its nucleus. Atoms containing the same number of protons but different numbers of neutrons are considered isotopes.

22. Refer to the table provided for Question 19:
    a. Calculate the mass number for each element in the table.
    b. Which element has two isotopes in the table?

23. A scientist announces the discovery of a new element. The only characteristic given in the report is the element's mass number of 266. Is this information sufficient, by itself, to justify the claim of the discovery of a new element? Explain.

24. How does the mass of an electron compare to the masses of a proton and a neutron?

25. How many protons and neutrons are needed for each magnesium isotope in this table?

ISOTOPES OF MAGNESIUM			
Isotope Symbol	Mass Number	Number of Protons	Number of Neutrons
Mg-24	24	a	b
Mg-25	25	c	d
Mg-26	26	e	f

The properties of an element are determined largely by the number and arrangement of electrons in its atoms.

26. Which are more likely to lose electrons, metallic elements or nonmetallic elements?

27. Noble gas elements rarely lose or gain electrons. What does this indicate about their chemical reactivity?

28. Predict whether each of the following elements would be more likely to form an anion or a cation: (*Note:* Anions are *negatively* charged; cations are *positively* charged.)
    a. Na   c. F   e. O   g. Sn
    b. Ca   d. Cu   f. Li   h. I

Connecting the Concepts

29. Which pair is more similar chemically? Defend your choice:
    a. copper metal and copper(II) ions
                or
    b. oxygen with mass number 16 and oxygen with mass number 18

30. The diameter of a magnesium ion (Mg^{2+}) is 156 pm (picometers, where 1 pm = 10^{-12} m); the diameter of a strontium ion (Sr^{2+}) is 254 pm. Estimate the diameter of a calcium ion (Ca^{2+}).

31. Three kinds of observations that may indicate a chemical change appear in the following list. However, a physical change may also result in each observation. Describe a possible chemical cause and a possible physical cause for each observation:
    a. change in color
    b. change in temperature
    c. formation of a gas

32. Identify the element that is described by each of the following statements:

   a. This element is a nonmetal. It forms anions with a 1– charge. It is in the same period as the metals used in a penny.

   b. This element is a metalloid. It is in the same period as the elements found in table salt.

33. Compare your use of the Snake River data to solve the fish-kill mystery in Unit 1 to Mendeleev's use of element data to create the periodic table.

34. Mendeleev arranged elements in his periodic table in order of their atomic weights. In the modern periodic table, however, elements are arranged in order of their atomic numbers. Cite two examples from the periodic table for which these two schemes would produce a different ordering of adjacent elements.

Extending the Concepts

35. How is mercury different from other metallic elements? Using outside resources, describe some applications that take advantage of the unique properties of metallic mercury.

36. Depending on how iron is heated and cooled, it can either be hard and brittle or malleable. Explain how the same metal can have both characteristics.

37. Construct a graph of the price per gram of an element versus its atomic number for each of the first 20 elements. Can the current cost of those elements be regarded as a periodic property? Explain. (*Hint:* Use a chemical supply catalog or the Web to locate the current price of each element.)

38. Classify the components of one or more pieces of jewelry you might possibly wear as being composed of metals, nonmetals, or metalloids.

Earth's Mineral Resources

Among Earth's resources, metals—and the minerals from which they are extracted—have long been used by humans. Those uses have ranged from tool making, energy transmission, and construction to works of art, decoration, and coin making. In this section, you will explore properties and uses of minerals and metals. Using copper as a case study, you will learn about Earth's mineral resources and how some minerals are converted to useful metals.

B.1 SOURCES AND USES OF METALS

Earth's Mineral Resources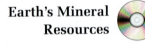

Human needs for resources—whether to create a new coin, make clothing, construct a space-vehicle rocket engine, or fertilize food crops—must all be met by chemical supplies currently present on Earth. These supplies of resources are often cataloged by where they are found. The table in Figure 2.15 indicates the chemical makeup of Earth.

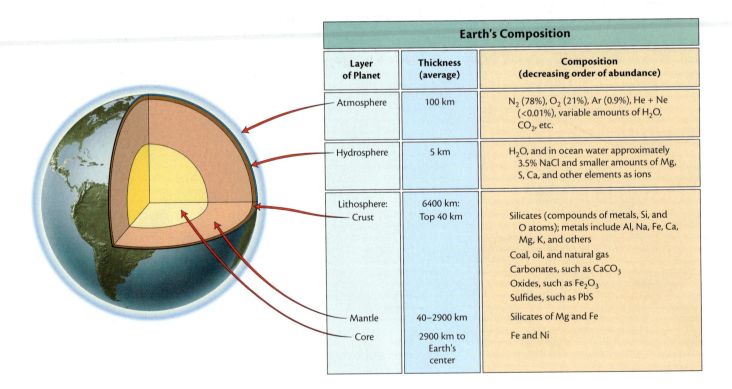

Earth's Composition		
Layer of Planet	Thickness (average)	Composition (decreasing order of abundance)
Atmosphere	100 km	N_2 (78%), O_2 (21%), Ar (0.9%), He + Ne (<0.01%), variable amounts of H_2O, CO_2, etc.
Hydrosphere	5 km	H_2O, and in ocean water approximately 3.5% NaCl and smaller amounts of Mg, S, Ca, and other elements as ions
Lithosphere: Crust	6400 km: Top 40 km	Silicates (compounds of metals, Si, and O atoms); metals include Al, Na, Fe, Ca, Mg, K, and others Coal, oil, and natural gas Carbonates, such as $CaCO_3$ Oxides, such as Fe_2O_3 Sulfides, such as PbS
Mantle	40–2900 km	Silicates of Mg and Fe
Core	2900 km to Earth's center	Fe and Ni

Figure 2.15 *Earth's composition. From what layer do most resources come that support human activities?*

Earth's atmosphere, hydrosphere, and outer layer of the lithosphere supply resources for all human activities. The **atmosphere** provides nitrogen, oxygen, neon, and argon. From the **hydrosphere** come water and some dissolved minerals. The **lithosphere,** which is the solid part of Earth, provides the greatest variety of chemical resources. For example, petroleum and metal-bearing ores are found in the lithosphere. An **ore** is a naturally occurring rock or mineral that can be mined and from which it is profitable to extract a metal or other material. An ore contains a mixture of components. Of these, **minerals** are the most important. They are naturally occurring solid compounds containing the element or group of elements of interest.

The deepest mines on Earth barely scratch the surface of its crust. If Earth were the size of an apple, all accessible resources of the lithosphere would be located within the apple's skin. From this thin band of soil and rock, we obtain the major raw materials needed to build homes, automobiles, appliances, computers, DVDs, and sports equipment—in fact, all manufactured objects.

As you can see from Table 2.3 (page 136), many of Earth's resources are not uniformly distributed. There is no relationship between a nation's supply of these resources and either its land area or its population. A particular region may be the predominant supplier of certain metals to industry. For example, Africa holds most of the world's known reserves of chromium (95%), cobalt (52%), and manganese (80%).

The growth of the United States as a major industrial nation has been facilitated, in part, by the quantity and diversity of its chemical resources. Yet, in recent years, the United States has imported increasing quantities of some chemical resources. For example, about 73% of the nation's tin (Sn) is imported. See Figures 2.16 and 2.17 (page 137) for examples of minerals and mining.

The greatest challenge regarding mineral resources is deciding on the wisest uses of available supplies. For example, is it worthwhile to mine a certain metallic ore at a particular site? The answer to this depends on several considerations, such as:

▶ the quantity of useful ore found at the site
▶ the percent of metal in the ore
▶ the type of mining and processing needed to extract the metal from its ore
▶ the distance between the mine and metal-refining facilities and markets
▶ the metal's supply-versus-demand status
▶ the environmental impact of the mining and metal processing

Copper, one of the materials you might think of using in your coin, provides an example of a vital chemical resource. In this section, you will first consider worldwide sources of copper and how copper-bearing materials are converted to pure copper. Later, you will explore some possible substitutes for this resource.

> Another meaning of mineral refers to *dietary minerals.* These are elements required by living organisms (see Unit 7).

| Table 2.3

PRODUCTION OF SELECTED METALS WORLDWIDE, 2002

Metal	Nation	Percent Production	Production (10³ metric tons)	World Total Production (10³ metric tons)
Aluminum	China	17	4300	25 900
	Russia	13	3347	
	Canada	10	2709	
	United States	10	2707	
	Australia	7	1836	
Copper	Japan	13	1499	11 500
	China	13	1490	
	Chile	13	1439	
	Russia	7	860	
	United States	6	683	
Iron ore	Brazil	21	125 300	592 700
	Australia	19	113 548	
	China	13	76 200	
	India	8	51 200	
	Russia	8	49 000	
Lead	United States	22	1380	6390
	China	20	1250	
	Germany	6	390	
	United Kingdom	6	370	
	Japan	4	290	
Nickel	Russia	23	310	1340
	Australia	16	211	
	Canada	13	178	
	Indonesia	9	122	
	New Caledonia	7	100	
Silver	Mexico	14	2.8	20
	Peru	14	2.7	
	China	13	2.5	
	Australia	10	2	
	United States	7	1.4	
Tin	China	33	93	278
	Indonesia	19	53	
	Peru	13	36	
	Malaysia	11	30	
	Thailand	8	21	
Zinc	China	24	2100	8910
	Canada	9	793	
	Japan	8	671	
	Australia	6	572	
	Republic of Korea	6	510	

Figure 2.16 *Some minerals among these samples are mined and processed to yield useful metals.*

Figure 2.17 *Mining metal ores has economic and environmental consequences. What impact might mining have on a local community?*

Copper is one of the most familiar and widely used metals in modern society. Among all the elements, it is second only to silver in electrical conductivity. This property and copper's relatively low cost, corrosion resistance, and **ductility** (ease of being drawn into thin wires) make copper the world's most common metal for electrical wiring. Copper is also used to produce brass, bronze, and other alloys; a variety of copper-based compounds; jewelry; and works of art. Table 2.4 summarizes copper's physical and chemical properties.

The first copper ores mined were relatively rich in copper—from 35% to 88%. Such ores are no longer available; however, ores less rich in copper can be used. In fact, it is now economically possible to mine ores containing less than 1% copper. Copper ore is chemically processed to produce metallic copper, which is then transformed into a

| Table 2.4

PROPERTIES OF COPPER	
Malleability and ductility	High
Electrical conductivity	High
Thermal conductivity	High
Chemical reactivity	Relatively low
Resistance to corrosion	High
Useful alloys formed	Bronze and brass, for example
Color and luster	Reddish, shiny

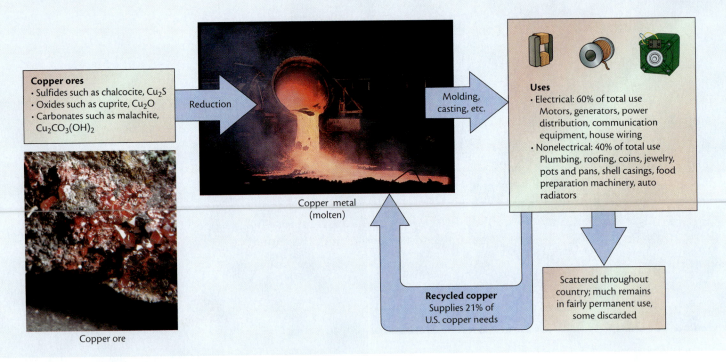

Copper ores
• Sulfides such as chalcocite, Cu_2S
• Oxides such as cuprite, Cu_2O
• Carbonates such as malachite, $Cu_2CO_3(OH)_2$

Copper ore

Reduction

Copper metal (molten)

Molding, casting, etc.

Uses
• Electrical: 60% of total use
  Motors, generators, power distribution, communication equipment, house wiring
• Nonelectrical: 40% of total use
  Plumbing, roofing, coins, jewelry, pots and pans, shell casings, food preparation machinery, auto radiators

Recycled copper
Supplies 21% of U.S. copper needs

Scattered throughout country; much remains in fairly permanent use, some discarded

Figure 2.18 *The life cycle of copper includes mining copper ores, reducing the ore to obtain the metal, fashioning the metal for final use, and then either recycling the metal or discarding it. What effect does recycling have on the energy these processes require?*

variety of useful materials. Figure 2.18 summarizes the life cycle of copper from sources to common uses to waste products.

Earth's accessible deposits of this valuable resource are destined to become depleted. Will future developments increase or decrease the need for copper? What copper substitutes are available? The following activity will help you address these questions.

Making Decisions

B.2 PRODUCTION VERSUS USE

Table 2.3 (page 136) provides information about the worldwide production of several selected metals. Note that the quantity of each metal produced varies. For example, nearly 100 times more aluminum is produced worldwide than tin. Why is this? Is aluminum more abundant than tin, or is there more demand and use for aluminum than for tin? How do the physical and chemical properties of these metals affect their production and eventual use?

Your teacher will divide the class into several groups. Each group will research one of the metals from Table 2.3 and will share what it learns with the class. Table 2.3 provides some initial information about the production quantity of each metal, as well as locations where each metal is produced.

As you learn more about your selected metal, use library and Internet resources to answer the following questions:

1. Investigate the cycle of production, consumption, and use for your metal (such as the copper life cycle in Figure 2.18).
   Generate a list of *at least* three current uses for your metal.

Research the history of your metal, as well. Perhaps, in years past, it replaced another metal for certain uses or was itself replaced. Note such information.

2. Investigate the metal's properties (both physical and chemical). Create a summary table similar to Table 2.4 (page 137).

3. For each use of your metal from the list created in Question 1, identify particular properties that make your metal an appropriate choice. If there were a replacement of your metal or if your metal replaced another metal for a particular use, what role did your metal's properties play in the transition?

4. How abundant is your metal? Are the uses and demand for your metal adequately met by the production of your metal, as noted in Table 2.3 (page 136)?

Investigating Matter

B.3 CONVERTING COPPER

Introduction

You have seen many chemical reactions in your lifetime. Some, such as a fireworks display, are memorable. Others, such as the slow process of rusting, are far less dramatic. Have you ever stopped to think about what happens to the atoms involved in those reactions? Are the materials that made up the fireworks still there after they are launched into the sky and ignited? What about the iron that turns into rust?

 Lab Video: Converting Copper

During this investigation, you will work with a powdered sample of elemental copper. As you observe its chemical behavior, think about why its properties make copper a good candidate for use in a coin.

Before starting, read the procedure to learn what you will need to do, note safety precautions, and plan necessary data collecting and observations. You will record several masses (in grams) throughout this investigation; it will be important to distinguish among them. The left column of your data table should be used to identify the objects for which mass is determined (for example, *crucible alone* or *crucible and copper before heating*). In the right column, you will record the measured masses.

Procedure

1. Construct a data table suitable for recording the observations and measurements you will make in this investigation.

2. Measure and record the mass of a clean, empty crucible. Add approximately 1 g copper powder to the crucible. Record the mass of the crucible with copper powder in it within the nearest 0.1 g. Find the actual mass of copper powder by subtracting the mass of the empty crucible from this value. Record the mass of copper powder.

3. Which properties of copper can you directly observe? Record your observations of the copper powder.

4. Set the crucible on a hot plate. The crucible lid should be left slightly ajar.

5. Set the hot plate on High.

6. Heat the crucible and its contents for two minutes. Use tongs to place the crucible on a ring-stand base or other heat-resistant surface. Hold the crucible with tongs while you *gently* use a spatula (a hot crucible is quite fragile) to break up the solid in the crucible to expose as much remaining copper metal as possible. (**Caution:** *Avoid touching the hot crucible or the hot-plate surface.*)

7. Continue heating for about 10 min more, removing the crucible from the hot plate and breaking up the solid with a spatula every one to two minutes.

8. When you have finished Step 7, turn off the hot plate and allow the crucible and its contents to cool to room temperature. Answer Questions 1 and 2 while you are waiting.

9. After the crucible and its contents have cooled, determine their mass. Use this value and the mass of the empty crucible to calculate the mass of the contents. Record these values in your data table.

> You will use your copper product later, during Investigating Matter C.11.

10. Label a clean, 100-mL beaker with your name, your class period, and the mass of copper product obtained. Transfer your product to the 100-mL beaker. Store the labeled beaker and product as indicated by your teacher.

11. Wash your hands thoroughly before leaving the laboratory.

Questions

1. a. Describe changes you observed as you heated the copper.
   b. Did the copper atoms remain in the crucible? Explain, using evidence from your observations.

2. a. Were the changes you observed physical changes or chemical changes?
   b. What observational evidence leads you to that conclusion?

3. a. How did the mass of the crucible contents change after you heated the copper?
   b. Explain why the mass of the crucible contents changed in that manner.

Relative Reactivities of Metals

B.4 METAL REACTIVITY

As you just observed, when copper metal is heated, it gradually reacts with oxygen gas in the air to produce a black substance. The equation is:

$$2\,Cu(s) \ + \ O_2(g) \ \longrightarrow \ 2\,CuO(s)$$

Copper   Oxygen   Copper(II) oxide

Although it reacts to form copper(II) oxide when heated, at room temperature the metal remains relatively unreactive in air. You are probably familiar with this fact from observing that copper wire and the copper surface on pennies do not turn black under normal conditions.

Magnesium metal also reacts with oxygen gas. However, unlike copper metal, magnesium heated in air ignites and produces a brief flash of light. See Figure 2.19. The equation for this reaction is:

$$2\,Mg(s) \quad + \quad O_2(g) \quad \longrightarrow \quad 2\,MgO(s)$$

Magnesium   Oxygen   Magnesium oxide

By contrast, gold (Au) does not react with any components of air, including oxygen gas. This is one reason gold is highly prized in long-lasting decorative objects, such as jewelry. Gold-plated electrical contacts, such as those used for automobile air bags and audio cable connectors, are very dependable because nonconducting oxides do not form on the gold-plated contact surfaces.

Observing how readily a certain metal reacts with oxygen provides information about the metal's chemical reactivity. If we rank elements in relative order of their chemical reactivities, the ranking is called an **activity series.** Based on what you have just learned about gold and magnesium and what you already know about copper, how would you rank the three metals in terms of their relative chemical reactivity?

> Two common compounds of copper and oxygen are CuO and Cu_2O. Because the name "copper oxide" could be applied to both, a Roman numeral is added to indicate copper's ionic charge. Copper(I) oxide is Cu_2O because it contains Cu^+ ions; copper(II) oxide is CuO; it contains Cu^{2+} ions.

> In writing chemical formulas for substances, the symbols for solid (s), liquid (l), gas (g), and aqueous solution (aq) are sometimes added. These symbols indicate the physical state of each reactant or product under conditions of the reaction.

Figure 2.19 *Magnesium and oxygen react so spectacularly that small samples of magnesium are used in some fireworks.*

B.5 RELATIVE REACTIVITIES OF METALS

Introduction

Lab Video: Relative Reactivites of Metals

In this investigation, you will observe the reactions of the metals copper, magnesium, and zinc with four different solutions. Each solution contains a particular metal cation. The solutions you will use are copper(II) nitrate, $Cu(NO_3)_2$ (containing Cu^{2+}); magnesium nitrate, $Mg(NO_3)_2$ (containing Mg^{2+}); zinc nitrate, $Zn(NO_3)_2$ (containing Zn^{2+}); and silver nitrate, $AgNO_3$ (containing Ag^+).

Before starting, read the procedure to learn what you will need to do, note safety precautions, plan necessary data collecting and observations, and design a suitable data table. Because this investigation involves combining various ionic solutions to determine their relative reactivities, your data table should clearly indicate which materials you combined and the observed results.

Procedure

Devise a systematic procedure that will allow you to observe the reaction (if any) between each metal and each of the four ionic solutions. You will conduct each reaction in a separate well of your wellplate, using 10 drops of 0.2 M solution and a small strip of metal. See Figure 2.20. How many different combinations of metals and solutions will you need to observe? How will you arrange things so you can complete your observations efficiently, yet keep track of which metal and which solution are in each well?

1. After considering the task described, construct a data table appropriate for recording the observations that you will make in this investigation. Keep the questions posed in the preceding paragraph in mind as you plan your data table.

Figure 2.20 *Metal strips and metal nitrate solutions should be arranged in an orderly manner in the wellplate.*

2. Obtain 1-cm strips of each of the metals to be tested. Clean the surface of each metal strip by rubbing it with sandpaper or emery paper. See Figure 2.21. Record your observations of each metal's appearance.

3. Begin your planned procedure. If you observe no reaction, write "NR" in your data table. If a reaction occurs, record the changes you observe. (***Caution:*** *Avoid allowing the* $AgNO_3$ *solution to come in contact with skin or clothing because it causes dark, nonwashable stains.*)

4. Dispose of all solid samples and wellplate solutions as directed by your teacher.

5. Wash your hands thoroughly before leaving the laboratory.

Questions

1. Which metal reacted with the most solutions?

2. Which metal reacted with the fewest solutions?

3. Assuming that you did not test silver metal, with which solutions (if any) would you expect silver metal to react? Explain your answer, citing evidence from your data and observations.

4. List the metals (including silver) in order, placing the most reactive metal first (the one reacting with the most solutions) and the least reactive metal last (the one reacting with the fewest solutions).

5. Refer to your "metal activity series" list from Question 4. Write a brief explanation of why the outside surface of a penny is made of copper instead of zinc.

6. a. Which of the four metals mentioned in this investigation might be an even better choice than copper for the outside surface of a penny? What observational evidence supports your conclusion?

b. Why do you think that the particular metal that you identified in Question 6a is not used as the outside surface of a penny?

7. Given your new knowledge about the relative chemical activities of these four metals,

a. which metal is *most* likely to be found in an uncombined, or "free," (metallic) state in nature?

b. which metal is *least* likely to be found chemically uncombined with other elements?

8. Reconsider your experimental design for this investigation:

a. Would it have been possible to eliminate one or more of the metal-solution combinations and still obtain all the information needed to create chemical activity ratings for the four metals?

b. If so, which combination or combinations could have been eliminated? Why?

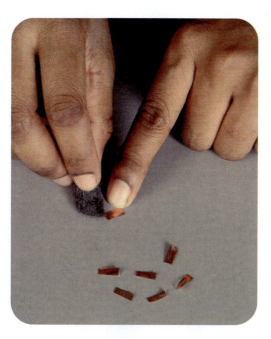

Figure 2.21 *Clean each metal strip with sandpaper or emery paper.*

Stone

Bronze

Iron

B.6 METALS: PROPERTIES AND USES

Humans have been described as toolmakers. Readily available stone, wood, and natural fibers were the earliest materials used in tools such as hammers, chisels, knives, spears, and grinding devices. The discovery that fire could transform materials in certain rocks into strong, malleable metals triggered dramatic advances in the growth of tool making and of civilization.

Gold and silver, found as free elements rather than in chemical combination with other elements, were probably the first metals humans used. They formed these metals into decorative and useful objects and, later, into coins. The metals' relative lack of reactivity made them excellent materials for those uses.

It is estimated that copper has been used for tools, weapons, utensils, and decorations for about 10 000 years. Bronze, an alloy of copper and tin, was developed about 3800 BC. Consequently, humans advanced from the Stone Age into the Bronze Age.

Eventually, early people developed *iron metallurgy,* the extraction of iron from its ores. This led to the start of the Iron Age more than 3000 years ago. In time, as humans learned more about chemistry and fire, they transformed various metallic ores into increasingly useful metals. Figure 2.22 portrays tools used in the three major ages of civilization.

ChemQuandary 1

DISCOVERY OF METALS

Copper, gold, and silver are far from being the most abundant metals on Earth. Aluminum, iron, and calcium, for example, are all much more plentiful. Why, then, were copper, gold, and silver among the first metallic elements discovered?

Copper crystals

METAL ACTIVITY SERIES

Element (in order of decreasing reactivity)	Metal Ions Found in Minerals	Process Used to Obtain the Metal	Metal
Lithium (Li)	Li^+	Pass direct electric current through the molten mineral salt (*electrometallurgy*)	Li(s)
Potassium (K)	K^+		K(s)
Calcium (Ca)	Ca^{2+}		Ca(s)
Sodium (Na)	Na^+		Na(s)
Magnesium (Mg)	Mg^{2+}		Mg(s)
Aluminum (Al)	Al^{3+}		Al(s)
Manganese (Mn)	Mn^{2+}, Mn^{3+}	Heat mineral with coke (C) or carbon monoxide (CO) (*pyrometallurgy*)	Mn(s)
Zinc (Zn)	Zn^{2+}		Zn(s)
Chromium (Cr)	Cr^{3+}, Cr^{2+}		Cr(s)
Iron (Fe)	Fe^{3+}, Fe^{2+}		Fe(s)
Lead (Pb)	Pb^{2+}	Heat (roast) mineral in air (*pyrometallurgy*) or find the metal free (uncombined)	Pb(s)
Copper (Cu)	Cu^{2+}, Cu^+		Cu(s)
Mercury (Hg)	Hg^{2+}		Hg(l)
Silver (Ag)	Ag^+		Ag(s)
Platinum (Pt)	Pt^{2+}		Pt(s)
Gold (Au)	Au^{3+}, Au^+		Au(s)

| **Table 2.5**

You have explored some chemistry of metals and know, for example, that copper metal is more reactive than silver but less reactive than magnesium. A more complete activity series is given in Table 2.5. The table also includes brief descriptions of common methods for retrieving each metal from its ore.

You can use such an activity series to predict whether certain reactions may occur. For example, you observed in Investigating Matter B.5 that zinc metal, which is more reactive than copper, reacted with copper ions in solution. However, zinc metal did not react with magnesium ions in solution. Why? Zinc is less reactive than magnesium. In general, a *more* reactive metallic element (higher in the activity series) will cause ions of a *less* reactive metallic element (lower in the activity series) to change to their corresponding metal.

B.7 TRENDS IN METAL ACTIVITY

Use Table 2.5 (page 145) and the periodic table (page 124) to answer the following questions:

Sample Problem: *Will Pb metal react with Ag⁺ ions?*
Yes; according to Table 2.5, lead is a more active metal than silver, so lead will reduce Ag^+ to Ag(s).

1. a. What trend in metallic reactivity is apparent as you move from left to right across a horizontal row (period) of the periodic table? (*Hint:* Compare the reactivity of sodium with that of magnesium and aluminum.)
   b. In which part of the periodic table are the most-reactive metals found?
   c. Which part of the periodic table contains the least-reactive metals?

2. a. Will iron (Fe) metal react with a solution of lead(II) nitrate, $Pb(NO_3)_2$?
   b. Will platinum (Pt) metal react with a lead(II) nitrate solution?
   c. Explain your answers to Questions 2a and 2b.

3. Use specific examples from the metal activity series in Table 2.5 in your answers to the following two questions:
   a. Are the least-reactive metals also the cheapest metals?
   b. If not, what other factor or factors might influence the market value of a metal?

B.8 MINING AND REFINING

As shown in Table 2.5 (page 145), most metals are not found free in nature. Instead, they are found combined with other elements within more stable compounds. These compounds must be mined from the lithosphere and refined to produce the free metal. The process of converting a combined metal (usually a metal ion) in a mineral to a free metal involves a particular kind of chemical change. For example, converting a copper(II) cation to an atom of copper metal requires the addition of two electrons.

Formation of Copper Metal: Reduction

In general, to convert metallic cations to atoms of pure metal, each cation must gain a particular number of electrons:

$$Cu^{2+} + 2e^- \longrightarrow Cu$$

Copper(II) ion   Copper metal

Chemists classify any chemical change in which a reactant can be considered to gain one or more electrons as a **reduction**. Thus, the conversion of copper(II) cations to copper metal is a reduction reaction, and we say that copper cations were **reduced**. You can convince yourself that this is a reduction reaction by examining the preceding equation.

Formation of Copper(II) Ions: Oxidation

Chemists classify the *reverse* reaction, in which an ion or other species can be considered to lose one or more electrons, as an **oxidation:**

$$\underset{\substack{\text{Copper} \\ \text{metal}}}{Cu} \longrightarrow \underset{\substack{\text{Copper(II)} \\ \text{ion}}}{Cu^{2+}} + 2e^-$$

Any reactant that appears to lose one or more electrons is said to be **oxidized.** In this case, a copper atom is oxidized to a copper(II) ion by loss of two electrons.

Oxidation–Reduction Reactions

Whenever one reactant loses electrons, another reactant must simultaneously gain them. In other words, oxidation and reduction reactions never occur separately. Oxidation and reduction occur together in what chemists call **oxidation–reduction reactions** or, to use a common chemical nickname, **redox reactions.**

You have already observed redox reactions in the laboratory. During Investigating Matter B.5 (pages 142–143), copper metal reacted with silver ions. The following equation shows the oxidation–reduction reaction you observed:

$$\underset{\substack{\text{Copper} \\ \text{metal}}}{Cu(s)} + \underset{\substack{\text{Silver ion} \\ \text{(colorless)}}}{2\,Ag^+(aq)} \longrightarrow \underset{\substack{\text{Copper(II) ion} \\ \text{(blue)}}}{Cu^{2+}(aq)} + \underset{\substack{\text{Silver} \\ \text{metal}}}{2\,Ag(s)}$$

Each metallic copper atom (Cu) was oxidized (converted to a Cu^{2+} ion by losing two electrons) and each silver ion (Ag^+ from $AgNO_3$ solution) was reduced (converted to an Ag atom by gaining one electron). What you *actually* observed in Investigating Matter B.5 was the formation of a solid (silver) and the solution becoming blue because of the formation of colored $Cu^{2+}(aq)$ ions; these observable changes indicated a redox reaction was occurring, even though you did not actually *see* the transfer of electrons.

In the same investigation, you found that copper ions could be recovered from a solution as metallic copper by allowing the copper ions to react with magnesium metal, an element more active than copper. Magnesium atoms were *oxidized;* copper ions were *reduced.* Do you see why?

$$\underset{\substack{\text{Copper(II) ion} \\ \text{(blue)}}}{Cu^{2+}(aq)} + \underset{\substack{\text{Magnesium} \\ \text{metal}}}{Mg(s)} \longrightarrow \underset{\substack{\text{Copper} \\ \text{metal}}}{Cu(s)} + \underset{\substack{\text{Magnesium ion} \\ \text{(colorless)}}}{Mg^{2+}(aq)}$$

Note that the total electrical charge on both sides of this equation is the same. Electrical charges—as well as atoms—must balance within a correctly written chemical equation. For instance, the net electrical charge on both the reactant side and the product side is 2+ in that equation.

In some circumstances, that reaction might be a useful way to obtain copper metal. However, as is often the case, the desired copper metal could only be obtained at the expense of using up another highly valuable material—in this case, magnesium metal.

> One easy way to remember this is OIL RIG: Oxidation Is Loss (of electrons), Reduction Is Gain (of electrons).

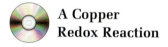 **A Copper Redox Reaction**

Many metallic elements are found in minerals in the form of cations because they combine readily with other elements to form ionic compounds. Obtaining a metal from its mineral requires energy and a source of electrons. A reactant that provides electrons is known as a **reducing agent.**

Using Redox Reactions to Obtain Pure Metals

Look again at Table 2.5 (page 145). The table lists several techniques that metallurgists use to reduce metal cations or, in other words, to supply one or more electrons to each cation. The specific technique chosen depends on the metal's reactivity and the availability of inexpensive reducing agents and energy sources.

Two approaches summarized in the table are electrometallurgy and pyrometallurgy. As the table suggests, *electrometallurgy* involves using an electrical current to supply electrons to metal ions, thus reducing them. This process is used when no adequate chemical reducing agents are available or when very high-purity metal is sought. *Pyrometallurgy,* the oldest ore-processing method, involves treating metals and their ores with heat (thermal energy), as in a blast furnace. Carbon (coke) and carbon monoxide are common reducing agents in pyrometallurgy: They provide electrons; metal ions are thus reduced to form metal atoms. An active metal can be used if neither of these reducing agents will do the job.

A third approach to obtaining metals from their ions is *hydrometallurgy,* which involves treating ores and other metal-containing materials with reactants in a water solution. You used such a procedure when you observed the reactivity of different metals in Investigating Matter B.5. Hydrometallurgy is used to recover silver and gold from old mine tailings (the mined rock left after most of the sought mineral is removed) by a process known as *leaching.* As higher-grade ores become scarcer, it will become economically feasible to use hydrometallurgy and other "wet processes" on metal-bearing minerals that can dissolve in water.

Modeling Matter

B.9 ELECTRONS AND REDOX PROCESSES

The processes of oxidation (the apparent loss of one or more electrons) and reduction (the apparent gain of one or more electrons) can be clarified by visual representations of these events. To develop such representations, you will consider atoms of each of the metals you encountered during Investigating Matter B.5 (page 142).

First, however, here is a review of some key details about an atom's composition. Magnesium (Mg), which is an active metal, formed Mg^{2+} ions in several reactions you observed during Investigating Matter B.5. The atomic number of Mg is 12, indicating that an electrically neutral atom of magnesium contains 12 protons and 12 electrons. Recall why those numbers must be equal for a neutral atom.

If magnesium is to form a Mg^{2+} ion, two negatively charged electrons must be removed from each magnesium atom. The bookkeeping involved can be summarized as

$$Mg \longrightarrow Mg^{2+} + 2\ e^-$$

Mg	Mg^{2+}	$2\ e^-$
12 protons (+)	12 protons (+)	
12 electrons (−)	10 electrons (−)	2 electrons (−)
Net charge: 0	Net charge: 2+	Net charge: 2−

To build a useful picture of this process in your mind, it is necessary to keep track of only the two electrons *released* by each magnesium atom, rather than monitoring all 12 of the atom's available electrons. (In fact, in any normal chemical reaction, a magnesium atom is not observed to release any of its other 10 electrons.)

Thus, for bookkeeping purposes, an atom of Mg will be depicted this way: Mg**:** (the element symbol with two dots attached). Each dot represents one of magnesium's two readily removable electrons. The symbol Mg represents the remaining parts of a magnesium atom, including its remaining ten electrons. The resulting expression for Mg is called an **electron-dot structure** or a **dot structure.** The equation for oxidation of Mg atoms can be represented this way in electron-dot terms:

$$Mg\text{:} \longrightarrow Mg^{2+} + 2\ e^-$$

> Lewis structure is another commonly used term for an electron-dot structure.

1. Construct a similar electron-dot expression for the change that occurred during Investigating Matter B.5 (page 142) when each of the following events took place:

   a. An atom of zinc, Zn, was converted to a Zn^{2+} ion. (*Hint:* Zn has two readily removable electrons.)
   b. A silver ion, Ag^+, was converted to a metallic silver atom, Ag(s).

2. Apply the definitions of *oxidation* and *reduction* to your two equations in Question 1, and label each reaction appropriately.

Now consider one of the complete reactions you observed during Investigating Matter B.5. When you immersed a sample of copper metal, Cu, in silver nitrate solution, $AgNO_3$, a blue solution containing Cu^{2+} formed, as well as crystals of solid Ag. The redox reaction that occurred is

$$Cu(s) + 2\,Ag^+(aq) \longrightarrow Cu^{2+}(aq) + 2\,Ag(s)$$

Using dot structures, this redox reaction can be represented as

$$Cu\text{:} + Ag^+ + Ag^+ \longrightarrow Cu^{2+} + Ag\cdot\ + Ag\cdot$$

3. Which reactant (Cu or Ag^+) is reduced?

4. Why is only one Cu atom needed for each two Ag^+ ions that react?

Each copper atom involved in this reaction loses two electrons. Thus, copper atoms must be oxidized in this redox reaction. It is clear from the dot structures that the two electrons lost by copper are gained by the two Ag^+ ions. So, Ag^+ is the *agent* that caused the removal of electrons from Cu, resulting in oxidation of Cu. The species involved in removing electrons from the oxidized reactant is called the **oxidizing agent**—in this case, Ag^+ ions.

5. a. Given the preceding explanation of an oxidizing agent, how would you define a *reducing agent*?

b. What must be the reducing agent in the reaction between $Cu(s)$ and Ag^+ ions?

6. Now consider the following reaction, which you observed during Investigating Matter B.5:

$$Zn(s) + Cu^{2+}(aq) \longrightarrow Zn^{2+}(aq) + Cu(s)$$

Draw an electron-dot representation of this reaction.

a. Which reactant is oxidized?

b. Which reactant is reduced?

7. Identify the following in the reaction represented in Question 6:

a. the oxidizing agent

b. the reducing agent

8. Consider both of the oxidation–reduction reactions you analyzed in this activity. What general features of an oxidation–reduction reaction would allow you to answer Questions 6 and 7 *without* drawing electron-dot representations?

9. Now consider a new oxidation–reduction reaction. Answer Questions 6 and 7 for the following equation:

$$Zn^{2+}(aq) + Mg(s) \longrightarrow Zn(s) + Mg^{2+}(aq)$$

10. Return to Investigating Matter B.5 and write an oxidation–reduction equation for the reaction of silver ions with magnesium atoms. Answer Questions 6 and 7 for this equation.

The chemical properties of elements can be understood in terms of their respective electron arrangements, as you have seen. In Section C, you will further explore the composition and properties of common materials composed of these elements.

SECTION B SUMMARY
Reviewing the Concepts

The resources for all human activities must be obtained from Earth's atmosphere, hydrosphere, and outer layer of its lithosphere. These resources are not uniformly distributed.

1. List two resources typically found in each of the three major "spheres" of Earth.

2. a. List and briefly describe three major parts of the lithosphere.
   b. Which layer serves as the main storehouse of chemical resources used in manufacturing consumer products?

3. Identify the nation that produces the most
   a. silver.　　b. copper.　　c. tin.

4. According to the information in Table 2.3 on page 136, which of these four nations—the United States, Australia, China, or Brazil—produces the largest masses of the eight listed resources in the table?

The feasibility of mining and extracting a mineral resource depends, in part, on how easily a particular metal can be processed and used, which depends on its chemical reactivity. Active metals are more difficult to process than are less-active metals and tend to corrode more quickly.

5. How do minerals differ from ores?

6. What factors determine the feasibility of mining a particular metallic ore at a certain site?

7. A nineteenth-century gold mine, inactive for over 100 years, has recently reopened for further mining. What factors may have influenced the decision to reopen the mine?

8. What is meant by referring to the quantity of "useful ore" at a site?

9. Why are active metals more difficult to process and refine than are less active metals?

10. Based on your results from Investigating Matter B.5 (page 142), which metals involved in that investigation would be the easiest to process? Why?

11. Why do most metals exist in nature as minerals rather than as pure metallic elements?

12. Which of these reactions is more likely to occur? Why? (Refer to Table 2.5, page 145.)
    a. calcium metal with chromium(III) chloride
    b. chromium metal with calcium chloride

13. Consider the following two equations. Which equation represents a reaction that is more likely to occur? Why?
    a. $Zn^{2+}(aq) + 2\ Ag(s) \longrightarrow Zn(s) + 2\ Ag^+(aq)$
    b. $2\ Ag^+(aq) + Zn(s) \longrightarrow 2\ Ag(s) + Zn^{2+}(aq)$

14. a. Why would it be a poor idea to stir a solution of lead(II) nitrate with an iron spoon? (Refer to Table 2.5, page 145.)
    b. Write a chemical equation to support your answer.

The processes of oxidation and reduction occur together, resulting in oxidation–reduction (redox) reactions.

15. Define oxidation and reduction in terms of electron transfer.

16. Write an equation for each of the following processes:
    a. the *reduction* of gold(III) ions to gold metal
    b. the *oxidation* of elemental vanadium to vanadium(IV) ions
    c. the *oxidation* of Cu^+ to Cu^{2+} ions

17. Identify each of the following equations as representing either an oxidation reaction or a reduction reaction:
    a. $Fe^{2+} + 2\ e^- \longrightarrow Fe$
    b. $Cr \longrightarrow Cr^{3+} + 3\ e^-$
    c. $Al^{3+} + 3\ e^- \longrightarrow Al$

18. Consider the following equation:

    oxidized reduced
    $$Zn(s) + Ni^{2+}(aq) \longrightarrow Zn^{2+}(aq) + Ni(s)$$

    a. Which reactant has been oxidized? Explain your choice.
    b. Which reactant has been reduced? Explain your choice.
    c. What is the reducing agent in this reaction?

19. Consider the following equation:

    $$2\ K(s) + Hg^{2+}(aq) \longrightarrow 2\ K^+(aq) + Hg(s)$$

    a. Which reactant has been oxidized? Explain your choice.
    b. Which reactant has been reduced? Explain your choice.
    c. What is the oxidizing agent in this reaction?

20. Write an equation for
    a. the oxidation of Al metal by Cr^{3+} ions.
    b. the reduction of Mn^{2+} ions by Mg metal.

Metal cations can be converted to metal atoms by electrometallurgy, pyrometallurgy, or hydrometallurgy.

21. Explain how each of the following processes converts metal cations to metal atoms:

    a. electrometallurgy
    b. pyrometallurgy
    c. hydrometallurgy

22. What processes would be most useful in obtaining the following elements from their ores?

    a. magnesium
    b. lead

Connecting the Concepts

23. How does access to mineral resources contribute to a nation's economic success? Explain.

24. In building ships, common practice is to attach a piece of magnesium to the hull to act as a "sacrificial anode." In terms of metal activity, explain how this helps to prevent the corrosion of other metal parts of the ship.

25. There are thousands of tons of gold in sea water. Explain why it is unlikely that sea water will ever be "mined" for gold.

26. At one time, food cans were made with tin and soldered with lead. What kinds of health hazards were posed by this design?

27. Is there any connection between the process used to reduce a metal cation and the position of that element on the periodic table?

Extending the Concepts

28. Although aluminum is a more reactive metal than is iron, it is often used to build outdoor products. Investigate why this use is feasible.

29. Find and describe an historical example of uneven distribution of mineral resources and its impact on relations among nations.

30. What conclusions about materials can be drawn from a study of substances used for currency in ancient civilizations? Explain your ideas by citing examples.

31. History documents that copper has been used by humans for ten thousand years, whereas aluminum has been used for only about 100 years. Suggest and explain some reasons for this time difference.

32. The reactive metal aluminum is often used in containers for acidic beverages. Investigate and describe the technology that makes this possible.

33. What is a *patina*? Explain its value both aesthetically and chemically.

Conserving Matter

In some chemical reactions, matter seems to be created—as a nail rusts, a new material appears; in other reactions, matter seems to disappear—a sheet of paper burns and apparently vanishes. In terms of fundamental particles, neither creation nor destruction of matter actually occurs. In this section, you will learn what happens to atoms in chemical reactions and how the atoms can be tracked through the use of chemical equations. This information will help you consider the fate of Earth's resources, as well as the fate of the materials and products developed from them.

C.1 KEEPING TRACK OF ATOMS

Think for a moment about what happens to molecules in gasoline as they burn in an automobile engine (Figure 2.23). The carbon and hydrogen atoms that make up these molecules react with oxygen atoms in air to form carbon monoxide (CO), carbon dioxide (CO_2), and water (H_2O). These products are released as exhaust and disperse in the atmosphere. Thus, atoms of carbon, hydrogen, and oxygen originally present in gasoline and air have not been destroyed; rather, they have become rearranged into different molecules.

In short, "using things up" means that materials are chemically changed, not destroyed. The **law of conservation of matter,** like all scientific laws, summarizes what has been learned by careful observation of nature: In a chemical reaction, matter is neither created nor destroyed. Molecules can be converted and decomposed by chemical processes; but atoms are forever. In a chemical reaction, matter—at the level of individual atoms—is always fully accounted for.

Because chemical reactions cannot create or destroy atoms, chemical equations representing such reactions must always be balanced. What does a "balanced equation" mean? Recall your introduction to chemical equations in Unit 1 (pages 34–37). Formulas for the reactants are placed on the left of the arrow; formulas for the products are placed on the right. In a **balanced chemical equation,** the number of atoms of each element is the same on the reactant and product sides.

Consider burning coal as an example. See Figure 2.24 (page 154). Coal is mostly carbon (C). If carbon burns completely, it combines with oxygen gas (O_2) to produce carbon dioxide (CO_2). Here is a representation of the atoms and molecules involved in this reaction:

Figure 2.23 *When a car's gas tank empties, where do the atoms in the gasoline go?*

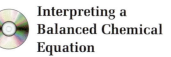 **Interpreting a Balanced Chemical Equation**

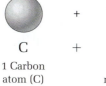 + →

C   + O_2 ⟶ CO_2

1 Carbon atom (C)   1 Oxygen molecule (O_2)   1 Carbon dioxide molecule (CO_2)

Note that the numbers of carbon and oxygen atoms on the reactant side equal the numbers of carbon and oxygen atoms on the product side. This indicates that the equation is balanced.

The representation of the coal-burning reaction shows that one carbon atom reacts with one oxygen molecule to form one carbon dioxide molecule:

$$C(s) \quad + \quad O_2(g) \quad \longrightarrow \quad CO_2(g)$$

Carbon   and   Oxygen   React to   Carbon
(in coal)   gas   produce   dioxide gas

The chemical reaction from Investigating Matter B.3 (page 139), where you heated powdered copper in air, provides another example. Copper metal (Cu) reacts with oxygen gas (O_2) to form copper(II) oxide (CuO). Look at the following representation:

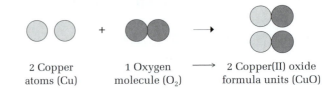

2 Copper   1 Oxygen $\longrightarrow$ 2 Copper(II) oxide
atoms (Cu)   molecule (O_2)   formula units (CuO)

Again, note that the numbers of copper and oxygen atoms on the reactant side equal the respective numbers of copper and oxygen atoms on the product side. Written as a chemical equation, the reaction is

$$2\,Cu(s) \quad + \quad O_2(g) \quad \longrightarrow \quad 2\,CuO(s)$$

Copper   and   Oxygen   React to   Copper(II)
metal   gas   produce   oxide

Figure 2.24 *When a pile of coal burns, what becomes of the atoms that made up the coal?*

It is standard procedure to imply but not write the coefficient "1."

Look at the periodic table to recall which elements are metals and which are nonmetals.

You may have noticed that numbers have been placed in front of the copper and copper(II) oxide formulas. A "one" is also implied for the O_2. These numbers are called *coefficients*. **Coefficients** indicate the relative number of units of each substance involved in the chemical reaction. Reading this equation from left to right, you would say, "Two copper atoms react with one oxygen molecule to produce two formula units of copper(II) oxide."

Why is the term *formula unit* used instead of *molecule*? Compounds of a metal and a nonmetal are ionic. (*Note:* It might be helpful to review pages 38–41 of Unit 1.) Ionic compounds are not found as individual molecules. Rather, they form large crystals made up of arrays of ions. Chemists use the term **formula unit** when referring to the smallest unit of an ionic compound.

Study Figure 2.25, which represents the copper–oxygen reaction in three ways. The photograph shows what you observed during Investigating Matter B.3, representing this reaction at a large-scale (macroscopic) level. The particulate-level drawing models individual atoms and molecules for the reactants, and formula units within the ionic crystal of the product, with the chemical equation representing this reaction symbolically. The ability to think about particles involved in chemical reactions and to represent reactions with symbols will help you link your observations to what you cannot see (that is, what happens with atoms, ions, and molecules) and to keep track of reactants and products.

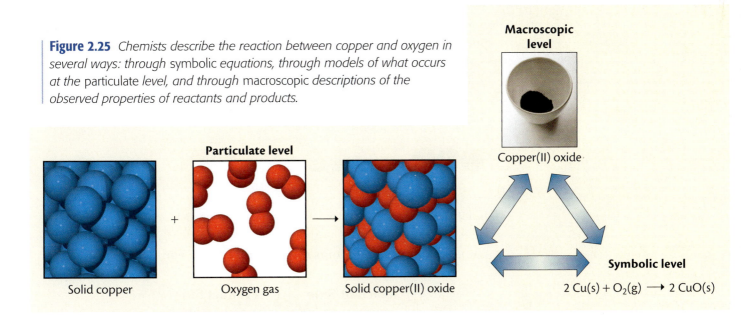

Figure 2.25 *Chemists describe the reaction between copper and oxygen in several ways: through symbolic equations, through models of what occurs at the* particulate *level, and through* macroscopic *descriptions of the observed properties of reactants and products.*

Macroscopic level

Copper(II) oxide

Particulate level

Solid copper   +   Oxygen gas   Solid copper(II) oxide

Symbolic level

$2 Cu(s) + O_2(g) \longrightarrow 2 CuO(s)$

In the following activity, you will practice recognizing and interpreting chemical equations.

Developing Skills

C.2 ACCOUNTING FOR ATOMS

Sample Problem: *The combustion reaction between propane (C_3H_8) and oxygen gas (O_2) is a common source of heat for campers, recreational-vehicle users, and others using tanks of propane fuel. See Figure 2.26. A chemical statement showing the reactants and products is*

$$C_3H_8(g) + O_2(g) \longrightarrow CO_2(g) + H_2O(g)$$

Interpret this reaction in terms of (a) words, (b) molecular models, and (c) an atom inventory. (d) Is this a balanced chemical equation?

> This chemical reaction produces thermal energy (heat), which can also be considered a product.

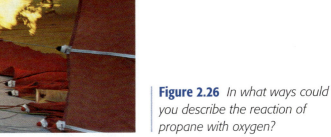

Figure 2.26 *In what ways could you describe the reaction of propane with oxygen?*

a. Interpreting this statement in words, "Propane gas reacts with oxygen gas to produce carbon dioxide gas and water vapor."

b. Using 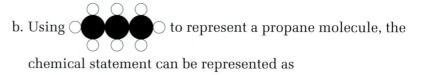 to represent a propane molecule, the chemical statement can be represented as

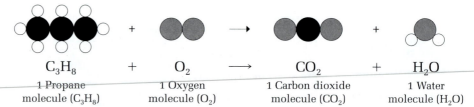

C_3H_8	+	O_2	$\longrightarrow$	CO_2	+	H_2O
1 Propane molecule (C_3H_8)		1 Oxygen molecule (O_2)		1 Carbon dioxide molecule (CO_2)		1 Water molecule (H_2O)

c. Counting the atoms on each side of the equation gives this atom inventory:

Reactant Side	Product Side
3 carbon atoms	1 carbon atom
8 hydrogen atoms	2 hydrogen atoms
2 oxygen atoms	3 oxygen atoms

d. The respective numbers of carbon, hydrogen, and oxygen atoms are different in reactants and products. Thus, the original statement is not a properly written chemical equation. That is, it is not balanced.

For each of the following chemical statements,

a. write an interpretation of the statement in words,
b. draw a representation of the chemical statement (some structures are provided),
c. complete an atom inventory of the reactants and products, and
d. decide whether the expression, as written, is balanced.

1. Many people use natural gas as a source of household heat. Natural gas contains methane (CH_4), which burns with oxygen gas (O_2) in air according to the equation

$$CH_4 + 2\,O_2 \longrightarrow CO_2 + 2\,H_2O$$

Let represent a methane molecule.

2. When an acid reacts with a metal, hydrogen gas and an ionic compound are often formed. An expression for hydrobromic acid (HBr) reacting with magnesium metal to form hydrogen gas and magnesium bromide ($MgBr_2$) is

$$HBr + Mg \longrightarrow H_2 + MgBr_2$$

Let ⬡⬡⬡ represent a formula unit of magnesium bromide ($MgBr_2$). (*Note:* Formula units are potentially misleading because, under ordinary circumstances, ionic compounds are not found as individual molecules, but as large crystals made

See page 61 in Unit 1 to view a representation of an ionic crystal structure.

up of many ions. Thus, this and similar representations of ionic compounds are technically inaccurate; however, for an atom inventory, they will suffice. Keep these limitations in mind.)

3. Hydrogen sulfide (H_2S) and metallic silver react in air to form silver sulfide (Ag_2S), commonly known as *silver tarnish,* and water:

$$4\ Ag + 4\ H_2S + O_2 \longrightarrow 2\ Ag_2S + 4\ H_2O$$

Let 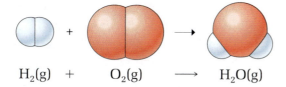 represent a hydrogen sulfide molecule and represent a formula unit of silver sulfide.

Try Questions 4 and 5 without drawing representations. (Why might this be a good decision?)

4. Wood or paper, composed mainly of cellulose, $C_6H_{10}O_5$, can burn in air to form carbon dioxide and water vapor:

$$C_6H_{10}O_5 + 6\ O_2 \longrightarrow 6\ CO_2 + 5\ H_2O$$

5. Nitroglycerin, $C_3H_5(NO_3)_3$, the active constituent of dynamite, decomposes explosively to form N_2, O_2, CO_2, and water.

$$2\ C_3H_5(NO_3)_3 \longrightarrow 3\ N_2 + O_2 + 6\ CO_2 + 5\ H_2O$$

> There are nine oxygen atoms in one $C_3H_5(NO_3)_3$ molecule. Can you see why?

C.3 NATURE'S CONSERVATION: BALANCED CHEMICAL EQUATIONS

The law of conservation of matter is based on the notion that atoms are indestructible. All chemical changes observed in matter can be interpreted as rearrangements among atoms. Correctly written (balanced) chemical equations represent such changes. In the preceding activity, you practiced recognizing balanced chemical equations. Now you will learn how to write them.

As an example, consider the reaction of hydrogen gas with oxygen gas to produce gaseous water. First, correctly write the reactant formula or formulas to the left of the arrow and the product formula or formulas to the right of the arrow, keeping in mind that hydrogen and oxygen gas are diatomic (*two*-atom) molecules.

> The law of conservation of matter holds under *normal* circumstances. You may be familiar with the equation $E = mc^2$, which indicates that small quantities of matter (mass) can be converted to energy within events such as nuclear reactions.

$$H_2(g) \quad + \quad O_2(g) \quad \longrightarrow \quad H_2O(g)$$

> For a complete listing of the elements found as diatomic molecules, see page 36.

Check the preceding expression by completing an atom inventory: Two hydrogen atoms appear on the left side and two on the right. Thus, hydrogen atoms are balanced. However, two oxygen atoms appear on the left side and only one on the right. Because oxygen is not balanced, the expression requires additional work.

Here is an *incorrect* way to complete the equation:

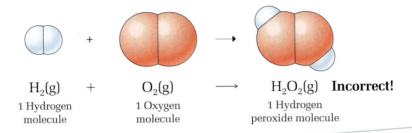

$H_2(g)$ + $O_2(g)$ ⟶ $H_2O_2(g)$ **Incorrect!**

1 Hydrogen molecule   1 Oxygen molecule   1 Hydrogen peroxide molecule

Although this chemical statement meets an atom-inventory check (there are two hydrogen and two oxygen atoms on both sides), this chemical expression is *quite wrong*. By changing the subscript of O from 1 to 2 in the product, the identity of the product has been changed from water (H_2O) to hydrogen peroxide (H_2O_2). Because hydrogen peroxide is *not* produced by this reaction, the chemical expression is incorrect. When balancing chemical equations, *subscripts* remain unchanged once correct formulas have been written for reactants and products. *Coefficients* (the numbers placed in front of formulas) must be adjusted to balance equations instead.

Additional hydrogen, oxygen, and water molecules must be added to the appropriate sides of the equation to balance the numbers of oxygen and hydrogen atoms. Another oxygen atom is needed on the product side to bring the number of oxygen atoms on both sides to two. Therefore, a water molecule is added:

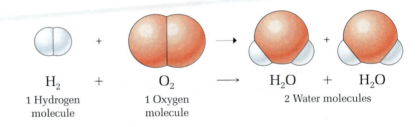

H_2 + O_2 ⟶ H_2O + H_2O

1 Hydrogen molecule   1 Oxygen molecule   2 Water molecules

As a result, two oxygen atoms appear on each side of the equation. Unfortunately, hydrogen atoms are no longer balanced: Now there are two hydrogen atoms on the left side and four atoms on the right. How can two hydrogen atoms be added to the reactant side? You are correct if you said by adding one hydrogen molecule to the left side:

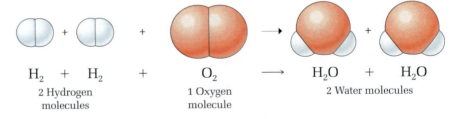

H_2 + H_2 + O_2 ⟶ H_2O + H_2O

2 Hydrogen molecules   1 Oxygen molecule   2 Water molecules

The atoms are now balanced. Count the atoms to verify that!

This information is usually summarized in a chemical equation:

$$2\ H_2(g) + O_2(g) \longrightarrow 2\ H_2O(g)$$

Under certain conditions, H_2 reacts with O_2 in a violent explosion. However, the reaction can be controlled and, in fact, it powers some types of rockets. See Figure 2.27. Used in fuel cells, this reaction can generate electricity.

Figure 2.27 *Some rockets are propelled by the thrust generated when liquid hydrogen and liquid oxygen react to form gaseous water.*

This equation can be interpreted as indicating, "Two molecules of hydrogen gas react with one molecule of oxygen gas to produce two molecules of water vapor."

The following additional suggestions may help you as you write correctly balanced chemical equations:

▶ If polyatomic ions, such as NO_3^- and CO_3^{2-} appear as both reactants and products, treat them as units, rather than balancing their atoms individually.

▶ If water is involved in the reaction, balance hydrogen and oxygen atoms last.

▶ Recount all atoms after you think an equation is balanced—just to be sure!

C.4 WRITING CHEMICAL EQUATIONS

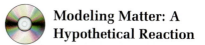 **Modeling Matter: A Hypothetical Reaction**

The reaction of methane gas (CH_4) with chlorine gas (Cl_2) occurs in sewage treatment plants and often in chlorinated water supplies (Figure 2.28). Common products are liquid chloroform ($CHCl_3$) and hydrogen chloride gas (HCl). A chemical statement describing this reaction follows. As you can see from a quick glance, the equation is not balanced:

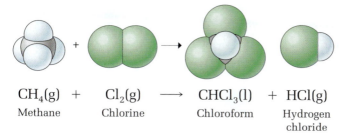

$$CH_4(g) + Cl_2(g) \longrightarrow CHCl_3(l) + HCl(g)$$
Methane   Chlorine   Chloroform   Hydrogen chloride

To complete this chemical equation, you can follow this line of reasoning: *One* carbon atom appears on each side of the arrow, so carbon atoms balance. Thus, the coefficients in front of CH_4 and $CHCl_3$ are regarded (at least for the moment) as "1." We will write it in explicitly, just as a reminder, but the coefficient 1 is removed from the final equation:

Figure 2.28 *A water-treatment plant, a portion of which is shown here, often uses chlorine to purify municipal water.*

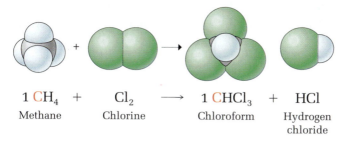

$$1\,CH_4 + Cl_2 \longrightarrow 1\,CHCl_3 + HCl$$
Methane   Chlorine   Chloroform   Hydrogen chloride

For convenience, the symbols (g) and (l) are removed. They will reappear in the final equation.

Four hydrogen atoms are on the left, but only two hydrogen atoms are on the right (one in $CHCl_3$, a second in HCl). To increase the number of hydrogen atoms on the product side, the coefficient of HCl can be adjusted. Because two more hydrogens are needed on the right, the number of HCl molecules must be changed from 1 to 3. This gives four hydrogen atoms on the right:

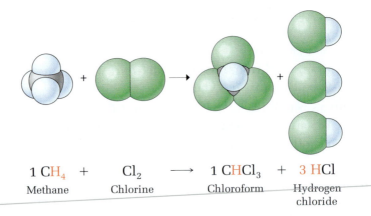

$$1 \text{ CH}_4 \ + \quad \text{Cl}_2 \quad \longrightarrow \quad 1 \text{ CHCl}_3 \ + \ 3 \text{ HCl}$$

Methane   Chlorine   Chloroform   Hydrogen chloride

Now both carbon and hydrogen atoms are balanced. What about chlorine? Two chlorine atoms appear on the left and six on the right side. These six chlorine atoms (three in $CHCl_3$, three in 3 HCl) must have come from three chlorine (Cl_2) molecules. Thus, 3 must be the coefficient of Cl_2.

The chemical equation appears to be balanced. An atom inventory verifies that the equation is complete and correct as written:

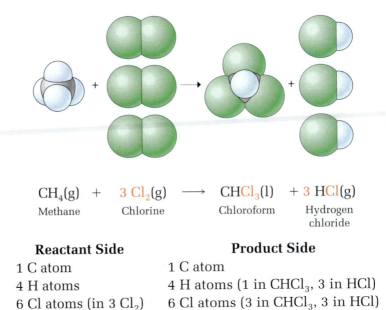

$$\text{CH}_4(g) \ + \ 3 \text{ Cl}_2(g) \quad \longrightarrow \quad \text{CHCl}_3(l) \ + 3 \text{ HCl}(g)$$

Methane   Chlorine   Chloroform   Hydrogen chloride

Reactant Side	**Product Side**
1 C atom	1 C atom
4 H atoms	4 H atoms (1 in $CHCl_3$, 3 in HCl)
6 Cl atoms (in 3 Cl_2)	6 Cl atoms (3 in $CHCl_3$, 3 in HCl)

Chloroform is one of the trihalomethanes (THMs) mentioned in Unit 1 (page 90).

Copy the following chemical expressions onto a separate sheet of paper, and balance each if needed. For Questions 1 through 4, draw a representation of your final equation to verify that it is balanced. Structures unfamiliar to you will be provided.

1. Two blast-furnace reactions are used to obtain iron from its ore:
   a. $C(s) + O_2(g) \longrightarrow 2 \text{ CO}(g)$
   b. $Fe_2O_3(s) + CO(g) \longrightarrow Fe(l) + 3 \text{ CO}_2(g)$

   Let ⬤ represent a formula unit of Fe_2O_3.

   Let ⬤⬤⬤ represent a molecule of CO_2.

2. The final step in the refining of a copper ore is:

$$CuO(s) + C(s) \longrightarrow Cu(s) + CO_2(g)$$

Let represent a formula unit of CuO.

3. Ozone (O_3) can decompose to form oxygen gas (O_2):

$$O_3(g) \longrightarrow O_2(g)$$

Let represent an ozone molecule.

4. Ammonia (NH_3) in the soil reacts continuously with oxygen gas (O_2):

$$NH_3(g) + O_2(g) \longrightarrow NO_2(g) + H_2O(l)$$

5. Copper metal reacts with silver nitrate solution to form copper(II) nitrate solution and silver metal:

$$Cu(s) + AgNO_3(aq) \longrightarrow Cu(NO_3)_2(aq) + Ag(s)$$

(*Hint:* Look at the formula for copper(II) nitrate, $Cu(NO_3)_2$. The subscript of two outside the parentheses indicates that this formula contains two nitrate (NO_3^-) anions. Therefore, one formula unit of $Cu(NO_3)_2$ contains one copper ion, two nitrogen atoms, and six oxygen atoms.)

6. Combustion of gasoline in an automobile engine can be represented by the burning of octane (C_8H_{18}):

$$C_8H_{18}(l) + O_2(g) \longrightarrow CO_2(g) + H_2O(g)$$

C.5 INTRODUCING THE MOLE CONCEPT

The Mole

In Question 2 of the previous activity, you analyzed the chemical equation

$$2\,CuO(s) + C(s) \longrightarrow 2\,Cu(s) + CO_2(g).$$

As you know, one interpretation of this equation is as follows: Two formula units of copper(II) oxide and one atom of carbon react to produce two atoms of copper and one molecule of carbon dioxide. Although correct, this interpretation involves such small quantities of material that a reaction on that scale would be completely unnoticed. Such information would not be very useful, for example, to a metal-processing company interested in how much carbon is needed to react with a large amount of copper(II) oxide to produce metallic copper.

Chemists have devised a counting unit called the **mole** (symbolized *mol*) that can help to solve the metal-processing company's problem. You are familiar with other common counting units, such as *pair* and *dozen*. The mole can be thought of as the chemist's *dozen*.

A *pair* of water molecules is two water molecules. One *dozen* water molecules is the same as 12 water molecules. One *mole* of water molecules is 602 000 000 000 000 000 000 000 water molecules. This special number—the number of particles or "things" in one mole—is more conveniently written as 6.02×10^{23}. Either way, this is a very large number!

The number 6.02×10^{23} is commonly known as Avogadro's number.

Figure 2.29 *One mole each of copper, table salt, and water. If these samples each represent one mole of substance, why do they occupy different volumes?*

The mole concept permits us literally to count by weighing.

To help you get a better idea of the size of a mole, consider this analogy: Imagine stringing one mole of paper clips (6.02×10^{23} paper clips) together and wrapping the paper-clip string around the world. It would circle the world about 400 trillion (4×10^{14}) times! Even if you connected a million paper clips each second, it would take you 190 million centuries to finish stringing together one mole of paper clips.

As large as one mole seems, however, drinking one mole of water molecules would leave you quite thirsty on a hot day. One mole of water is *less than one-tenth of a cup* of water—only 18 g (or 18 mL) of water. Because atoms and molecules are so very small, the amount of those tiny building blocks needed to make up the "stuff" we see is nearly countless.

That is the reason the mole is so useful for work in chemistry. It represents a number of atoms, molecules, or formula units large enough to be conveniently weighed or measured in the laboratory. Furthermore, the atomic weights of elements can be used to find the mass of one mole of any substance, a value known as the **molar mass** of a substance. Figure 2.29 shows one mole of several familiar substances.

Molar Mass

Because mass is a readily measured quantity, molar mass can be used to count particles by simply measuring the mass of a substance. Specific examples will help you to better understand this powerful idea. See Figure 2.30. Suppose you need to find the molar masses of carbon (C) and of copper (Cu). In other words, you want to know the mass of one mole of carbon atoms (6.02×10^{23} C atoms) and the mass of one mole of copper atoms (6.02×10^{23} Cu atoms). Rather than counting that collection of atoms onto a laboratory balance, you can quickly get the correct quantities from atomic-weight data. The atomic weight of each element is found on the periodic table. Carbon's atomic weight is 12.01 and copper's is 63.55. If the unit *grams* is attached to these numbers, the resulting values represent their molar masses:

$$1 \text{ mol C} = 12.01 \text{ g C} \qquad 1 \text{ mol Cu} = 63.55 \text{ g Cu}$$

In summary, the mass (in grams) of one mole of an element's atoms equals the numerical value of that element's atomic weight. Any element's molar mass can be easily found from the periodic table.

Figure 2.30 *One mole of carbon (left) and copper (right). What do the different samples have in common?*

The molar mass of a diatomic element, such as O_2, is *twice* the mass given in the periodic table. Can you explain why this is so?

The molar mass of a compound is simply the sum of the molar masses of its component atoms. For example, consider two compounds of importance to metal-processing companies interested in refining large amounts of copper—carbon dioxide, CO_2, and the copper-containing mineral malachite, $Cu_2CO_3(OH)_2$.

One mole of CO_2 molecules contains one mole of C atoms and two moles of O atoms. Adding the molar mass of carbon and twice the molar mass of oxygen gives

$$1 \text{ mol C} \times \frac{12.01 \text{ g C}}{1 \text{ mol C}} = 12.01 \text{ g C}$$

$$2 \text{ mol O} \times \frac{16.00 \text{ g O}}{1 \text{ mol O}} = 32.00 \text{ g O}$$

Molar mass of CO_2 = (12.01 g + 32.00 g) = 44.01 g CO_2

Recall that *mol* is the symbol for the unit *mole*.

The molar mass of malachite is found in a similar way. However, we must carefully count the atoms of each element in more complex compounds such as this.

One mole of malachite, $Cu_2CO_3(OH)_2$, contains 2 mol Cu, 1 mol C, 5 mol O, and 2 mol H. (Can you see where five oxygen atoms are found in the formula?)

$$2 \text{ mol Cu} \times \frac{63.55 \text{ g Cu}}{1 \text{ mol Cu}} = 127.1 \text{ g Cu}$$

$$1 \text{ mol C} \times \frac{12.01 \text{ g C}}{1 \text{ mol C}} = 12.01 \text{ g C}$$

$$5 \text{ mol O} \times \frac{16.00 \text{ g O}}{1 \text{ mol O}} = 80.00 \text{ g O}$$

$$2 \text{ mol H} \times \frac{1.008 \text{ g H}}{1 \text{ mol H}} = 2.016 \text{ g H}$$

Molar mass of $Cu_2CO_3(OH)_2$ = 221.1 g $Cu_2CO_3(OH)_2$

Developing Skills

C.6 MOLAR MASSES

Find the molar mass of each of the following substances:

1. nitrogen atoms, N
2. nitrogen molecules, N_2
3. sodium chloride (table salt), NaCl
4. sucrose (table sugar), $C_{12}H_{22}O_{11}$
5. chalcopyrite (a copper-containing mineral), $CuFeS_2$
6. magnesium phosphate, $Mg_3(PO_4)_2$
7. caffeine, $C_8H_{10}N_4O_2$
8. calcium hydroxyapatite (a mineral found in teeth), $Ca_{10}(PO_4)_6(OH)_2$
9. alunite (an aluminum mineral), $KAl_3(SO_4)_2(OH)_6$
   (*Hint:* Verify that this formula includes 14 mol of oxygen atoms.)

The term *one mole of nitrogen* is not helpful. It could refer to one mole of N atoms or to one mole of N_2 molecules. Thus, it is important to specify clearly which substance is involved.

C.7 EQUATIONS AND MOLAR RELATIONSHIPS

Why are chemical equations and molar masses useful in answering questions about large-scale reactions? The coefficients in a chemical equation show both the relative numbers of individual molecules (or formula units) of reactants and products and the relative numbers of moles of these same substances.

Consider again the copper-refining example. The mole "counting unit" makes it relatively easy to find the mass of carbon dioxide released during refining:

$$2 \, CuO(s) \quad + \quad C(s) \quad \longrightarrow \quad 2 \, Cu(s) \quad + \quad CO_2(g)$$

| 2 Formula units CuO | 1 Atom C | 2 Atoms Cu | 1 Molecule CO_2 |

Alternatively,

$$2 \text{ mol } CuO + 1 \text{ mol } C \longrightarrow 2 \text{ mol } Cu + 1 \text{ mol } CO_2$$

Thus, for every *two* moles of CuO that react, *one* mole of CO_2 is produced. The molar masses of all four substances can be used to interpret the equation in terms of the masses involved:

$$2 \, CuO(s) \quad + \quad C(s) \quad \longrightarrow \quad 2 \, Cu(s) \quad + \quad CO_2(g)$$

2 mol CuO	1 mol C	2 mol Cu	1 mol CO_2
79.55 g CuO + 79.55 g CuO	12.01 g C	63.55 g Cu + 63.55 g Cu	44.01 g CO_2
159.10 g CuO	12.01 g C	127.10 g Cu	44.01 g CO_2

> 2 mol CuO, containing 2 mol Cu atoms as well as 2 mol O atoms, has a mass of 159.10 g, which is twice the molar mass of CuO.

Compare the total mass of products to the total mass of reactants. This illustrates the *law of conservation of matter*. Note that the total mass of reactants (159.10 g + 12.01 g) equals the total mass of products.

Metal processing companies typically work with quantities much larger than several grams; therefore, they produce much more than one or two moles of material. See Figure 2.31. The refiners must rely on

Figure 2.31 *Following reduction from its oxide ore, molten aluminum metal is poured into a mold. The United States produces over two million metric tons of aluminum annually.*

their understanding of the molar relationships represented in equations such as the one above to complete the needed calculations involved in their large-scale reactions.

Sample Problem: *A refiner needs to convert 955.0 g CuO to pure Cu. What mass of C is needed for this reaction?*

The previous equation indicates that 1 mol C is needed to refine 2 mol CuO. But how many moles of CuO are represented by 955.0 g CuO? One mol CuO has a mass of 79.55 g, therefore

$$\frac{1 \text{ mol CuO}}{79.55 \text{ g CuO}} = \frac{x \text{ mol CuO}}{955.0 \text{ g CuO}}$$

$$955.0 \text{ g CuO} \times 1 \text{ mol CuO} = 79.55 \text{ g CuO(s)} \times x \text{ mol CuO}$$

$$x = 955 \text{ g } \cancel{\text{CuO}} \times \frac{1 \text{ mol CuO}}{79.55 \text{ g } \cancel{\text{CuO}}} = 12.01 \text{ mol CuO}$$

We started solving this sample problem with a proportion; however, in rearranging the proportion we created a **conversion factor.** The top and the bottom of this conversion factor—1 mol CuO/79.55 g CuO—both refer to the same number of particles, which are expressed in different units. This fraction allows us to convert from one unit, in this case, *grams,* to another unit, *moles,* by describing the relationship between the two units of measurement.

The refiner now knows that there are 12.01 mol CuO in 955.0 g CuO. How many moles of C are needed? Based on the equation on page 164, which states 1 mol C is required to react with 2 mol CuO, how many moles of C are needed to refine 12.01 mol CuO?

While it is quite possible to solve this problem using proportions, it is often faster and more efficient to create a conversion factor relating the quantities in question (which you would do, anyway, by solving the proportion expression).

In this case, the relevant conversion factor would be 1 mol C for every 2 mol CuO, based on the coefficients in the chemical equation. Notice, as in the previous example, we write the conversion factor to allow the appropriate units to divide and cancel:

$$12.01 \text{ } \cancel{\text{mol CuO}} \times \frac{1 \text{ mol C}}{2 \text{ } \cancel{\text{mol CuO}}} = 6.005 \text{ mol C}$$

Once this information is known, it is possible to find the mass of C needed to refine 955.0 g of CuO, again using an appropriate conversion factor:

$$6.005 \text{ } \cancel{\text{mol C}} \times \frac{12.01 \text{ g C}}{1 \text{ } \cancel{\text{mol C}}} = 72.12 \text{ g C}$$

Identify the conversion factor in the last calculation. Where did the information contained in the conversion factor come from? Once the molar masses and molar relationships involved are known, chemists can predict reactions involving any amount of material they may need or may wish to produce.

C.8 MOLAR RELATIONSHIPS

Base your answers to the following questions on this equation:

$$CuO(s) + 2\ HCl(aq) \longrightarrow CuCl_2(aq) + H_2O(l)$$

1. Find the molar mass of each reactant specified in the equation.
2. Find the mass (in grams) of each of the following substances:
   a. 1.0 mol HCl
   b. 5.0 mol HCl
   c. 0.50 mol CuO
3. Find the number of moles of substance that are represented by
   a. 941.5 g $CuCl_2$
   b. 201.6 g $CuCl_2$
   c. 73.0 g HCl
4. a. How many moles of CuO are needed to react with 4 mol HCl?
   b. How many moles of HCl are needed to react with 4 mol CuO?

C.9 COMPOSITION OF MATERIALS

One decision you will make when designing your coin is whether to use only one material or a combination of materials. If your design uses more than one material, you will need to specify how much of each material will be present in the coin. The percent by mass of each material found in an item such as a coin is called its **percent composition.**

In Section A, you learned that the composition of the U.S. penny has changed several times. During 1943, it was made of zinc-coated steel. After this date and up until 1982, the penny was made mostly of copper. Since 1982, U.S. pennies have been made primarily of zinc.

Sample Problem: *A post-1982 penny with a mass of 2.500 g is composed of 2.4375 g zinc and 0.0625 g copper. What is its percent composition?*

The percent composition of the penny can be found by dividing the mass of each constituent metal by the total mass of the penny and multiplying by 100%.

$$\frac{2.4375\ g\ zinc}{2.500\ g\ total} \times 100\% = 97.50\%\ zinc$$

$$\frac{0.0625\ g\ copper}{2.500\ g\ total} \times 100\% = 2.50\%\ copper$$

The idea of percent composition helps geologists describe how much metal or mineral is present in a particular ore. They can then evaluate whether the ore should be mined and how it should be processed.

Figure 2.32 *Ores of copper include (left to right) chalcocite, chalcopyrite, and malachite.*

A compound's formula indicates the relative number of atoms of each element present in the substance. For example, one common commercial source of copper metal is the mineral chalcocite, copper(I) sulfide (Cu_2S). Its formula indicates that the mineral contains twice as many copper atoms as sulfur atoms. The formula also reveals how much copper can be extracted from a certain mass of the mineral, which is an important consideration in copper mining and production.

Why are the ideas of molar mass and percent composition useful in determining how much copper can be obtained from copper-containing minerals and ores? Some copper-containing minerals are listed in Table 2.6 and shown in Figure 2.32. The percent of copper in chalcocite can be found, based on what you know about molar masses.

The formula for chalcocite indicates that one mole of Cu_2S contains two moles of Cu, or 127.10 g Cu and one mole of S, or 32.07 g S. The molar mass of Cu_2S is $(2 \times 63.55 \text{ g}) + 32.07 \text{ g} = 159.17 \text{ g}$. Therefore,

$$\% \text{ Cu} = \frac{\text{Mass of Cu}}{\text{Mass of Cu}_2\text{S}} \times 100\%$$

$$\frac{127.10 \text{ g Cu}}{159.17 \text{ g Cu}_2\text{S}} \times 100\% = 79.85\% \text{ Cu}$$

A second calculation indicates that Cu_2S contains 20.15% sulfur. The sum of the percent copper and the percent sulfur equals 100.00%. Why?

Knowing the percent composition of metal in a particular mineral helps us decide whether a particular ore should be mined; however, it is not the sole criterion. What else do we need to consider?

Suppose an ore contains the mineral chalcocite, Cu_2S. Because nearly 80% of this mineral is Cu (see the calculation), it seems likely that this ore is worth mining for copper. However, we must also consider the quantity of mineral actually contained in the ore. All other factors being equal, an ore that contains only 10% chalcocite would be a less desirable copper source than one containing 50% chalcocite. Thus, two factors must be taken into account when deciding on the quality of a particular ore source: (1) the percent mineral in the ore and (2) the percent metal in the mineral.

| Table 2.6 |

SOME COPPER-CONTAINING MINERALS	
Common Name	**Formula**
Chalcocite	Cu_2S
Chalcopyrite	$CuFeS_2$
Malachite	$Cu_2CO_3(OH)_2$

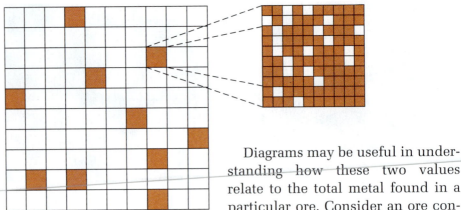

Figure 2.33 *The large square represents the total ore sample. (left) Each square within the large square represents 1% of the sample. The ten colored squares indicate an ore that contains 10% chalcocite. (right) 80% of each chalcocite square is colored to show the percent copper in chalcocite. Overall, only 8% of the total ore sample is copper.*

Diagrams may be useful in understanding how these two values relate to the total metal found in a particular ore. Consider an ore containing 10% chalcocite. Look at Figure 2.33. Suppose the larger box in Figure 2.33 represents a piece of this ore. According to this box, 10% of the ore is composed of the mineral chalcocite. (Ten squares are colored to show this.) You know that chalcocite itself is approximately 80% copper by mass. To represent this, 80% of each chalcocite box is colored in Figure 2.33 (one box is enlarged to show the detail). This representation may help estimate visually how much copper is in this particular ore. How might similar diagrams represent an ore that is 50% chalcocite?

Figure 2.34 *Hematite (Fe₂O₃, top) and magnetite (Fe₃O₄, bottom) are two iron-containing ores. If all other factors were equal, which ore would you mine for iron? Why?*

Developing Skills

C.10 PERCENT COMPOSITION

1. The chemical formula for azurite, a copper-containing mineral, is $Cu_3(CO_3)_2(OH)_2$. Complete an atom inventory for this compound. Use the material contained in Developing Skills C.2 (page 155) as a model.

2. Use Table 2.6 (page 167) to answer the following. In your calculations, assume that each mineral is present at the same concentration in an ore. Also assume that copper metal can be extracted from a given mass of any ore at the same cost. (*Note:* The percent copper in chalcocite was calculated on page 167.)

   a. Calculate the percent copper in chalcopyrite.
   b. Calculate the percent copper in malachite.
   c. Which of these minerals—chalcocite, chalcopyrite, or malachite—could be mined most profitably?

3. Suppose you have the choice either to mine ore that is 20% chalcocite (Cu_2S) or ore that is 30% chalcopyrite ($CuFeS_2$). Assuming all other factors are equal, which would you choose? Draw diagrams similar to those in Figure 2.33 to support your answer.

4. Two common iron-containing minerals are hematite (Fe_2O_3) and magnetite (Fe_3O_4). (Refer to Figure 2.34.) If you had the same mass of each mineral, which sample would contain the larger mass of iron? Support your answer with calculations.

C.11 RETRIEVING COPPER

Introduction

While completing Investigating Matter B.3 (pages 139–140), you heated metallic copper, producing a black powder, copper(II) oxide (CuO). Because atoms are always conserved in chemical reactions, the original copper atoms must still exist. In this investigation, you will attempt to recover those atoms of metallic copper.

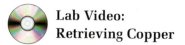

Lab Video: Retrieving Copper

Before starting, read the procedure to learn what you will need to do, note safety precautions, and plan necessary data collecting and observations. Prepare a table in your lab notebook where you can record the various masses measured in this investigation, and clearly label what those data represent.

Procedure 👓

Part I: Separating Copper(II) Oxide (CuO) from the Sample

Most likely, during Investigating Matter B.3, not all the original copper powder reacted with oxygen gas when you heated the copper in air. Some copper metal probably is still mixed in with the black copper(II) oxide. The first steps now involve separating this mixture into copper and copper(II) oxide. To do this, you will add dilute hydrochloric acid (HCl) to the black powder. Copper metal does not react with hydrochloric acid, so it will remain as a solid. The black copper(II) oxide, however, reacts with hydrochloric acid to produce copper(II) chloride (CuCl$_2$) and water, as shown in the following equation:

$$CuO(s) \quad + \quad 2\,HCl(aq) \quad \longrightarrow \quad CuCl_2(aq) \quad + \quad H_2O(l)$$

Copper(II) oxide   Hydrochloric acid   Copper(II) chloride   Water

1. Obtain your beaker containing the black powder from Investigating Matter B.3. Look closely at its contents. Is the material uniform throughout? If not, why not?

2. Construct a data table appropriate for recording the data you will collect in this investigation.

3. Add 50 mL of 1 M HCl to the beaker containing the copper oxide mixture. Record your observations of the solution. ⚠ (*Caution: 1 M hydrochloric acid may damage your skin. If some HCl does spill on your skin, ask another student to notify your teacher immediately. Begin rinsing the affected area with tap water immediately.*)

4. Gently heat the mixture to about 40 °C on a hot plate, and maintain at that temperature for 15 min, stirring every few minutes with a glass rod.

5. Remove the beaker from the hot plate. Turn off the hot plate. Allow any unreacted copper metal still remaining to settle to the bottom of the beaker. Then slowly decant the liquid into a second, empty 100-mL beaker.

> Based on what you learned about mixtures on page 28, how would you classify the sample in your beaker?

> The symbol M, as in 1 M HCl, stands for molar concentration. A 1 M solution contains one mole of solute (in this case, HCl) dissolved in one liter of solution.

Figure 2.35 *Transferring copper metal from a beaker to filter paper.*

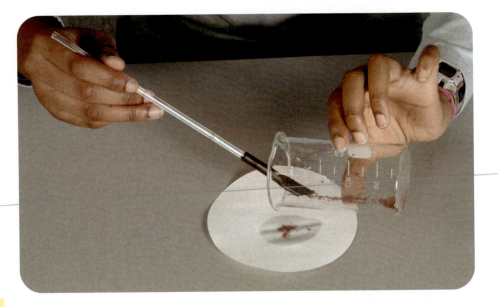

To **decant,** let the solid material settle to the bottom of the beaker, and gently pour the liquid into a separate container. Pour carefully so that solid material doesn't transfer over with the liquid.

6. Wash the solid copper remaining in the first beaker several times by swirling it gently with approximately 10 mL of distilled water. Slowly decant the liquid into the second beaker, adding it to the liquid you collected in Step 5. Set aside the collected liquid contained in the second beaker (approximately 60 mL) for Step 9.

7. Measure and record the mass of a piece of filter paper. Transfer the solid copper to the paper, as shown in Figure 2.35, and allow it to dry overnight.

8. After the sample and filter paper have dried, find the mass of the solid copper. This represents the portion of the original copper powder that failed to react to form CuO. Record this mass in your laboratory notebook.

Part II: Converting Copper(II) Chloride ($CuCl_2$) to Copper (Cu)

The final step in this investigation is to convert the dissolved copper(II) chloride ($CuCl_2$) to copper metal. Perhaps you already have an idea how to do this. Recall the metal activity series that you devised during Investigating Matter B.5 (pages 142–143). By adding solid metal samples to solutions that contain other metal ions, you compared the chemical activity of each metal. Review your observations from that earlier investigation. Which of the tested metals are more active than copper? Try to predict the result of placing one of those more active metals in your copper(II) chloride solution. Check your prediction by completing these steps.

9. Obtain a watch glass that can completely cover the top of the beaker containing copper(II) chloride ($CuCl_2$) solution from Step 6. For each gram of copper powder that you started with from Investigating Matter B.3, add about one gram of zinc metal to the $CuCl_2$ solution. See Figure 2.36.

Figure 2.36 *Adding zinc metal to a copper(II) chloride solution. How does the resulting chemical reaction relate to your earlier consideration of the activity series?*

10. Immediately cover the beaker with the watch glass, and allow it to stand for several minutes. Record your observations.

11. After the reaction has subsided, remove the watch glass and gently dislodge the solid copper that has formed on the surfaces of the zinc pieces.

12. Continue to dislodge the copper from the zinc until you are convinced that the zinc has stopped reacting with the solution. (How can you decide?) Then, add 10 mL of 1 M HCl to the beaker and carefully remove any large pieces of solid zinc from the beaker with forceps. Follow your teacher's instructions for discarding the solid zinc. Replace the watch glass on the beaker. Record your observations.

> Note the caution about hydrochloric acid in Step 3.

13. After a few minutes, carefully decant as much of the liquid as possible into another empty beaker. Follow your teacher's instructions for discarding the liquid.

14. Wash the solid copper several times with distilled water.

15. Transfer the copper to a preweighed piece of filter paper, and allow the copper to dry overnight.

16. After the sample and filter paper have dried, find the mass of copper metal. This represents the copper that you recovered from the copper(II) chloride solution.

17. Follow your teacher's instructions for disposing of waste materials.

18. Wash your hands thoroughly before leaving the laboratory.

Questions

1. During Investigating Matter B.3, not all of the original copper powder reacted when you heated it in air.

   a. What observational evidence leads you to think that the reaction was incomplete?
   b. How would you revise the procedure so that more copper(II) oxide could form?

2. During Investigating Matter B.3 (page 139),

   a. what mass of the original powdered copper sample reacted when you heated it? (*Hint:* Refer to the original mass of copper you used during this investigation and the mass of copper residue found in Step 8 to calculate this.)
   b. what percent of the total copper sample reacted?

3. In the reaction between copper(II) chloride ($CuCl_2$) solution and zinc metal, in Investigating Matter B.5 (page 142), each Cu^{2+} ion gained two electrons to form an atom of copper metal. Each zinc metal atom lost two electrons to form a Zn^{2+} ion:

   a. Write a balanced chemical equation that represents this process. (*Hint:* To review how, turn to pages 148–150.)

b. Based on the chemical equation you wrote in Question 3a, identify
   i. the reactant that was *oxidized.*
   ii. the reactant that was *reduced.*
   iii. the *reducing* agent.
   iv. the *oxidizing* agent.

4. Adding HCl to CuO, in Investigating Matter B.3 (page 139), resulted in the formation of a blue solution. This color is due to the presence of Cu^{2+}(aq) ions. Consult your observations when answering the following questions:

   a. Describe what happened to the solution color after you added zinc (see Steps 9 and 10) in this investigation.
   b. What caused the changes you observed in the solution?
   c. How can the color of the solution be used to indicate when the zinc metal has removed the Cu^{2+} ions from the solution?

5. To recover Cu metal from the $CuCl_2$ solution, you had to use other resources:

   a. What resources were "used up" in this recovery process?
   b. Where (to what location) did each resource finally go?

C.12 CONSERVATION IN THE COMMUNITY

Depleting Resources

In some ways, Earth is like a spaceship. The resources "on board" are all that are available to the ship's inhabitants. Some resources—such as fresh water and air, fertile soil, plants, and animals—can eventually be replenished by natural processes. These are called **renewable resources.** As long as natural cycles are not disturbed too much, supplies of renewable resources can be maintained indefinitely. Other materials—such as metals, natural gas, coal, and petroleum—are considered **nonrenewable resources** because they cannot be readily replenished. If atoms are always conserved, why do some people say that a resource may be "running out"? Can a resource actually "run out"?

The answer can be found by first remembering that *atoms* are conserved in chemical processes, but *molecules* might not be. For example, the current production of new petroleum molecules in nature is much, much slower than the current rate at which petroleum molecules are being burned to release thermal energy, as well as carbon dioxide, water, and other molecules. Thus, the total inventory of petroleum on Earth is declining, but the total number of carbon and hydrogen atoms on Earth remains constant.

A resource—particularly a metal—can be depleted in another way. As you have learned, profitable mining depends on finding an ore with at least some minimal metal content. This minimum level depends on the metal and its ore: from as low as 1% for copper or 0.001% for gold to as high as 30% for aluminum.

> Resources such as petroleum are regarded as nonrenewable resources because they can be formed only over millions of years.

Once ores with high metal content are depleted, lower grade ores with less metal content are processed. Meanwhile, atoms of the metal that were previously concentrated in rich deposits in limited parts of the world gradually become spread out (dispersed) over wider areas of Earth. Eventually, the mining and extraction of certain metals may become prohibitively expensive, thus making the metal too costly for general use. At that time, for practical purposes, the supply of that resource can be considered depleted.

Conserving Resources

Can such depletion scenarios be avoided? Can Earth's mineral resources be conserved? One strategy for conservation is to slow the rate at which resources are used. This can be best addressed by considering the four Rs—*rethinking, reusing, replacing,* and *recycling.*

Part of this strategy includes rethinking personal and societal habits and practices involving resource use. Such rethinking can involve decisions such as whether it is better to use paper or plastic bags in grocery stores or, as is practiced in some nations, if it is even better to consider reusing your grocery bags.

Rethinking can take the form of reexamining old assumptions, identifying resource-saving strategies, and perhaps uncovering new solutions to old problems. A fundamental part of rethinking resource conservation and management is to consider the most direct option—source reduction. That simply means decreasing the amount of resources used. If fewer resources are used now, the more remain available for future generations.

Another approach is replacing a resource by finding substitute materials with similar properties, preferably from renewable resources. In addition, some manufactured items can be refurbished or repaired for reuse, rather than sent to a landfill. Common examples include car parts and printer cartridges, both of which can often be reused after rebuilding or refilling. Finally, certain items can be recycled, or gathered for reprocessing. Such recycling allows the resources present in the items to be used again. Figure 2.37 (page 174) illustrates the steps involved in recycling aluminum cans.

However, even recycling can create environmental problems. For example, aluminum scrap, excluding cans, may contain up to 1% magnesium. The magnesium must be removed before the aluminum is processed into other products. One conventional method to remove magnesium requires adding chlorine gas, which requires special handling in the workplace and can contribute to air pollution.

Dealing with Waste

Ultimately, some items will be discarded if they are no longer wanted or needed. Each person in this country throws away an average of about 2 kg (4 lb) of unwanted materials (waste) daily. Some products, such as yesterday's newspaper, become waste after they fulfill their initial

Minimum profitability levels for mining depend not only on the metal's abundance in the ore, but also on the metal's chemical activity. Less active metals (such as gold) are often found in their native, or uncombined, state. In compounds, less active metals are more readily released from their compounds than are more active metals (such as aluminum). Thus, mining richer ores of active metals is usually more profitable.

Green chemistry is an approach that aims to eliminate pollution in industrial settings by preventing it from happening in the first place through enactment of one or more of the four Rs.

In some nations, most glass bottles are cleaned and refilled, not discarded.

Collect used aluminum cans

Use aluminum cans for beverages and other products

Deliver to aluminum recycling center

Process cans at recycling center
• remove nonaluminum materials
• shred aluminum into chips
• melt aluminum chips in furnace
• pour molten aluminum into ingot molds

Manufacture new aluminum cans

Deliver aluminum ingots for aluminun-can manufacture

Figure 2.37 *Why does recycling aluminum require less energy than producing aluminum from its ore?*

purpose. Others, such as telephones and computers, become waste when they are discarded in favor of newer models. Combined, materials discarded by U.S. citizens and businesses represent about two kilograms (over four pounds) per person per day.

As you can see in Figure 2.38, the largest fraction of municipal solid waste in the U.S. is paper. While recycling reduces this fraction somewhat, the market for recycled paper is limited. Paper that is not recycled leaves a high proportion of combustibles as waste, which has made *waste-to-energy plants* an increasingly attractive option. Approximately 100 waste-to-energy plants currently operate in the United States, burning about 97 000 tons of solid waste per day. Each ton of garbage that serves as "fuel" in these plants produces about a third of the energy released by a similar quantity of coal.

Although waste-to-energy plants produce some fly ash and solid residues, such plants can allow the use of materials that otherwise would be disposed of as part of an unwanted product. In addition, waste-to-energy plants tend to increase recycling, both on-site and in the communities in which they are located.

Recycling, landfills, and combustion for energy production are three options for the final step in the life cycle of a material—a topic you will now consider as you prepare to select a material for your coin design.

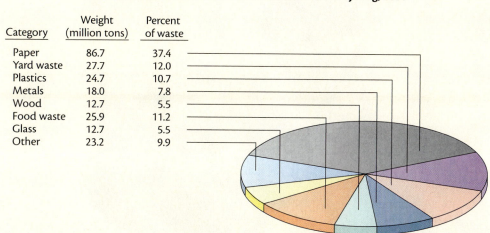

Total Waste Generated in the United States before Recycling, 2000

Category	Weight (million tons)	Percent of waste
Paper	86.7	37.4
Yard waste	27.7	12.0
Plastics	24.7	10.7
Metals	18.0	7.8
Wood	12.7	5.5
Food waste	25.9	11.2
Glass	12.7	5.5
Other	23.2	9.9

Total generation: 231.9 million tons

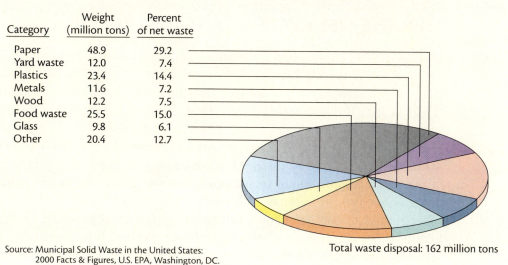

Net Waste Generated in the United States after Recycling, 2000

Category	Weight (million tons)	Percent of net waste
Paper	48.9	29.2
Yard waste	12.0	7.4
Plastics	23.4	14.4
Metals	11.6	7.2
Wood	12.2	7.5
Food waste	25.5	15.0
Glass	9.8	6.1
Other	20.4	12.7

Source: Municipal Solid Waste in the United States: 2000 Facts & Figures, U.S. EPA, Washington, DC.

Total waste disposal: 162 million tons

Figure 2.38 *Composition of the U.S. municipal waste stream before and after recycling.*

C.13 RETHINKING, REUSING, REPLACING, AND RECYCLING

Deciding between paper, plastic, or your own reusable grocery bags is just one of many decisions you face on a daily basis regarding material use and conservation.

1. Consider when your family is finished reading the daily newspaper (Figure 2.39). What choice would you decide to make for disposing of the newspaper?

2. Why did you decide on that choice?
   a. What are *benefits* of your decision?
   b. What are *drawbacks* of your decision?

Figure 2.39 *What are some options for disposing of a newspaper?*

C.14 THE LIFE CYCLE OF A MATERIAL

In designing a new product for human use, we must consider the full life cycle of the materials involved. A material's *life cycle* has several distinct stages. Raw materials are first obtained and then refined and synthesized into the desired material. That material is then used to make the product designed for a particular use. When the product is no longer useful, the materials may be recovered and reused, or they may become scattered in landfills. Figure 2.40 illustrates this general life cycle.

In every step of the cycle, energy and resources are used. Investigating Matter B.3 (Converting Copper) and Investigating Matter C.11 (Retrieving Copper) are good examples of this fact. Recall that thermal energy (from the hot plate) and chemical resources (hydrochloric acid and zinc metal) first were used to convert the copper metal to other copper-containing substances, and then they were also used to recover

the copper metal. Because energy and resource use affect economics as well as the environment, each step in the life cycle of a material becomes a factor to consider when we design a new product.

The next activity will allow you to model this process for a familiar material—copper.

Figure 2.40 *A material's life cycle. Paper, for example, has a life cycle with several distinct stages.*

Developing Skills

C.15 COPPER LIFE-CYCLE ANALYSIS

Imagine that you are involved in designing copper water pipes. Use Figure 2.18 (page 138) and Figure 2.40 to conduct a life-cycle analysis for the copper metal in your pipes. Based on these figures, make a list of the steps involved in the life cycle of the copper used to make the pipes. Consider how each step will affect your final copper water-pipe design by answering the following questions:

1. Which steps in the life cycle of copper consume energy? For each step, explain the particular energy needs.

2. Which life-cycle steps (such as the reduction of a mineral in an ore to produce the metal) require the use of other materials?

3. How will you obtain the copper for your water pipes?

4. People may one day want to replace their copper water pipes with pipes composed of a new material.
   a. What steps would be needed to recycle the copper?
   b. What potential uses can you think of for the recycled copper?

5. Consider your answers to the previous questions.
   a. How does each decision influence the cost of your copper water pipes?
   b. How might an effective national copper recycling program affect the eventual cost of copper water pipes?

Making Decisions

C.16 DESIGNING A COIN

By now, you have a good idea of some properties your new half-dollar coin should possess. Perhaps you have even narrowed the range of materials you are considering. To prepare for the final phase of your coin design, answer the following questions:

1. Look at the list of required and desired physical and chemical properties that you developed on page 127 of Section A. Revise and update your list using what you have learned since completing that activity.

2. Coins are traditionally made of metals.
   a. What alternative materials might you consider?
   b. What are advantages or disadvantages of these alternatives?

3. Construct a table similar to the one on the next page. Enter all needed and desirable properties from your revised list in Question 1. Then list the materials you are considering for your coin. If you have not already done so, begin researching the properties of these materials. Indicate whether each material meets each criterion. For example, you might check the box if it does, and place an "x" in the box if it does not.

	Materials				
	A	**B**	**C**	**D**	**E**
Necessary properties	'				
Desirable properties					

SAMPLE FOR REFERENCE ONLY

4. Consider the design of your coin:
   a. What images are you considering for the obverse (heads) and reverse (tails) sides of the coin?
   b. What anticounterfeiting features will you include in your design?
   c. How large should this coin be?
   d. What should be its mass?

5. Consider the cost of materials in your design:
   a. What would happen if the materials in the coin were worth more than the face value of the coin (fifty cents)?
   b. What materials, if any, have you already ruled out because of this concern?

6. What other costs should you consider as you design your coin?

7. Consider the life cycle of your coin:
   a. How will the coin materials be obtained and processed?
   b. How long do you expect each coin to stay in circulation?
   c. What will happen to the materials when used coins are removed from circulation?
   d. How might the impact on the environment be addressed at each stage of the coin's life cycle?

SECTION C SUMMARY
Reviewing the Concepts

The atoms that compose matter are neither created nor destroyed in chemical reactions.

1. State the law of conservation of matter.

2. What is a scientific law?

3. Why are expressions such as "using up" and "throwing away" misleading, if the law of conservation of matter is taken into account?

4. Complete atom inventories to decide if each of these chemical expressions is balanced.

   a. the preparation of tin(II) fluoride, a component of some toothpastes (called *stannous fluoride* in some ingredient lists):

   $$Sn(s) \ + \ HF(aq) \ \longrightarrow \ SnF_2(aq) \ + \ H_2(g)$$

   | Tin metal | Hydrofluoric acid | Tin(II) fluoride | Hydrogen gas |

   b. the synthesis of carborundum for sandpaper:

   $$SiO_2(s) \ + \ C(s) \ \longrightarrow \ SiC(s) \ + \ CO(g)$$

   | Silicon dioxide (sand) | Carbon | Silicon carbide (carborundum) | Carbon monoxide |

   c. the reaction of an antacid with stomach acid (hydrochloric acid):

   $$Al(OH)_3(s) \ + \ 3 \ HCl(aq) \ \longrightarrow \ AlCl_3(aq) \ + \ 3 \ H_2O(l)$$

   | Aluminum hydroxide | Hydrochloric acid | Aluminum chloride | Water |

Coefficients in a chemical equation indicate the relative number of units of each reactant and product involved. Subscripts in a substance's formula may not be changed to balance an equation.

5. Consider this equation:

   $$N_2(g) + 3 \ H_2(g) \longrightarrow 2 \ NH_3(g)$$

   a. What is the coefficient for hydrogen gas?
   b. What is the coefficient for NH_3 gas?
   c. What is the coefficient for nitrogen gas?

6. For each of the following processes, draw a representation of the chemical statement, balance the representation, and verify your answer.

   a. preparing tungsten from one of its minerals:

   $$\_\_WO_3 + \_\_H_2 \longrightarrow \_\_W + \_\_H_2O$$

   b. heating lead(II) sulfide in air:

   $$\_\_PbS + \_\_O_2 \longrightarrow \_\_PbO + \_\_SO_2$$

   c. rusting (oxidation) of iron metal:

   $$\_\_Fe + \_\_O_2 \longrightarrow \_\_Fe_2O_3$$

7. Balance each of these chemical expressions.

   a. preparing phosphoric acid (used in making soft drinks, detergents, and other products) from calcium phosphate and sulfuric acid:

   $$\_\_Ca_3(PO_4)_2 + \_\_H_2SO_4 \longrightarrow \_\_H_3PO_4 + \_\_CaSO_4$$

   b. completely burning gasoline:

   $$\_\_C_8H_{18} + \_\_O_2 \longrightarrow \_\_CO_2 + \_\_H_2O$$

8. A student is asked to balance this chemical expression:

   $$Na_2SO_4 + KCl \longrightarrow NaCl + K_2SO_4$$

   The student decides to balance it this way:

   $$Na_2SO_4 + K_2Cl \longrightarrow Na_2Cl + K_2SO_4$$

   a. Complete an atom inventory of the student's answer. Are the atoms conserved?
   b. Did the student create a properly balanced chemical equation? Explain.
   c. If your answer to Question 8b is *no*, write a correctly balanced equation.

One mole of substance contains 6.02×10^{23} particles. The molar mass of a substance can be calculated from the atomic weights of its component elements.

9. If you could spend a billion dollars (1×10^9 dollars) per second, how many years would it take to spend one mole of dollars?

10. Find the molar mass of each of the following substances:

   a. oxygen gas, O_2
   b. ozone, O_3
   c. limestone, $CaCO_3$
   d. a typical antacid, $Mg(OH)_2$
   e. aspirin, $C_9H_8O_4$

11. How can samples of 63.6 g copper metal and 23.0 g sodium metal, with different masses, volumes, and densities, both correctly represent 1.00 mol of substance?

12. A major advantage of the mole concept is that it enables a chemist to "count by weighing." If one mole of potassium metal has a mass of 39.1 g,

   a. how many atoms are in 39.1 g potassium?
   b. how many atoms are in 19.55 g potassium?
   c. how many atoms are in 3.91 g potassium?
   d. how many atoms are in 1.0 g potassium?

The coefficients in a balanced chemical equation indicate the molar relationships among all reactants and products.

13. For the equation

   $3\ PbO(s) + 2\ NH_3(g) \longrightarrow 3\ Pb(s) + N_2(g) + 3\ H_2O(l)$,

   a. how many moles NH_3 are needed to react with 9 mol PbO?
   b. how many moles N_2 are produced by the reaction of 10 mol NH_3?
   c. how many moles Pb are produced from 5 mol PbO?

14. For the equation in Question 13,

   a. how many moles (maximum) N_2 can be produced from 34.0 g NH_3?
   b. what mass Pb can be produced from the complete reaction of 3.0 mol PbO?
   c. what maximum mass N_2 can be produced from 34.0 g NH_3?
   d. what mass PbO, which fully reacts, will produce 415 g Pb?

The percent composition of a substance can be calculated from the relative number of atoms of each element in the substance's formula.

15. In carbon dioxide, two-thirds of the atoms are oxygen atoms; however, the percent oxygen by mass is not 67%. Explain.

16. Find the percent metal (by mass) in each of the following compounds:

   a. Ag_2S
   b. Al_2O_3
   c. $CaCO_3$

17. A 50.0-g sample of ore contains 5.00 g lead(II) sulfate, $PbSO_4$:

   a. What is the percent lead (Pb) in $PbSO_4$?
   b. What is the percent $PbSO_4$ in the ore sample?
   c. What is the percent Pb in the total ore sample?
   d. Use a diagram to represent the proportions of lead and lead(II) sulfate in the ore.

> Resources are either renewable or nonrenewable. Resources can be conserved by recycling, controlling the rate of use, or replacing them with substitute resources.

18. a. What is the difference between *reusing* and *recycling*?
    b. Give two examples of each, other than those presented in the textbook.

19. In addition to those found in the textbook, list four examples of
    a. renewable resources.
    b. nonrenewable resources.

20. Classify each use as either *recycling* or *reusing*:
    a. storing water in used juice bottles for an emergency.
    b. converting plastic milk containers into fibers used to weave clothing fabric.
    c. packing breakable items with shredded newspapers.

21. How would the life cycle of a light bulb compare to that of a newspaper? Consider material sources and disposal and recycling.

Connecting the Concepts

22. How is a scientific law, such as the law of conservation of matter, different from a law created by the government?

23. What additional information would you need to be able to count 1000 nails by weighing them?

24. In Investigating Matter B.3 and C.11, you converted and retrieved copper through a series of chemical reactions:
    a. Compare the physical properties of (i) copper metal, (ii) copper(II) oxide, and (iii) copper(II) chloride.
    b. Describe a chemical process that produced one of the substances listed in Question 24a.
    c. Student results often show less than 100% recovery of the initial copper sample. Does this suggest that the law of conservation of matter is not always true? Explain.

25. The Earth has been compared to a space station.
    a. In what ways is this analogy useful?
    b. In what ways is it misleading?

26. Chromium minerals are found at three different mine sites in these forms:
    a. site 1: chromite, $FeCr_2O_4$
    b. site 2: crocoite, $PbCrO_4$
    c. site 3: chrome ochre, Cr_2O_3

    Based only on percent composition of these minerals, at which site would chromium mining be most feasible?

27. Describe at least two benefits of discarding less waste material and recycling more of it.

28. One method to produce chromium metal includes, as a final step, the reaction of chromium(III) oxide with silicon at high temperature:

    $$2\ Cr_2O_3(s) + 3\ Si(s) \longrightarrow 4\ Cr(s) + 3\ SiO_2(s)$$

    a. How many moles of each reactant and product are shown in this chemical equation?
    b. What mass (in grams) of each reactant and product is indicated by this equation?
    c. Show how this equation illustrates the law of conservation of matter.

Extending the Concepts

29. Why is aluminum metal more easily produced from recycled aluminum cans than from aluminum contained in clay, bauxite, or aluminum oxide ore?

30. Atoms that presently make up your body once may have been part of the body of a Tyrannosaurus rex, or even Alexander the Great or Cleopatra. Explain how this is possible.

31. Investigate the amount and composition of wastes in your community. Construct a pie chart like those shown in Figure 2.38 (page 175).

    a. How does your community compare to the nation?
    b. What steps could be taken to reduce the quantity of discards in your community?

32. In your laboratory experiences, you may encounter some compounds called *hydrates*. Examples of hydrates include
    (i) $Na_2S_2O_3 \cdot 5H_2O$,
    (ii) $CaSO_4 \cdot 2H_2O$,
    (iii) $Na_2CO_3 \cdot 10H_2O$:

    a. Why are these compounds called hydrates?
    b. Calculate the molar mass of each listed hydrate.
    c. Calculate the percent composition of water in each hydrate.
    d. Although hydrates contain significant amounts of water, they are found as solid substances at room temperature. How is this possible?

Materials: Designing for Desired Properties

As Earth's chemical resources continue to be extracted and used, issues of scarcity, cost, and environmental, economic, or political factors sometimes motivate people to consider alternatives. Three related options are to find, modify, or create new materials to serve as substitutes. An ideal substitute satisfies three requirements: It is plentiful, inexpensive, and, of course, its useful properties match or exceed those of the original material. Alternative materials seldom meet all these conditions completely. As a result, we must weigh the benefits and burdens involved in each option. This is true whether the design challenge involves a new coin, a better microwave oven, or a new computer memory device.

D.1 STRUCTURE AND PROPERTIES: ALLOTROPES

Structures and Properties: Allotropes

You have learned in your study of chemistry that elements are the fundamental building blocks of matter. You also know that each element—identified by a name, symbol, and particular numbers of electrons and protons in its atoms—displays characteristic properties. Elements can be classified as metals, nonmetals, and metalloids by such properties. You are aware that some elements, such as magnesium, are reactive and others, such as krypton, are virtually inert. Given all of that, try solving the following puzzle.

ChemQuandary 2

A CASE OF ELUSIVE IDENTITY

Imagine that you receive samples of three solids and are told that each is a pure element.

You examine the first sample. It is a black solid that feels soft and a bit "greasy" to the touch. The sample leaves black marks when you rub it on paper. It is a useful lubricant and a fairly good conductor of electricity. A gram of the material sells for less than a penny.

You then inspect the second sample. It is a colorless, glasslike solid. The sample leaves deep scratch marks when rubbed on the table; in fact, it is among the hardest substances known. It is useful as an abrasive and as a coating for saw blades. The sample is an electrical insulator, which means that it does not conduct electricity. Depending on the quality of the solid sample, it can cost $50 per gram or more than $20 000 per gram.

(continued)

(continued from page 184)

Finally, you consider the third sample. It is a fine, powdery solid made up, it is claimed, of the roundest-shaped molecules known. Although it has been found in prehistoric layers of Earth's crust, the substance was only discovered in 1985. The cost of this substance is dropping; however, in pure form, it currently costs between $170 and $3000 per gram, depending on the chemical supplier and the amount purchased at one time. That price—gram-for-gram—is currently higher than the cost of gold.

These are, indeed, three distinctly different substances. Yet, you are told that all three samples are exactly the *same* element.

How can atoms of one element make up such different materials? What additional information do you need to explain this? Which element is being described?

Before you attempt to answer these questions, it will be helpful for you to know that chemists recognize the three samples just described as allotropes of the same element. *Allotropes* are different forms of an element that each have distinctly different physical or chemical properties.

To be considered allotropes, the forms of the element must be in the same state. Solid iron (Fe) and molten iron have distinctly different properties; however, because they are in different states, they are not allotropes. Oxygen exists as two allotropes, O_2 and O_3. O_2, commonly referred to as *oxygen gas,* is the more common form, making up 20% of the air you breathe. O_3, called *ozone,* is a gaseous substance found in the upper layers of our planet's atmosphere, as well as in some of the air pollution associated with cities.

It is possible that some information in ChemQuandary 2 provided enough clues for you to identify the element involved. If you guessed carbon (C), you are correct. The identities of each of the three carbon allotropes can now be revealed. Look at Figure 2.41 (page 186).

The first sample, a soft, black solid, is the carbon allotrope known as *graphite.* You encounter graphite either directly or indirectly nearly every day. It is a major component of pencil lead. Due in part to its conductive properties, it is also found in many batteries.

The second sample is carbon in the form of *diamond.* Its melting point is among the highest of all elements. Both natural and synthetic diamonds exist. The cost of a diamond depends in part on its quality and optical characteristics. When carefully selected and cut, natural diamonds have high decorative value, even though their chemical formula, C(s), is the same as the formula for the graphite in common pencil lead.

The third sample is one of the many recently identified allotropes of carbon, a group of ball-like structures, and even tube-like structures, known as *fullerenes.* The particular fullerene described here is *buckminsterfullerene,* a 60-carbon hollow sphere (C_{60}), which resembles a soccer ball. Other fullerene molecules have formulas such as C_{70}, C_{240}, and C_{540}. Structures composed of a "rolled-up" layer

Because graphite and diamond are both pure carbon, they can burn in oxygen to produce carbon dioxide gas and thermal energy. Despite that, diamonds (and even graphite) are not recommended as a heat source. The estimated bill would obviously be astronomically high!

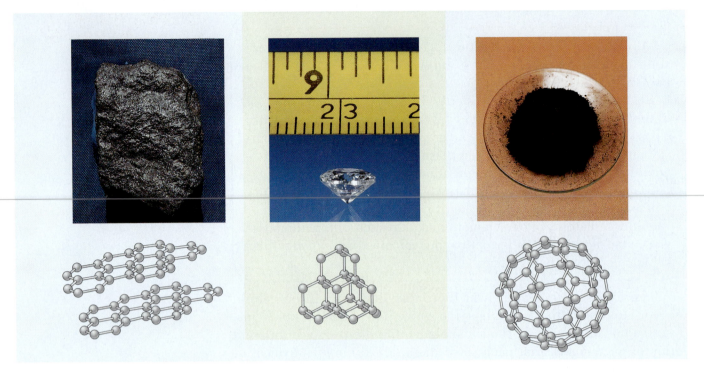

Figure 2.41 *Allotropic forms of carbon—their appearance and structures. From left to right, graphite, diamond, and buckminsterfullerene. The three-dimensional diamond and graphite structures continue in all directions.*

of carbon atoms that form hollow tubes are called *nanotubes*, which are represented in Figure 2.42.

How do chemists account for the different properties of allotropes? The explanation lies in how the atoms of the element are linked and organized—that is, in the structure of the substances. Although the three allotropes of carbon are all composed only of carbon atoms, they have distinctly different atomic arrangements. The three allotrope models in Figure 2.41 illustrate this fact. Figure 2.43 shows another example of one of these allotropes, that of a diamond formed through synthetic rather than natural processes.

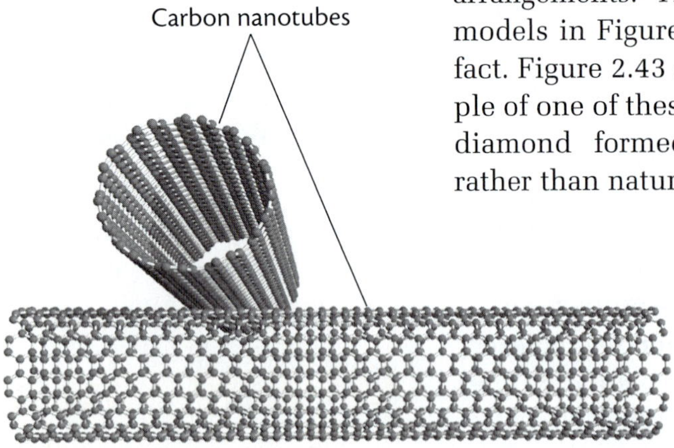

Carbon nanotubes

Figure 2.42 *Representation of a nanotube—another form of carbon. Nanotubes are very strong. They also have conductive properties that may prove useful in nanoelectronic devices.*

Figure 2.43 *Using high temperature and pressure, it is possible to convert graphite to diamonds.*

D.2 LINKING PROPERTIES TO STRUCTURE

Based on the structural models shown in Figure 2.41, answer the following questions about some properties of carbon allotropes. The following example has been answered for you.

Sample Problem: *What feature of diamond's structure may account for its property as a hard, rigid substance, one that can scratch most other materials?*

Notice that every carbon atom in the interior of the diamond model is linked to four other carbon atoms by chemical bonds. These bonds have the effect of holding each carbon atom in place in a rigid, three-dimensional structure. Each carbon–carbon linkage in diamond is a covalent bond involving a shared pair of electrons. Thus, bending or denting a diamond crystal would be extremely difficult because it would involve breaking a network of strong chemical bonds that locks each carbon atom in a fixed position. This accounts for diamond's rigidity and hardness.

1. How might the actual structure of diamond help explain why it is sometimes found in the form of large, single crystals?

2. What feature of graphite's structure might account for its usefulness as a lubricant?

3. Why are fullerenes powdery as solids rather than composed of large-scale chunks?

4. A molecule of buckminsterfullerene (C_{60}) can be regarded as a hollow sphere. Chemists have demonstrated that it is possible to place an atom of another element inside the sphere. Can you think of any practical applications for carrying atoms of another element inside fullerene molecules?

D.3 ENGINEERED MATERIALS

Ceramics

Throughout history—first by chance and more recently guided by science and technology—humans have greatly extended the array of materials available for their use. Chemists and materials scientists have learned to modify the properties of matter by physically blending or chemically combining two or more substances. Sometimes they desire only slight changes in a material's properties. At other times, chemists create new materials with properties different from those of starting substances.

The black "lead" in a pencil is mainly graphite, a natural form of the element carbon. You have just learned that pure carbon can display very different properties depending on how the atoms are bonded together in three-dimensional space. Thus, carbon in the form of graphite has properties different from those of carbon in diamonds or fullerenes.

When first discovered, graphite was mistakenly identified as a form of lead, and the name stuck.

Hard pencil lead (such as No. 4) produces very light lines on paper. Why is that so different from the properties of soft pencil lead (such as No. 1), which makes very broad, easily smudged lines? The properties of pencil lead are controlled by the amount of clay mixed with the graphite. Increasing the quantity of clay produces harder pencil lead, which means that less graphite can be rubbed off onto paper.

Examples abound of other materials designed or modified to meet specific needs. Clay, one of the most plentiful materials on this planet, is composed mainly of the mineral kaolinite ($Al_2O_3 \cdot 2SiO_2 \cdot 2H_2O$), along with magnesium, sodium, and potassium ions. Early humans found that they could mix clay with water and then mold and heat the mixture to create useful products, such as pottery and bricks.

In more recent times, researchers have used newly developed techniques and materials to produce *engineered ceramics*. Figures 2.44a and 2.44b compare the sources, processing, and products of conventional ceramics with newer, stronger engineered ceramics.

Figure 2.44a *Conventional ceramic products, such as bricks and pottery, are made from clay. For example, kaolinite clay (top left) can be formed by hand or machine and fired in a kiln.*

What properties of conventional ceramics made them useful in pottery and bricks? Characteristics such as hardness, rigidity, low chemical reactivity, and resistance to wear were certainly desirable. This durability has made ceramics an attractive option for medical applications, such as ceramic hip replacements. The main attractions of ceramics for future use, however, are their high melting points and their strength at high temperatures. Indeed, such ceramics have become attractive substitutes for steel in some applications. For example, diesel or turbine engines made of ceramics can operate at higher temperatures than can metal engines. Such high-temperature engines run with increased efficiency, thus reducing fuel use.

The major problem for researchers to solve is that ceramics are brittle. They can fracture if exposed to rapid temperature changes, such as during hot engine cool down. To avoid cracks in ceramic engine components (and resulting engine failure), scientists and engineers have developed low-thermal-expansion materials and precise manufacturing methods that carefully control the microstructure of the materials. As these manufacturing practices become more economical, ceramics are likely to be more widely used in high-temperature applications.

Figure 2.44b Engineered ceramics, many of which are designed to be higher melting, stronger, or less brittle than conventional ceramics, are used in computer circuit board resistors and in huge jet-engine turbines, both shown here.

Plastics

These synthetic substances are composed of carbon-atom chains and rings with hydrogen and other atoms attached. Plastics generally are less dense than metals and can be designed to be "springy" or resilient in situations where metals might become dented. They have already replaced metals for many uses, such as plastic automobile bumpers and even automobile bodies.

The properties of certain plastics can be customized. For example, polyethylene can be tailored to display relatively soft and pliable properties (as in a squeeze bottle for water) or crafted to be hard and brittle, almost glasslike in its behavior. Unfortunately, most plastics are made from petroleum, a nonrenewable resource already in great demand as a fuel.

Optical fibers developed by chemists, physicists, and engineers are well on their way to replacing conventional copper wires in phone- and data-transmission lines. See Figure 2.45. Voice or electronic messages are sent through these thin, specially designed glass tubes of very pure silicon dioxide (SiO_2) as pulses of laser light. As many as 50 000 phone conversations or data transmissions can take place simultaneously in one glass fiber the thickness of a human hair. A typical 72-strand optical fiber ribbon can carry over a million messages. The fiber's larger carrying capacity and noise-free characteristics outweigh its higher initial cost.

Chemists, chemical engineers, and materials scientists continue to find new and better alternatives to traditional materials for a variety of applications. Such custom tailoring at the molecular level is possible because of chemical knowledge—knowledge of how the atomic composition of materials affects their observable properties and behavior.

Now it's your turn to consider possible uses of some alternative materials.

Figure 2.45 *Fiber optic material transmits considerably more information using light than copper wires can transmit using electrical signals.*

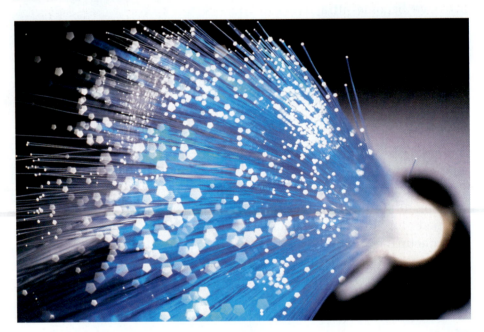

Making Decisions

D.4 ALTERNATIVES TO METALS

1. a. Select four uses of copper from the list contained in Figure 2.18 (page 138).
    b. For each use, suggest an alternative material that could serve the same purpose. Consider both conventional materials and possible new materials.

2. a. Suggest some common metallic items that might be replaced by ceramic or plastic versions.
    b. i. In what ways would the ceramic or plastic versions be preferable?
       ii. In what ways might the original metallic versions be preferable?

3. Suppose silver became as common and inexpensive as copper. In what uses would silver most likely replace copper? Explain.

D.5 STRIKING IT RICH

Introduction

Seeing is believing—or so it is said. In this investigation, you will change the appearance of some pennies through chemical and heat treatment.

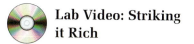

Lab Video: Striking it Rich

Before starting, read the procedure to learn what you will need to do, note safety precautions, and plan necessary data collecting and observations. You will record observations at several points in the investigation.

Procedure

1. Construct a data table appropriate for recording the data you will collect in this investigation.

2. Obtain three pennies. Use steel wool to clean each penny until it is shiny. Record the appearance of the pennies.

3. Set aside one of the cleaned pennies to serve as a *control;* that is, as an untreated sample that can be compared later to the other two treated coins.

4. Weigh a 2.0-g to 2.2-g sample of granulated zinc (Zn) or zinc foil. Place it in a 250-mL beaker.

5. Use a graduated cylinder to measure 25 mL of 1 M zinc chloride ($ZnCl_2$) solution. Add the solution to the beaker that contains the zinc metal. (***Caution:*** *1 M zinc chloride solution can damage skin. If any accidentally spills on you, ask a classmate to notify your teacher immediately; wash the affected area with tap water immediately.*)

6. Cover the beaker with a watch glass and place it on a hot plate. Gently heat the solution until it just begins to bubble, then lower the temperature just to sustain gentle bubbling. Do not allow the solution to boil vigorously or become heated to dryness. (***Caution:*** *Note the warning in Step 5 about the zinc chloride solution.*)

7. Using forceps or tongs, carefully lower two clean pennies into the solution in the beaker. To avoid causing a splash, do not drop the coins into the solution. Put the watch glass on the beaker and keep the solution boiling gently for two to three minutes. You should notice a change in the appearance of the pennies during this time.

8. Using forceps or tongs, remove the two coins from the beaker. Rinse them under running tap water, then gently dry them with a paper towel. Set one treated coin aside for later comparisons and use the other treated coin in Step 9.

9. Set a hot plate to medium-high heat. Using the tongs, place the second treated penny directly on the hot-plate surface and count to five. Use the tongs to flip the coin and count to five again. Repeat until you observe a color change. See Figure 2.46.

Figure 2.46 *Heating the treated penny on a hot plate.*

10. Pick up the coin from the hot plate surface with the tongs, rinse the heated coin under running tap water, and gently dry it with a paper towel. Record your observations.

11. Observe and compare the appearances of the three pennies. Record your observations.

12. When finished, discard the used zinc chloride solution and the used zinc as directed by your teacher.

13. Wash your hands thoroughly before leaving the laboratory.

Questions

1. a. Compare the color of the three coins—untreated (the control), heated in the zinc chloride solution only, and heated in the zinc chloride solution and then on a hot plate.
   b. Do the treated coins appear to be composed of metals other than copper? If so, explain.

2. If someone claimed that a precious metal was produced in this investigation, how would you decide whether the claim was correct?

3. Identify at least two practical uses for metallic changes similar to those you observed in this investigation.

4. a. What happened to the copper atoms originally present in the treated pennies? Provide evidence to support your conclusion.

   b. Do you think the treated pennies could be converted back to ordinary coins? If so, what procedures would you use to accomplish this? (**Caution:** *No laboratory work should be performed without your teacher's approval and direct supervision.*)

D.6 COMBINING ATOMS: ALLOYS & SEMICONDUCTORS

The investigation you just completed demonstrated how metallic properties can be modified by creating an **alloy**, a solid combination of atoms of two or more metals. The immersion of a penny in hot zinc chloride solution produced a silvery alloy of zinc and copper called *brass*. When you heated the penny on the hot plate, the zinc and copper atoms mixed more. The overall mixing of zinc and copper atoms to form brass is depicted in Figure 2.47. The resulting solid solution has a different proportion of zinc and copper atoms and is another form of brass. Most brass materials have a golden color, unlike either copper or zinc. Brass is also harder than copper metal. Some other common alloys with familiar names are listed in Table 2.7 (page 194).

It is clear that one way to modify the properties of a particular metal is to form it into an alloy, just as you did when you produced a gold-colored penny. Often the results of alloying metals are unexpected, as you are about to discover.

Combinations of Elements: Alloys

(a)
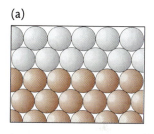

(b)
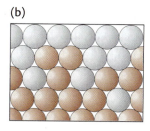

(c)

Figure 2.47 *Forming a brass outer layer onto a penny (a–c). Brass samples contain mostly copper and anywhere between 10% to 40% zinc (by mass).*

> ### ChemQuandary 3
>
> #### FIVE CENTS' WORTH
> A U.S. nickel is composed of an alloy of nickel and copper. Based on your familiarity with the appearance of that common five-cent coin, you might be surprised to learn it is composed of more copper than nickel! Specifically, each U.S. five-cent coin contains only 25% nickel and 75% copper.
>
> *In terms of its composition, what would be a more accurate name for a nickel?*
>
>
>
> What does this suggest about the difference between an alloy (in this case, a "solid solution" of copper and nickel atoms) and a simple mixture of powdered copper and powdered nickel?

COMMON ALLOY COMPOSITIONS AND USES

Alloy and Composition	Examples (composition in mass percent)	Comments
Brass Copper and zinc	Red brass 90% Cu, 10% Zn Yellow brass 67% Cu, 33% Zn Naval brass 60% Cu, 39% Zn, 1% Sn	Properties of brass vary with the proportion of copper and zinc used and with the addition of small amounts of other elements. Brass is used in plumbing and lighting fixtures, rivets, screws, and ships.
Bronze Primarily copper, with tin, zinc, and other elements	Coinage bronze 95% Cu, 4% Sn, 1% Zn Aluminum bronze 90% Cu, 10% Al Hardware bronze 89% Cu, 9% Zn, 2% Pb	Bronze is harder than brass. Its properties depend on the proportions of its components. Bronze is used in bearings, machine parts, telegraph wires, gunmetal, coins, medals, artwork, and bells.
Steel Primarily iron, with carbon and small amounts of other elements	Steel 99% Fe, 1% C Nickel steel 96.5% Fe, 3.5% Ni Stainless steel 90–92% Fe, 0.4% Mn, <0.12% C, Cr (trace)	Properties of steel are often determined by carbon content. High-carbon steel is hard and brittle; low- or medium-carbon steel can be welded and shaped. Steel is used in automobile and airplane parts, kitchen utensils, plumbing fixtures, and architectural decoration.
Other common alloys	Pewter 85% Sn, 6.8% Cu, 6% Bi, 1.7% Sb	Pewter is often used in figurines and other decorative objects.
	Mercury amalgams 50% Hg, 20% Ag, 16% Sn, 12% Cu, 2% Zn	Mercury amalgams have often been used for dental fillings.
	14-carat gold 58% Au, 14–28% Cu, 4–28% Ag	14-carat gold is popular in jewelry.
	White gold 90% Au, 10% Pd	White gold is also principally used in jewelry.

| Table 2.7

In addition to alloys composed of "solid solutions" of two or more metals, such as steel, other useful alloys include some well-defined compounds; that is, they have a constant, definite ratio of metallic atoms. One example is Ni_3Al, a low-density, strong metallic alloy of nickel and aluminum used as a component of jet aircraft engines. A very hard chromium–platinum alloy, Cr_3Pt, forms the basis of some commercial razor blade edges. And a special group of alloys, including the niobium–tin compound Nb_3Sn, displays **superconductivity**—the ability to conduct an electric current without any electrical resistance— if cooled to a sufficiently low temperature.

Thus, one strategy for modifying the properties of metals is to produce alloys, which have properties that differ from those of the component elements. In fact, dramatic changes in the properties of a substance are possible when an extremely small amount of an element—as small as one atom per million atoms—is intermingled within a solid substance.

The metalloid silicon belongs in a class of materials called **semiconductors.** What does this term mean? You should recall that some elements (metals) are good conductors of electricity, whereas other elements (nonmetals) are not. Semiconductors, as the name implies, lie somewhere in between. In addition to silicon (Si), elements and compounds with semiconducting properties include germanium (Ge), gallium arsenide (GaAs), and cadmium sulfide (CdS). Locate the elements identified as semiconductors on the periodic table (page 124).

In the crystal structure of pure silicon, each atom is bonded to four other atoms. As a result, silicon's electrons are incapable of moving through the crystal. In other words, crystals of pure silicon display poor electrical conductivity at normal temperatures.

Although it sounds strange, adding certain impurities to pure silicon dramatically enhances its semiconducting properties. The impurities are atoms of other elements such as phosphorus, arsenic, aluminum, or gallium. This process, called **doping,** creates a situation in the solid silicon that allows charge carriers—either electrons or tiny regions of positive charge—to become mobilized within the crystal. This greatly improves silicon's semiconductor characteristics.

Most semiconductor devices, such as transistors and integrated circuits used in computers and other electronics, are based on solid crystals of silicon that contain intentionally added impurities. See Figure 2.48. Those added atoms are responsible for our present silicon-based technological era.

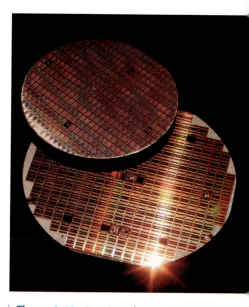

Figure 2.48 *Semiconductor devices like these allow computers to process digital information.*

Silicon Valley in California got its name from the element that is vital to many computer-related companies located there.

D.7 MODIFYING SURFACES

In Section A, you learned that high copper prices in the early 1980s forced the United States to use zinc as the primary metal in pennies. Copper covers the coin's surface. This copper exterior not only preserves the traditional appearance of the coin, it also protects it from corrosion. Using one material to protect the surface of another, less durable material is by no means new or unusual. A layer of protective material can enhance the performance or appearance of a manufactured product, while allowing a less expensive or more available material to be used for most of the item. As you continue reading, you will learn about several types of surface treatments, both new and old, and how they can enhance materials and products.

Coatings

The surface treatment that is probably most familiar to you is a **coating.** Coatings include paints, varnishes, and shellacs that are commonly applied to homes, cars, and many other products. These materials are generally applied late in the manufacturing or construction process and are physically attached to the surface to be protected. Although coatings sink into pores of the base material, no chemical bonding between the coating and the base takes place.

Paint, a typical coating, contains a *pigment* and a *solvent,* and may also contain a resin or other components, depending on intended uses. The paint pigment provides color, the solvent allows application of the paint in liquid form, and the resin provides the desirable protective properties. Although paints have existed for centuries, growing concerns about the solvents released in their application and the heavy metals in their pigments have spurred development of new formulations and methods of application.

Of particular interest is the technique of *powder coating,* which eliminates the need for a solvent. See Figure 2.49. In this method, all coating components are blended and ground to an extremely fine powder. The product to be coated is then cleaned to ensure good adhesion. The powder is mixed with compressed air, pumped into spray guns, and given an electrical charge. The product is electrically grounded, allowing it to be coated by the charged powder. After coating, the product is heated to "cure" the powder into a tough shell. Many products, including bicycle and auto parts, are powder coated to create long-lasting, attractive surfaces.

Figure 2.49 *This motor scooter's metal parts were powder coated to resist corrosion.*

Electroplating

The process used to cover one metal with another involves using direct-current (DC) electricity, which causes redox reactions to occur between the metal and a metal-ion solution. This general process is known as **electroplating.** As you have already learned, cations of most metallic elements can be reduced. This fact is exploited in electroplating. For example, metal bumpers on trucks are often made of steel. In wet or snowy climates, the exposed steel would quickly corrode or rust. Manufacturers protect the steel by coating it with chromium and nickel. See Figure 2.50. Coins, too, have been plated with nickel to resist corrosion.

Writing **half-reaction** equations, which separately represent the reduction and oxidation parts of a redox process, can be helpful in understanding the plating process. The reduction half-reaction for plating nickel metal on steel is written this way:

$$Ni^{2+}(aq) + 2\ e^- \longrightarrow Ni(s)$$

Where do the electrons in this equation come from? Electroplating requires a power source, usually a battery or power supply when the reaction is conducted in the laboratory. The power source transports electrons to the **cathode** (where reduction occurs). The ultimate source of electrons in any electrochemical process is the **anode.** In electroplating, the anode is usually made of the metal to be plated. As metal cations are reduced (removed from the solution) and deposited on the object attached to the cathode, metal atoms at the anode are oxidized and dissolve into the plating solution. Figure 2.51 illustrates this process.

Thus, the other half-reaction for the plating system is an oxidation reaction, the reverse of the half-reaction shown above. Based on what you know about metal reactivities, would you expect this system of reactions to proceed on its own without any external power source? Although systems that can plate metals without applying electric current have been developed, they must include a chemical reducing agent, which is sometimes more expensive than electric current.

Figure 2.51 *Electroplating is the result of an oxidation–reduction process. The new coating is a layer of metal atoms deposited by reducing metal ions from the electroplating solution.*

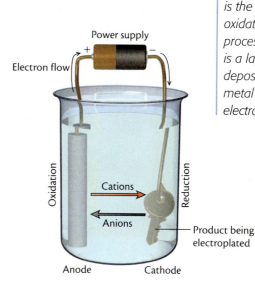

Power supply

Electron flow

Oxidation

Reduction

Cations

Anions

Product being electroplated

Anode   Cathode

All That's Gold Doesn't Glitter . . . Or Does It?

Did you ever wonder where the gold metal that makes up a high-school ring comes from? Or who is responsible for locating and recovering the material? John Langhans may have helped. He's a metallurgist with Barrick Goldstrike Mines Inc., a large mining company in Elko, Nevada, that finds and recovers gold.

John's job is not an easy one. The normal concentration of gold in Earth's crust is only about 15 parts per billion (ppb). During the 1800s and early 1900s, miners looked for areas where gold was concentrated in the form of nuggets. More refined mining techniques enabled miners to locate deposits of finely divided gold. Such deposits account for most of today's gold mines.

Gold is often associated with certain types of rocks; quartz is an example. It is relatively easy to recover gold from these rock deposits. In the past decade, however, many of these deposits have been mined out, and the remaining gold is much harder to process.

When associated with sulfide minerals, gold is known as *refractory ore*. More sophisticated equipment and techniques must be employed to recover this gold. By using knowledge of chemistry and highly efficient recovery methods, John and his coworkers can extract gold from this material.

One method John's company uses is "autoclaving." In this process, the sulfide minerals are cooked under fairly high temperatures and pressures to dissolve the sulfides and expose the gold. The autoclaves work in a manner similar to that of a pressure cooker. You may be familiar with this piece of cookware if it is used to prepare meals in your home. Once the gold is exposed, a dilute basic solution containing cyanide ions (CN^-) is added to the ore to dissolve the gold. A gold cyanide complex $[Au(CN)_2]^-$ forms when gold atoms come in contact with the solution:

$$4\,Au + 8\,CN^- + 2\,H_2O + O_2 \longrightarrow 4\,[Au(CN)_2]^- + 4\,OH^-$$

Later, the gold is recovered through a process called *electrowinning*. Electrowinning uses an electrical current to transfer the gold onto steel wool. Finally, technicians heat the gold-plated steel wool to liquefy the gold, and pour the liquid gold into bar molds.

> John supervises laboratory personnel whose job it is to research and develop new methods for recovering gold . . .

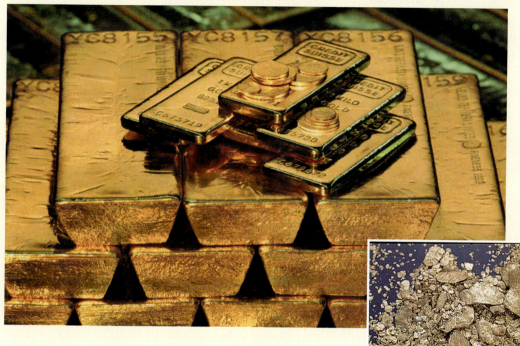

John supervises laboratory personnel whose job it is to research and develop new methods for recovering gold that are both safe and economical. He and his team also monitor the processing operations of mines to ensure that the environment is not degraded as a result of those operations.

Some Questions to Ponder . . .

Class rings are not 100% pure gold. Because pure gold is very soft, it is mixed with other metals (copper, silver, nickel, or zinc) to increase hardness and durability. The greater the amount of added metal, the lower the karat value. Pure gold is 24 karat. Most gold-containing rings are probably 12 karat, or about 50% gold.

Because of the specialized processes that Barrick Goldstrike uses, the company is able to profitably recover gold from refractory ore, even at grades as low as 5 grams gold per metric ton ore.

1. If a 12-karat-gold class ring has a mass of 11.2 g, how many metric tons of ore were processed in order to produce the gold to make the ring?

2. What is the value of the gold in the class ring if the market value of gold is about $10.50/g? (You may want to use the current actual price of gold, which is available in the business section of many newspapers or on the World Wide Web.)

3. During the 1800s, gold miners looked for gold in the form of nuggets. What property of gold allows it to be found in this form?

Unlike coatings, platings are usually bonded to the surface of the bulk material. This is desirable because it keeps the metal finish firmly attached to the object. After several layers of atoms have been deposited, the plating has the properties of the plating metal; therefore, it can impart the properties for which it was selected. What about the first few layers of atoms? Metallurgists have known for years that these layers tend to have properties distinct from both the plating and the base materials. This knowledge helped lead to the development of another type of surface modification—thin films.

Thin Films

Films are materials distinguished by the incredibly small thickness of their deposition on a surface. Modern *thin films* may be only one or two atoms or molecules thick! Films made of metal compounds, such as carbides and oxides, were originally developed to harden the edges of cutting tools. They are also used to decorate or protect many metal products. Alternating layers of these materials can enhance certain desired properties of the metal.

Thin metal films find many applications in the optics and electronics industries. For example, tungsten films connect circuits on microelectronic devices. Gold films can be deposited on glass or other surfaces to produce high-quality mirrors. Semiconductors also include thin films in their design.

Scientists are investigating a model provided by nature to develop new thin-film approaches. For example, seashells are gradually formed through the deposition of layers of inorganic and organic substances. Similar approaches are being used to create layers of engineered ceramics on surfaces.

The most common organic thin films are made of polymers, a subject you will explore in more detail in Unit 3. For now, you should know about the existence of polymer films because they are used to protect currency and documents from counterfeiting and could be similarly used to protect coins. Polymers coat many products you use daily. Some common examples are automobile glass and the UV-protective and antiglare films on sunglasses.

Figure 2.52 *This automobile paint appears to change color depending on the angle from which it is viewed.*

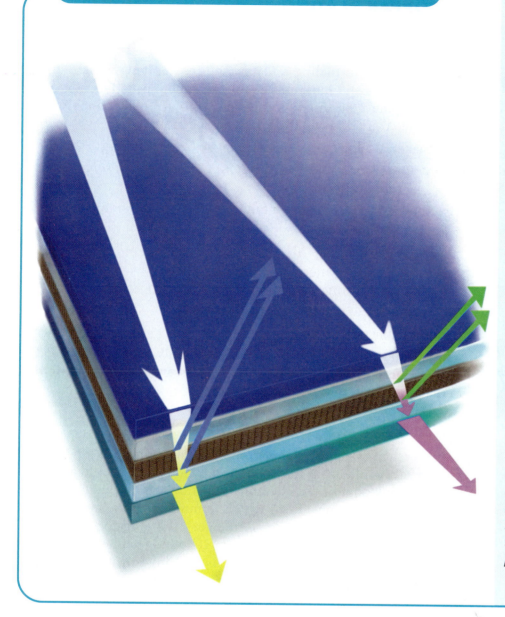

Figure 2.53 *Color-changing paint. This diagram shows a cross-section of a transparent mica-based pigment in color-changing paint. Mica (brown layer) is thinly coated on both sides with titanium dioxide (TiO₂, light-blue layers). The pigment is then covered with a clear lacquer coating (top and bottom layers).* **Left:** *White light travels through the lacquer coating. As it encounters each TiO_2-mica interface, blue light waves reflect and reinforce each other (called constructive interference). Viewed at this angle, the paint appears blue.* **Right:** *Changing the angle at which white light contacts the pigment affects the distance that it travels through each layer. This causes a different wavelength of light (shown as green) to reflect and constructively interfere at the two TiO_2-mica interfaces, while other colors of light destructively interfere. At this viewing angle, the paint appears green. The mica and TiO_2 coating thicknesses can be altered to produce different color variations.*

A novel use of polymer thin films is in color-changing paint. Scientists have observed that the colors of butterflies result from the interference patterns of light caused by tiny fibers in their wings. Seeking to duplicate this effect, researchers have used spun polyester, liquid-crystal polymers, and mica-based films to develop car paint that appears to change color depending upon the angle from which it is viewed. Figure 2.52 illustrates this feature, and Figure 2.53 shows how the color-changing paint works.

The development of thin films is a good example of productive interaction of science and technology. As new discoveries are made, new applications result. Demands for better methods and more desirable properties lead to additional basic research and subsequent applications.

D.8 COPPER PLATING

Introduction

**Lab Video:
Copper Plating**

As you just learned, plating a metal requires applying direct current, the presence of metal ions, and a suitable anode, usually made of the metal to be plated. In this investigation, you will plate copper onto a metal object. You will need the following materials:

- beaker, U-tube or transparent plastic (Tygon) tubing
- copper-plating solution (**Caution:** *Copper-plating solutions are hazardous and corrosive.*)
- object to be electroplated (an iron nail, a nickel or quarter, a steel key, pencil lead)
- copper strips
- a 1.5-V or 9-V battery or other low-voltage power supply (**Caution:** *Always be careful when working with electricity.*)
- wire leads with alligator clips
- voltmeter or CBL kit (optional)

Before starting, read the procedure to learn what you will need to do, note safety precautions, and plan necessary data collecting and observations.

Procedure

Set up a system to deposit copper onto a metallic object. Keep in mind which metal should act as the *anode* and which metal should act as the *cathode.* Think about how electrons will flow—from which terminal will the electrons come and to which terminal will the electrons go? Test your setup, taking care not to immerse the alligator clips in the plating solution, and record your results. If nothing happens, have your teacher check your setup. When you finish the investigation, be sure to follow your teacher's instructions for disposing of wastes. Wash your hands thoroughly before leaving the laboratory.

Questions

1. What was the anode in this oxidation–reduction process? Explain how you came to this conclusion, based on observational evidence from the investigation.
2. Write an equation for the reaction that occurred at the anode.
3. What was the cathode?
4. Write an equation for the reaction that occurred at the cathode.

5. Would the metal used as the cathode affect the result of electroplating? Explain.

6. Do you think this method would be useful for large-scale copper plating? Why or why not?

D.9 MORE COIN DESIGN CONSIDERATIONS

In D.6 and D.7 (pages 193–201), you learned about using alloys and surface modification to create various products. Would you consider using either option for your coin design?

Perhaps you have thought about using one of the alloys described in Table 2.7 (page 194). You might be considering using a surface modification technique that has been described.

1. What properties of alloys make them an attractive option in coin design?

2. Why might simply modifying the coin's surface be preferable?

3. List the benefits and the drawbacks of using alloys versus using surface modification. Consider factors such as chemical reactivity, durability, and cost.

SECTION D SUMMARY
Reviewing the Concepts

Allotropes of an element have distinctly different physical or chemical properties. These variations are explained by differences in the arrangement and bonding of the element's atoms.

1. What is an allotrope?

2. Name two elements other than carbon that form allotropes.

3. A diamond, a chunk of coal, and your pencil lead contain the same substance:
   a. How are their properties different?
   b. Why are their properties different?
   c. What accounts for the differences in the cost of these items?

The properties of matter may be modified by physically blending or chemically combining two or more substances.

4. How do engineered materials differ from natural materials?

5. List two advantages and two disadvantages of using engineered ceramics in high-temperature applications.

6. Describe two examples of properties that can be modified in plastics to make them useful for new applications.

An alloy possesses properties that differ, sometimes significantly, from the properties of its constituent elements.

7. What is an alloy?

8. Give examples of two alloys you use regularly. (*Hint:* See Table 2.7, page 194.)

9. What nonmetal is a component of both steel and stainless steel? (*Hint:* See Table 2.7, page 194.)

10. Give the formula, use, and an important physical property of an alloy that is also a well-defined compound.

A poorly conductive material can sometimes be transformed into a semiconductor by adding a small amount of a particular impurity.

11. Describe the periodic table location of elements that behave as semiconductors.

12. List three elements commonly used for doping semiconductors.

13. What is the primary use of the products of semiconductor technology?

> The surface properties of a material can be modified through application of coatings, thin films, or electro-plated metals.

14. Give three examples of coatings.

15. What desirable properties do coatings provide?

16. Describe two ways that electroplating is used in industry.

17. List two applications of thin-film technology.

Connecting the Concepts

18. Classify each major method of coating a surface—coating, plating, and thin films—as either a chemical change or a physical change. Explain your answers.

19. In Investigating Matter D.8 (page 202), electricity was used to convert Cu^{2+} ions to $Cu(s)$.

   a. Was this chemical change an oxidation or a reduction?

   b. Rather than obtaining electrons from electricity, you could use another metal to provide electrons. Based on the metal activity series (Table 2.5, page 145), name two metals that could be used for this purpose.

20. Can a rusty car bumper be protected from further rusting by electroplating? Explain.

21. New materials can successfully substitute for metals in some products. Explain why each of the following replacements can be made:

   a. Ceramics replace steel in turbine engines.

   b. Plastics replace steel in automobile bumpers.

   c. Optical fibers replace copper in phone wires.

Extending the Concepts

22. Obtain information about various allotropic forms of the element sulfur.

   a. Draw models depicting the major sulfur allotropes.

   b. Compare the properties of these allotropes.

23. Research and report on new applications of carbon allotropes known as *fullerenes*.

24. Research how *doping* works, and, specifically, how it allows electrons to move more easily within semiconductors.

25. Investigate recent advances and potential applications in the field of superconductivity.

MAKING MONEY

SELECTING THE BEST COIN DESIGN

When a new product is created, the design process often includes proposals from several individuals or teams. A panel of experts or consumers then chooses the best proposal. Because each school can submit only one design for the new coin, you will follow a similar process in selecting the best proposal. Each team will present its coin design and findings to the class, which will serve as the selection committee. The following information summarizes the essential features expected of your coin design proposal:

Coin Model

Include a model or a detailed drawing of the proposed coin. If you create a model, it does not have to be made of the specified material, but it should resemble the desired material. Ensure that all models and designs are at least five times the actual coin size. In addition, specify the actual size, thickness, and mass of the proposed coin.

Material Specifications

Describe the material or materials selected for the coin. Present your rationale for choosing those materials. Include an analysis of the necessary and desirable properties of the selected material or materials, as well as strategies that will discourage counterfeiting.

Material Sources

Provide details on the source or sources of the raw materials needed to produce the new coin.

Life-Cycle Analysis

Present an analysis of the life cycle of the proposed coin. How long do you expect the coin to last in general circulation? Will the material or materials that make up the coin be recycled or reused? If so, how?

Cost Analysis

Present an analysis of estimated production costs for the new coin.

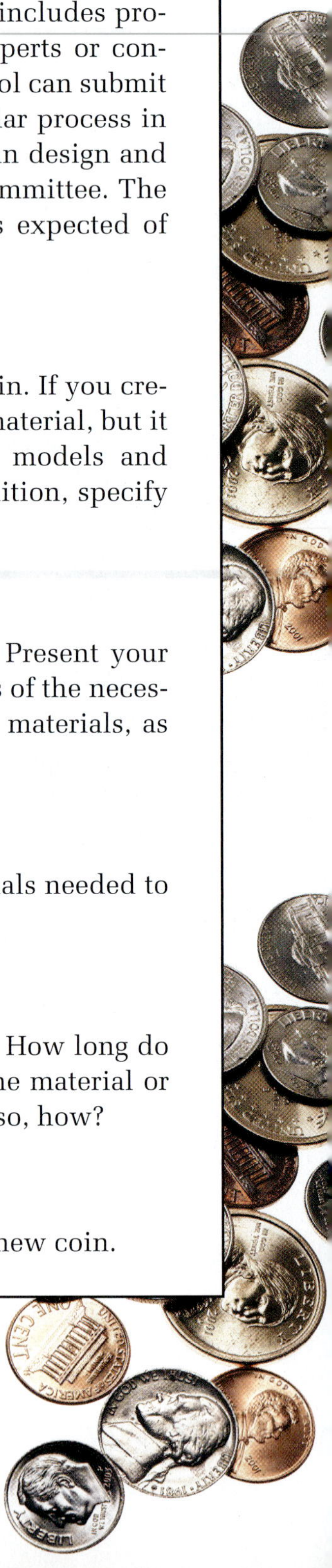

LOOKING BACK AND LOOKING AHEAD

As you come to the end of this unit, pause and reflect on what you have learned so far. You have learned some working language of chemistry, such as symbols, formulas, and equations; laboratory techniques; and major ideas in chemistry, such as periodicity, the law of conservation of matter, and atomic theory.

This knowledge can help you better understand some societal issues. Central among these concerns are the use and management of Earth's chemical resources, which include water, metals, petroleum, food, and air.

You also have explored other issues that enter into policy decisions about emerging technological challenges. Chemistry often plays a crucial role in helping people recognize and resolve such issues, as you have learned.

However, many problems are far too complex for a simple technological fix. Issues of policy are not usually "either/or situations," but they often involve weighing many considerations. As a voting citizen, you will deal with issues that require some scientific understanding. Tough decisions may be needed. The remaining chemistry units that you study will continue to prepare you for those responsibilities.

Unit 3 deals with petroleum, a nonrenewable chemical resource.

UNIT 3

Petroleum: Breaking and Making Bonds

WHAT are important chemical and physical properties of hydrocarbons?

WHY are hydrocarbons commonly used as fuels?

WHY are carbon-based molecules so versatile as chemical building blocks?

WHAT properties are important to consider in finding substitutes for petroleum?

Interest is growing in alternative-energy-powered transportation. Why? What advantages and disadvantages do petroleum alternatives offer? How can the global supply of petroleum best be used? Turn the page to learn more about this energy-rich resource.

HYBRIS HYBRID VEHICLE TV COMMERCIAL: WORKING SCRIPT

Text/Script	Image/Sound
Clean . . . **Comfortable . . .** **Convenient . . .**	(Video image of driver and three passengers riding in a *Hybris*.) (The words "clean," "comfortable," and "convenient" appear, in turn, as the announcer speaks them. "Clean" appears at the top of the screen; "Comfortable" appears in the middle; "Convenient" appears at the bottom. The "C" moves from the top to the bottom as the words appear.)
Those three words describe *Hybris,* the new breakthrough in personal vehicles. *Clean . . .* Unlike older petroleum-fueled vehicles, the *Hybris* hybrid-powered automobile is virtually emission free. In fact, it is considered an Ultra Low-Emissions Vehicle by the EPA. Why? Because the *Hybris's* gasoline engine is combined with an energetic electric motor to deliver maximum power and efficiency. Excess energy generated by the gasoline engine is stored in rechargeable batteries to later power the electric motor. You can drive knowing that you're helping, not hurting, the environment.	(Show the *Hybris* traveling on a road.) (*"Clean"* appears on the screen and moves from left to right.) (Show the *Hybris* decelerating at a stoplight and then, after sitting at a stoplight, accelerating next to nonhybrid vehicles.)
Comfortable . . . You and three others can easily ride in the *Hybris.* And—thanks to the electric motor that takes over in city driving—the ride can be so quiet that driving the *Hybris* seems more like a walk in the park.	 (*"Comfortable"* appears on the screen and moves from the top left corner to the bottom right corner.)

Text/Script	Image/Sound
Convenient . . .	(*"Convenient"* appears on the screen and moves from top center to bottom center.)
Imagine not having to stop as often at gas stations.	(The *Hybris* drives past a gasoline station.)
The *Hybris* uses nearly all the energy in each drop of gasoline. In fact, every time you step on the brake pedal, you store energy that will save you big money.	
With an EPA-rated mileage ranging from 50 to 56 miles per gallon, driving your *Hybris* hybrid-powered automobile is easy on your pocketbook—as little as five cents per mile to operate. Compare that to more than eleven cents per mile for conventional gasoline-burning vehicles.	(*Hybris* shown in a variety of settings—beach, mountains, recreation areas.) (Image of a piggy bank being shaken with coins, mostly pennies, falling onto a table.)
Help conserve petroleum resources! Visit your *Hybris* dealer today.	 (View of a *Hybris* at a dealer showroom with several people examining it. The *Hybris* logo then appears below the vehicle.)

Petroleum: What Is It?

You live in a world of newly developed products and materials. Whether on billboards, on radio, on television, or in magazines, each advertisement attempts to sell a product by educating its audience about the product's unique features. The product may be faster, lighter, easier to use, newer, or great tasting—to sample a few of many common product claims. The advertisement for the *Hybris* is no exception; it highlights several energy- and fuel-related features of the new vehicle that will (or so it is claimed) "help conserve petroleum resources."

Figure 3.1 *Much time and energy will be expended to refine this mixture of petroleum before it is viable as a fuel or as a source of "builder molecules." What about petroleum makes it such an attractive resource?*

The word *petroleum* is quite familiar to you. But do you know what petroleum *is* or what it is *made of*? See Figure 3.1. Can you explain what properties make petroleum useful for both burning and building? In this section, you will explore characteristics of some key compounds found in petroleum. Specifically, you will focus on their physical properties, their molecular structures, and how atoms bond to form these compounds.

A.1 WHAT IS PETROLEUM?

Crude oil was used as lamp oil in ancient times.

Petroleum is a vital chemical world resource. Petroleum pumped from underground is called **crude oil.** This liquid varies from colorless to greenish-brown to black, and can be as fluid as water or as resistant to flow as soft tar.

We cannot use crude oil commercially in its natural state. Instead, we transport it by pipeline, ocean tanker, train, or barge to oil refineries, where it is separated into simpler mixtures. See Figure 3.2. Some

Figure 3.2 *The largest oil tanker in the world can carry up to 4.1 million barrels (over 150 billion gallons) of crude oil. Such ships are so big, onboard crew members often need bicycles to travel from one point to another.*

mixtures are ready to use, while others require further refinement. Refined petroleum is chiefly a mixture of various **hydrocarbons,** which are molecular compounds that only contain atoms of hydrogen and carbon. You can readily see how this class of compounds got its name.

Burning petroleum provides nearly half of the total annual U.S. energy needs. Most petroleum is used as a fuel. Converted to gasoline, petroleum powers millions of U.S. automobiles, each traveling an average of 14 000 miles annually. Other petroleum-based fuels provide heat to homes and to office buildings, deliver energy to generate electricity, and power diesel engines and jet aircraft.

Petroleum's other major use is as a raw material from which a stunning array of familiar and useful products is manufactured—from CDs, sports equipment, clothing, automobile parts, plastic charge cards, and carpeting to prescription drugs and artificial limbs. Figure 3.3 shows some petroleum-based products.

Based on your experiences with petroleum fuels and products, what percent of petroleum would you estimate is used for burning? What percent of petroleum would you estimate is used for producing other useful substances? Can you identify other uses of petroleum? The answers in the next paragraph may surprise you.

What did you predict for the percent of petroleum used for burning? Did you predict 50%? Was your prediction 60%? Astonishingly, 89% of all petroleum is burned as fuel. Only about 7% is used for producing new substances, such as medications and plastics. The remaining 4% is used as lubricants, road-paving materials, and an assortment of miscellaneous products. In fact, for every gallon of petroleum used to produce useful products, more than five gallons are burned for energy.

What happens to molecules contained in petroleum when they are burned or used in manufacturing? As in all chemical reactions, the atoms become rearranged to form new molecules.

When hydrocarbons burn, they react with oxygen gas to form carbon dioxide (CO_2) gas and water vapor. (See equations below representing the burning of two typical hydrocarbons.) These gases then disperse in the air. The hydrocarbon fuel is used up; it will take millions of years for natural processes to replace it. Thus, petroleum is a *nonrenewable resource,* as were the minerals you studied in Unit 2.

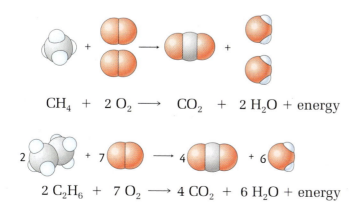

$$CH_4 + 2 O_2 \longrightarrow CO_2 + 2 H_2O + energy$$

$$2 C_2H_6 + 7 O_2 \longrightarrow 4 CO_2 + 6 H_2O + energy$$

> The word *petroleum* comes from the Latin words *petr-* ("rock") and *oleum* ("oil").

Figure 3.3 *It is likely you use petroleum-based products in almost every area of your life. How many can you identify?*

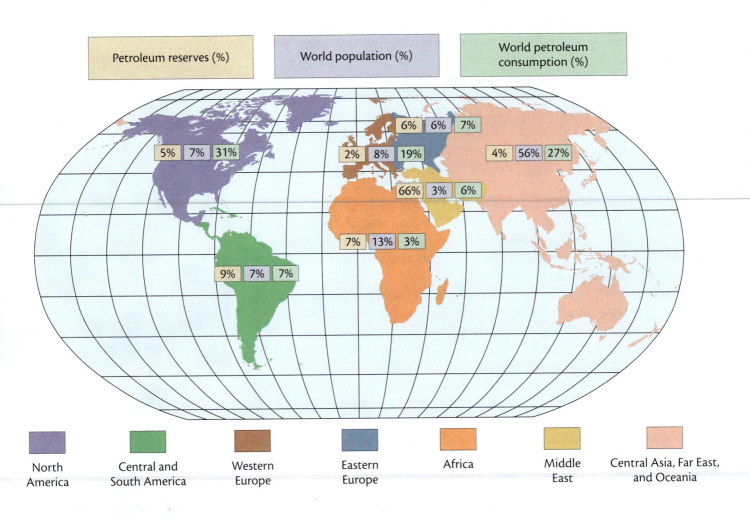

Legend:
- North America
- Central and South America
- Western Europe
- Eastern Europe
- Africa
- Middle East
- Central Asia, Far East, and Oceania

	Petroleum Reserves		Population		Petroleum Consumption	
	10^9 Barrels	Percent	Millions	Percent	10^3 Barrels per day	Percent
North America	54	5.3	421	6.7	23 843	30.5
Central and South America	96	9.3	434	6.9	5238	6.7
Western Europe	17	1.7	485	7.8	14 698	18.8
Eastern Europe	58	5.7	386	6.2	5257	6.7
Africa	77	7.4	831	13.3	2675	3.4
Middle East	686	66.4	178	2.9	5043	6.5
Central Asia, Far East, and Oceania	44	4.2	3508	56.2	21 452	27.4
Totals	1032	100.0	6243	100.0	78 206	100.0

Figure 3.4 *Distribution of world's petroleum reserves, population, and consumption of petroleum.*

Figure 3.5 *Two employees of the Kuwait Oil Company work on a crude oil flowline at Maqwa Oil Field, 50 kilometers south of Kuwait City.*

Like other resources, petroleum is not uniformly distributed around the world. The worldwide distributions are summarized in Figure 3.4. The North American petroleum reserves are only about 5% of the world's known supply. Central Asia, the Far East, and Oceania account for 56% of the world's population, but this region has only about 4% of the world's petroleum reserves. In contrast, approximately 66% of the world's known crude oil reserves are located in Middle Eastern nations (see Figure 3.5).

You have just learned what petroleum is, how we use it, and where we find it. Petroleum is a complex mixture of hydrocarbons that must be refined or separated into simpler mixtures to become useful. In the following investigation, you will learn about this basic separation process as you study a simple mixture of two liquids.

Recall that you used several separation techniques to clean up the foul-water mixture in Unit 1.

Investigating Matter

A.2 SEPARATION BY DISTILLATION

Introduction

You know that you can often separate substances by taking advantage of their different physical properties. One physical property commonly used to separate liquids is their density. However, density differences will work only if substances are insoluble in each other, which is not the case with petroleum; its components are soluble in each other. Another physical property chemists often use is *boiling point*. The separation of liquid substances according to their differing boiling points is called **distillation.**

As you heat a liquid mixture containing two components, the component with the lower boiling point will vaporize first and leave the distillation flask. That component will then condense back to a liquid as it passes through a condenser—all before the second component begins to boil. See Figure 3.6. Each condensed liquid component, called the **distillate,** can thus be collected separately.

 Lab Video: Separation by Distillation

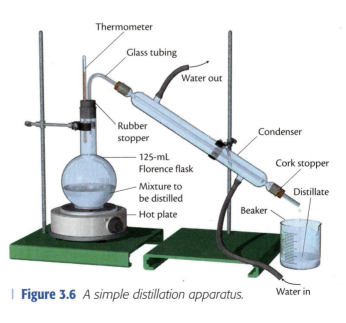

Thermometer
Glass tubing
Water out
Rubber stopper
Condenser
125-mL Florence flask
Cork stopper
Mixture to be distilled
Distillate
Hot plate
Beaker
Water in

| **Figure 3.6** *A simple distillation apparatus.*

| Table 3.1 |

POSSIBLE COMPONENTS OF DISTILLATION MIXTURE			
Substance	Formula	Boiling Point (°C)	Appearance with I_2
2-Propanol (rubbing alcohol)	C_3H_8O	82.4	Bright yellow
Acetone	C_3H_6O	56.5	Yellow to brown
Water	H_2O	100.0	Colorless to light yellow
Cyclohexane	C_6H_{12}	80.7	Magenta

The boiling points listed in Table 3.1 are based on normal sea-level atmospheric pressure.

In this investigation, you will use distillation to separate a mixture of two liquids. Then you will identify the two substances in the mixture by comparing the observed distillation temperatures with the boiling points of several possible compounds listed in Table 3.1.

The following procedure provides guidance about when you should collect and record data (consider Step 8). Create a data table with columns to record time and temperature data. Remember to include proper units for each data column. Before you begin, carefully read the procedure to learn what is involved, note safety precautions, and plan for your data collecting and observations.

Procedure

(**Caution:** *This distillation should only be completed with a hot plate or other electric heating source. The presence of open flames near the distillation apparatus represents a fire hazard.*)

1. Construct data tables to record your observations and measurements.

2. Assemble an apparatus similar to that shown in Figure 3.6. Label two beakers *Distillate 1* and *Distillate 2*.

3. Using a clean, dry, graduated cylinder, measure a 50-mL sample of the distillation mixture. Pour the mixture into the distillation flask and add a boiling chip.

4. Record your observations of the starting mixture.

5. Connect the flask to a condenser, shown in Figure 3.6 (page 215). Ensure that the hoses are attached to the condenser and to the water supply as shown. Position the Distillate 1 beaker at the outlet of the condenser so that it will catch the distillate, as shown in Figure 3.7.

6. Ensure that all connections are tight and will not leak.

7. Turn on the water to the condenser, and then turn on the hot plate to start gently heating the flask. (**Caution:** *The substances, other than water, are volatile and highly flammable. Be sure that no flames or sparks are in the area.*)

8. Record the temperature every minute until the first drop of distillate falls into the beaker. Then continue to record the

Figure 3.7 *This distillation apparatus is used to separate a mixture of two liquids. The component with the lowest boiling point vaporizes first, converts back to a liquid in the condenser, and collects (here) in a beaker.*

temperature every 30 seconds. Continue to heat the flask and collect the distillate until the temperature begins to rise again. At this point, replace the Distillate 1 beaker with the Distillate 2 beaker.

9. Continue heating and recording the temperature every 30 seconds until the second substance just begins to distill. Record the temperature at which the first drop of the second distillate falls into the beaker. Collect 1 to 2 mL of the second distillate. (*Caution: Do not allow all of the liquid to boil from the flask.*)

10. Turn off the hot plate and allow the distillation apparatus to cool. While the apparatus is cooling, test the relative solubility of solid iodine (I_2) in Distillate 1 and Distillate 2 by adding a few crystals of iodine to each beaker and stirring. Record your observations. (*Caution: Iodine is corrosive on contact. It will stain skin or clothing.*)

11. Disassemble and clean the distillation apparatus, and dispose of your distillates as directed by your teacher.

12. Wash your hands thoroughly before leaving the laboratory.

Data Analysis

Plot your data on a graph of time (*x*-axis) versus temperature (*y*-axis). As you observed, heating the liquid mixture raised its temperature. However, once the first component began to boil and vaporize from the mixture, the temperature of the liquid remained steady until that component completely distilled. Continued heating then caused the temperature to rise again, this time until the second component began to boil and distill. Because the liquid temperature does not change appreciably during distillation of a particular component, those graph-line portions should appear flat (horizontal).

> Features of a correctly drawn graph are given on pages 80–81.

Questions

1. a. Using your graph, identify the temperatures at which Distillate 1 and Distillate 2 were collected.
   b. How well do the horizontal plateaus in your graph match the temperatures at which you collected the first drops of each distillate?

2. Use data in Table 3.1 to identify each distillate sample.

3. Combine your data with the data of students who distilled the same mixture:

   a. Examine the combined data, and find the average temperature (*mean*) and the most frequently observed temperature (*mode*) for each of the two plateaus you observed.
   b. All laboratory teams investigating the same mixture did not observe the same distillation temperatures. Describe some factors that may contribute to this inconsistency.

> As you learned in Unit 1 (pages 13–14), the *mean* and *median* are useful values for describing "middle" characteristics for a set of related data.

4. In which of the distillates was solid iodine more soluble? What observational evidence leads you to that conclusion?

5. What laboratory tests could you complete to decide whether the liquid left behind in the flask is a mixture or a pure substance?

6. Of the substances listed in Table 3.1 (page 216), which two would be most difficult to separate by distillation? Why?

7. a. How would a graph of time vs. temperature look for the distillation of a mixture of all *four* substances listed in Table 3.1 (page 216)?

   b. Sketch the predicted graph and describe its features.

A.3 PETROLEUM REFINING

> Although the names given to various fractions and their boiling ranges may vary somewhat, crude oil refining always has the same general features.

Unlike the simple laboratory mixture you separated in the preceding investigation, crude oil is a mixture of many compounds. Separating such a complex mixture requires applying distillation techniques in large-scale oil refining. The refining process does not separate each compound in crude oil. Rather, it produces several distinctive mixtures, called **fractions.** This process is known as **fractional distillation.** See Figure 3.8. Compounds in each fraction have a particular range of boiling points and specific uses.

Figure 3.9 illustrates fractional distillation (fractionation) of crude oil. First, the crude oil is heated to about 400 °C in a furnace and then pumped into the base of a distilling column (fractionating tower), which is usually more than 30 m (100 ft) tall. Many components of the heated crude oil vaporize. The temperature of the distilling column is highest at the bottom, and it decreases toward the top. Trays arranged at appropriate heights inside the column collect the various condensed fractions.

During distillation, the vaporized molecules move upward in the distilling column. The smaller, lighter molecules have the lowest boiling points and either condense high in the column or are drawn off the top of the tower as gases. Fractions with larger molecules have higher boiling points and are more difficult to separate from one another; therefore, they require more thermal energy (heat energy) to vaporize. These larger molecules condense back to liquid in trays lower in the column. Substances with the highest boiling points never do vaporize. These thick (*viscous*) liquids—called *bottoms*—drain from the column's base. Each arrow in Figure 3.9 indicates the name of a particular fraction and its boiling-point range.

Figure 3.8 *These fractionating towers contain many different levels of condensers to cool the oil vapor as it rises. Temperatures range from about 400 °C (at the base) to 40 °C (at the top).*

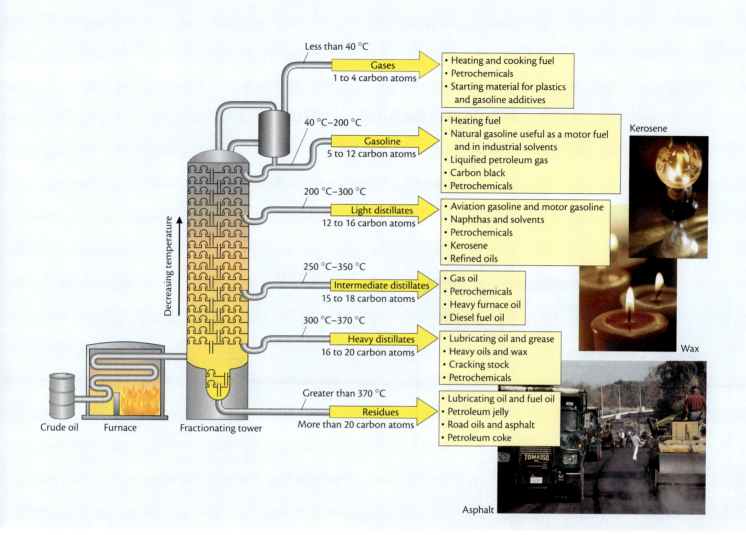

Figure 3.9 Fractional distillation (fractionation) of crude oil.

As you learn more about characteristics of the fractions obtained from petroleum, think about how people use petroleum-based products in both traditional and hybrid vehicles.

A.4 EXAMINING PETROLEUM'S MOLECULES

Petroleum's gaseous fraction contains compounds with low boiling points (less than 40 °C). These small hydrocarbon molecules, which contain from one to four carbon atoms, have low boiling points because they are only slightly attracted to each other or to other molecules in petroleum. Forces of attraction between molecules are called **intermolecular forces.** As a result of their weak intermolecular forces, these small hydrocarbon molecules readily separate from each other and rise through the distillation column as gases.

Petroleum's liquid fractions—including gasoline, kerosene, and heavier oils—consist of molecules that have from five to about twenty carbon atoms. Molecules with even more carbon atoms are found in the greasy

Just as "interstate highway" refers to a road that runs *between* states, intermolecular forces act *between* molecules.

fraction—the bottoms—that does not vaporize. These viscous compounds have the strongest intermolecular forces among all substances found in petroleum. It is not surprising that they are solids at room temperature.

Now complete the following activity to learn more about a key physical property of hydrocarbons—their boiling points.

Developing Skills

A.5 HYDROCARBON BOILING POINTS

Chemists often gather and analyze data about the physical and chemical properties of substances. These data can be organized in many ways, but the most useful techniques are those that uncover trends or patterns among the data. The development of the periodic table is an example of this approach.

You can examine patterns among boiling points of some hydrocarbons to make useful predictions.

| Table 3.2

BOILING POINTS OF SELECTED HYDROCARBONS		
Hydrocarbon	Boiling Point (°C)	Molecular Formula
Butane	−0.5	C_4H_{10}
Decane	174.0	$C_{10}H_{22}$
Ethane	−88.6	C_2H_6
Heptane	98.4	C_7H_{16}
Hexane	68.7	C_6H_{14}
Methane	−161.7	CH_4
Nonane	150.8	C_9H_{20}
Octane	125.7	C_8H_{18}
Pentane	36.1	C_5H_{12}
Propane	−42.1	C_3H_8

Use the data found in Table 3.2 to answer the following questions:

1. a. In what pattern or order are the data in Table 3.2 organized?
    b. Is this a useful way to present the information? Explain.

2. Suppose that you were searching for a trend or a pattern among these boiling points.

    a. Propose a more useful way to arrange these data that might suggest how the molecules of each hydrocarbon interact.
    b. Reorganize the data table based on your idea.

Now use your reorganized data table to answer the following questions:

3. Which substance or substances are gases at room temperature (22 °C)?

4. If a substance is a gas at 22 °C, what can you infer about the boiling point of that substance?

5. Which substance or substances boil between 22 °C (room temperature) and 37 °C (body temperature)?

6. What can you infer about intermolecular forces among decane molecules compared to those forces among butane molecules?

"Infer" means to reach a conclusion based on evidence and reasoning.

A.6 CHEMICAL BONDING

Hydrocarbons and their derivatives are the focus of a branch of chemistry known as **organic chemistry.** These substances are called *organic compounds* because early chemists thought that living systems—plants or animals—were needed to produce them. However, chemists have known for more than 150 years how to make most organic compounds without any assistance from living systems. In fact, reactants other than petroleum can be used to produce organic compounds. You will learn about some of these important reactants in Section C.

In hydrocarbon molecules, carbon atoms are joined to form a backbone called a **carbon chain.** Hydrogen atoms are attached to the carbon backbone. Carbon's versatility in forming bonds helps explain the abundance of different hydrocarbon compounds, as you will soon learn. Hydrocarbons can be regarded as the starting point of an even larger number of compounds that contain atoms of other elements attached to a carbon chain.

> The carbon chain forms a framework to which a wide variety of other atoms can be attached.

Electron Shells

How are atoms of carbon or other elements held together in compounds? The answer is closely related to the arrangement of electrons in atoms. You know that atoms are made up of neutrons, protons, and electrons. In addition, you know that neutrons and protons are located in the small, dense, central region of the atom, the *nucleus.* Electrons reside in separate energy levels in space surrounding the nucleus. Similar energy levels are grouped into **shells,** each of which can hold only a certain maximum number of electrons. For example, the first shell surrounding the nucleus of an atom has a capacity of two electrons. The second shell can hold a maximum of eight electrons.

Consider an atom of helium (He), the first member of the noble gas family. See Figure 3.10 (left). A helium atom has two protons (and two neutrons) in its nucleus, and two electrons occupying the first, or innermost, shell. Because two is the maximum this shell can hold, the shell is completely filled.

The next noble gas, neon (Ne), has atomic number 10. See Figure 3.10 (right). Each electrically neutral neon atom contains 10 protons and 10 electrons. Two electrons occupy and completely fill the first shell. The remaining eight electrons fill the second shell. In neon, each filled shell has reached its electron capacity.

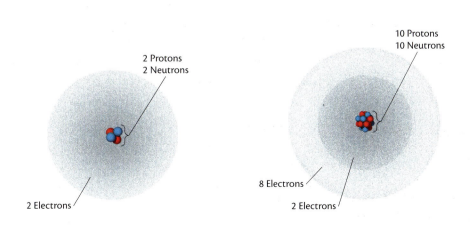

2 Protons
2 Neutrons

2 Electrons

10 Protons
10 Neutrons

8 Electrons

2 Electrons

Figure 3.10 *Atomic structure of helium (left). Note that the nucleus is composed of two protons and two neutrons. Two electrons are located within—and fill—the first electron shell, depicted here by an electron cloud.*
Atomic structure of neon (right). Note that the nucleus is composed of ten protons and ten neutrons. Two electrons are located—and fill—the first electron shell, and eight electrons are located within—and fill—the second electron shell.

Figure 3.11 *One product of the reaction between sodium (Na) and water is hydrogen gas. So much heat is produced in this violent reaction that the hydrogen gas often undergoes combustion accompanied by flames.*

Helium (He, atomic number 2) and neon (Ne, atomic number 10) are not chemically reactive; their atoms do not combine with each other or with atoms of other elements to form compounds. By contrast, sodium (Na) atoms—with atomic number 11 and one more electron than neon atoms—are extremely reactive (Figure 3.11). Chemists explain sodium's reactivity as due to its tendency to lose that one electron from its third, unfilled shell. Doing so leaves 10 electrons, giving the atom the same electron occupancy as a neon atom. Fluorine (F) atoms each have nine electrons—one fewer than neon atoms—and are also extremely reactive. Their reactivity is due to their tendency to gain an additional electron, thus achieving a filled outer shell of electrons.

Noble-gas elements are essentially unreactive because their separate atoms have filled electron shells. All but helium have eight electrons in their outer shell; helium needs only two electrons to attain its first-shell maximum. A useful key to understanding the chemical behavior of many elements is to recognize that atoms with *filled electron shells* are particularly stable; that is, they tend to be chemically unreactive.

Covalent Bonds

Chemical Bonding

In most substances composed of nonmetals, atoms achieve filled electron shells by *sharing* electrons. Typically, only **valence electrons,** which are the electrons within an atom's unfilled, outer shell, participate in bonding. As you will see, the sharing of two or more valence electrons between two atoms—called a **covalent bond**—allows both atoms to fill their outer shells completely.

A hydrogen molecule (H_2) provides a simple example of electron sharing. Each hydrogen atom contains only one electron, so one more electron is needed to fill the first shell. Two hydrogen atoms can accomplish this if they each share their single electron. If an electron is represented by a dot (·), then the formation of a hydrogen molecule can be depicted as shown below.

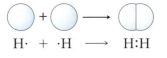

$$H\cdot \ + \ \cdot H \ \longrightarrow \ H\!:\!H$$

The chemical bond formed between two atoms that share a pair of electrons is called a **single covalent bond** (see Figure 3.12). Through such sharing, both atoms achieve the stability associated with completely filled electron shells. A carbon atom, atomic number 6, has six electrons—two in its first shell and four more in the second shell. To fill the second shell to its capacity of eight, four more electrons are needed. These electrons can be obtained through covalent bonding.

Figure 3.12 *Model of a hydrogen (H₂) molecule.* *A hydrogen molecule is held together by a single covalent bond. Each hydrogen atom in the molecule shares two electrons, thereby filling each atom's outer electron shell. Electrical attractions between shared electrons and each atom's positively charged nucleus (depicted by the high cloud density between the two nuclei) hold the H₂ molecule together.*

Consider the simplest hydrocarbon molecule, methane (CH_4). In this molecule, each hydrogen atom shares its single electron with a valence electron of the carbon atom. A representation of this arrangement is:

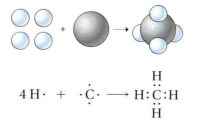

$$4\,H\cdot \;+\; \cdot\overset{\displaystyle\cdot}{\underset{\displaystyle\cdot}{C}}\cdot \;\longrightarrow\; H\!:\!\overset{\displaystyle H}{\underset{\displaystyle H}{C}}\!:\!H$$

As in the formula for a hydrogen molecule, dots surrounding each element's symbol represent the valence electrons for that atom. Structures such as those on page 223 are **electron-dot formulas,** also known as **Lewis dot structures** or, simply, **Lewis structures.** The two electrons in each covalent bond "belong" to both bonded atoms. Dots placed between the symbols of two atoms represent electrons that are shared by those atoms.

When determining the electrons associated with each atom, each shared electron in a covalent bond is "counted" twice, once for each element. For example, count the dots surrounding each atom in methane (see below). Note that each hydrogen atom has a filled outer electron shell with two electrons. The carbon atom also has a filled outer electron shell with eight electrons. Each hydrogen atom is associated with one pair of electrons; the carbon atom has four pairs of electrons, or eight electrons.

For convenience, each pair of electrons in a covalent bond can be represented by a line drawn between the symbols of each atom. This yields another common representation of a covalently bonded molecule, called a **structural formula.**

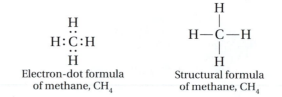

Electron-dot formula
of methane, CH_4

Structural formula
of methane, CH_4

Although you can draw flat, two-dimensional pictures of molecules on paper, assembling three-dimensional models gives a far more accurate representation. Such models help predict a molecule's physical and chemical behavior. The following investigation provides an opportunity for you to assemble and use molecular models.

Investigating Matter

Modeling Alkanes

A.7 MODELING ALKANES

Introduction

In this investigation, you will assemble models of several simple hydrocarbon molecules. See Figure 3.13. Your goal is to associate the three-dimensional shapes of these molecules with the names, formulas, and pictures used to represent them on paper.

There are several useful ways to construct models of molecules; each model has its own advantages and disadvantages (Figure 3.14). Two common types of molecular models are shown in Figure 3.15. Most likely, you will use *ball-and-stick models.* Each ball represents an atom, and each stick represents a single covalent bond (one shared electron pair) connecting two atoms.

Of course, molecules are not composed of ball-like atoms located at the ends of stick-like bonds. Experimental evidence shows that atoms are in contact with each other, much like that depicted in *space-filling*

Figure 3.13 *Due to the extremely small size of atoms and molecules, people often use model kits to visualize the three-dimensional geometry of hydrocarbons and other molecular structures.*

models. However, ball-and-stick models are still useful because they can clearly represent the bonding and geometry of molecules.

Look again at the Lewis dot structure and the structural formula for methane (CH_4) on page 224. Methane, the simplest hydrocarbon, is the first member of a series of hydrocarbons known as **alkanes.** Each carbon atom in an alkane forms *single* covalent bonds with four other atoms. You will explore alkanes in this investigation.

Before you begin, carefully read the procedure to learn what is involved and to plan for your data collecting and observations.

Procedure

1. Assemble a three-dimensional model of a methane (CH_4) molecule. Compare your model to the Lewis dot formula and the structural formula on page 224. Note that the angles defined by bonds between atoms are not 90°, as you might have expected by looking at the structural formula. If you were to build a close-fitting box to surround a CH_4 molecule, the box would be shaped like a pyramid with a triangle as its base. This three-dimensional shape is called a **tetrahedron.**

Figure 3.14 *Scientists use many types of physical and mental models to better understand concepts related to atoms and molecules.*

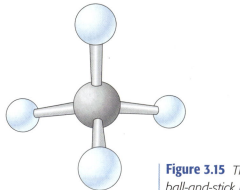

Figure 3.15 *Three-dimensional CH_4 models: ball-and-stick (left); and space-filling (right).*

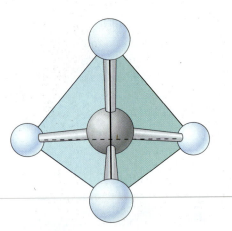

Figure 3.16 *The tetrahedral shape of a methane molecule.*

The tetrahedral geometry that you observe results from the fact that similar electrical charges repel. The four pairs of electrons in the bonds surrounding the carbon atom repel each other and arrange themselves to be as far away from one another as possible. In this spatial arrangement, they point to the corners of a tetrahedron, as shown in Figure 3.16. The angle formed between the C—H bonds is 109.5°, a value that has been verified experimentally.

2. Convert your three-dimensional model into a two-dimensional drawing, similar to the one shown in Figure 3.16, that conveys the tetrahedral structure of methane. Shade the carbon atom to distinguish it from the hydrogen atoms.

3. Assemble models of a two-carbon alkane molecule and a three-carbon alkane molecule. Recall that each carbon atom in an alkane is bonded to four other atoms.

   a. How many hydrogen atoms are in a two-carbon alkane molecule?

   b. How many hydrogen atoms are in a three-carbon alkane molecule?

   c. Draw a ball-and-stick model, similar to the one shown in Figure 3.15 (page 225), of the three-carbon alkane. Shade the carbon atoms to distinguish them from the hydrogen atoms.

   d. Draw Lewis dot structures and structural formulas for the two- and three-carbon alkanes.

| Table 3.3

			SOME MEMBERS OF THE ALKANE SERIES	
			Alkane Formulas	
Name	**Total Carbon Atoms**	**Boiling Point (°C)**	**Molecular Formula**	**Condensed Formula**
Methane	1	−161.7	CH_4	CH_4
Ethane	2	−88.6	C_2H_6	CH_3CH_3
Propane	3	−42.1	C_3H_8	$CH_3CH_2CH_3$
Butane	4	−0.5	C_4H_{10}	$CH_3CH_2CH_2CH_3$
Pentane	5	36.1	C_5H_{12}	$CH_3CH_2CH_2CH_2CH_3$
Hexane	6	68.7	C_6H_{14}	$CH_3CH_2CH_2CH_2CH_2CH_3$
Heptane	7	98.4	C_7H_{16}	$CH_3CH_2CH_2CH_2CH_2CH_2CH_3$
Octane	8	125.7	C_8H_{18}	$CH_3CH_2CH_2CH_2CH_2CH_2CH_2CH_3$
Nonane	9	150.8	C_9H_{20}	$CH_3CH_2CH_2CH_2CH_2CH_2CH_2CH_2CH_3$
Decane	10	174.0	$C_{10}H_{22}$	$CH_3CH_2CH_2CH_2CH_2CH_2CH_2CH_2CH_2CH_3$

4. **Molecular formulas** specify the number of each atom type within a molecule. The molecular formulas of the first two alkanes are CH_4 and C_2H_6. Identify the molecular formula for the third alkane.

Examine your three-carbon alkane model and the structural formula you drew for it. Note that the middle carbon atom is attached to *two* hydrogen atoms, but the carbon atom at each end is attached to *three* hydrogen atoms. This molecule can be represented as $CH_3-CH_2-CH_3$, or $CH_3CH_2CH_3$. Formulas such as these provide convenient information about how atoms are arranged in molecules. For many purposes, such **condensed formulas** are more useful than molecular formulas, such as C_3H_8.

Consider the molecular formulas of the first few alkanes: CH_4, C_2H_6, and C_3H_8. Given the pattern represented by that series, try to predict the formula of the four-carbon alkane. The general molecular formula of all open-chain alkane molecules can be written as C_nH_{2n+2}, where n is the number of carbon atoms in the molecule. Therefore, even without assembling a model, you can predict the formula of a five-carbon alkane: If $n = 5$, then $2n + 2 = 12$, and the resulting formula thus is C_5H_{12}.

The names of the first ten alkanes are given in Table 3.3 and some examples of uses of these compounds are shown in Figure 3.17 and Figure 3.18. As you can see, each name is composed of a prefix, followed by -*ane* (designating an alk*ane*). The prefix indicates the number of carbon atoms in the backbone carbon chain. *Meth-* means one carbon atom, *eth-* means two, *prop-* means three, and *but-* means four. For alkanes with five to ten carbon atoms, the prefixes are derived from Greek—*pent-* for five, *hex-* for six, and so on.

5. Disassemble your molecular models and replace all parts in their container.

Figure 3.17 *Propane often serves as an emergency energy source during power outages. Here, workers weigh filled tanks to determine how much propane they contain.*

n can be any positive integer.

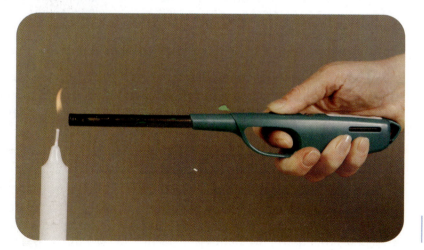

Figure 3.18 *Butane is a liquid commonly used as fuel in lighters.*

Questions

6. Write structural formulas for butane and pentane.

7. a. Name the alkanes with the following condensed formulas:
   i. $CH_3CH_2CH_2CH_2CH_2CH_2CH_3$
   ii. $CH_3CH_2CH_2CH_2CH_2CH_2CH_2CH_2CH_3$
   b. Write molecular formulas for the two alkanes in Question 7a.

8. a. Write the formula of an alkane containing 25 carbon atoms.
   b. Did you decide to write the molecular formula or the condensed formula for this compound? Why?

9. Find the molar mass of pentane. (*Hint:* Remember to take subscripts into account, as you first learned to do in Unit 2, page 163.)

10. Name the alkane molecule that has a molar mass of
   a. 30 g/mol.   b. 58 g/mol.   c. 114 g/mol.

Developing Skills

A.8 TRENDS IN ALKANE BOILING POINTS

In Investigating Matter A.2, you used the technique known as *distillation* that separates liquid mixtures according to the boiling points of the components. You also know that the fractions of petroleum are separated based on their boiling points. Why do fractions with the largest molecules have the highest boiling points? Why are the smallest molecules found in fractions with the lowest boiling points? In this activity, you will explore this trend in alkane boiling points.

Using data for the alkanes found in Table 3.3 (page 226), prepare a graph of boiling points. The *x*-axis scale should range from 1 to 13 carbon atoms (even though you will initially plot data for 1 to 10 carbon atoms). The *y*-axis scale should extend from $-200\ °C$ to $+250\ °C$.

1. Label the axes appropriately and plot the data. Draw a best-fit line through your data points according to the following guidelines:

   ▶ The line should follow the trend of your data points.
   ▶ The data points should be roughly equally distributed above and below the line.
   ▶ The line should not extend past your data points.

   Figure 3.19 shows an example of a best-fit line.

2. Estimate the average change in boiling point (in °C) when one carbon atom and two hydrogen atoms ($-CH_2-$) are added to a particular alkane chain.

   The pattern of boiling points among the first ten alkanes allows you to predict the boiling points for other alkanes.

   a. Using your graph, estimate the boiling points of undecane ($C_{11}H_{24}$), dodecane ($C_{12}H_{26}$), and tridecane ($C_{13}H_{28}$). To do this, follow the trend of your graph line by extending a dashed line from the graph line you drew for the first ten alkanes. This procedure is called *extrapolation*. Then read your predicted boiling points for C_{11}, C_{12}, and C_{13} alkanes on the *y*-axis.

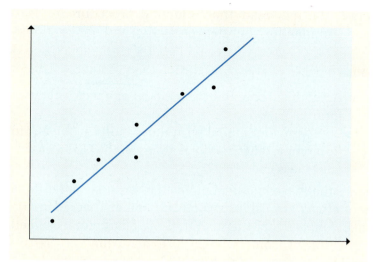

   b. Compare your predicted boiling points to actual values provided by your teacher.

3. You learned that a substance's boiling point depends in part on its *intermolecular forces,* which are the attractions among its molecules. For the alkanes you have studied, what is the relationship between the strength of these attractions and the number of carbon atoms in each molecule?

A.9 ALKANES REVISITED

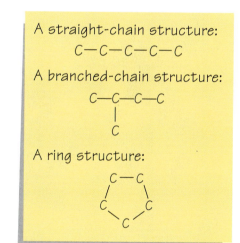

Investigating Matter

Introduction

The alkane molecules you have considered so far are **straight-chain alkanes,** where each carbon atom is only linked to one or two other carbon atoms.

   In alkanes with four or more carbon atoms, other arrangements of carbon atoms are possible. In **branched-chain alkanes,** one carbon atom can be linked to three or four other carbon atoms. An alkane composed of four or more carbon atoms can have either a straight-chain structure or a branched-chain structure. In this investigation, you will use ball-and-stick molecular models (see Figure 3.20) to explore such variations in alkane structures—variations that can lead to different properties.

A straight-chain structure:

$$C—C—C—C—C$$

A branched-chain structure:

$$C—C—C—C$$
$$|$$
$$C$$

A ring structure:

Figure 3.20 Students working together to construct molecular models.

Before starting, read the following procedure to learn what is involved and plan for your data collecting and observations.

Procedure

1. a. Assemble a ball-and-stick model of a molecule with the formula C_4H_{10}.
   b. Compare your model with those built by others. How many different arrangements of atoms in the C_4H_{10} molecule did your class construct?

Molecules that have identical molecular formulas but different arrangements of atoms are called **structural isomers.** By comparing models, convince yourself that there are only two structural isomers of C_4H_{10}. The formation of isomers helps to explain the very large number of compounds composed of carbon in chains or rings.

2. a. Draw a Lewis dot structure for each C_4H_{10} isomer.
   b. Write a structural formula for each C_4H_{10} isomer.

3. As you might expect, alkanes containing larger numbers of carbon atoms also have larger numbers of structural isomers. In fact, the number of different isomers increases rapidly as the number of carbon atoms increases. For example, chemists have identified three pentane (C_5H_{12}) isomers, as shown in Table 3.4. Try building these structural isomers.

4. Now consider possible structural isomers of C_6H_{14}, hexane.

   a. Working with a partner, draw structural formulas for as many different C_6H_{14} isomers as possible. Compare your structures with those drawn by other groups.
   b. How many different C_6H_{14} isomers were found by your class?

5. Build models of one or more C_6H_{14} isomers, as assigned by your teacher.

   a. Compare the three-dimensional models built by your class with corresponding structures drawn on paper.
   b. Based on your careful examination of the three-dimensional models, how many different C_6H_{14} isomers are possible?

PENTANE ISOMERS AND THEIR BOILING POINTS

Structural Formula	Boiling Point (°C)
$CH_3 - CH_2 - CH_2 - CH_2 - CH_3$	36.1
$CH_3 - CH - CH_2 - CH_3$ $\quad\quad\; \vert$ $\quad\quad CH_3$	27.8
$\quad\quad\; CH_3$ $\quad\quad\;\; \vert$ $CH_3 - C - CH_3$ $\quad\quad\;\; \vert$ $\quad\quad\; CH_3$	9.5

| Table 3.4

A.10 BOILING POINTS OF ALKANE ISOMERS

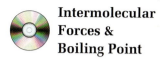

Intermolecular Forces & Boiling Point

You have already observed that boiling points of straight-chain alkanes are related to the number of carbon atoms in their molecules. Increased intermolecular forces are associated with the greater molecule-to-molecule contact possible for larger alkanes.

1. Boiling points for several isomers of pentane and octane are listed in Table 3.4 and Table 3.5. For each set of isomers, how does the boiling point change as the extent of carbon-chain branching increases? Based on your conclusions, assign each of these boiling points to one of the following C_7H_{16} isomers: 98.4 °C, 92.0 °C, 79.2 °C.

   a. CH_3—CH_2—CH_2—CH_2—CH_2—CH_2—CH_3
   b. CH_3—CH_2—CH—CH_2—CH_2—CH_3
   　　　　　　　　|
   　　　　　　　CH_3

   c. CH_3—CH_2—CH_2—$\overset{\overset{\displaystyle CH_3}{|}}{\underset{\underset{\displaystyle CH_3}{|}}{C}}$—$CH_3$

2. Here is the structural formula of a C_8H_{18} isomer:

   CH_3—CH_2—CH_2—$\overset{\overset{\displaystyle CH_3}{|}}{\underset{\underset{\displaystyle CH_3}{|}}{C}}$—$CH_2$—$CH_3$

   a. Compare this isomer to each C_8H_{18} isomer listed in Table 3.5. Predict whether this isomer has a higher or lower boiling point than each of the other listed C_8H_{18} isomers.
   b. Would the C_8H_{18} isomer shown here have a higher or lower boiling point than each of the three C_5H_{12} isomers depicted in Table 3.4? Why?

3. How do you explain the boiling point trends that you investigated in this activity?

SOME OCTANE ISOMERS AND THEIR BOILING POINTS	
Structural Formula	**Boiling Point (°C)**
CH_3—CH_2—CH_2—CH_2—CH_2—CH_2—CH_2—CH_3	125.7
CH_3—CH_2—CH_2—CH_2—CH_2—CH—CH_3 with CH_3 below	117.7
CH_3—CH—CH_2—C—CH_3 with CH_3 groups	99.2

| Table 3.5

A.11 FUELS AND CLIMATE

Different parts of the nation experience very different climates. Automobile fuel used in Maine during its cold winter months differs from fuel used in Arizona during hot summer months (Figure 3.21).

1. Why do differing climates require different automobile fuels? What practical reasons are there (in terms of transport, storage, and transfer of fuel) for tailoring fuel to a specific climate?

2. Suggest a physical property that petroleum engineers should consider when supplying gasoline for two such disparate climates.

3. Based on your answer to Question 2, what aspects of the components of fuel used in a Maine winter would differ from the components of fuel used during a summer in Arizona? (*Hint:* Think in terms of molecular structures.)

Chemists and chemical engineers use information about molecular structures and boiling points to separate the complex mixture known as petroleum into a variety of useful substances, many of which you are quite familiar with. In Section B, you will learn how bonding helps explain the use of petroleum as a fuel.

Figure 3.21 *How would the properties of the gasoline being pumped in these two situations vary?*

SECTION A SUMMARY
Reviewing the Concepts

Petroleum (crude oil), a nonrenewable resource that must be refined prior to use, consists of a complex mixture of hydrocarbon molecules.

1. What is a hydrocarbon?

2. What does it mean to *refine* a natural resource?

3. What characteristics of petroleum make it a valuable resource?

4. What is the likelihood of discovering a pure form of petroleum that can be used directly as it is pumped from the ground? Explain your answer.

5. What is meant by saying that oil is *crude*?

Petroleum is a source of fuels that provide thermal energy. It is also a source of raw materials for the manufacture of many familiar and useful products.

6. On average, the United States uses about 20 million barrels of petroleum daily:

   a. What is the average number of barrels of petroleum used daily in the United States for building (nonfuel) purposes?

   b. How many barrels of petroleum, on average, are burned as fuel daily in the United States?

7. Name several fuels obtained from crude petroleum.

8. a. List four household items made from petroleum.

   b. What materials could be substituted for each of these four household items if petroleum were not available to make them?

9. List several products that might not be widely and easily available if petroleum supplies were to dwindle.

The distribution of crude oil reserves does not necessarily correspond to areas of high petroleum use.

The following two questions refer to Figure 3.4, page 214.

10. a. Which world region has the most petroleum reserves relative to its population?

    b. Which region has the least petroleum reserves relative to its population?

11. a. Which regions consume a greater proportion of the world's supply of petroleum than they possess?

    b. Which regions consume a smaller proportion of the world's supply of petroleum than they possess?

Liquid substances can often be separated according to their differing boiling points in a process called distillation.

12. Under what conditions could density be used to separate two different liquids?

13. Referring to Table 3.1 (page 216), a mixture of which two of the substances listed would be the easiest to separate from each other by distillation? Explain your reasoning.

14. Sketch the basic setup for a laboratory distillation. Label the key features of your sketch.

15. Referring to Table 3.1 (page 216), sketch a graph of the distillation of a mixture of acetone and water. Label its key features.

Fractional distillation of crude oil produces several distinctive and usable mixtures (fractions). Each fraction contains molecules of similar sizes, boiling points, and intermolecular forces.

16. How does fractional distillation differ from simple distillation?

17. Petroleum fractions include light, intermediate, and heavy distillates and residues. List three useful products derived from each of these three fractions.

18. Where in a distillation tower—top, middle, or bottom—would you expect the fraction with the *highest* boiling point range to be removed? Why?

19. After fractional distillation, each fraction is still a mixture. Suggest a way to further separate the components of each fraction.

20. Rank the following straight-chain hydrocarbons from their lowest boiling point to their highest: hexane (C_6H_{14}), methane (CH_4), pentane (C_5H_{12}), and octane (C_8H_{18}). Explain your rankings in terms of intermolecular forces.

The atoms in hydrocarbons and in other molecules are held together by covalent bonds.

21. What is a covalent bond?

22. Why do atoms with filled outer electron shells not form covalent bonds?

23. It has been suggested that a covalent bond linking two atoms is like two dogs tugging on the same sock. Explain how this analogy describes the way that shared electrons hold together atoms in a covalent bond.

Molecules can be represented by Lewis dot structures, structural formulas, or molecular formulas.

24. What does each dot in a Lewis dot structure represent?

25. What is the advantage of using a dash instead a pair of electron dots to represent a covalent bond?

26. a. What information does a structural formula convey that a molecular formula does not?

b. In what ways is a structural formula an inadequate representation of an actual molecule?

27. Choose a branched six-carbon hydrocarbon molecule

a. Draw a Lewis dot structure to represent its structure.

b. Draw a structural formula for the same molecule.

28. Each carbon atom has six total electrons. Why, then, does the electron-dot representation of a carbon atom show only *four* dots?

29. Use the general molecular formula to write the molecular formula for an alkane containing

    a. 9 carbons.   c. 10 carbons.
    b. 16 carbons.   d. 18 carbons.

30. Calculate the molar mass of each alkane listed in Question 29.

31. Name and give the molecular formula for the alkane with a molar mass of

    a. 44 g/mol.
    b. 72 g/mol.

32. What does -*ane* imply about the carbon–carbon bonding in hexane?

33. Are the following three molecules isomers of one another? Explain your answer.

$$CH_3 \quad\quad CH_3 \quad\quad CH_3$$
$$|\quad\quad\quad\quad |\quad\quad\quad\quad |$$
$$CH_2{-}CH_2{-}CH_2 \quad\quad CH_2$$
$$|$$
$$CH_2{-}CH_2{-}CH_3$$

$$CH_3{-}CH_2{-}CH_2{-}CH_2{-}CH_3$$

34. Draw structural formulas for at least three structural isomers of C_9H_{20}.

35. What is the shortest-chain alkane that can demonstrate isomerism?

36. An unbranched hydrocarbon molecule can be represented as a linear chain or as a zig-zag chain. Explain in what way both representations are correct.

37. a. Draw two hexane structural isomers, one a straight-chain molecule and the other a branched-chain molecule.
    b. Which of the two isomers you drew would have the lower boiling point? Explain your choice.

38. Which of each pair of the following hydrocarbon molecules would have the lower boiling point? In each case, describe your reasoning.

    a. a short, straight chain or a long, straight chain
    b. a short, branched chain or a long, branched chain
    c. a short, branched chain or a long, straight chain

Connecting the Concepts

39. Why is petroleum considered a nonrenewable resource?

40. Using a Venn diagram, distinguish fractional distillation from simple distillation.

41. Simple distillation is never sufficient to separate two liquids completely. Explain.

42. In a fractionating tower, petroleum is generally heated to 400 °C. What would happen if it were heated only to 300 °C?

43. The molar masses of methane (16 g/mol) and water (18 g/mol) are similar. At room temperature, methane is a gas and water is a liquid. Explain this difference in terms of their intermolecular forces.

44. The traditional unit of volume for petroleum is the *barrel*, which contains 42 gallons. Assume that those 42 gallons of petroleum provide 21 gallons of gasoline. How many barrels of petroleum does it take to operate an automobile for one year, assuming the vehicle travels 10 000 miles and gets 27 miles per gallon of gasoline?

45. Which mixture would be easier to separate by distillation—a mixture of pentane and straight-chain octane or a mixture of pentane and a branched-chain octane isomer? Explain the reasoning behind your choice.

46. Explain why thermal energy is added at one point and removed at another point in the process of distillation.

Extending the Concepts

47. Is it likely that the composition of crude oil found in Texas is the same as that of crude oil found in Kuwait? Explain your answer.

48. Gasoline's composition is varied by oil companies for use in different parts of the nation and for use in different seasons. What factors help determine the composition of gasoline blended for different seasons?

49. What kind of petroleum trade relationship would be expected between North America and the Middle East? If other world regions become more industrialized and if global petroleum supplies decrease, how might the North America–Middle East trade relationship change?

50. How would the hydrocarbon boiling points listed in Table 3.2 (page 220) change if they were measured under increased atmospheric pressure? (*Hint:* Although butane is stored as a liquid inside of a butane lighter, it escapes through the lighter's nozzle as a gas.)

51. The two isomers of butane have different physical properties, as illustrated by their different boiling points. They also have different chemical properties. Explain how isomerism may contribute to their differences in chemical behavior.

52. What properties of petroleum make it an effective lubricant?

53. One problem with diesel fuel is that it solidifies at cold temperatures. How can its formulation be altered to prevent this problem?

Petroleum: An Energy Source

For almost 5000 years, humans have used petroleum in small amounts. Only in the last 150 years, roughly since the first U.S. oil well was drilled in 1859, has society become much more dependent on this nonrenewable resource. See Figure 3.22. In the following activity, you will begin to understand just how much everyday life is influenced by petroleum's role as an energy source—not just in powering automobiles and other vehicles, but in energizing modern society itself.

Figure 3.22 *In 1859, Edwin Drake struck oil at the world's first successful oil well, in Titusville, PA. Oil was already refined and used for many purposes at that time. Drake developed an efficient way to retrieve petroleum, leading to an oil boom in that region.*

ChemQuandary 1

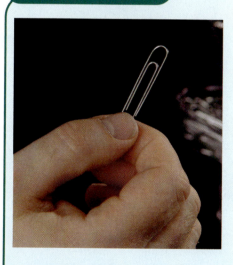

ENERGY TRACING

Examine a paper clip. It is composed of a metallic alloy that could have a "materials history" written about it— a story revealing the origins of the paper clip's material, similar to those materials you studied in Unit 2.

What is less obvious, perhaps, is that this same paper clip can also be analyzed in terms of its energy history. Complete an "energy trace" for a paper clip by following these steps. (a) Decide what general events must have occurred to produce the paper clip and to deliver it to your school. (b) From where did the energy for each event come? In other words, what was the source of the needed energy? (c) In your view, is the total cost of a box of paper clips related more to the materials used (the metals in the alloy) or to the energy required? Explain your reasoning.

After completing an "energy trace" for a paper clip, consider how much more complicated and energy-consuming it is to produce the *Hybris* or any other automobile.

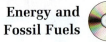

Energy and Fossil Fuels

B.1 ENERGY AND FOSSIL FUELS

Fossil fuels are believed to originate from biomolecules of prehistoric plants and animals. See Figure 3.23. The energy released by burning these fuels represents energy originally captured from sunlight by prehistoric green plants during photosynthesis. Thus, **fossil fuels**—petroleum, natural gas, and coal—can be thought of as forms of buried sunshine.

Most evidence indicates that fossil fuels originated from living matter in ancient seas some 500 million years ago. These species died and eventually became covered with sediments. Pressure, heat, and microbes converted what was once living matter into petroleum, which became trapped in porous rocks. It is likely that some petroleum is still being formed from sediments of dead matter. However, such a process is far too slow to consider petroleum a renewable resource.

Figure 3.23 *Fossil fuels are thought to have originated from living matter; this illustration depicts a forest in the Upper Carboniferous or Pennsylvanian division of the Paleozoic Era.*

Fossil-fuel energy is comparable in some ways to the energy stored in a wind-up toy race car. The "winding-up" energy that was originally supplied tightened a spring in the toy. Most of that energy, stored within the coiled spring, is a form of **potential energy,** which is energy of position (or condition). As the car moves, the spring unwinds, providing energy to the moving parts. Energy related to motion is called **kinetic energy.** Thus, the movement of the car is based on converting potential energy into kinetic energy. Eventually, the toy "winds down" to a lower-energy, more stable state and stops.

In a similar manner, **chemical energy,** which is another form of potential energy, is stored within bonds in chemical compounds. When an energy-releasing chemical reaction takes place (such as when fuel burns), bonds break and reactant atoms reorganize to form new bonds. The process yields products with different and more stable bonding arrangements of their atoms. That is, the products have less potential energy (chemical energy) than did the original reactants. Some of the energy stored in the reactants has been released in the form of heat and light.

The combustion, or burning, of methane (CH_4) gas illustrates such an energy-releasing reaction. It can be summarized this way:

Energy from Combustion

> Methane is a major component of natural gas.

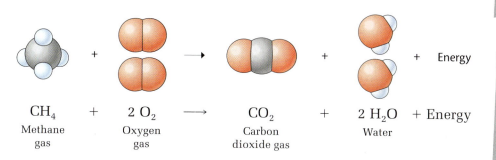

$$CH_4 + 2 O_2 \longrightarrow CO_2 + 2 H_2O + Energy$$

Methane gas   Oxygen gas   Carbon dioxide gas   Water

> This simplification allows us to more easily analyze the energy involved. The actual reaction is more complicated.

The reaction releases considerable **thermal energy** (heat). In fact, laboratory burner flames are based primarily on that reaction—conducted, of course, under very controlled conditions. To gain a better understanding of the energy involved, just imagine that the reaction takes place in two simple steps: One step involves bond breaking and one step involves bond making.

In the first step, suppose that all the chemical bonds in one CH_4 molecule and two O_2 molecules are broken. The result of this bond-breaking step is that separated atoms of carbon, hydrogen, and oxygen are produced. All such bond-breaking steps are energy-requiring processes, or **endothermic** changes. In an endothermic change, energy must be added to "pull apart" the atoms in each molecule. Thus, energy appears as a *reactant* in Step 1, as shown in the following chemical equation:

> An endothermic reaction can proceed only if energy is continuously supplied.

Step 1

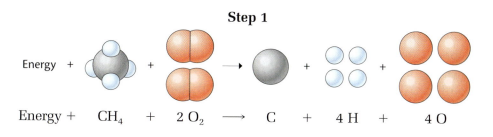

$$Energy + CH_4 + 2 O_2 \longrightarrow C + 4 H + 4 O$$

To complete the methane-burning reaction, suppose that the separated atoms now join to form the new bonds needed to make the product molecules: one CO_2 molecule and two H_2O molecules. The formation of chemical bonds is an energy-releasing process, or an **exothermic** change. Because energy is given off, it appears as a *product* in Step 2, as shown in the following chemical equation:

Step 2

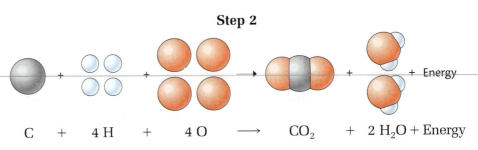

$$C \quad + \quad 4\,H \quad + \quad 4\,O \quad \longrightarrow \quad CO_2 \quad + \quad 2\,H_2O + Energy$$

When methane burns, the energy released in forming carbon–oxygen bonds in CO_2 and hydrogen–oxygen bonds in H_2O is greater than the energy used to break the carbon–hydrogen bonds in CH_4 and the oxygen–oxygen bond in O_2. That is why this overall chemical change is exothermic. The complete energy "accounting summary" for burning methane is shown in Figure 3.24.

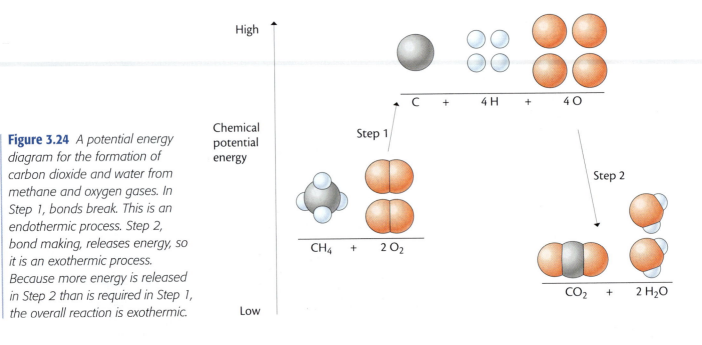

Figure 3.24 *A potential energy diagram for the formation of carbon dioxide and water from methane and oxygen gases. In Step 1, bonds break. This is an endothermic process. Step 2, bond making, releases energy, so it is an exothermic process. Because more energy is released in Step 2 than is required in Step 1, the overall reaction is exothermic.*

Whether an overall chemical reaction is exothermic or endothermic depends on how much energy is added (endothermic process) in bond breaking and how much energy is given off (exothermic process) in bond making. If more energy is given off than is added, the overall change is exothermic. However, if more energy is added than is given off, the overall change is endothermic.

Figure 3.25 Bicycling is an example of continuous conversions between potential and kinetic energy.

In general, if a process converts potential energy into kinetic energy, then the reverse process converts kinetic energy back to potential energy. For example if you wind the spring of a model race car, you are converting your energy of motion (winding) into energy stored in the spring. A biker or a skateboarder uses the transformation of potential energy into kinetic energy to complete jumps and other tricks (Figure 3.25).

Likewise, if a particular chemical reaction is exothermic (releasing thermal energy), then the reverse reaction is endothermic (converting thermal energy into potential energy). For example, burning hydrogen gas—involving the formation of water—is exothermic. The energy released by the formation of H—O bonds in water is greater than that required to break H_2 and O_2 bonds:

$$2\ H_2 + O_2 \longrightarrow 2\ H_2O + \text{Energy}$$

Therefore, the separation of water into its elements—the reverse reaction—must be endothermic, equal in energy used to that released when water is formed from gaseous H_2 and O_2. Electrical energy is expended and is converted into potential energy that is stored, again, in chemical bonds. See Figure 3.26.

$$\text{Energy} + 2\ H_2O \longrightarrow 2\ H_2 + O_2$$

> If a rock rolling downhill is an exothermic process, then pushing the same rock back uphill must be an endothermic process.

Figure 3.26 An electric current provides energy to decompose water, producing hydrogen (right test tube) and oxygen (left test tube) gas. This process is referred to as the "electrolysis" of water. As follows from the formula of water (H_2O), hydrogen and oxygen gas are produced in a 2 to 1 ratio.

B.2 ENERGY CONVERSION

Scientists and engineers have increased the usefulness of energy released from burning fuels through devices that convert thermal energy into other forms of energy. In fact, much of the energy you use daily goes through several conversions before it reaches you.

Consider the energy-conversion steps involved in the operation of a hair dryer, illustrated in Figure 3.27. What detailed, step-by-step "energy story" can you devise using these illustrations?

You probably noted that in the first step, stored chemical energy (potential energy) in a fossil fuel, Figure 3.27(a), is released in a power-plant furnace, producing thermal (heat) energy, as shown in Figure 3.27(b). The thermal energy then converts water in the boiler to steam that spins the turbines (mechanical energy), generating electrical energy as shown in Figure 3.27(c). Thus, the power plant converts thermal energy produced in the furnace to mechanical energy, which is a form of kinetic energy, in the turbines. It is then converted

(a)
(b)
(c)
(d)
(e)

Figure 3.27 *Many steps are involved in the processes by which the chemical energy stored in the bonds of hydrocarbon molecules is eventually converted to the thermal energy a hair dryer produces.*

to electrical energy, as shown in Figure 3.27(d). When electricity reaches the hair dryer, it is converted back to thermal energy to dry your hair, and it is also converted to mechanical energy, Figure 3.27(e), as a small fan blade spins to blow the hot air. Some sound energy is also produced, as any hair-dryer user knows!

Despite all the changes involved, no energy is consumed or "used up" in any of these steps. The form of energy just changes from *chemical* to *thermal* to *mechanical* to *electrical.* This concept is summarized by the **law of conservation of energy,** which states that energy is neither created nor destroyed in any mechanical, physical, or chemical processes.

B.3 AUTOMOBILE ENERGY CONVERSIONS

Developing Skills

In one sense, an automobile—whether powered by electricity, gasoline, or even solar energy—can be considered a collection of energy-converting and energy-powered devices. Imagine that you are an "energy detective," and follow some typical energy conversions in an automobile. Use Figure 3.28 to answer the following questions:

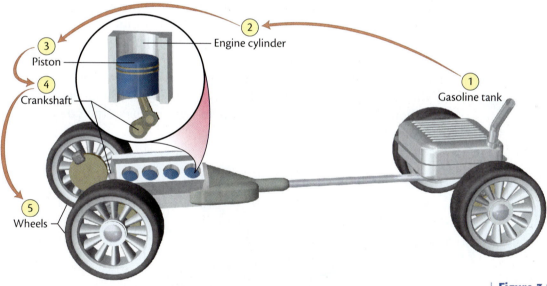

Figure 3.28 *What energy conversions are necessary to propel a car? Trace the energy conversions, starting with the gas tank and ending with movement of the wheels.*

1. Name the type of energy involved in each step required to cause the wheels to propel a conventional automobile forward.

2. Do you think that all the energy originally stored in gasoline can be traced to the rotating wheels? Explain.

B.4 ENERGY EFFICIENCY

Petroleum is neither limitless nor inexpensive. One way to maximize the benefits from available supplies of petroleum-based fuels is to reduce the total number of energy conversions the fuel undergoes. We can also seek ways to increase the efficiency of energy-conversion devices.

Unfortunately, devices that convert chemical energy to thermal energy and then to mechanical energy are typically less than 50% efficient. Solar cells, which convert solar energy to electrical energy, and fuel cells, which convert chemical energy to electrical energy, hold long-term promise for either replacing petroleum or increasing the efficiency of its use.

Although energy-converting devices definitely have increased the usefulness of petroleum and other fuels, some of these devices have problems associated with their use. For example, they may sometimes generate potentially harmful by-products. (You will learn more about this issue later in the unit.) More fundamentally, some useful energy is always "lost" whenever energy is converted from one form to another. That is, no energy conversion is totally efficient; some energy, usually liberated as heat energy, always becomes unavailable to do useful work.

Consider an automobile with 100 units of chemical energy stored in the mixture of molecules that make up gasoline in its fuel tank. See Figure 3.29. Even a well-tuned automobile converts only about 25% of that chemical energy (potential energy) to useful mechanical energy (kinetic energy). The remaining 75% of gasoline's chemical energy is lost to the surroundings as heat (thermal energy). The following activity will allow you to realize what this means in terms of gasoline consumption and expense.

You will learn more about fuel cells later in this unit on pages 290–291.

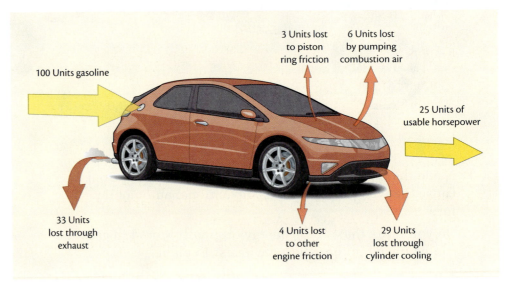

100 Units gasoline

3 Units lost to piston ring friction

6 Units lost by pumping combustion air

25 Units of usable horsepower

33 Units lost through exhaust

4 Units lost to other engine friction

29 Units lost through cylinder cooling

Figure 3.29 *Energy use in a moving automobile. Note that only about 25% of the potential energy in the gasoline is used to power the vehicle.*

B.5 ENERGY CONVERSION EFFICIENCY

How much gasoline in a car's gas tank is actually used by the engine? How much is wasted—and at what cost?

Sample Problem: *Assume that your family drives 225 miles each week and that the car can travel 23.0 miles on one gallon of gasoline. How much gasoline does the car use in one year?*

This question can be answered by attaching the units to all values, then multiplying and dividing them as though they were arithmetic expressions.

For example, the relevent information can be expressed as:

$$\frac{225 \text{ mi}}{1 \text{ wk}} \quad and \quad \frac{23.0 \text{ mi}}{1 \text{ gal}}$$

or, if needed, as inverted expressions:

$$\frac{1 \text{ wk}}{225 \text{ mi}} \quad and \quad \frac{1 \text{ gal}}{23.0 \text{ mi}}$$

Calculating the desired answer also involves using information you already know—there are 52 weeks, for example, in one year. You also know that the desired answer must have units of "gallons per year" (gal/yr). When care is taken to ensure that units are multiplied and divided to produce gal/yr, the following expression is formed:

$$\frac{225 \text{ mi}}{1 \text{ wk}} \times \frac{1 \text{ gal}}{23.0 \text{ mi}} \times \frac{52 \text{ wk}}{1 \text{ yr}} = 509 \text{ gal/yr}$$

Now answer the following questions:

1. Assume that a conventional automobile averages 23.0 miles per gallon of gasoline and travels 11 000 miles annually:

   a. How much fuel will be burned in one year?
   b. If gasoline is $3.00 per gallon, what would be spent in one year?

2. Assume that a *Hybris* hybrid-powered automobile averages 50.0 miles per gallon of gasoline and travels 11 000 miles annually.

   a. How much gasoline will the *Hybris* burn in one year?
   b. If gasoline is $3.00 per gallon, what would be spent on fuel annually?

3. Assume the automobile in Question 1 uses only 25.0% of the energy released by burning gasoline.

   a. How many gallons of gasoline are wasted each year due to energy conversion inefficiency?
   b. How much does this wasted gasoline cost at $3.00 per gallon?

4. Suppose a new car travels 70.0 miles on one gallon of gasoline with a 40.0% efficient engine:

   a. How much fuel is saved annually compared to cars in Questions 1 and 3?
   b. How much fuel is wasted annually due to energy inefficiency?

Oil's Well That Ends Well:
The Work of a Petroleum Geologist

Susan Landon is an independent petroleum geologist. Using her knowledge of chemistry, geology, geography, biology, and mathematics, she analyzes rocks and other geologic and geographic features for indications of likely oil and natural-gas reserves.

Susan searches for the folded or faulted terrain of Earth that often contains petroleum. Hydrocarbons (oil or natural gas) may be trapped in reservoirs in these areas. Using data about a given geographic area—such as satellite images, aerial photographs, topographic maps, and information about the rocks—Susan predicts the quantity of oil or natural gas

> A geologist can examine the chips . . . and identify the presence of oil.

that may have been generated from organic matter in the rocks, and where it might occur.

If signs are good for a substantial petroleum deposit, Susan and her team work to find investors to fund further exploration. Such exploration often includes more detailed research, analysis, and drilling. During drilling, Susan and her colleagues examine rock samples and analyze data regarding electrical and other physical properties of the rocks encountered. Based on the findings, the group recommends either completing the well or plugging it.

In Susan's opinion, a good liberal arts background with a major in geology or a related science is excellent preparation for advanced studies in geology.

Wells? Now There's a Deep Subject!

Although there are many techniques for predicting the presence of underground oil or natural gas, the only way to know for sure is to drill a well.

The process of drilling a well begins with a drilling rig consisting of a derrick—a large, vertical structure made of metal that holds and supports a drilling pipe. Rotated by a motor, the drilling pipe ends in a bit that digs and scrapes down through soil and rock. Drilling fluid (usually a special mixture of water and other materials that is called "mud") is pumped down through the pipe to cool the bit and carry chips of rock back to the surface.

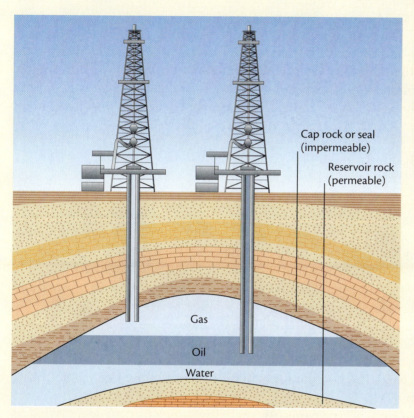

Cap rock or seal
(impermeable)

Reservoir rock
(permeable)

Gas

Oil

Water

A geologist can examine the chips and other material brought to the surface to determine the rock type and identify the presence of oil.

Although oil or natural gas sometimes flows to the surface spontaneously as a result of underground pressure, the more common procedure is to install a pump. Pumped from the well, the oil or natural gas is transported by truck or pipeline to a refinery. At the refinery, the oil or natural gas is processed and delivered to commercial and private consumers, who convert it into everything from polypropylene socks to carpeting to household heat.

B.6 COMBUSTION

Lab Video
"Combustion"

Introduction

You strike a match and a hot, yellow flame appears. If you bring the flame close to a candlewick, the candle ignites and burns (Figure 3.30). These events are so commonplace that you probably do not think about the complex chemical reactions at work.

Candle burning involves chemical reactions of the wax, which is composed of long-chain alkanes, with oxygen gas at elevated temperatures. Although many chemical reactions are involved in burning (or *combustion*), chemists simplify the process by usually focusing on the overall changes. For example, the complete burning of one component of candle wax, $C_{25}H_{52}$, can be summarized this way:

$$C_{25}H_{52}(s) \ + \ 38\,O_2(g) \longrightarrow 25\,CO_2(g) \ + \ 26\,H_2O(g) \ + \ Energy$$

| Paraffin wax (alkane) | Oxygen gas | Carbon dioxide gas | Water vapor |

As you already know, a burning candle gives off energy—the reaction is exothermic. Thus, less energy must be stored in the product molecules (in the carbon-to-oxygen bonds of carbon dioxide gas and the hydrogen-to-oxygen bonds of water vapor) than was originally stored in the reactant molecules (in the carbon-to-hydrogen bonds of wax and the oxygen-to-oxygen bonds of oxygen gas).

Fuels provide thermal energy as they burn. But how much energy is released? How can we measure the quantity of released energy? In this investigation, you will measure the *heat of combustion* of a candle (paraffin wax) and compare this quantity with known values for other hydrocarbons.

Before you begin, carefully read the procedure to learn what is involved, note safety precautions, and plan for your data collecting and observations. In this laboratory investigation, you will record data before and after the candle is burned. The first (left) column of your data table can be used to record each data item collected. The two middle columns can be used for values recorded before and after burning the candle. A fourth (right) column can be used to report the difference between the *before* and *after* values that you noted.

Procedure

1. Determine the mass of the tea candle. Record its value.

2. Carefully measure (to the nearest milliliter) about 100 mL of chilled water. (The chilled water, provided by your teacher, should be 10 to 15 °C colder than room temperature.) Pour the 100-mL sample of chilled water into an empty soft-drink can.

Figure 3.30 *As with all combustion reactions, thermal energy is released when candles burn. What other form(s) of energy result from this process?*

3. Set up the apparatus as shown in Figure 3.31, but do not light the candle yet! Adjust the can so that the top of the candlewick is about 2 cm from the bottom of the can.

4. Measure both the room temperature and the water temperature to the nearest 0.1 °C. Record these values.

5. Place the candle under the can of water. Light the candle. As the water heats, stir it gently. (**Caution:** *Do not stir with the thermometer—use a stirring rod.*)

6. As the candle burns and becomes shorter, you may need to lower the can so that the flame remains just below the bottom of the can. (**Caution:** *Lower the can with great care.*)

7. Continue heating the water until its temperature rises as far above room temperature as it was below room temperature at the start. (For example, if the water temperature were 15 °C before heating and room temperature is 25 °C, you would heat the water to 35 °C, 10 °C higher than room temperature.)

8. When the desired temperature is reached, extinguish the candle flame.

9. Continue stirring the water until its temperature stops rising. Record the highest temperature reached by the warmed water.

10. Determine and record the mass of the cooled candle, including all wax drippings.

11. Wash your hands thoroughly before leaving the laboratory.

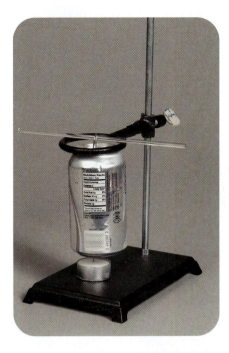

Figure 3.31 *Apparatus for determining heat of combustion.*

Data Analysis

A characteristic property of a material is the quantity of heat needed to raise the temperature of one gram of the material by one degree Celsius. This value is called the **specific heat capacity** of the material. The specific heat capacity of liquid water is about 4.2 J/(g·°C) (joules per gram per °C). This means that each 4.2 J of energy absorbed by one gram of water will increase its temperature by one degree Celsius.

Sample Problem: *Suppose a 10.0-g water sample is heated from 25.0 °C to 30.0 °C, an increase of 5.0 °C. How much thermal energy must have been added to the water?*

The answer can be reasoned this way: It takes 4.2 J to raise the temperature of 1 g water by 1 °C. In this example, however, there is 10.0 times more water and a temperature increase that is 5.0 times greater. Thus, 10.0 × 5.0, or 50 times more thermal energy is needed. Therefore, the specific heat must be multiplied by 50 to obtain the answer: 50 × 4.2 J = 210 J. It takes 210 J to increase the temperature of 10.0 g water by 5.0 °C.

Specific heat capacity is sometimes shortened to the term *specific heat*.

The precise specific heat capacity of water is 4.184 J/g. However, a rounded-off value of 4.2 J/g is adequate.

| Table 3.6

HEATS OF COMBUSTION FOR SELECTED HYDROCARBONS

Hydrocarbon	Formula	Heat of Combustion (kJ/g)	Molar Heat of Combustion (kJ/mol)
Methane	CH_4	55.6	890
Ethane	C_2H_6	52.0	1560
Propane	C_3H_8	50.0	2200
Butane	C_4H_{10}	49.3	2859
Pentane	C_5H_{12}	48.8	3510
Hexane	C_6H_{14}	48.2	4141
Heptane	C_7H_{16}	48.2	4817
Octane	C_8H_{18}	47.8	5450

The quantity of thermal energy given off when a certain amount of a substance burns is called the **heat of combustion.** See Table 3.6. The heat of combustion can be expressed as the thermal energy released when either one gram or one mole of substance burns. If the amount of substance burned is one mole, the quantity of thermal energy involved is called the **molar heat of combustion.**

Using your laboratory data, you can calculate the heat of combustion of paraffin wax:

1. Calculate the mass of water heated. (*Hint:* The density of liquid water is 1.0 g/mL. Thus, each milliliter of water has a mass of 1.0 g.)

2. Calculate the total rise in the temperature of the water.

3. Calculate how much thermal energy was used to heat the water sample. Use values from the two preceding steps to reason out the answer, as illustrated in the earlier sample problem.

4. Calculate the total mass of paraffin wax burned.

5. Calculate the heat of combustion of paraffin, expressed in units of
   a. joules per gram (J/g) of paraffin.
   b. kJ/g of paraffin.

   (*Hint:* Assume that all the energy released by the burning paraffin wax is absorbed by the water.)

6. Calculate the molar heat of combustion of paraffin, expressed in units of kJ/mol. (*Hint:* First calculate the thermal energy released when one mole of paraffin burns. Because one mole of paraffin ($C_{25}H_{52}$) has a mass of 352 g, the *molar* heat of combustion will be 352 times greater than the heat of combustion expressed in units of kJ/g.)

1 kJ = 1000 J

Questions

Your teacher will collect your heat of combustion data, expressed in units of kilojoules per gram (kJ/g) of paraffin. Use the combined results of your class to determine the *best estimate* for this value. Will you decide to use the average value of the class results or the median value? Why? Use your selected value to answer the following questions:

1. How does your experimental heat of combustion (in kJ/g) for paraffin wax, $C_{25}H_{52}$, compare to the accepted heat of combustion for propane, C_3H_8? See Table 3.6.

2. How do the molar heats of combustion (in kJ/mol) for paraffin and propane compare?

3. Explain the differences in your answers to Questions 1 and 2 for paraffin and propane. (*Hint:* Keep in mind the calculation you completed in Step 6.)

4. In your view, which hydrocarbon—paraffin or propane—is the better fuel? Explain your answer, citing evidence from your observations and data.

5. In calculating the heat of combustion of paraffin, you assumed that all thermal energy from the burning wax went to heating the water:
   a. Was this a good assumption? Explain.
   b. What other laboratory conditions or assumptions might cause errors in your calculated values?

B.7 USING HEATS OF COMBUSTION

With abundant oxygen gas and complete combustion, the burning of a hydrocarbon can be described by the equation

Hydrocarbon + Oxygen gas ⟶ Carbon dioxide + Water + Thermal energy

"Thermal energy" is written as a product of the reaction because energy is released when a hydrocarbon burns (Figure 3.32). The combustion of a hydrocarbon is a highly exothermic reaction.

The equation for burning ethane (C_2H_6) is

$$2 C_2H_6 + 7 O_2 \longrightarrow 4 CO_2 + 6 H_2O + \underline{?} \text{ kJ thermal energy}$$

To complete this equation, the correct quantity of thermal energy involved must be included. Table 3.6 indicates that ethane's molar heat of combustion is 1560 kJ/mol. That is, burning one mole of ethane releases 1560 kilojoules of energy. According to the chemical equation above, two moles of ethane ($2 C_2H_6$) are burned. Of course, the total thermal energy must correspond to the amounts of all other reactants and products involved. Thus, the total thermal energy released will be *twice* that quantity released when 1 mole of ethane burns. Therefore, the complete combustion equation for ethane is:

$$2 C_2H_6 + 7 O_2 \longrightarrow 4 CO_2 + 6 H_2O + 3120 \text{ kJ}$$

Figure 3.32 *Natural gas (often used as a fuel to heat homes and for cooking) is a mixture of hydrocarbons. Its primary component is methane, the smallest of the hydrocarbon molecules, consisting of only one carbon and four hydrogen atoms.*

As you found in Investigating Matter B.6, and as suggested in Table 3.6 (page 250), heats of combustion can also be expressed as the energy involved when one gram of hydrocarbon burns (kJ/g). That information is useful in finding out how much energy is released when a certain mass of fuel is burned.

Sample Problem: *How much thermal energy would be produced by burning 12.0 g octane, C_8H_{18}?*

Table 3.6 indicates that burning 1.00 g octane releases 47.8 kJ. Burning 12.0 times *more* octane produces 12.0 times *more* thermal energy, or

$$12.0 \times 47.8 \text{ kJ} = 574 \text{ kJ}$$

The calculation can also be written this way:

$$12.0 \text{ g octane} \times \frac{47.8 \text{ kJ}}{1 \text{ g octane}} = 574 \text{ kJ}$$

ChemQuandary 2

A BURNING PROBLEM: PART I

Examine the heat of combustion and the molar heat of combustion data summarized in Table 3.6 (page 250). What do you notice about the trend in the heat of combustion for hydrocarbons expressed as kilojoules per gram (kJ/g) compared with the trend expressed as kilojoules per mole (kJ/mol)? The units *grams* and *moles* both specify the *quantity* of a substance. What difference in these two quantities would explain the trends you observed in Table 3.6?

Moles express amount; grams express mass.

Assuming these trends apply to larger hydrocarbon molecules, predict the heat of combustion for decane, $C_{10}H_{22}$, expressed as kilojoules per gram (kJ/g) decane and kilojoules per mole (kJ/mol) decane. Which prediction was easier to make, and why might this be so?

Developing Skills

B.8 HEATS OF COMBUSTION

This activity will give you a better understanding of the energy involved in burning hydrocarbon fuels. Use Table 3.6 (page 250) to answer these questions.

Sample Problem: *How much energy (in kilojoules) is released by completely burning 25.0 mol hexane, C_6H_{14}?*

According to Table 3.6, the molar heat of combustion of hexane is 4141 kJ. This means that 4141 kJ of energy is released as 1.00 mol hexane burns. Therefore, burning 25.0 times more hexane will liberate 25.0 times more energy:

$$25.0 \text{ mol } C_6H_{14} \times \frac{4141 \text{ kJ}}{1 \text{ mol } C_6H_{14}} = 104\ 000 \text{ kJ}$$

Burning 25.0 mol hexane would thus release 104 000 kJ of thermal energy.

1. Write a chemical equation that includes thermal energy for the complete combustion of the following alkanes:

   a. propane   c. octane
   b. butane   d. decane

2. a. How much thermal energy is produced by burning two moles of octane?
   b. How much thermal energy is produced by burning one gallon of octane? (A gallon of octane occupies a volume of about 3.8 L. The density of octane is 0.70 g/mL.)
   c. Suppose a car operates so inefficiently that only 16% of the thermal energy from burning fuel is converted to useful "wheel-turning" (mechanical) energy. How many kilojoules of useful energy would be stored in a 20.0-gallon tank of gasoline? (Assume that octane burning and gasoline burning produce the same results.)

3. The molar heat of combustion of carbon contained in coal is 394 kJ/mol C.

   a. Write a chemical equation for burning the carbon contained in coal. Include the thermal energy produced.
   b. Gram for gram, which is the better fuel—carbon or octane? Explain your answer using calculations.
   c. In what applications might coal replace petroleum-based fuel?
   d. Describe one application in which coal would be a poor substitute for petroleum-based fuel.

B.9 ALTERING FUELS

As automobile use grows throughout the world, the demand for gasoline continues to increase rapidly. Because the gasoline fraction in a barrel of crude oil normally represents only about 18% of the total, researchers have been anxious to find a way to increase this yield. One promising method has been based on the discovery that it is possible to alter the structures of some of petroleum's hydrocarbon molecules so that 47% of a barrel of crude oil can be converted to gasoline. This chemical technique deserves further attention.

Chemists and chemical engineers are adept at modifying or altering available chemical resources to meet new needs. Such alterations might also involve converting less-useful materials to more-useful products, or converting a low-demand material into high-demand materials.

Cracking

Recall that the gasoline fraction obtained from crude oil refining includes hydrocarbons with 5 to 12 carbons per molecule. (See Figure 3.9, page 219.)

By 1913, chemists had devised a process for converting larger molecules in kerosene into smaller, gasoline-sized molecules by heating the kerosene to 600 to 700 °C. The process of converting large hydrocarbon molecules into smaller ones through the application of thermal energy (heat) and a catalyst is known as **cracking.** Today, more than a third of all crude oil undergoes cracking. The process has been improved by adding catalysts. A **catalyst** increases the speed of a chemical reaction by participating in it; however, the catalyst is not used up. Catalytic cracking is more energy efficient because it occurs at a lower temperature, 500 °C rather than 700 °C, which is made possible by the catalyst's lowering the energy requirements of reacting molecules.

Through cracking, a 16-carbon molecule, for example, might be changed into two 8-carbon molecules:

$$C_{16}H_{34} \longrightarrow C_8H_{18} + C_8H_{16}$$

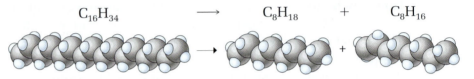

In practice, molecules with up to about 14 carbon atoms can be produced through cracking. Molecules with 5 to 12 carbon atoms are particularly useful in gasoline, which remains the most important commercial product of refining. Some C_1, C_2, C_3, and C_4 molecules produced in cracking are immediately burned, keeping the temperature high enough for more cracking to occur.

Octane Rating

Gasoline is composed mainly of straight-chain alkanes, such as hexane (C_6H_{14}), heptane (C_7H_{16}), and octane (C_8H_{18}). In most automobiles, the gasoline–air mixture is compressed in cylinders just prior to being ignited by spark plugs. This compression can be enough to heat the alkanes to the point where they burn before the spark plug ignites them.

Modern, well-tuned engines rarely knock under normal conditions.

The premature burning causes engine "pinging," or "knocking," as the piston bangs backwards against the crankshaft at the wrong time and may contribute to engine problems. Branched-chain alkanes burn more satisfactorily, being less likely to undergo combustion due to compression in engine cylinders; they do not ping as much. The structural isomer of octane shown here has excellent combustion properties in automobile engines. This octane isomer is known chemically as 2,2,4-trimethylpentane. Can you see how the name of the molecule relates to its structure? For convenience, this substance is frequently referred to by its common name, *isooctane:*

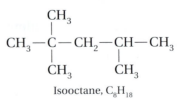

Isooctane, C_8H_{18}

As you probably know, gasoline is sold in a variety of grades—and at corresponding prices. A common reference standard for gasoline quality is the *octane scale.* On this scale, isooctane, the branched-chain hydrocarbon you just learned about, is assigned an octane number of 100. Straight-chain heptane (C_7H_{16}), a fuel with very poor engine performance, is assigned an octane number of zero. Gasoline samples can be rated in comparison with isooctane and heptane.

The **octane rating** for a particular fuel is determined by testing the fuel's burning efficiency under two conditions: in a free-running engine and in an engine under load, as in towing or passing conditions (Figure 3.33). The results of these two tests are then averaged to obtain the octane rating. The higher the octane rating of a gasoline becomes, the better its antiknock characteristics. Octane ratings in the high 80s and low 90s (87, 89, 92) are quite common, as a survey of nearby gasoline pumps, like those pictured in Figure 3.34, will reveal.

Figure 3.33 *Researchers use test engines to determine the octane rating of fuel.*

Figure 3.34 *Octane ratings posted on gasoline pumps.*

From the 1920s until the 1970s, the octane rating of gasoline was increased at low cost by adding a substance such as tetraethyl lead, $(C_2H_4)_4Pb$, to the fuel. This additive enhanced the burning efficiency of straight-chain gasoline molecules and added about three points to "leaded" fuel's octane rating. Unfortunately, lead from the treated gasoline was discharged into the atmosphere along with other vehicle exhaust products. Lead is harmful to the environment; therefore, lead-based gasoline additives are no longer used in the United States.

Assigning an octane number of 100 to isooctane is arbitrary and does not imply that it has the highest possible octane number. In fact, several fuels burn more efficiently in engines than isooctane does; they are assigned octane numbers higher than 100.

Oxygenated Fuels

The phaseout of lead-based gasoline additives in the United States meant that alternative octane-boosting supplements were required. A group of additives called **oxygenated fuels** now are blended with gasoline to enhance its octane rating. The molecules of these additives contain oxygen in addition to carbon and hydrogen.

Although oxygenated fuels actually deliver less energy per gallon than regular gasoline hydrocarbons do, their economic appeal stems from their ability to increase the octane number of gasoline while reducing exhaust-gas pollutants. In most conditions, oxygenated fuels encourage more complete combustion, producing lower emissions of air pollutants, such as carbon monoxide (CO).

A common oxygenated fuel is *methanol* (methyl alcohol, CH_3OH), which is added to gasoline at distribution locations. In addition to its octane-boosting properties, methanol can be made from natural gas, coal, corn, or wood—a contribution toward conserving nonrenewable petroleum resources. A blend of 10% ethanol (ethyl alcohol, CH_3CH_2OH) and 90% gasoline, sometimes called *gasohol,* can be used as an oxygenated fuel in nearly all modern automobiles without engine adjustments or problems.

ChemQuandary 3

A BURNING PROBLEM: PART II

Methanol (CH_3OH) and ethanol (CH_3CH_2OH), sometimes produced from corn, are used as gasoline additives or substitutes. The heat of combustion values for methanol and ethanol are 23 kJ/g and 30 kJ/g, respectively. Consider the chemical formulas of methanol and ethanol. Gram for gram, why are the heats of combustion of methanol and ethanol considerably lower than those of any hydrocarbons considered so far? See Table 3.6 (page 250).

Methyl tertiary-butyl ether (MTBE), with an octane rating of 116, was initially introduced in the late 1970s as an octane-boosting fuel additive. In the 1990s, MTBE became the most common oxygenated fuel additive in gasoline. At its peak, annual U.S. production of MTBE increased to more than 4 billion gallons (about 16 gallons per person), making it a top-ten industrial chemical substance.

By the late 1990s, however, evidence began to mount of contamination of groundwater and drinking water supplies due to MTBE seeping from defective underground gasoline storage systems. MTBE dissolves readily in water and is difficult to remove in water-treatment processes. The unpleasant taste and odor that MTBE imparts to water, even at concentrations below those regarded as a public health concern, triggered consumer complaints. The EPA credited MTBE with substantial reductions in emissions of air pollutants from gasoline-powered vehicles in the 1990s. However, in light of these more recent concerns, policies to reduce or to ban the use of MTBE as an oxygenated fuel are under active consideration.

Other octane-boosting strategies involve altering the structures of hydrocarbon molecules in petroleum. This works because branched-chain hydrocarbons burn more satisfactorily than do straight-chain hydrocarbons. (Recall isooctane's octane number compared with that of straight-chain heptane.) Straight-chain hydrocarbons are converted to branched-chain hydrocarbons by a process called **isomerization.** During isomerization, hydrocarbon vapor is heated with a catalyst:

$$CH_3-CH_2-CH_2-CH_2-CH_2-CH_3 \xrightarrow[\text{Catalyst}]{\text{Heat}} CH_3-CH-CH-CH_3$$

$$C_6H_{14}(g) \qquad\qquad\qquad C_6H_{14}(g)$$
Straight-chain   Branched-chain
isomer   isomer

The branched-chain alkanes produced by isomerization are blended with C_5 to C_{12} molecules obtained from cracking and distillation, producing a high-quality gasoline. Although cracked and isomerized molecules improve how gasoline burns, they also increase its cost. One reason for this increase is the extra fuel needed to produce such gasoline.

B.10 FUEL FOR TRANSPORTATION

Making Decisions

Now that you have examined the importance of petroleum as a fuel, particularly for transportation (Figure 3.35), it is time to revisit the television ad for the *Hybris* that opened this unit. Later, you will have the opportunity to critically review and design an advertisement for an alternate-powered vehicle.

Use what you have learned about petroleum to answer the following questions:

1. In what ways does the manufacture of a *Hybris* hybrid-powered car use petroleum as a fuel?

2. What alternatives to petroleum can you suggest for each petroleum use that you listed in Question 1?

3. How do the alternatives that you suggested in Question 2 compare to petroleum in terms of energy efficiency?

4. What direct, nonfuel uses of petroleum or its fractions are involved in the daily operation of a *Hybris*?

5. What alternatives to petroleum can you suggest for each direct, nonfuel use that you listed in answering Question 4?

6. Write one or two paragraphs that evaluate other assertions made in the *Hybris* advertisement.

Section C will enrich your knowledge of the "building" role of petroleum, focusing on petroleum as a source of an impressive array of useful compounds and products. This information will help you further decide about the accuracy of the claims in the *Hybris* TV advertisement and will also guide your design of an advertisement for an alternative-fuel vehicle.

Figure 3.35 *Petroleum, precursor to the fuels used by these vehicles, is a nonrenewable resource. What alternative fuels can be devised to power vehicles such as these?*

SECTION B SUMMARY
Reviewing the Concepts

Chemical energy, a form of potential energy, is stored within the bonds in chemical compounds.

1. From a chemical viewpoint, why is petroleum sometimes considered "buried sunshine"?

2. Define and give one example of
   a. potential energy.   b. kinetic energy.

3. In terms of chemical bonds, what happens during a chemical reaction?

4. Based on its structural formula, which has more potential energy, a molecule of methane or a molecule of butane? Explain your answer.

5. Classify each of the following as primarily a demonstration of kinetic energy or potential energy:
   a. a skateboard positioned at the top of a hill
   b. a charged battery in a flashlight that's turned off
   c. a rolling soccer ball
   d. gasoline in a parked car
   e. water flowing over a waterfall

The energy change in a chemical reaction equals the difference between the energy required to break reactant bonds and the energy released in forming product bonds. Reactions are classified as either exothermic or endothermic.

6. Why is energy required to break chemical bonds?

7. For each of the following events, determine whether the reaction is exothermic or endothermic. Explain your answers in terms of bond breaking and bond making:
   a. burning wood in a campfire
   b. cracking large hydrocarbon molecules
   c. digesting a candy bar

8. Burning a candle is an exothermic reaction. Explain this fact in terms of the quantity of energy stored in the bonds of the reactants compared with the quantity of energy stored in the bonds of the products.

9. Using Figure 3.24 on page 240 as a model, draw a potential energy diagram that illustrates the energy change when hydrogen gas reacts with oxygen gas to produce water and thermal energy.

Thermal energy can be converted into other forms of energy. Although energy can neither be created nor destroyed, useful energy is lost each time energy is converted from one form to another.

10. State the law of conservation of energy.

11. Match the following events (a–d) with the energy conversions (i–iii) that are involved:
    a. riding a bike
    b. illuminating a lightbulb
    c. producing electricity in a wind-powered generator
    d. running a gasoline-powered lawn mower
       i.   mechanical to electrical energy
       ii. chemical to mechanical energy
       iii. electrical to light energy

12. In powering an automobile, 25% of the energy is said to be useful. What happens to the other 75% of the energy?

13. Explain what *energy conversion efficiency* means.

14. One gallon of gasoline produces about 132 000 kJ of energy when burned. Assume that an automobile is 25% efficient in converting this energy into useful work:
    a. How much energy (in kJ) is "wasted" when a gallon of gasoline burns in an automobile?
    b. What happens to this wasted energy?

> When a hydrocarbon burns completely, it reacts with oxygen gas from the air to liberate thermal energy and produce carbon dioxide gas and water vapor.

15. Write a balanced chemical equation, including the quantity of thermal energy (refer to Table 3.6, page 250), for the complete combustion of

    a. pentane.   b. propane.   c. hexane.

16. The combustion of acetylene, C_2H_2 (used in a welder's torch), can be represented as

    $$2\ C_2H_2 + 5\ O_2 \longrightarrow 4\ CO_2 + 2\ H_2O + 2512\ kJ$$

    a. What is the molar heat of combustion of acetylene in kilojoules per mole?
    b. If 12 mol acetylene burns fully, how much thermal energy will be produced?

17. a. List two factors that would help you decide which hydrocarbon fuel to use in a particular application.
    b. Explain how you would use each factor in making your decision.

18. When candle wax (a mixture of hydrocarbons) burns, it seems to disappear. What actually happens to the wax?

19. *Water gas* (a 50–50 mixture of CO and H_2) is made by the reaction of coal with steam. Because the United States has substantial coal reserves, water gas might serve as a substitute fuel for natural gas (composed mainly of methane, CH_4). Water gas burns according to this equation:

    $$CO + H_2 + O_2 \longrightarrow CO_2 + H_2O + 525\ kJ$$

    a. How does water gas compare to methane in terms of the thermal energy produced when fully burned?
    b. If a water gas mixture containing 10.0 mol CO and 10.0 mol H_2 were completely burned in O_2, how much thermal energy would be produced?

> The thermal energy (measured in joules) required to raise the temperature of one gram of a material by one degree Celsius is its specific heat capacity.

20. In a laboratory activity, a student team measures the heat released by burning heptane (C_7H_{16}). Using the following data, calculate the molar heat of combustion of heptane in kJ/mol:

    • mass of water   179.2 g
    • initial water temperature   11.6 °C
    • final water temperature   46.1 °C
    • mass of heptane burned   0.585 g

21. How much energy is released when a 5600-g sample of water cools from 99 °C to 28 °C? The specific heat capacity of water is 4.18 J/(g·°C).

> Larger molecules in petroleum's kerosene fraction (C_{12} to C_{16}) can be converted to smaller molecules useful in the gasoline fraction (C_5 to C_{12}). The speed of the process, called cracking, is increased by using a catalyst.

22. What is a catalyst?

23. Write one possible balanced equation showing how each of the following molecules could be cracked into molecules for the gasoline fraction:

    a. $C_{17}H_{36}$ b. $C_{18}H_{38}$

24. In terms of energy efficiency, why are catalysts used during the cracking process?

25. Explain why only small amounts of catalysts are needed to crack large amounts of petroleum.

26. Why do we no longer use tetraethyl lead as a gasoline additive?

27. Compare the molecular structures of octane and isooctane.

28. How does the addition of oxygenating compounds affect a fuel's octane rating?

29. List two ways to increase a fuel's octane rating.

Connecting the Concepts

30. In a laboratory activity, a student completely burns 4.2 g ethanol (C_2H_5OH). The molar heat of combustion of ethanol is 1366 kJ/mol.

    a. How much thermal energy is released?
    b. The thermal energy from this reaction is used to warm a 468-g sample of water in a calorimeter. The water temperature changes from 21 °C to 89 °C. How much thermal energy is absorbed by the water? The specific heat capacity of water is 4.18 J/(g·°C).
    c. Compare the quantity of heat released by this reaction to the quantity of heat absorbed by the calorimeter. What accounts for the difference?

31. What modifications would have to be made to the laboratory procedure in Section B.6 (pages 248–251) if a liquid or gaseous fuel were used?

32. Why is the energy stored in petroleum more useful than solar energy for most applications? Give at least three reasons in your explanation.

Extending the Concepts

33. MTBE was once widely used as an oxygen source for automobile fuels. Investigate the chemical properties of this additive and the controversy surrounding its use. Report on its current status.

34. Referring to Figure 3.29 (page 244), suggest specific engine-design modifications that might increase the proportion of energy available to propel the car.

35. Write a word equation for photosynthesis. Include energy in your equation.

36. Using Figure 3.24 on page 240 as a model, draw a potential energy diagram for photosynthesis.

37. Compare the equation for burning hydrocarbons to the equation for photosynthesis. How are these two reactions related?

Petroleum: A Building-Material Source

Just as an architect uses knowledge of available construction materials in designing a building, a chemist—a "molecular architect"—uses knowledge of available molecules in designing new molecules (Figure 3.36). Architects must know about the structures and properties of common building materials. Likewise, chemists must understand the structures and properties of their raw materials. The "builder molecules"—that small portion of petroleum not used for energy, lubricants, and road tar—are available for building by molecular architects. In this section, you will explore the structures of common hydrocarbon builder molecules and some materials made from them.

Figure 3.36 *When designing a new molecule, chemists often use computer modeling programs to make predictions about its structure and properties. This is particularly helpful when considering how a new biomolecule will interact with other substances inside the human body.*

C.1 CREATING NEW OPTIONS: PETROCHEMICALS

Until the early 1800s, all objects and materials used by humans were either created directly from wood or stone or crafted from metals, glass, and clays. Available fibers included cotton, wool, linen, and silk. All medicines and food additives came from natural sources. Celluloid (from wood) and shellac (from animal materials) were the only sources for commercially produced **polymers,** which are large molecules typically composed of 500 to 20 000 or more repeating units of simpler molecules known as **monomers.**

Today, many common objects and materials created by the chemical industry are unlike anything seen or used by citizens of the 1800s or even the mid-1900s. See Figure 3.37. Many of these differences are due to the use of compounds produced from oil or natural gas, called **petrochemicals.** Some petrochemicals, such as detergents, pesticides, pharmaceuticals, and cosmetics, are used directly. Most petrochemicals, however, serve as raw materials in producing other synthetic substances, particularly a wide range of polymers.

Synthetic polymers include paint components, fabrics, rubber, insulating materials, foams, adhesives, molding, and structural

Figure 3.37 *Two street scenes— one from the early 1900s (top) and one from the early 2000s (bottom). What similarities and differences do you notice in the use of materials?*

Double bonds will be discussed in more detail on page 266.

materials. Worldwide production of petroleum-based polymers is more than four times that of aluminum products.

The astonishing fact is that it takes relatively few *builder molecules* (small-molecule compounds) to make thousands of new substances, including many polymers. One builder molecule is *ethene,* C_2H_4, a hydrocarbon compound commonly called *ethylene.* The structural formula for ethene is shown in the following equation. The two carbon atoms in an ethene molecule share *two* pairs of electrons. This produces an arrangement of electrons called a **double covalent bond.** Because of the high reactivity of its double bond, ethene is readily transformed into many useful products. A simple example—the formation of ethanol (ethyl alcohol) from water and ethene—illustrates how ethene reacts:

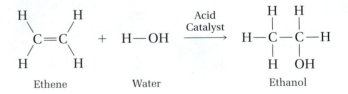

Ethene   Water   Ethanol

In this reaction, the water molecule (HOH) "adds" to the double-bonded carbon atoms by adding an H to one carbon atom and an OH group to the other carbon atom. This type of chemical change is called an **addition reaction.**

Ethene can also undergo an addition reaction with itself. Because the added ethene molecule also contains a double bond, another ethene molecule can be added, and so on. This creates a long-chain substance called *polyethene,* commonly known as *polyethylene,* a polymer consisting of 500 to 20 000 or more repeating units of ethene (ethylene) monomer. The chemical reaction that produces polyethene can be represented this way:

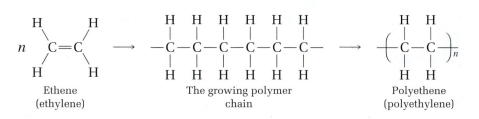

| Ethene (ethylene) | The growing polymer chain | Polyethene (polyethylene) |

Polymers formed by reactions such as this are called—sensibly enough—**addition polymers.** Polyethene is commonly used in grocery bags and packaging. The United States produces millions of kilograms of polyethene annually.

We can make a great variety of addition polymers from monomers that closely resemble ethene. The most common variation is to replace one or more hydrogen atoms in ethene with an atom or atoms of another element. In the following examples, note the replacements for hydrogen atoms in each monomer and polymer:

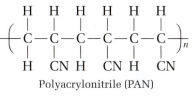

Vinyl chloride   Polyvinyl chloride (PVC)

Acrylonitrile   Polyacrylonitrile (PAN)

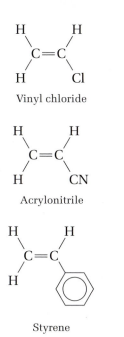
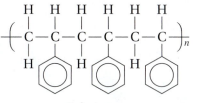

Styrene   Polystyrene

Perhaps you wonder why two names are shown for the same substance— ethene/ethylene and polyethene/polyethylene. Chemists have developed a system for naming substances that makes it easy to communicate clearly with each other. However, some substances were given names long before this system was universally accepted. Some "common names" of substances, such as ethylene and polyethylene, continue in wide use.

The hexagon-shaped ring depicted in styrene and polystyrene represents benzene, a substance discussed later on page 271.

(a)

(b)

(c)

Figure 3.38 *Products commonly made from (a) polyvinyl chloride, (b) polypropylene, and (c) polystyrene.*

The atoms that compose the monomers determine the properties of the resulting polymer. Thus, polyethene and polyvinyl chloride are fundamentally different materials. However, as with many modern materials, the properties of a polymer are often altered to meet a variety of needs and to produce many products. See Figure 3.38.

C.2 POLYMER STRUCTURE AND PROPERTIES

Modeling Matter

Polymers— 3D Structures & Properties

Unmodified, the arrangement of covalent bonds in long, stringlike polymer molecules causes the molecules to coil loosely. A collection of polymer molecules (such as those in a sample of molten polymer) can intertwine, much like strands of cooked spaghetti. In this form, the polymer is flexible and soft.

1. Using a pencil line on paper to represent a linear polymer, draw a collection of loosely coiled polymer molecules.

Ductility refers to the ability of a material to be drawn out, as into thin strands. For most polymers like the ones you just drew, flexibility and ductility depend on temperature. When the material is warm, the polymer chains can slide past one another easily. The polymer becomes more rigid when it cools. We use such polymers, classified as *thermoplastics,* in many everyday products, such as soft drink bottles, milk bottles, and plastic bags.

The flexibility of a polymer can also be enhanced by adding molecules that act as internal lubricants among the polymer chains. For example, untreated polyvinyl chloride (PVC) is used in rigid pipes and house siding. With added lubricant molecules, polyvinyl chloride becomes flexible enough for use in such consumer goods as raincoats and inflatable pool toys.

The reactions that form polymer chains can also take place perpendicular to the main chain, forming *side chains.* These polymers are called **branched polymers.** The branched form of polyethylene is shown in the following illustration. The extent of branching can be controlled by adjusting reaction conditions.

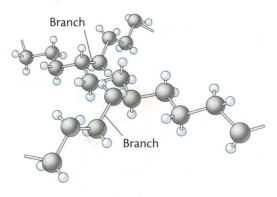

Branch

Branch

2. Draw at least two different models of branched-chain polymers. Try to vary your representations—the forms of branched polymers can differ greatly.

Branching changes the properties of a polymer by affecting the ability of chains to slide past one another and by altering intermolecular forces.

Another way to alter the properties of polymers is through **cross-linking.** Polymer rigidity can be increased if the polymer chains are cross-linked so that they can no longer move or slide readily. You can see this for yourself if you compare the flexibility of a plastic soda bottle with that of its screw-on cap. Polymer cross-linking is much greater in the cap. The following illustration shows a cross-linked form of polyethene.

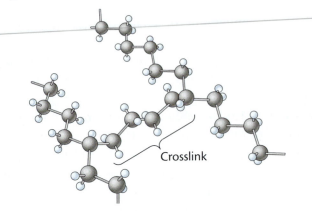

Crosslink

Rubber-Formation & Chemical Structure

3. Draw several linear polymer chains that have been cross-linked.

4. Draw several cross-linked, branched polymers. Then describe how cross-linked *branched* polymers compare to cross-linked *linear* polymers.

Chemists can control polymer strength and toughness. To do this, the polymer chains are arranged so that they lie in the same general direction. The aligned chains are stretched until they uncoil. Polymers that remain uncoiled after this treatment make strong, tough films and fibers. Such materials include polyethene, which is used in everything from garbage bags to artificial ice rinks, and polypropylene, which is used in bottles and some carpeting.

5. Draw several aligned linear polymer chains.

6. Draw several aligned and cross-linked linear polymer chains.

C.3 BEYOND ALKANES

Carbon atoms are versatile building blocks. Carbon can form bonds with other atoms in several different ways. As you learned in Section A, each carbon atom in an alkane molecule is bonded to four other atoms. Compounds such as alkanes are called **saturated hydrocarbons** because each carbon atom forms as many single covalent bonds as it can.

In some hydrocarbon molecules, carbon atoms bond to three other atoms, not four. Members of this series of hydrocarbons are called **alkenes.** The simplest alkene was briefly introduced to you (see page 262)—ethene, C_2H_4.

The carbon–carbon bonding that characterizes alkenes is a double covalent bond. In a double bond, four electrons (two electron pairs) are shared between the bonding partners. Alkenes, which contain carbon–

carbon double bonds, are described as **unsaturated hydrocarbons.** Not all carbon atoms are bonded to their full capacity with four other atoms. Because of their double bonds, alkenes are more chemically reactive—and therefore better builder molecules—than are alkanes.

The **substituted alkenes** make up another class of builder molecules. In addition to carbon and hydrogen, these molecules contain one or more other atoms, such as oxygen, nitrogen, chlorine, or sulfur. Adding atoms of other elements to hydrocarbon structures significantly changes their chemical reactivity.

Even molecules composed of the same elements can have quite different properties. For example, ethanol and dimethyl ether both contain carbon, hydrogen, and oxygen atoms. In fact, they share the same molecular formula, C_2H_6O. However, ethanol, also called grain alcohol, is liquid at room temperature, while dimethyl ether—often used as a refrigerant—is a gas. The dramatic differences in their properties and uses result from particular arrangements of atoms in these two molecules, as shown by the following structures.

> Dimethyl ether is a propellant in many aerosol spray cans and recently has been successfully used as a diesel-fuel substitute.

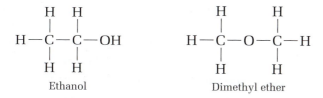

Ethanol   Dimethyl ether

As you know, the strength of intermolecular forces is related to molecular structure. In this case, attractions among ethanol molecules, in which oxygen is bonded to hydrogen, are stronger than attractions among dimethyl ether molecules. This helps to explain why ethanol is liquid and dimethyl ether is gaseous at room temperature.

Investigating Matter

C.4 THE BUILDERS

Introduction

This investigation, in which you will use models to represent various arrangements of atoms, will help you become more familiar with alkenes and their polymers. Be prepared to record your observations and answers to questions in your laboratory notebook.

Procedure

Part I: Alkenes

1. Examine the following Lewis-dot formula and structural formula for ethene, C_2H_4. Confirm that each atom has attained its filled outer shell of electrons:

Lewis dot formula   Structural formula CH_2CH_2 or C_2H_4
                                              Molecular formula

The names of alkenes follow a pattern much like that of the alkanes. The first three alkenes are *ethene, propene,* and *butene.* The same prefixes that you learned for alkanes are used to indicate the number of carbon atoms in the molecule's longest carbon chain. However, each alkene name ends in *-ene* instead of *-ane.*

2. Recall that the alkane general formula is $C_nH_{2n + 2}$. Examine the molecular formulas of ethene (C_2H_4) and butene (C_4H_8). What general formula for alkenes do these molecular formulas suggest?

3. Assemble models of an ethene (C_2H_4) molecule and an ethane (C_2H_6) molecule. See Figure 3.39. Compare the arrangements of atoms in the two models. Rotate the two carbon atoms in ethane about their single C—C bond. Then try a similar rotation with ethene.

   a. What do you observe?
   b. Can you build a molecule in which you can complete a rotation about a C=C double bond?
   c. Write a general rule to summarize your findings.

4. Build a model of butene (C_4H_8). Remember that alkenes must contain a double bond. Compare your model to those made by others.

   a. How many different arrangements of atoms in a C_4H_8 molecule appear possible? Each arrangement represents a different substance—another example of isomers.
   b. Which structural formulas in Figure 3.40 correspond to models built by you and your classmates?

As with alkanes, we name alkenes according to their longest carbon chain. The carbon atoms are numbered, beginning at the end of the chain closest to the double bond. The name of each isomer starts with the number assigned to the first double-bonded carbon atom. Look again at the butene isomer structures in Figure 3.40 and confirm these naming rules with 1-butene and 2-butene.

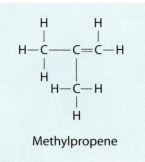

1-Butene, or simply butene

2-Butene

Methylpropene

Figure 3.40 *Three isomers of butene, C_4H_8.*

5. Do each of the following pairs represent isomers of the same substance, or do they both represent the same substance?

a. CH_2=CH—CH_2—CH_3 and CH_3—CH_2—CH=CH_2

b. CH_2=$\underset{\underset{CH_3}{|}}{C}$—$CH_3$ and CH_3—$\underset{\underset{CH_2}{||}}{C}$—$CH_3$

6. How many isomers of propene (C_3H_6) are there? Support your answer with appropriate structures.

7. Do the following two structures represent isomers or the same substance? Explain.

$$\begin{array}{cc} CH_2-CH_2 \\ |\qquad\ | \\ CH_2\quad CH_2 \\ \diagdown\ \diagup \\ CH_2 \end{array} \qquad \begin{array}{c} CH_3-CH-CH_2 \\ |\qquad\ | \\ CH_2-CH_2 \end{array}$$

8. Assemble a model of a hydrocarbon molecule containing a *triple* covalent bond. Base this on your knowledge of hydrocarbon molecules with either single or double bonds between carbon atoms. Your completed model represents a member of the hydrocarbon series known as **alkynes.** Given your understanding of the naming of alkanes and alkenes, write structural formulas for

a. ethyne, commonly called *acetylene* (Figure 3.41).
b. 2-butyne.

9. Are alkynes saturated or unsaturated hydrocarbons? Explain.

Part II: Compounds of Carbon, Hydrogen, and Singly Bonded Oxygen

10. Assemble as many different molecular models as possible, using all nine of the following atoms:

▶ 2 carbon atoms (each forming four single bonds)
▶ 6 hydrogen atoms (each forming a single bond)
▶ 1 oxygen atom (forming two single bonds)

Figure 3.41 *Ethyne (commonly called acetylene) is the primary fuel in blowtorches. When burned, it generates the high temperatures needed to weld and to cut through metal.*

11. On paper, draw a structural formula for each compound you assembled in Step 10, indicating how the nine atoms are connected. Compare your structures with those built by other classmates. Be sure that you are satisfied that all possible structures have been produced, then answer these questions:

   a. How many distinct structures did you identify?
   b. Are all of these structures isomers? Explain.

12. Each structure you have identified possesses distinctly different properties.

   a. Recalling that "like dissolves like," which molecule should be most soluble in water?
   b. Which molecule should have the highest boiling point? Why?

> Remember that boiling points are related to strengths of intermolecular forces.

Part III: Alkene-Based Polymers

In this part of the investigation, you and your classmates will use models to simulate the formation of several addition polymers.

13. Build models of two ethene molecules.

14. Using the information on page 263 as a guide, combine your two models into a **dimer**, a two-monomer structure. How did you have to modify the monomer structure to accomplish this?

15. Combine your dimer with that of another lab team. Continue this process until your class has created one long-chain structure. Although your resulting molecular chain is not yet long enough to be regarded as a model of a polymer, you have simulated the processes involved in creating a typical addition polymer.

16. Repeat Steps 13 and 14, first by replacing ethene with vinyl chloride and then by replacing ethene with propene. Then find out whether you can follow the same steps for styrene. (See page 263 for the molecular structures of these substances.)

17. Assume the structures you built in Steps 15 and 16 became significantly larger. Give the name of each resulting polymer.

18. Are the chains you built *linear* or *branched*? How would this affect the properties of the polymer?

19. Build a polyethene chain that includes some cross-linking.

   a. How does cross-linking change the behavior of the model, if at all?
   b. How would this cross-linking change the observed properties of the polymer?

C.5 MORE BUILDER MOLECULES

So far, you have examined only a small part of the inventory of builder molecules available to chemical architects. Now you will explore two classes of compounds in which carbon atoms are joined in *rings* rather than in chain structures.

As a first step to doing this, picture a straight-chain hexane molecule:

$$CH_3—CH_2—CH_2—CH_2—CH_2—CH_3$$

Next, remove one hydrogen atom from the carbon atom at each end. Then imagine those two carbon atoms bonding to each other. The result is the molecule known as *cyclohexane:*

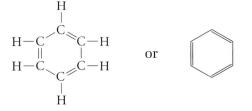

Hexane   Cyclohexane

Cyclohexane, obtained by petroleum refining, is a starting reactant for making nylon, a familiar petrochemical polymer. Cyclohexane is representative of the **cycloalkanes,** which are saturated hydrocarbons made up of carbon atoms joined in rings.

Another class of hydrocarbon builder molecules is **aromatic compounds.** Aromatic compounds have chemical properties distinctly different from those of the cycloalkanes and their derivatives. The structural formula of benzene (C_6H_6), which is the simplest aromatic compound, is shown in the following illustration. In the representation on the right, each "corner" of the six-carbon (hexagonal) ring represents a carbon atom with its hydrogen atom:

Although chemists who first investigated benzene proposed these structures, the chemical properties of the compound did not support their models. Recall that carbon–carbon double bonds (C=C) are usually very reactive. But, in terms of its reactivity, benzene behaves as though it does not contain any such double bonds. A deeper understanding of chemical bonding was needed to explain benzene's puzzling structure.

Substantial experimental evidence indicates that all carbon–carbon bonds in benzene are identical. Thus, its structure is not well represented by alternating single and double bonds. Instead, chemists often represent a benzene molecule in the following way:

The inner circle represents equal sharing of bonding electrons among all six carbon atoms. The hexagonal ring represents the bonding of six carbon atoms to each other. "Each corner" in the hexagon ring is the location of one carbon atom and hydrogen atom, thus accounting for benzene's formula, C_6H_6.

Nylon-Formation & Chemical Structure

The pleasant odor of the first aromatic compounds that were discovered prompted their descriptive name.

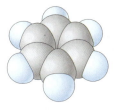

Benzene, C_6H_6

Although only small amounts of aromatic compounds are found in petroleum, fractionation and cracking produce large amounts of these aromatic compounds. Gasoline contains benzene and other aromatic compounds as octane-number enhancers, but they are used primarily as builder molecules. Entire chemical industries (dye and drug manufacturing, in particular) are based on the unique chemistry of aromatic compounds (Figure 3.42).

C.6 BUILDER MOLECULES CONTAINING OXYGEN

In assembling molecular models with C, H, and O atoms, you probably discovered one of the following compounds:

$$CH_3{-}OH \qquad CH_3{-}CH_2{-}OH$$

Methanol (methyl alcohol)   Ethanol (ethyl alcohol)

Figure 3.42 *Aromatic compounds are the primary builder molecules for dyes such as these.*

As you can see, each molecule has an $-OH$ group attached to a carbon atom. This general structure is characteristic of a class of compounds known as **alcohols.** The $-OH$ group is recognized by chemists as one type of **functional group**—an atom or group of atoms that imparts characteristic properties to organic compounds. If the letter R is used to represent all the rest of the molecule other than the functional group, then the general formula of an alcohol can be written as

$$R{-}OH$$

Any alcohol

> The $-OH$ alcohol functional group should not be confused with the hydroxide anion, OH^-, found in ionic compounds such as sodium hydroxide, NaOH.

In this formula, the line indicates a covalent bond linking the oxygen of the $-OH$ group with the adjacent carbon atom in the molecule. In methanol (CH_3OH), the letter R would represent CH_3-; in ethanol (CH_3CH_2OH), R would represent CH_3CH_2-.

The formulas and structures of two common alcohols follow. What would R represent in each compound?

$$CH_3CH_2CH_2OH$$

1-Propanol

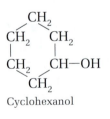

Cyclohexanol

Two other classes of oxygen-containing organic compounds, **carboxylic acids** and **esters,** are versatile builder molecules. The functional group in each of these classes of compounds contains two oxygen atoms, as shown at the top of the next page:

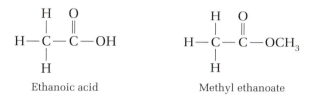

Carboxylic acid   Ester

Note that both classes of compounds have one oxygen atom double-bonded to a carbon atom and a second oxygen atom single-bonded to the same carbon atom. *Ethanoic acid* (a carboxylic acid that is better known as *acetic acid*) and *methyl ethanoate* (an ester more commonly called *methyl acetate*) are examples of these two classes of compounds:

Ethanoic acid   Methyl ethanoate

Some familiar examples that include or involve the use of alcohols, carboxylic acids, and esters are shown in Figure 3.43. These builder molecules can be modified by adding other functional groups that include nitrogen, sulfur, or chlorine atoms. The rich variety of functional groups greatly expands the types of molecules that chemists can ultimately build.

Figure 3.43 *Examples that include or involve (a) carboxylic acids and (b) esters and alcohols.*

(a) carboxylic acids

(b) esters and alcohols

C.7 CONDENSATION POLYMERS

Earlier in this section, you learned that many polymers can be formed from alkene-based monomers through addition reactions. Not all polymers are formed in this way, however. Natural polymers, such as proteins, starch, cellulose (in wood and paper), and synthetic polymers, including familiar nylon and polyester, are also formed from monomers. However, unlike addition polymers, these polymers form with the loss of simple molecules, such as water, when monomer units join. The term **condensation reaction** applies to this second type of polymer-making process, and the resulting product is called a **condensation polymer.** Here is a simple representation of this process:

$$-R_1-O-H \ + \ H-O-R_2- \ \longrightarrow \ -R_1-O-R_2- \ + \ H-OH$$

Monomer 1 \qquad\qquad Monomer 2 \qquad\qquad Condensation polymer \qquad Water

> Recall that R stands for the "rest" of the molecule, or everything other than the functional group.

One common condensation polymer is polyethylene terephthalate (PET). This polymer is often used in large soft-drink containers. It has many other applications—as thin film for videotape and as Dacron for clothing, surgical tubing, and fiberfill. More than two million kilograms of PET are produced each year in the United States. Some examples of products made from PET are shown in Figure 3.44.

| **Figure 3.44** *Common uses of PET.*

While some PET may be recycled, material containing high quantities of additives, such as dyes or other polymers, is either disposed of in a landfill or is incinerated. Another option is a DuPont process that breaks down PET into its monomers, which are then repolymerized into high-quality PET products. This technology decreases the need for new petroleum-based builder molecules as well as decreases the quantity of PET discarded in landfills.

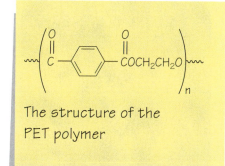

The structure of the PET polymer

Condensation reactions can be used to make small molecules as well as polymers. In the next investigation, you will use condensation reactions to produce several esters. These reactions illustrate how you can chemically combine organic compounds to create new, useful substances.

C.8 CONDENSATION

Introduction

In this investigation, you will produce several products through the condensation reaction of an organic acid (an acid derived from a hydrocarbon) with an alcohol. The esters you will produce have familiar, pleasing fragrances. Many perfumes contain esters; the characteristic aromas of many herbs and fruits arise from esters in the plants.

Lab Video: Condensation

Organic acids are also called *carboxylic acids*. See page 273.

One example of the formation of an ester is the production of methyl acetate from ethanoic acid (acetic acid) and methanol in the presence of sulfuric acid as a catalyst:

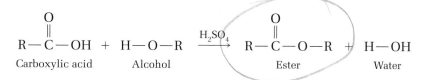

To emphasize the roles of functional groups in the formation of an ester, a general equation can be written using *R* notation:

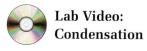

Note how the functional groups of the acid and alcohol combine to form a water molecule, while the remaining atoms join to form an ester molecule.

Using a chemical process similar to the one in the preceding reaction, you will now produce the ester known as *methyl salicylate.*

Before starting, read the procedure to learn what you will need to do, note safety precautions, and plan necessary data collecting and observations. Because this investigation involves the reactions of various alcohols and carboxylic acids, your data table should clearly communicate which substances you combined and the observed properties of the resulting esters.

Procedure

1. Prepare a hot-water bath by adding about 50 mL tap water to a 100-mL beaker. Place the beaker on a hot plate, and heat the water until it is near boiling. Be sure that no flames or sparks are in the area. (*Note:* Use a hot plate, *not* a Bunsen burner, to heat the water bath.)

2. Obtain a small, clean test tube. Place 10 drops of methanol into the tube. Next add 0.1 g salicylic acid. Then add 2 drops of concentrated sulfuric acid to the tube. (**Caution:** *Concentrated sulfuric acid will cause burns to skin or fabric. Add the acid slowly and very carefully. Other than water and sulfuric acid, the substances used in this investigation are volatile and flammable.*)

3. As you dispense these reagents, note their odors. (**Caution:** *Do not directly sniff any reagent fumes—some may irritate or burn nasal passages.*) Record any odors you happen to note.

4. Place the test tube in the hot-water bath you prepared in Step 1, as shown in Figure 3.45.

5. Keep the tube in the hot water, and do not spill the contents. Note any color changes. Continue heating for five minutes.

6. If you have not noticed an odor from the test tube after five minutes, remove the test tube from the water bath, hold the test tube away from you with the tongs, and wave your hand across the top of the test tube to waft any vapors toward your nose. See Figure 3.46. Record your observations about the odor of the product. Compare your observations with those of other class members.

7. Repeat Steps 1 through 6 using 20 drops of pentyl alcohol (not methanol), 20 drops of glacial (concentrated) acetic acid (not salicylic acid), and 2 drops of concentrated sulfuric acid. (**Caution:** *Avoid inhaling any glacial acetic acid fumes.*)

8. Now repeat Steps 1 through 6, this time using 20 drops of octyl alcohol, 20 drops of glacial (concentrated) acetic acid, and 2 drops of concentrated sulfuric acid.

9. Dispose of your products as directed by your teacher.

10. Wash your hands thoroughly before leaving the laboratory.

Figure 3.45 *Proper setup for heating substances with a water bath.*

Figure 3.46 *To avoid burning or irritating your nasal passages, gently "waft" vapors toward your nose with your hand.*

Questions

1. Write the molecular formulas of the acid and alcohol from which you produced methyl salicylate. (*Hint:* You may need to consult a chemistry reference book.)

2. Write a chemical equation for the formation of methyl salicylate.

3. Repeat Questions 1 and 2 for the second and third esters you produced.

4. Describe the odors of the three esters produced in this laboratory investigation.

5. Classify each of the following substances as a carboxylic acid, an alcohol, or an ester:

a. $CH_3-CH_2-CH_2-CH_2-OH$

b. $CH_3-\overset{\overset{\textstyle O}{\|}}{C}-O-CH_2-CH_3$

c. $CH_3-\underset{\underset{\textstyle CH_2-\overset{\overset{\textstyle O}{\|}}{C}-OH}{|}}{CH}-CH_2-CH_3$

d. $CH_3-\overset{\overset{\textstyle O}{\|}}{C}-OH$

C.9 BUILDER MOLECULES IN TRANSPORTATION

Making Decisions

You have surveyed how some petroleum components are used as building blocks for countless everyday objects. You can easily find examples of products built from petroleum-based materials in your surroundings, products that were unknown even a few decades ago. See Figure 3.47. Synthetic polymers have many advantages over traditional materials, including favorable strength-to-weight ratios and recyclability.

In the TV ad that opened this unit (see page 210), a new automobile was described as helping to conserve petroleum resources. Is this possible? Is it plausible? In this activity, you will consider this claim and evaluate how builder molecules are used in automobile construction.

Figure 3.47 *How many petroleum-based components do you see?*

1. List several automobile components commonly made of polymers.

2. The TV advertisement claimed that driving the *Hybris* will "help conserve petroleum resources."

   a. Does that claim refer to petroleum-based builder molecules used in the manufacture of the *Hybris*?

   b. Would the components on your list from Question 1 possibly differ between a *Hybris* and a traditional automobile? Why?

3. Try to think of a replacement material (not another polymer or petrochemical) for each component on your list.

4. Decide whether any disadvantages are associated with each replacement material you have suggested. List the two biggest disadvantages for each component, or write "none" if the replacement material provides similar properties at a comparable cost. A table such as the one shown here will help you organize your ideas and answers.

Summary For Question 4

Automobile Component	Petroleum Based? (yes or no)	Potential Replacement Material	Disadvantages of Replacement Material

5. Advocates of petroleum conservation typically emphasize increasing an automobile's energy efficiency, rather than replacing auto components made from petroleum's polymer-based builder molecules. Why might that be?

SECTION C SUMMARY
Reviewing the Concepts

The chemical combination of small, repeating molecular units (monomers) forms large molecules (polymers). Polymers can be designed or altered to produce materials with desired flexibility, strength, and durability.

1. How many repeating units are found in each of these structures?
   a. a monomer
   b. a dimer
   c. a trimer
   d. a polymer

2. Draw pictures of polymers with and without cross-links. How does cross-linking alter the properties of a polymer?

3. List four examples of natural polymers and four examples of synthetic polymers.

4. What structural features make the properties of one polymer different from those of another?

5. List and explain two methods for altering the characteristics of a polymer.

Unsaturated hydrocarbons that contain one or more double bonds are called alkenes. Unsaturated hydrocarbons that contain a triple bond are called alkynes.

6. Why is the term *unsaturated* used to describe the structures of alkenes and alkynes?

7. Rank the following in order of decreasing chemical reactivity: alkyne, alkane, and alkene. Explain your answer.

8. Why is an unsaturated hydrocarbon likely to be more reactive than a saturated hydrocarbon?

9. Write the molecular formula for a compound named
   a. octane.
   b. pentyne.
   c. decene.

10. How does the presence of a double or triple bond affect the ability of a hydrocarbon molecule to rotate about its carbon chain?

11. Draw a Lewis dot structure and a structural formula for each of the following molecules:
    a. propane
    b. propene
    c. propyne

12. Why is it impossible to have a sample of *methene,* a one-carbon alkene?

Addition reactions involve the chemical combination of monomers to form a polymer.

13. Use structural formulas to illustrate how propylene (propene), shown below, can polymerize to form polypropylene:

$$
\begin{array}{c}
\quad\ \ \text{H}\ \ \text{H}\ \ \text{H} \\
\quad\ \ |\ \ \ |\ \ \ | \\
\text{H} - \text{C} - \text{C} = \text{C} - \text{H} \\
\quad\ \ | \\
\quad\ \ \text{H}
\end{array}
$$

14. Explain why alkanes cannot be used to make polymers.

15. What change occurs in the monomer structure during polymerization?

Ring compounds include cycloalkanes and aromatic compounds.

16. In what ways are cycloalkanes different from aromatic compounds?

17. Draw a structural formula and write the molecular formula for cyclopentane.

18. How did the term *aromatic compounds* originate?

19. Why is the circle-within-a-hexagon representation of a benzene molecule a better model than a hexagon with alternating double bonds?

20. What do the six "corners" in the hexagon-shaped benzene representation signify?

Functional groups—such as alcohols, carboxylic acids, and esters—impart characteristic properties to organic compounds.

21. a. Write the structural formula for a molecule containing three carbon atoms that represents
    i.   an alcohol.
    ii. an organic acid.
    iii. an ester.
    b. Circle the functional group within each structural formula.
    c. Name each compound.

22. a. What does the R stand for in ROH?
    b. What type of compound would this be?

23. What functional group is responsible for the pleasing odors of many herbs, fruits, and perfumes?

Condensation reactions involve the chemical combination of two larger molecules with the loss of a small molecule.

24. Acetic acid, CH_3COOH, and butyric acid, $CH_3(CH_2)_2COOH$, are common reactants in condensation reactions. Their structural

$$CH_3-\overset{\overset{\displaystyle O}{\|}}{C}-OH \text{ and}$$

formulas are

$$CH_3-CH_2-CH_2-\overset{\overset{\displaystyle O}{\|}}{C}-OH.$$ Predict the products of a condensation reaction between each of these two acids and each of the alcohols in the following list. Provide the name and structural formula for the product. For example, the product of a condensation reaction between acetic acid and ethanol is *ethyl acetate*. (*Hint:* To organize your response, construct a table with two columns—one for each acid—and three rows—one for each alcohol.)
    a. methanol (CH_3OH)
    b. ethanol (C_2H_5OH)
    c. propanol (C_3H_7OH)

25. Why is the word *condensation* used to describe the reaction that forms esters? *[handwritten: → because water is produced as a product]*

26. Name three examples of polymers formed by condensation reactions.

Connecting the Concepts

27. Identify the missing molecule (represented by a question mark) in each of the following equations. (*Hint:* If you are uncertain about an answer, start by completing an atom inventory. Remember that the final equation must be balanced.)

a. CH_2=CH_2 + ? $\longrightarrow$ CH_3—CH_2Br

b. CH_3—$\overset{\overset{\displaystyle H}{|}}{C}$=$CH_2$ + ? $\longrightarrow$ CH_3—$\overset{\overset{\displaystyle}{|}}{\underset{\underset{\displaystyle OH}{|}}{CH}}$—$CH_3$

c. C_6H_{13}—$\overset{\overset{\displaystyle O}{\|}}{C}$—OH + ? $\longrightarrow$

C_6H_{13}—$\overset{\overset{\displaystyle O}{\|}}{C}$—$O^-Na^+$ + HOH

28. Assemble models of benzene and cyclohexane. Describe differences between the two molecular structures.

29. Architects have a saying, "form follows function." Does this saying apply to chemistry or is the opposite—"function follows form"—more correct? Explain your answer.

30. List five products manufactured from cyclic compounds.

31. How does adding a double or a triple bond change the shape of the carbon backbone in a hydrocarbon?

32. Draw structural formulas for both a saturated form of C_4H_8 and an unsaturated form of C_4H_8.

Extending the Concepts

33. Considering that petroleum is a natural resource, why aren't all petroleum-based polymers classified as natural?

34. Research and write about one example in which substituting a different functional group in the drug's molecular structure has had a dramatic effect on its medicinal properties.

35. Organic molecules that contain alcohol or carboxylic acid functional groups are often more soluble in water than are similarly sized hydrocarbon molecules. Explain.

36. Investigate several different polyesters. How are they formed? What explains their differences in properties? Give two structural formulas for polyesters. List five everyday products made of polyester.

37. C_2H_6O is the formula for two isomers; one has a boiling point of 78 °C and the other has a boiling point of −22 °C:

a. Draw the structural formula for each isomer.
b. Which of the two isomers has the higher boiling point? Explain your choice.

Energy Alternatives to Petroleum

Petroleum is a nonrenewable resource; its total available inventory on Earth is finite. Thus, people must eventually find other energy sources to meet needs of modern society. In this section, you will explore alternatives to petroleum and also learn about other emerging ways to power personal and commercial vehicles.

> The energy used in one week in the United States is approximately 2×10^{18} J. That is equivalent to the energy released by a hydrogen-bomb explosion, a severe earthquake, or by burning 2×10^{10} gallons of gasoline. By contrast, the radiant energy received by Earth *daily* from the Sun is approximately 1×10^{26} J; which is nearly 100 million times more.

D.1 ENERGY: PAST AND PRESENT

The Sun is our planet's primary energy source. Through photosynthesis, radiant energy from the Sun is stored as chemical potential energy. That is, green plants use the Sun's radiant energy to convert carbon dioxide and water (molecules involving chemical bonds with low potential energy) into carbohydrates and oxygen gas (molecules involving chemical bonds that have high potential energy). Thus, green plants convert solar energy into chemical potential energy, which is stored in the bonds of carbohydrate and oxygen molecules (Figure 3.48). When animals ingest and digest the plants and combine the carbohydrates with oxygen in a complex stepwise process, this chemical energy is released and used by the animals to form other organic molecules. Organic molecules found in plants and animals are called **biomolecules.**

> Even though energy is required in digestion to break the chemical bonds in carbohydrates and oxygen, the digestion of plants is, overall, exothermic. More energy is released by forming products with low-energy bonds than is required to break higher-energy bonds in the reactants. These ideas were introduced in Section B.1 (introduced on pages 238–241).

Figure 3.48 *Through photosynthesis, solar energy is converted into chemical energy stored in biomolecules of these plants. The challenge is to find the cleanest, most efficient approach to harnessing this chemical potential energy.*

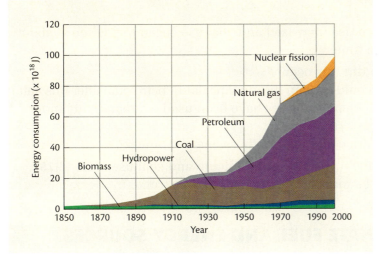

Figure 3.49 *Annual U.S. consumption of energy from various sources (1850–2000).*

Solar energy and energy stored in biomolecules are key energy sources for life on Earth. Since the discovery of fire, human use of stored solar energy held in the chemical bonds that make up wood, coal, and petroleum has had a major influence on civilization's development. In fact, the forms, availability, and cost of energy greatly influence how—and even where—people live.

In the past, abundant supplies of inexpensive energy were available. Until about 1850, wood, water, wind, and animal power satisfied the United States' slowly growing energy needs. Wood, then the predominant energy source, was readily available to most people for heating, cooking, and lighting. People used water, wind, and animal power, all forms of energy whose primary source was the Sun, for transportation and to power machinery and industrial processes. With industrialization and population growth, the demand for energy increased and the primary fuel sources changed.

Figure 3.49 illustrates how U.S. energy sources have changed since 1850. In the next activity, you will explore how energy supplies and fuel use have shifted in the United States during the past 150 years.

In 1850, the population of the United States was 23 million. By 2005, the U.S. population had increased to nearly 300 million.

Developing Skills

D.2 FUEL SOURCES OVER THE YEARS

As you can see from Figure 3.49, there has been a definite shift away from biomass (mainly wood, but also ethanol and waste) toward fossil fuels—coal, petroleum, and natural gas. Use Figure 3.49 to answer the following questions:

1. a. Find the dates of the period during which biomass (mainly wood) was used to meet at least half of the nation's total energy needs.
   b. What were the chief modes of travel during that period?
   c. What factors might explain the declining use of biomass after that period?
   d. What main energy source did people use to replace biomass?

2. Compared with other energy sources, only a small quantity of petroleum was used as fuel before the 1920s. What do you think petroleum's main uses might have been before that time?

3. Petroleum use increased and coal use began to decline at about the same time.

   a. When did this occur?
   b. What could explain the growing use of petroleum after this date?
   c. Which energy sources have been used comparatively more since 1975?
   d. What are major uses of these energy sources?

4. a. Describe the trends in petroleum and coal use since 1975.
   b. What factors could account for these trends?

D.3 ALTERNATE FUEL AND ENERGY SOURCES

Everyday life in the United States requires considerable quantities of energy. As you just learned, the energy sources used in this country have changed over time. As energy demands have accelerated, the nation increasingly has relied on nonrenewable fossil fuels—coal, petroleum, and natural gas. What is the future for fossil fuels, particularly petroleum?

The United States is a mobile society. Some 70% of the petroleum consumed in the United States is used for transportation. Although efforts to revitalize and improve public transportation systems merit attention, most experts predict our nation's citizens will continue to rely on personal vehicles well into the foreseeable future. And remember, even energy-conserving mass-transit systems must have a fuel source. What options, then, does chemistry offer to extend, supplement, or even replace petroleum as a primary energy source?

Oil Shale

One metric ton of oil shale typically contains the equivalent of 80 to 330 L of oil.

Petroleum from tar sands and oil shale rock is an option with some promise. See Figure 3.50. Major deposits of **oil shale** are located west of the Rocky Mountains. These rocks contain *kerogen*, which is partially formed oil. When the rocks are heated, kerogen decomposes into a material quite similar to crude oil.

Figure 3.50 *The people in this photo are searching for fossils within a pile of oil shale. When heated, oil shale will decompose into a liquid similar to crude oil. However, it takes several steps to remove impurities present in the oil shale.*

Unfortunately, vast quantities of sand and rock must be processed to recover this fuel. Moreover, current extraction methods use the equivalent of half a barrel of petroleum to produce every barrel of shale oil. Enormous volumes of water are also needed for processing, which poses a problem where water is scarce.

Coal Liquefaction

Because known coal (see Figure 3.51) reserves in the United States are much larger than known reserves of petroleum, another possible alternative to petroleum is a liquid fuel produced from coal. The technology to convert coal to liquid fuel (and also to convert coal to builder molecules) has been available for decades, having been used in Germany since the mid-1900s. Current coal-to-liquid-fuel technology is quite well developed here in the United States. However, the present cost of mining and converting coal to liquid fuel (commonly known as *coal liquefaction*) is considerably greater than that of producing the same quantity of fuel from petroleum. But if petroleum costs increase sufficiently, obtaining liquid fuel from coal—itself a nonrenewable resource—may become a more attractive option.

Figure 3.51 *Trains often transport coal, an important source of energy.*

One disadvantage of coal as an energy source is the air pollution generated when it burns. You will learn more about this in Unit 4.

Renewable Petroleum Replacements

Petroleum replacement candidates are not limited to other fossil fuels; currently there is rising interest in transitioning to renewable fuel sources. Certain plants, including some 2000 varieties of the genus *Euphorbia,* capture and store solar energy as hydrocarbon compounds, rather than as carbohydrates. These compounds may prove to be extractable and usable as a petroleum substitute.

Another such petroleum substitute under consideration is **biodiesel,** a fuel that can be burned in diesel engines. One major benefit of biodiesel is that any source of plant or animal fat can be converted into biodiesel. Although biodiesel must be refined for use in automobiles, it would offer a "green" strategy for producing a renewable petroleum substitute. Currently, biodiesel is sold blended with petroleum-based diesel and is seldom used as a pure fuel on its own. In the coming Investigating Matter D.4, you will have the opportunity to synthesize crude biodiesel from common vegetable cooking oil.

Figure 3.52 *Wind moves these turbines—generating electricity without burning fossil fuels.*

Other Energy Strategies

To meet some U.S. energy requirements, it is possible to move away from petroleum-based technology altogether. Alternative energy sources currently in use or under investigation include hydropower (water power), *nuclear fission* and *fusion, solar energy, wind energy* (Figure 3.52), burning *biomass*, and *geothermal energy*. Alternate approaches include constructing more energy-efficient buildings, vehicles, and machines, as well as using alternative fuels. In addition, reducing energy use—through such actions as carpooling, using public transportation, and turning off unneeded appliances—helps to conserve fossil fuels and to reduce energy needs overall. All of these steps are intended to further reduce our need to burn petroleum.

D.4 BIODIESEL FUEL

Introduction

Biodiesel is a mixture of methyl esters of fatty acids. It is made from any source of fat or oil. In this investigation, you will use ordinary vegetable cooking oil. As you will see, enough fuel can be produced by this procedure to be burned, although it is not pure enough to be used as a fuel in a car or truck. See Figure 3.53.

This laboratory synthesis is based on a chemical reaction that produces biodiesel and glycerol. Cooking oil is initially mixed with methanol; potassium hydroxide is added as a catalyst. The products separate into two layers, with biodiesel on top. The biodiesel is separated, washed, and then is ready for further testing.

Before starting, read the following procedure to learn what you will need to do, note safety precautions, and plan all necessary data collecting and observations. In setting up your data table, note that Step 17 calls for completing a second trial.

Figure 3.53 *This person makes fuel for the biodiesel-powered 1974 automobile from used cooking oil donated by a local restaurant.*

Procedure

1. After reading the procedure and the questions that follow, construct a data table appropriate for recording the data you will collect in this investigation.

2. Using a graduated cylinder, measure 100 mL of vegetable oil and pour it into a 250-mL beaker.

3. Carefully add 15 mL of methanol to the liquid in the beaker. (**Caution:** *Methanol is flammable and toxic.*)

4. Slowly add 1 mL of 9 M potassium hydroxide (KOH) to the liquid in the beaker. (**Caution:** *Potassium hydroxide is corrosive.*)

5. Taking turns with your partner, stir the mixture for 10 minutes. As you take turns stirring, set up the apparatus described in Step 13.

6. Allow the mixture to sit for two minutes so that it separates into two layers.

7. Using a Beral pipet, carefully remove the top layer (biodiesel) and place it into a clean 100-mL beaker.

8. Wash the biodiesel fluid by adding 10 mL of distilled water to it. Stir vigorously for several seconds.

9. Allow the mixture to sit for two minutes so that it separates into two layers.

10. Using the same Beral pipet as before, carefully remove the top layer (biodiesel) and place it into a clean, previously weighed 100-mL graduated cylinder.

11. Measure and record the mass and volume of biodiesel that you collected.

12. Carefully pour 5 mL of biodiesel onto a wad of steel wool (about the diameter of a quarter coin) in a metallic sample cup, as shown in Figure 3.54. Measure and record the total mass of the cup and its contents.

Figure 3.54 *Biodiesel being poured onto steel wool in a metallic sample cup.*

13. Set the metal sample cup underneath a soft-drink can, similar to the calorimeter setup shown in Figure 3.31, page 249. Measure 100.0 mL of chilled water provided by your teacher and pour the water into the soft-drink can.

14. Measure and record both the room temperature and the water temperature to the nearest 0.1 °C.

15. Ignite the biodiesel. Allow it to warm the water to the same number of degrees Celsius *above* room temperature as the number of degrees Celsius it started *below* room temperature.

16. Extinguish the flame by blowing on it. Allow the cup to cool; record the cup's total mass.

17. Repeat Steps 12 through 16.

Questions

1. What is the density of the biodiesel that you produced?

2. In producing the biodiesel sample, what changes did you observe in the characteristics of the starting materials compared to those of the final products?

3. What was the purpose of washing the biodiesel fluid in Step 8?

4. Compare the mass of biodiesel obtained (the yield) in your laboratory production of fuel with that of other laboratory teams.

5. All students followed the same laboratory procedure in producing biodiesel. What can account for the differences in biodiesel yield noted in Question 4 among teams?

6. Calculate the heat of combustion for your sample of biodiesel.

7. Gasoline produces 48 kJ/g of thermal energy when burned. How does your biodiesel compare to gasoline?

Making Decisions

D.5 BIODIESEL AS A PETROLEUM SUBSTITUTE

Consider these data, and then answer the following questions.

- ▶ 41 pounds of soybeans are required to make one gallon of biodiesel
- ▶ 1 bushel of soybeans weighs 60 pounds
- ▶ 38.1 bushels of soybeans, on average, are produced per acre (Figure 3.55)
- ▶ U.S. agriculture uses nearly 990 million acres (47.3% for crops, 52.6% for livestock) and has 74 million surplus acres
- ▶ biodiesel releases 123 540 kJ per gallon
- ▶ gasoline releases 120 560 kJ per gallon
- ▶ the production of one gallon of biodiesel or gasoline requires about 24 900 kJ

Figure 3.55 *Soybeans are one possible source of biodiesel. What benefits of biodiesel can you identify?*

Questions

1. Approximately 30 billion gallons of diesel fuel are used in the United States each year.
   a. How many pounds of soybeans would be required to produce enough biodiesel to meet this need?
   b. How many bushels of soybeans is this?
   c. How many acres would be required to grow the quantity of soybeans needed to produce enough biodiesel to meet all U.S. diesel needs?

2. Could the United States grow enough soybeans to replace all fossil diesel fuel with soybean-based biodiesel? Explain.

3. Create a list of pros and cons of using biodiesel as a petroleum replacement.

D.6 ALTERNATE-FUEL VEHICLES

As you now know, personal vehicles consume a significant portion—about 50%—of petroleum burned for fuel in the U.S. Because there is a limited supply of petroleum as a resource and emissions are produced by petroleum-burning engines, alternate-fuel vehicles are being developed, tested, and used. What are some of these fuels, and how are they used to power the vehicles? What are advantages and disadvantages of various alternative fuels? The overview that follows will help you prepare your own automobile advertisement.

Compressed Natural Gas

Most passenger vehicles and buses can be converted to dual-fuel vehicles that run on either natural gas or gasoline. Natural gas, mainly methane (CH_4), is produced either from gas wells or during petroleum processing. Compressed and stored in high-pressure tanks, this product is commonly known as **compressed natural gas (CNG).** A refillable CNG tank, capable of powering an automobile up to 300 miles, can be comfortably installed in a car's trunk. Many CNG-powered vehicles are operating worldwide, particularly in government and mass transit fleets. See Figure 3.56.

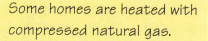

Some homes are heated with compressed natural gas.

Among the advantages of CNG are wide availability and an 80% decrease (compared to gasoline) in carbon monoxide (CO) and nitrogen oxide (NO_x) emissions. However, refueling systems require a compressor, which increases the vehicle cost by $2000 to $4000. Also, there is a higher fire risk resulting from a collision of a CNG-powered vehicle.

Figure 3.56 *Natural gas–powered buses line up in front of Freeborn Hall at the University of California, Davis. The buses replaced the 1960-era diesel buses on campus; one will run on a mixture of natural gas and hydrogen—the first of its kind in the nation.*

Fuel-Cell Power

Fuel Cells

One developing option for generating electricity to power vehicles is the **fuel cell.** It did not become a practical energy source until the 1960s, when fuel cells were designed and used in the U.S. Space Program. Any fuel containing hydrogen (such as methanol or natural gas) can be used in a fuel cell.

As shown in Figure 3.57, one common form of fuel cell converts oxygen gas and hydrogen fuel into electrical energy and water, which is its only emission. From a chemical viewpoint, such an operating fuel cell represents another way to release and harness the chemical potential energy stored in the chemical bonds that hold hydrogen and oxygen molecules together:

$$2 \, H_2 + O_2 \longrightarrow 2 \, H_2O + \text{Electrical energy (and some thermal energy)}$$

The fuel cell involves platinum electrodes that catalyze the removal of electrons from hydrogen atoms. In one common type of fuel cell called a *proton exchange membrane* (PEM) fuel cell (see Figure 3.58), electrodes are in contact with a polymer that allows hydrogen ions to pass through, but not hydrogen molecules. The electrons flow from the fuel cell through an external circuit (where they do useful work), returning to the fuel cell at the other electrode. See Figure 3.58. That second electrode catalyzes the reaction of oxygen gas (the oxidant) with hydrogen ions and electrons to produce water molecules, thus completing the electrical circuit. In the PEM hydrogen–oxygen fuel cell, the net products are water molecules and electrical energy.

Fuel cells require no electrical recharging and eliminate or substantially reduce release of air pollutants. They are more efficient than are internal-combustion engines; they can obtain more useful power from a given quantity of fuel than can be obtained by burning it. However, challenges remain in developing fuel handling and processing options, and reducing fuel-cell manufacturing costs and operating costs.

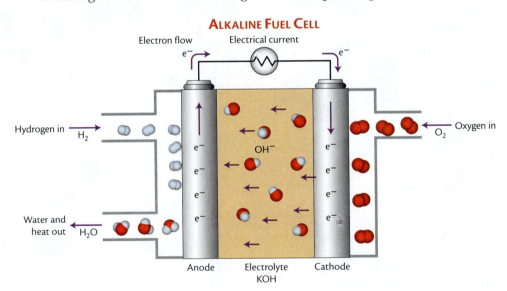

Figure 3.57 *The reaction* $2 \, H_2 + O_2 \longrightarrow 2 \, H_2O$ + *electrical energy takes place inside this fuel cell.*

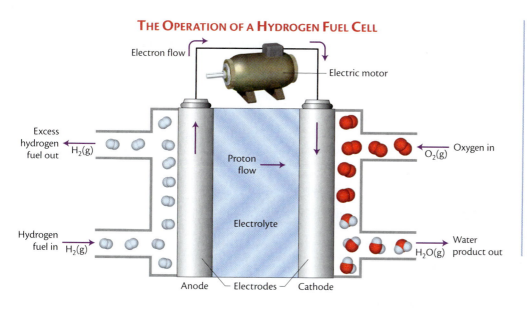

THE OPERATION OF A HYDROGEN FUEL CELL

Electron flow

Electric motor

Excess hydrogen fuel out — $H_2(g)$

Proton flow

Oxygen in — $O_2(g)$

Electrolyte

Hydrogen fuel in — $H_2(g)$

Water product out — $H_2O(g)$

Anode — Electrodes — Cathode

Figure 3.58 *Schematic diagram of a proton exchange membrane (PEM) fuel cell. Hydrogen ions form when electrons are removed from H_2 at the anode. The electrons flow through the circuit to the cathode, doing useful work along the way (note the electric motor in the circuit). The H^+ ions (protons) flow through polymer to the cathode, where they combine with oxygen gas and electrons to produce water.*

The expense (in terms of high energy costs) required to obtain high-purity hydrogen (H_2) for the type of fuel cell just described represents a major challenge. Currently, there is no commercial-scale energy-efficient way known to produce and distribute hydrogen gas in large enough quantities to make large-scale fuel cell use commercially viable. All current sources of hydrogen gas require petroleum-based hydrocarbons as a source of both the hydrogen gas and the energy required to produce it.

Hybrid Gasoline–Electric Power

Some car designers believe that a *gasoline–electric vehicle* will best meet consumer needs while reducing emissions and fuel costs. See Figure 3.59. *Hybrid gasoline–electric vehicles,* or **hybrid vehicles,** like the fictitious *Hybris* featured in this unit, come equipped with a gasoline-burning engine as well as a battery-powered electric motor. The batteries are recharged while driving, partially through conversion of the car's kinetic energy into stored chemical potential energy whenever the vehicle's brakes are applied.

As you learned in Section B, traditional gasoline-powered automobiles are unable to convert all the chemical energy stored in the bonds of gasoline and oxygen gas into useful mechanical energy. Figure 3.29 (page 244) documented that only 25% of the original chemical energy is converted into usable horsepower, while the other 75% is lost to the surroundings as thermal energy.

Figure 3.59 *In the past few years several commercial hybrid vehicles have entered the U.S. automobile market.*

A hybrid vehicle, however, is able to store some of the energy that would otherwise be "lost" and save it for later use. (See Figure 3.60.) In essence, chemical potential energy—originally stored in gasoline and oxygen—is converted into kinetic energy as the car accelerates. When the car possesses more kinetic energy than is required—for example, as it goes downhill or when brakes are applied—excess kinetic energy is converted back into chemical potential energy stored in the car's battery. (By comparison, when brakes are applied in traditional gasoline-

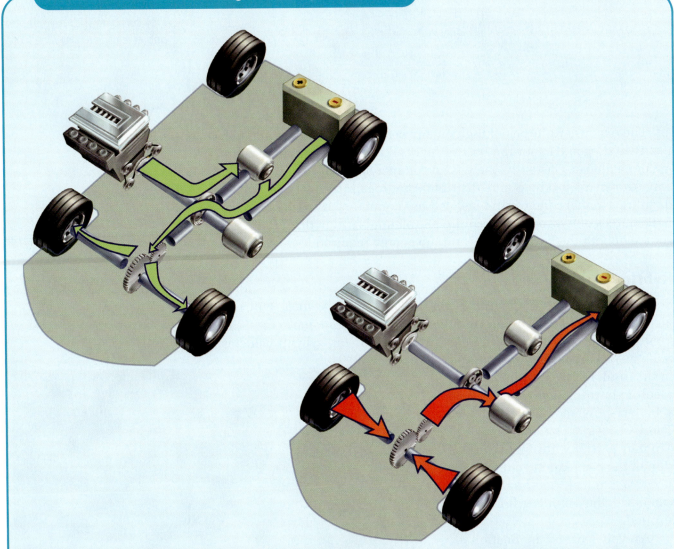

Scientific American Working Knowledge Illustration

Figure 3.60 *Hybrid car. Hybrid gasoline-electric vehicles represent alternatives to traditional gasoline-powered vehicles. Left: As the hybrid car accelerates (green arrows), it draws power from both the gasoline engine and electric storage battery. In some models, the electric battery provides all the needed power at low speeds, which also lowers gasoline consumption. Right: When brakes are applied (red arrows), some kinetic energy is transformed into chemical potential energy stored in the car's electric battery. This energy is used when the car again accelerates; in traditional gasoline-powered vehicles, all of this energy would be lost as heat through friction by the brakes.*

powered vehicles, the car's kinetic energy is converted into wasted thermal energy through friction as the brakes pads slow the car.) Additionally, hybrid cars are engineered to switch over from the gasoline engine to the electric motor whenever possible, such as when idling at a red light. This increases the car's overall energy efficiency; nearly 40% of the gasoline's chemical energy is converted into useful kinetic energy.

Hybrid vehicles typically achieve over 40 miles per gallon and can travel more than 650 miles between fueling stops. Although these vehicles produce the same kinds of emissions as fossil-fuel-burning vehicles, the quantity produced over a given distance is smaller.

Making Decisions

D.7 EVALUATING ALTERNATE-FUEL OPTIONS

Based on what you have just learned about alternate-fuel vehicles, create a table that will easily enable you to compare the three power options in terms of *fuel type, availability, efficiency, cost, environmental impact, renewability*, and *ease of refueling*. Work as a group to enter appropriate information in the table. Refer to the information in your completed table to answer the following questions:

1. While driving each of the three types of alternative-fuel vehicles, under what circumstances do they still rely on petroleum as a fuel? Explain.

2. Which of the three alternative-fuel vehicle options involve completely renewable fuel sources? Explain.

3. Which of the three options has the greatest potential for a completely renewable fuel source in the future? For each option you name, propose an approach that might support the transition from nonrenewable fuel to renewable fuel.

4. The *Hybris* advertised at the start of this unit is a hybrid vehicle. Why is this type of alternative-fuel vehicle currently the best-selling vehicle of the three alternate-fuel vehicle types?

5. Which of the three options do you consider the most promising overall for future use? Be prepared to list and explain the factors upon which you based your decision.

> The three power options are fuel cells, compressed natural gas (CNG), and hybrid gasoline-electric.

SECTION D SUMMARY
Reviewing the Concepts

Throughout human history, society has relied on a variety of biomolecules as energy sources.

1. What is a *biomolecule*?
2. Refer to Figure 3.49 (page 283). Identify the primary source of energy in the United States
   a. in 1900. b. in 1940. c. since 1975.
3. List three fuels that contain biomolecules used for energy.

4. What is the source of the energy stored in biomolecules?
5. Although petroleum has been used for thousands of years, only in relatively recent history has it become a major energy source. List three technological factors that can explain this.

Since petroleum is a nonrenewable resource, it will be eventually necessary to develop alternative energy sources.

6. List one advantage and one disadvantage of using each of the following alternate fuel sources:
   a. oil shale
   b. coal
   c. biodiesel

7. List three alternative energy sources that do not involve biomolecules.
8. What percent of petroleum consumed in the United States is used for transportation?

Diesel-fuel substitutes can be produced from renewable resources.

9. What is the primary functional group in biodiesel molecules?
10. Why was potassium hydroxide used in making biodiesel in Investigating Matter D.4 (Step 4, page 286)?

11. What are three advantages of biodiesel over petroleum-based diesel fuel?

Several types of alternative-fuel vehicles are being developed, tested, and used.

12. List an advantage and a disadvantage of each of the following alternative power sources for vehicles:
    a. compressed natural gas
    b. hydrogen fuel cell
    c. hybrid gasoline–electric
13. What are the emission products of fuel-cell powered vehicles?

14. What is the most common source of the hydrogen gas used in fuel cells?
15. Name the energy transformation that occurs in a hybrid vehicle when
    a. the vehicle is accelerated.
    b. the vehicle's brakes are applied.

Connecting the Concepts

16. Some energy authorities recommend exploring ways to use more renewable energy sources such as hydroelectric, solar, and wind power as replacements for nonrenewable fossil fuels.
    a. Why might this be a useful policy?
    b. Which of these renewable sources are least likely to replace fossil fuels at this time? Why?

17. If there is a dramatic increase in petroleum prices on the world market, explain which alternative source of energy you would expect to serve as the most practical substitute
    a. immediately.
    b. in ten years.

18. Suggest three ways individuals could reduce their total consumption of petroleum products.

19. We often consider alternative energy sources to be pollution free. Choose two alternative energy sources and evaluate them in terms of their possible impact on the environment.

20. Using Figure 3.49 on page 283 as a guide, make a pie chart showing the percent of all fuel sources used in
    a. 1900.   b. 2000.

21. Some people who do not own a car believe that they do not consume any fossil fuel. Evaluate their claim.

22. When biodiesel burns, it emits CO_2, like any other hydrocarbon-based fuel. Explain why it is not considered as a *net* contributor to atmospheric CO_2.

Extending the Concepts

23. Predict how your community would respond to a proposal for installing wind- or solar-power devices on a large scale to provide electrical power. Consider both advantages and disadvantages that might be considered in a town meeting.

24. Analyze how the relatively low cost and abundant availability of petroleum have affected the search for and development of alternative energy sources.

25. Identify and evaluate some state or federal programs that encourage development or use of alternative energy sources.

26. Experts disagree about how long fossil-fuel supplies will last. Research and evaluate some of their current opinions.

Putting It All Together

GETTING MOBILE

If you are an average U.S. high school student, you have already viewed about a half-million television commercials, many of them for automobiles. As alternative-fuel vehicles become more available, advertisements similar to the one that opened this unit will become more common.

To make informed, intelligent consumer decisions, you should analyze information conveyed in such advertisements. You can further develop your analytical skills by devising and defending your own product claims.

You will now draft and produce your own automobile advertising message. After presenting your commercial message, you will answer questions posed by other students.

DESIGNING AND EVALUATING VEHICLE ADS

Each student team will create and present an advertisement featuring an imaginary but plausible vehicle that uses a particular type of fuel. You can choose from these power options: biodiesel, CNG, hybrid gasoline-electric, or hydrogen–oxygen fuel cell. You will present the commercial message to the class for analysis and comment, based on ideas introduced and discussed in your chemistry class.

Each advertising message must meet certain specifications. Use these specifications to guide the development of your advertisement, as well as to evaluate the presentations of your classmates.

Timing

Your message should take no more than one minute of "air time."

Scientific Claims

All scientific claims must be accurate. Because the type of fuel (energy source) represents the unique feature of your vehicle, you should highlight and explain it briefly. If appropriate, compare your vehicle to those that depend solely on petroleum for fuel. Include implications of emissions released by your vehicle, fuel efficiency, and how, other than in its possible use as a fuel, petroleum is related to your vehicle. (*Hint:* Review your earlier responses to Making Decisions D.7 (page 293).)

Presentation

In addition to presenting a commercial message with accurate claims, your goal is to present an organized, visually stimulating, and motivating advertisement. Remember that any commercial message should interest a potential buyer in the product.

LOOKING BACK AND LOOKING AHEAD

This unit has illustrated once again that chemical knowledge can help inform personal and community decisions related to resource use and possible replacement options. Thus far in your *ChemCom* studies, you have focused on three types of resources: water, minerals, and petroleum. The next unit explores another kind of resource—the atmosphere—the thin, virtually invisible envelope of gases that surrounds Earth's surface. As you will learn, human interactions with the atmosphere have provoked important questions and concerns.

UNIT 4

Air: Chemistry and the Atmosphere

HOW does the composition of Earth's atmosphere affect atmospheric properties and behavior?

HOW does solar radiation interact with the atmosphere to influence conditions on Earth?

WHAT are major causes and consequences of acid rain?

HOW can air pollution be minimized?

The principal of Riverwood High School has just announced a new school bus-idling policy. You will investigate this decision for an article in the *Riverwood High Gazette.* What are the benefits and drawbacks of this policy? Turn the page to learn more about your reporting assignment.

NEW SCHOOL BUS-IDLING POLICY

The principal at Riverwood High School recently made the following announcement over the school intercom:

> "For our students' well-being and as a contribution to general environmental quality, effective next month we will require that all school bus drivers shut off their bus engines while waiting for students to board buses in the school parking lot or whenever buses are parked for several minutes."

Reporting Assignment

As a reporter for the *Riverwood High Gazette,* you have been asked to investigate the new school bus-idling policy. Your editor asked another reporter to interview the principal, as well as several students, teachers, and parents to find out what Riverwood High School community members think.

INTERVIEW NOTES

Luis Rodriguez (9th grade student): "If the buses aren't running, it could get pretty cold while we wait for them in the wintertime. How bad can bus exhaust actually be for us?"

Kayla Johnson (12th grade student): "I don't know why school bus idling would cause any problems. In science class we learned that burning gasoline, or any other hydrocarbons, produces just carbon dioxide and water vapor. They're already in the atmosphere, anyway."

Isaiah Martin (school parent): "I think it's a great idea. My son takes the bus, and I'm glad the principal and other school officials are concerned about his health. Plus, I've heard that bus exhaust creates acid rain, ozone depletion, global warming, and many other major problems."

Ms. Newton (social studies teacher): "I'm concerned about how much money this might cost the school district. I've heard that starting and stopping an engine is more expensive than letting it run. I realize there are benefits, but are they really worth it?"

Ms. Powers (principal): "The school board and I discussed this issue in depth before approving this policy. We learned that the city of Riverwood is establishing a similar policy for municipal buses. From what city officials told us, exhaust from buses negatively affects both the environment and public health. Students are our top priority and we feel this is a positive step in protecting their well-being."

Looking Ahead

Now that you have some idea of what a small sampling of people think about controlling school bus idling, your editor has asked you to create a special feature for the next edition of the *Riverwood High Gazette.* Your primary goal is to give the Riverwood High community the facts about this policy, including its possible benefits and drawbacks.

This unit will provide you with useful background on air quality and gas behavior that you will need to complete this assignment. As a reporter, you have been given considerable freedom in the design of the feature. However, your editor has also stated several requirements for the article. Your feature must

- ▶ address the particular concerns and opinions expressed by those who were interviewed.
- ▶ be scientifically accurate.
- ▶ support claims with information from credible sources.
- ▶ include diagrams that are attractive and informative.
- ▶ cite references for all the sources that were used to obtain your information or graphics (other than your textbook).

SECTION A

Gases in the Atmosphere

You have been asked to inform the Riverwood High community about the benefits and drawbacks of the new school bus-idling policy. In their interviews, Kayla Johnson, Isaiah Martin, and Principal Powers implied that bus exhaust changes the composition of the atmosphere. Is this true? To address this question, you will need some information. To start, you will find out what gases are naturally present in the atmosphere and how those gases behave under various conditions. The activities and information provided in Section A will help you learn more about atmospheric gases.

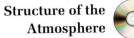

Structure of the Atmosphere

> The word *troposphere* is based on the Greek words *tropos*, which means turn, and *sphaira*, which means sphere.

A.1 STRUCTURE OF THE ATMOSPHERE

Most of the atmosphere's mass and all of its weather are located within 10 to 15 km of Earth's surface (see Figure 4.1). This is a very small distance compared to the size of Earth; if Earth were roughly the size of a basketball, the portion of the atmosphere called the **troposphere** would be relatively thinner than the basketball's pebbled texture that covers its surface. Gases mix continuously in the troposphere; therefore, its composition is reasonably uniform around the world. Table 4.1 lists the main components of the troposphere. Chemical analysis reveals that ancient air trapped in glacial ice has about the same chemical makeup as does our current atmosphere (Figure 4.2). This finding suggests that there has been

Figure 4.1 *Just how thin is the atmosphere that surrounds us? Looking up from Earth, the skies can seem endless. This view highlights the narrow strip of atmosphere that separates Earth from space.*

Figure 4.2 *Analysis of gases trapped in glacial ice shows that the composition of the atmosphere has not changed much since ancient times.*

COMPONENTS OF DRY TROPOSPHERIC AIR

Substance	Formula	Percent of All Gas Molecules
Major components		
Nitrogen	N_2	78.08
Oxygen	O_2	20.95
Minor components		
Argon	Ar	0.93
Carbon dioxide	CO_2	0.033
Trace components		
Neon	Ne	0.0018
Ammonia	NH_3	0.0010
Helium	He	0.0005
Methane	CH_4	0.0002
Krypton	Kr	0.0001

Table 4.1 *This chart identifies gases normally present in the tropospheric air closest to Earth's surface. The pie chart (right) illustrates the relative amounts of each gaseous substance contained in this air.*

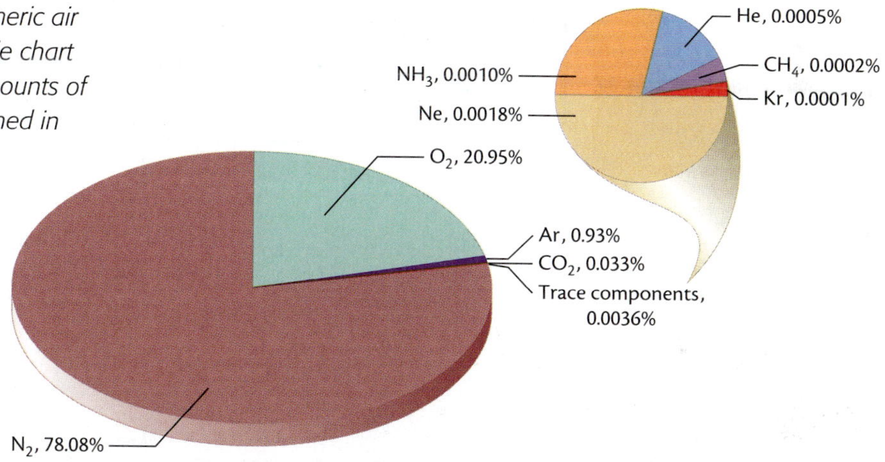

He, 0.0005%
CH_4, 0.0002%
Kr, 0.0001%
NH_3, 0.0010%
Ne, 0.0018%
O_2, 20.95%
Ar, 0.93%
CO_2, 0.033%
Trace components, 0.0036%
N_2, 78.08%

little change, except for carbon dioxide concentrations, in tropospheric air over a very long time.

Air is a mixture of gases. Nitrogen is the most abundant gas, followed by oxygen and substantially lower amounts of argon and carbon dioxide. In addition to the gases listed in Table 4.1, air samples can contain up to 5% water vapor, although in most locations the water vapor range is from 1 to 3%. Other gases naturally present in air at concentrations below 0.0001% (1 ppm) include hydrogen gas (H_2), xenon (Xe), ozone (O_3), nitrogen oxides (NO and NO_2), carbon monoxide (CO), and sulfur dioxide (SO_2). Human activity and natural phenomena, such as volcanic eruptions, can alter the concentrations of some trace gases and add other substances to air. This may lead to decreased air quality, as you will learn later in this unit.

The amount of water vapor in the air determines its humidity.

The only components of the atmosphere visible from space are condensed water vapor in the form of clouds and, at times, traces of colored nitrogen oxide (NO_x) pollutants near urban areas.

Suppose it were possible for you to fly from Earth's surface up into the farthest regions of the atmosphere. What do you think you would encounter as you traveled 5, 10, or 50 km away from Earth? Would the air temperature change? Would you be surrounded by the same mixture of gas molecules as your altitude increased? Although you cannot take a trip such as this right now, you can find answers to these and other questions about the atmosphere. Technology allows scientists to measure a range of air characteristics at different altitudes. In the following activity, you will analyze this type of data.

Developing Skills

A.2 GRAPHING ATMOSPHERIC DATA

Table 4.2 summarizes atmospheric data gathered at various altitudes. Use the information provided to answer the questions that follow:

The unit mmHg, millimeters of mercury, is a common unit of gas pressure (see page 312).

ATMOSPHERIC DATA

Altitude (km)	Air Temperature (°C)	Air Pressure (mmHg)	Mass (g) of 1-L Air Sample	Molecules in 1-L Air Sample
0	20	760	1.20	250×10^{20}
5	−12	407	0.73	150×10^{20}
10	−45	218	0.41	90×10^{20}
12	−60	170	0.37	77×10^{20}
20	−53	62	0.13	27×10^{20}
30	−38	18	0.035	7×10^{20}
40	−18	5.1	0.009	2×10^{20}
50	2	1.5	0.003	0.5×10^{20}
60	−26	0.42	0.0007	0.2×10^{20}
80	−87	0.03	0.00007	0.02×10^{20}

| Table 4.2

1. Predict the shape of the graph line for a *y* versus *x* plot of
   a. air temperature versus altitude.
   b. air pressure versus altitude.

2. Prepare graphs according to the following instructions. Your teacher will indicate whether you can use graphing software or graphing calculators for these tasks:
   a. Plot air temperature versus altitude data.
   b. Plot air pressure versus altitude data.
   c. Draw a best-fit line through the plotted points for each graph. (*Note:* The graph line may be straight or curved.)
   d. Does the shape of either graph differ from your prediction in Question 1? If so, how?

3. Compare ways that air temperature and air pressure change with increasing altitude.

   a. Which follows a more regular pattern?
   b. Explain this behavior.

4. Based on the data, would you expect air pressure to rise or fall if you traveled from sea level (0 km) to

   a. Pike's Peak (4301 m above sea level)?
   b. Death Valley (86 m below sea level)?

5. a. Suppose you gathered 1-L samples of air at several altitudes. How would the following values change with altitude?
      i. mass of the air sample
      ii. total molecules in the air sample

   b. If you were to plot the air-sample mass versus the number of molecules contained in the sample, what would the graph look like?
   c. Why?

6. Scientists often characterize the atmosphere as composed of four general layers. In order, they are the *troposphere* (nearest Earth's surface), the *stratosphere,* the *mesosphere,* and the *thermosphere* (outermost layer).

   a. Mark both graphs from Question 2 with lines at appropriate altitudes to indicate where you think the general transitions between the atmospheric regions might be.
   b. How confident are you in marking the mesosphere–thermosphere transition on your graph? Explain.

Death Valley

Investigating Matter

A.3 PROPERTIES OF GASES

Introduction

Kayla Johnson said in her interview (see page 300), quite correctly, that gases such as carbon dioxide and water vapor are naturally present in the atmosphere. Because atmospheric gases are generally colorless, odorless, and tasteless, you might doubt that they are forms of matter; they seem to be "nothing." However, gases do have definite physical and chemical properties, just as the materials in the other two states of matter—solids and liquids—do. In this laboratory investigation, you will complete activities that illustrate some properties of air.

Before starting this investigation, read the procedure carefully. Predict what you think will happen in each of the nine activities; write down your predictions. Also, note safety precautions and plan necessary data collecting and observations.

Properties of Gases

Lab Video: Exploring Properties of Gases

Procedure

Five stations, A through E, have been set up around the laboratory. At each station you will complete the activities indicated for that station. These activities can be done in any order; that is, work at Station D can be completed before Station B activities, and so on. Follow these general instructions:

- ▶ Reread the procedure.
- ▶ Review your prediction.
- ▶ Complete the activity.
- ▶ Record your observations.
- ▶ Restore the station to its original condition.

When you have completed your work at all five stations, answer the questions that follow.

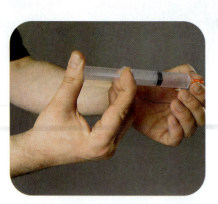

Figure 4.3 *Manipulating a sealed syringe (Activity 1).*

Station A

Activity 1

1. Draw some air into a syringe.
2. Seal the tip by placing the cap on the *open* end.
3. Holding the cap in place, gently push the plunger down with your thumb, as shown in Figure 4.3.
4. Release the plunger.
5. Record your observations.

Activity 2

1. Inflate and tie off two new balloons so that they are approximately the same size; about the size of a grapefruit.
2. Use tongs to submerge one inflated balloon in an ice–salt water bath.
3. Use tongs to submerge the second inflated balloon in a container of hot tap water.
4. Record your observations.

Station B

Activity 3

1. Inflate a balloon to approximately the size of a grapefruit and tie off the end.
2. Place the inflated balloon on a balance, using a piece of tape to hold it in place.
3. Record the mass of the balloon (and attached tape).

4. Remove the balloon from the balance. Use a pin to gently puncture the balloon near its neck, and release most of the gas contained within the balloon.

5. Place the deflated balloon on the balance, with the tape still attached, and record its mass.

6. Record your observations.

Activity 4

1. As shown in Figure 4.4, insert the rounded end of a new, uninflated balloon part way into an empty soft-drink bottle, stretching the balloon's neck over the mouth of the bottle.

2. Try to blow up the balloon so that it fills the bottle.

3. Remove and discard the used balloon.

4. Record your observations.

Figure 4.4 *Placement of a balloon in a bottle (Activity 4).*

Station C
Activity 5

1. Lower an empty drinking glass, with its open end facing downward, into a larger container of water.

2. With the open end still under water, slowly tilt the drinking glass.

3. Record your observations.

Activity 6

1. Fill a test tube to the rim with water.

2. Cover the test tube opening with a piece of stiff plastic.

3. Press down the plastic to make a tight seal with the mouth of the test tube.

4. While continuing to press the plastic to the test tube, invert the test tube above a sink or a pan.

5. Without causing any jarring, gently remove your hand from the piece of plastic.

6. Repeat the process with the test tube half-full of water.

7. Record your observations.

Station D

Activity 7

1. Fill a test tube to the rim with water.
2. Cover the test tube mouth with a piece of plastic wrap.
3. While continuing to press the plastic wrap to the mouth of the test tube, invert the test tube and partially immerse it in a container of water.
4. Remove the piece of plastic wrap.
5. Move the test tube up and down, keeping its lower (open) end under water.
6. Repeat the process with the test tube half-full of water.
7. Record your observations.

Activity 8

1. Locate the plastic bottle with a small hole in its side.
2. Cover the hole in the side of the bottle with your finger.
3. Fill the bottle with water.
4. Replace the cap tightly.
5. Holding the bottle over a sink, remove your finger from the hole.
6. Still holding the bottle over the sink, remove the cap.
7. Record your observations.

Station E

Activity 9

1. Place about 10 mL of water in a clean, empty aluminum soft-drink can.
2. Place the can on a hot plate and bring the water to a rapid boil.
3. Using tongs to handle the soft-drink can, quickly remove the can from the heat and immediately invert it into a container of ice water, as shown in Figure 4.5.
4. Record your observations.

Figure 4.5 *Use tongs to quickly invert a can containing boiling water into a container of ice water (Activity 9).*

Questions

1. Which of the previous activities suggest that air is composed of matter? Explain, using observational evidence to support your answer.
2. For each activity, briefly describe how well your predictions corresponded with the actual results you observed, and propose explanations for any differences between your predictions and the observed results of the activities.

3. For any three activities in this investigation,
   a. describe your observations in detail.
   b. explain the role of air in the activity.
   c. draw particle models that show the interactions between the gas particles and the other particles of matter.

4. Describe another investigation or experience that suggests that
   a. air is matter.
   b. air exerts pressure.

5. Complete this investigation at home or in the school lunchroom: Put one end of a straw in a glass of water. Hold another straw outside the glass. Place the ends of both straws in your mouth and try to drink the water through the straw in the glass. See Figure 4.6.

Describe what you observed when you tried this investigation. Based on your observations, what makes it possible to drink liquid through a straw?

Figure 4.6 *What will happen when you try to drink water through the straw inside the glass?*

A.4 PRESSURE

In everyday language, the word *pressure* can have many meanings. Perhaps you use it to mean that you feel too busy or that you feel forced to behave in certain ways. The greater the pressure, the more "boxed in" you feel. To scientists, pressure also refers to force and space, but in quite different ways.

In science, **pressure** refers to the force applied divided by the surface area upon which the force is applied:

$$\text{Pressure} = \frac{\text{Force}}{\text{Area}}$$

You can see from the equation that pressure is *directly* proportional to the force applied, and it is *inversely* proportional to the area upon which the force acts.

Think back to a common experience: Someone steps on your foot. The pressure (or maybe even pain) that you feel depends on the two variables involved in the previous equation—*force* and *area*. You surely can feel the difference between a light-weight person or a heavy person stepping on your foot, assuming that both people have the same size foot and wear similar shoes. The larger the force, in this case the weight, the more pressure that is exerted. If applied weight (force) is the same, then different shoe types (area) will affect the pressure (Figure 4.7).

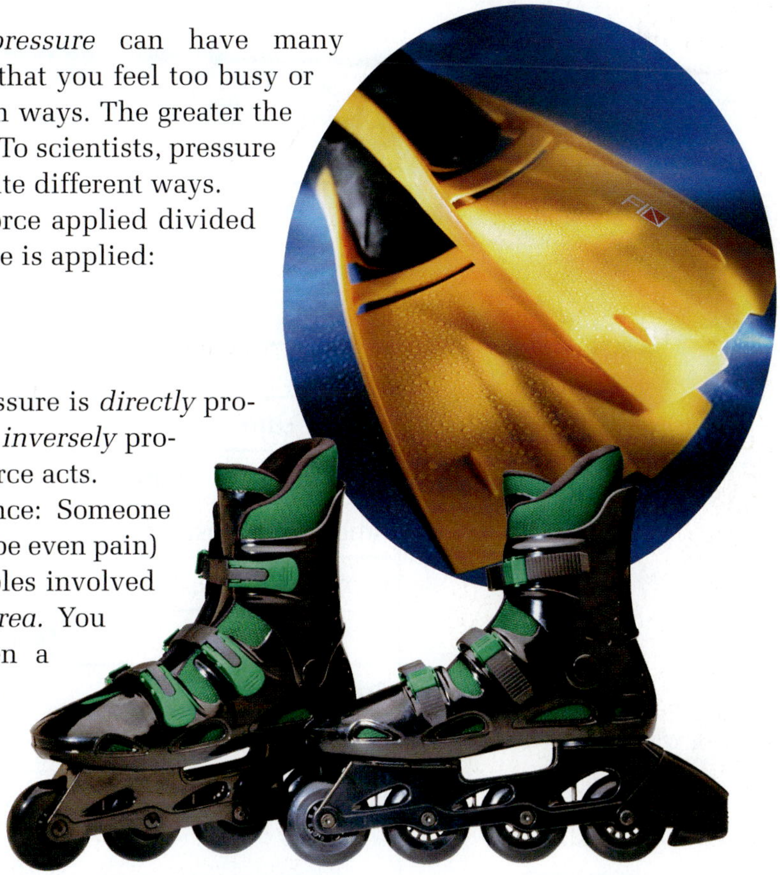

Figure 4.7 *If someone were to accidentally step on your foot, which "shoes" would you prefer that the person wore? Explain your choice in terms of the concept of pressure.*

In the same way that someone exerts pressure on your foot, gas molecules exert pressure on the walls of their container. You observed this in several recent laboratory investigations (see pages 306–308). For instance, in Activity 6 (page 307), you may have possibly concluded that air pressure holds the piece of plastic against the water in an inverted test tube. How much pressure did the air exert on the piece of plastic, and how could you measure this pressure?

You probably are familiar with several units used to report pressure. For example, U.S. weather forecasters report barometric air pressure in *inches of mercury.* When you check air pressure in car or bicycle tires, most likely your tire gauge reads *pounds per square inch,* or psi. Other pressure units you may have already heard of are *millimeters of mercury* (mmHg), and *atmospheres* (atm). With so many different units available for measuring and reporting pressure, how do scientists communicate with one another?

Scientists have agreed to use certain units when communicating results. This system of units, called the **International System of Units (SI**), allows scientists from around the world to communicate clearly with each other. Table 4.3 lists some SI units. Note that some units, such as *length* and *mass,* are **base units;** they express the fundamental

SI UNITS FOR SELECTED PHYSICAL QUANTITIES

The Seven Base Units of SI		
Physical Quantity	**Name**	**Symbol**
Length	meter	m
Mass	kilogram	kg
Time	second	s
Electric current	ampere	A
Thermodynamic temperature	kelvin	K
Amount of substance	mole	mol
Luminous intensity	candela	cd
Some Derived Units		
Area	square meter	m^2
	square centimeter	cm^2
Volume	cubic meter	m^3
	cubic centimeter	cm^3
Force	newton	N (equal to $kg \cdot m/s^2$)
Pressure	pascal	Pa (equal to N/m^2)

| Table 4.3

physical quantities of the modernized metric system. Others, such as *area, force, volume,* and *pressure,* are **derived units;** they are formed by mathematically combining two or more base units.

Recall that pressure can be expressed as *force per area.* It is useful to understand the SI units of force and area before learning about the SI derived unit for pressure. The SI derived unit for force is the **newton** (**N**). To visualize one newton of force, imagine holding a personal-sized bar of soap (with a mass slightly greater than 100 g) in your open hand. That bar of soap "pushes" downward on your open hand with a force of about one newton (1 N).

The **force** (or weight) of any object depends on its mass and gravitational attraction. **Area** refers to the total surface of an object. For a square or a rectangular surface, area is calculated by multiplying the lengths of its two adjacent sides. The SI base unit of length is the **meter** (m); thus, area is expressed as a derived unit—m × m, or square meters (m²).

When a one-newton (1-N) force is exerted upon an area of one square meter (1 m²), the pressure equals one *pascal* (1 Pa). The **pascal** (Pa) is the SI derived unit of pressure. One pascal (1 Pa) is a relatively small pressure; it is roughly the pressure exerted downward on a slice of bread by a thin layer of soft butter spread on its top surface. Because a pascal is so small, the *kilopascal* (kPa), a unit 1000 times larger than the pascal, is commonly used to express many pressures.

Pascal is pronounced so it rhymes with rascal. Some automobile tires specify inflation pressure in kilopascals (kPa), as well as pounds per square inch (psi).

A.5 APPLICATIONS OF PRESSURE

Developing Skills

Using what you have learned about pressure, answer the following questions:

1. Figure 4.8 shows two bricks of the same mass lying on the ground.

   a. Is each brick exerting the same total *force* upon the ground? Explain.
   b. Is each brick exerting the same *pressure* upon the ground? Explain.

2. Calculate the pressure, expressed in pascals, that the brick on the right side of the photo exerts upon the ground. It is exerting a force of 18 N, and the dimensions of the brick surface touching the ground are 9.3 cm × 5.5 cm. (Remember, there are 100 cm in 1 m.)

3. You need to chop some wood to build a fire. You quite obviously reach for an ax rather than a hammer to complete this task. Explain your choice in terms of the concept of pressure.

Figure 4.8 *Two similar bricks lie on the ground. Are they exerting the same pressure upon the ground?*

A.6 ATMOSPHERIC PRESSURE

You have just learned what pressure means to a scientist and how it can be expressed in quantitative units. Perhaps you are wondering how pressure relates to the gases in the atmosphere that are of concern at Riverwood High School. As you discovered in Investigating Matter A.3, the atmosphere exerts a force on every object within the atmosphere. (Figure 4.9). On a typical day at sea level, air exerts a force of about 100 000 N on each square meter of your body, resulting in air pressure of about 100 kPa. This pressure is also roughly equal to one *atmosphere* (1 atm), which is another unit commonly used by scientists and one with which you will become familiar. One atmosphere is equal to 101.3 kPa.

Pressure can also be expressed in units of *mmHg*, or *inches of mercury*. Such units suggest that air pressure can be measured as the height of a column of mercury. How is this possible? Recall that in part of a laboratory investigation that you completed (Activity 7, page 308), you covered a test tube filled with water and inverted it into a container of water. You then uncovered the test tube. What did you observe? What force supported the weight of the column of water in the test tube?

Now imagine repeating that laboratory investigation with a taller test tube and once again with an even taller test tube. If the test tube were tall enough, at a certain height water would no longer fill the tube entirely when it was inverted in a container of water. This puzzling effect was first observed in the mid-1600s. Scientists discovered that 1 atm of air pressure could support a column of water only as tall as about 10.3 m (33.9 ft). If the investigation was tried with even taller inverted test tubes, the water column still would reach only the 10.3-m level.

A *barometer* (a device that measures atmospheric pressure) based on a tube filled with water would be much too tall to be useful. Therefore, scientists replaced the water with mercury, a liquid 13.6 times denser than water. The resulting mercury column, illustrated by the barometer depicted in Figure 4.10, is shorter than the water column by a factor of 13.6. Thus, at 1 atm pressure, the mercury column has a height of 760 mm; 1 atm = 760 mmHg.

Figure 4.9 *Tornados are powerful reminders of atmospheric pressure.*

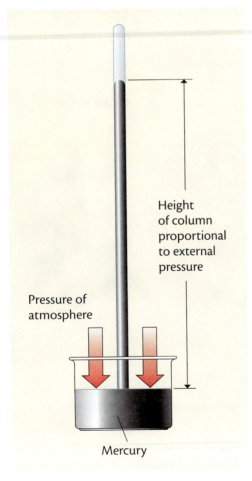

Height of column proportional to external pressure

Pressure of atmosphere

Mercury

Figure 4.10 *A mercury barometer. On a typical day at sea level, the atmosphere will support a column of mercury 760 mm high. The pressure unit of atmospheres is related to pressure in millimeters of mercury: 1 atm = 760 mmHg.*

A.7 ATOMS AND MOLECULES IN MOTION

As you get up on a cold winter morning, you look out the window to see clouds of exhaust coming from cars, trucks, and buses driving past your home. Your hand recoils from the hot spray coming from the shower head; it is too uncomfortable for you to enjoy your morning shower. Turning on too much cold water, however, leads to a chilling experience.

Your experiences with gases, liquids, and solids are all at a sensory level; that is, you observe matter and its changes through your senses of sight, smell, touch, taste, and hearing. In so doing, you arrive at some generalizations about the world around you. Making observations is the first step in "doing science." A more difficult task is crafting a theory that explains all observations and that also predicts the outcomes of investigations yet to be completed.

One such theory concerns the motion of atoms and molecules within all states of matter. What do you already know about the three states of matter from observing them? You know that a *solid* has a definite shape and that its structure is rigid. Experience also tells you that a *liquid* flows and, unlike a solid, depends on a container to define its shape. A sample of *gas*, likewise, does not have a definite shape. If you have blown up a balloon and then let go of it without tying it off, you also know that air rapidly leaves the balloon and the balloon changes shape.

The atoms, molecules, or ions that make up solids are held closely to one another. Although the particles in solids vibrate about a position, they are tightly packed and cannot move past one another. The molecules and ions within liquids are more mobile, they can move past each other; however, intermolecular attractive forces prevent them from moving too far apart. The arrangement of particles in solids and liquids are depicted in Figure 4.11.

> Scientific laws describe the behavior of events in nature, but they do not provide explanations for the observed behavior. Scientific theories and models attempt to offer how-type explanations of phenomena in the natural world.

> As you learned in Unit 3 (page 219), intermolecular forces are attractive forces between molecules.

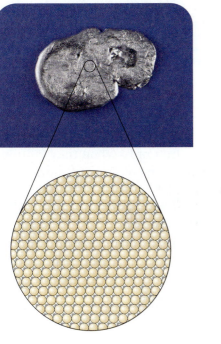 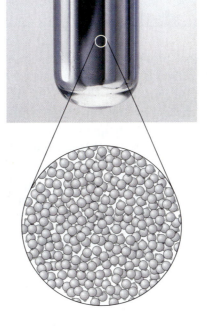

Figure 4.11 *You can observe the properties of solids (zinc, left) and liquids (mercury, right) using your senses. Models help explain how atoms interact with each other and their surroundings.*

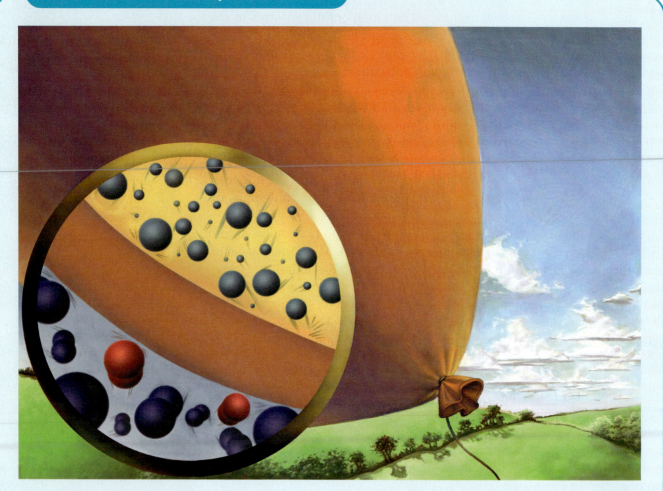

Figure 4.12 *Particulate-level model of gases.* A helium-filled balloon soars above the landscape. In magnification, helium atoms are depicted within the balloon, He(g). Nitrogen molecules (N_2, blue) and oxygen molecules (O_2, red), representing 98% of air, are shown outside the balloon. What does this model suggest about spacing and movement of gas particles compared to those in liquids and solids?

Most molecules in the gaseous state are not as strongly attracted to each other. In contrast to particles in solids and liquids, particles in gases are very far apart. In fact, their size is negligible compared to the great distances that separate them. Gas molecules move in straight-line paths at very high speeds and change direction only when they collide with each other or with another object. Collisions of gaseous molecules with the container walls produce gas pressure. Figure 4.12 illustrates the motion of particles in a gas sample. Compare this model of gases to the models of solids and liquids in Figure 4.11 (page 313).

Gas molecules move at average speeds that depend on how much kinetic energy they possess. Recall from Unit 3 (page 239) that kinetic energy, which sometimes is called *energy of motion*, is the energy associated with any moving object. Its value depends on both the mass of

Figure 4.13 *A technician in Wheat Ridge, CO, repairs a hail-damaged trunk panel. Was this damage more likely caused by pea-sized hail or golf ball-sized hail?*

the moving object and its velocity. Traveling at the same velocity, a more massive object has greater kinetic energy than does a less massive object (see Figure 4.13). Thus, a baseball has much greater kinetic energy than a ping-pong ball when both are traveling at the same speed. If two moving objects have equal mass, the object moving faster has greater kinetic energy. This helps to explain the difference in damage to a car bumper when it taps the wall of a parking garage at 10 km/h compared to a collision with a parked car at 30 km/h.

10 km/h = 6.3 mph

The behavior of moving gas molecules can be explained by the idea of kinetic energy. This explanation is based on several postulates, some of which you have just considered. The following list summarizes these postulates:

A postulate is a statement accepted as the basis for further reasoning and study.

▶ Gases consist of tiny particles (molecules) whose size is negligible compared with the great distances that separate the molecules from each other. On average, the separation between gaseous molecules at 1 atm pressure and 0 °C is about 11 times greater than the diameter (size) of the molecules.

▶ Gas molecules are in constant, random motion. They collide with each other and with the walls of their container or surrounding objects. Gas pressure is caused by molecular collisions with the container walls.

▶ Molecular collisions are *elastic*. This means that although individual molecules in a gas sample may gain or lose kinetic energy, there is no gain or loss in total kinetic energy from all of these collisions.

▶ At a given temperature, gas molecules have a range of kinetic energies. However, the *average* kinetic energy of the molecules is constant and depends only on the temperature of the gas sample. Therefore, molecules of different gases at the same temperature have equal average kinetic energies. As gas temperature increases, the average velocity and kinetic energy of the gas molecules also increase.

These postulates serve as the basis of the **kinetic molecular theory (KMT)** of gases, which can be used to explain various observations of gas behavior. The postulates can also be used to make predictions. For example, the theory can predict what will happen to the pressure of a gas sample if its volume is doubled, assuming no change in temperature. Read on to see how.

 **Gas Laws**

A.8 PRESSURE–VOLUME BEHAVIOR OF GASES

Earlier, when you pushed down on a sealed syringe filled with gas, you observed that a gas sample can easily be compressed. A gas sample can be compressed much more easily than can a liquid or solid sample. In other words, the volume of a gas sample is easily changed when an external pressure is applied to it.

Consider your experience with the syringe. The more pressure you applied to the plunger, the smaller the volume of gas held inside the syringe became. Consequently, the pressure of the gas sample trapped below the plunger increased. Think of it this way: As the syringe volume within which the gas particles are confined continues to become smaller, the particles collide with each other and with the syringe walls more frequently. The net force exerted by all of those collisions upon the interior walls of the syringe, or pressure, increases.

Assume that the pressure of gas inside the syringe was 1 atm before you started pushing in on the plunger (at this point, the atmosphere was exerting the only pressure on the plunger). If, by pushing on the plunger, the volume of gas in the syringe were then lowered to *one-half* of its original volume, the pressure of the gas sample would be *doubled.* If the gas volume in the syringe were reduced to *one-fourth* of its original value, the gas pressure would become *four times larger.* In each case, the gas pressure inside the syringe increases as the gas volume decreases (see Figure 4.14). Figure 4.15 illustrates this relationship between gas pressure and its volume.

> You observed that gas trapped in a syringe could be compressed in Activity 1, page 306.

Figure 4.14 *A student pushes a plunger into a closed syringe. What's keeping him from pushing it in all the way?*

Figure 4.15 *As pressure increases, the volume of gas trapped in the syringe decreases proportionally.*

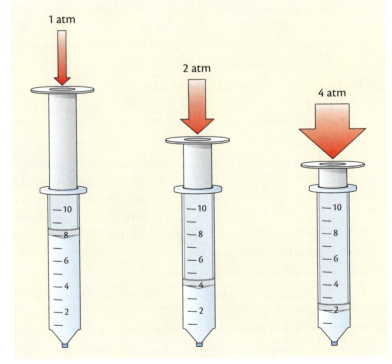

The Pressure–Volume Relationship

How can kinetic molecular theory help explain the gas behavior just described? Gas molecules are in constant, random motion; gas pressure is caused by molecules colliding with the walls of their container.

The relationship between the volume and the pressure of a gas sample at constant temperature can be described in several ways. The changes in volume and pressure predicted by KMT for a particular gas sample at constant temperature can be described with words or pictures, such as those shown in Figures 4.15 and 4.16. These changes can also be described graphically, as shown in Figure 4.17. Note the shape of the curve; this indicates an *inverse relationship* between gas pressure and volume.

This relationship is useful for predicting the new gas pressure or volume resulting from a change in one of these two variables. Again, consider the syringe shown in Figure 4.15.

Sample Problem: *You know that initially the gas sample occupies a volume of 8.0 mL and exerts a pressure of 1.0 atm. How would the pressure of the gas sample change if its volume were increased to 10.0 mL?*

Take a few moments to think about this question, using your own experiences with the syringe and the illustrations in Figure 4.15 to guide your thinking. If you increase the volume of the syringe by pulling out on the plunger, what will happen to the gas pressure within the syringe? Will the new pressure be less than or greater than the initial pressure? Did you reason that the new pressure would be *lower* than 1.0 atm? By now, you probably have gained an understanding of the relationship between gas volume and pressure that can guide you in predicting general results of such changes. Now you are ready to learn how to make *quantitative* predictions about this type of gas behavior.

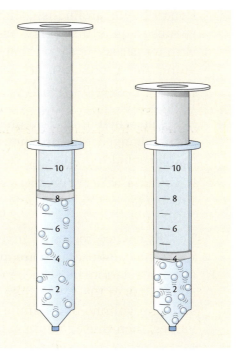

Figure 4.16 *As the volume of a sample of gas is reduced, the number of molecular collisions with the container wall—and thus the gas pressure—increases proportionally. The gas molecules in each syringe have been greatly magnified to depict the relative distances among them, and thus to account for the more frequent collisions with the container wall (and, thus, higher gas pressure) in the syringe on the right.*

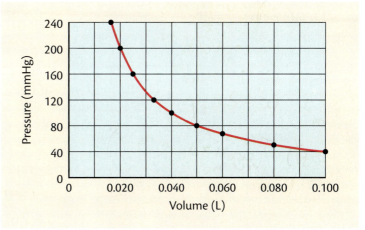

Figure 4.17 *The volume (V) of a gas sample, maintained at constant temperature and amount, is inversely proportional to its pressure (P). Therefore P × V is constant. A plot of pressure versus volume for any gas sample at constant temperature and amount will be similar to this one.*

Boyle's Law

Robert Boyle, an English scientist who studied gases, was the first to propose a quantitative law based upon the volume–pressure relationship for gases. He found that, for a sample of gas at constant temperature, the product of its measured pressure (P) and volume (V) remains the same, or $P \times V = k$. This relationship is known as **Boyle's law.**

Even when either the pressure or volume of that same sample of gas is changed, the product of pressure multiplied by volume will still be equal to that same value (k) if the temperature and amount of gas are kept constant:

$$P_2 \times V_2 = k$$

> An inverse relationship is a mathematical relationship, where an *increase* in one variable results in a *decrease* of another variable. However, the product of the two variables remains the same.

Boyle's law expresses an inverse relationship. In fact, when changes in volume and pressure are made to a gas sample at constant temperature, a relationship can be expressed between the initial and final conditions of the gas sample. If P_1 and V_1 represent the initial pressure and volume of a sample of gas and P_2 and V_2 represent its final values, then this equation can be written as

$$P_1 \times V_1 = P_2 \times V_2$$

or

$$P_1 V_1 = P_2 V_2$$

This equation, which is an expanded expression of Boyle's law, will help you to decide whether your prediction from page 317 was, in fact, correct.

$$P_1 = 1.0 \text{ atm} \longrightarrow P_2 = ? \text{ atm}$$

$$V_1 = 8.0 \text{ mL} \longrightarrow V_2 = 10.0 \text{ mL}$$

$$P_1 V_1 = P_2 V_2$$

$$1.0 \text{ atm} \times 8.0 \text{ mL} = P_2 \times 10.0 \text{ mL}$$

$$P_2 = \frac{(1.0 \text{ atm})(8.0 \text{ mL})}{10.0 \text{ mL}} = 0.80 \text{ atm}$$

The final gas pressure (0.80 atm) is less than the initial gas pressure: As gas volume in the syringe in Figure 4.15 (page 316) *increases,* the gas pressure *decreases.* Does this match your original reasoning?

In the following activity, you will explain several common observations based on what you know about the relationship between gas pressure and volume. You will also have the opportunity to apply Boyle's law to solve several pressure–volume gas problems.

A.9 PREDICTING GAS BEHAVIOR: PRESSURE–VOLUME

1. Explain each of the following observations:

   a. Even if they have ample supplies of oxygen gas, airplane passengers experience discomfort when the cabin undergoes a drop in air pressure.
   b. New tennis balls are sold in pressurized containers.
   c. After descending from a high mountain, the capped, half-filled plastic water bottle from which you drank while standing at the summit now appears dented or slightly crushed.

2. You buy helium gas in small pressurized cans to inflate party balloons. The can label indicates that the container delivers 7100 mL of helium gas at 100.0 kPa pressure. The volume of the gas container is 492 mL:

   a. Do you think that the initial pressure of helium gas inside the can is greater or less than 100.0 kPa? Explain.
   b. Calculate the initial pressure of helium gas inside the container.
   c. Was your prediction in Question 2a correct?

3. Two glass bulbs are separated by a closed valve (see Figure 4.18). The 0.50-L bulb on the left contains a gas sample at a pressure of 6.0 atm. The 1.7-L bulb on the right is evacuated; it contains no gas:

   a. Draw a model of gas molecules in the two glass bulbs before and after the middle valve is opened.
   b. Explain your model in terms of kinetic molecular theory.
   c. Predict, in general, what will happen to the total volume of the gas sample if you open the middle valve. Explain.
   d. Predict, in general, what will happen to the total pressure of the gas sample if you open the middle valve. Explain.
   e. Calculate the actual pressure of the gas sample after the middle valve is opened.

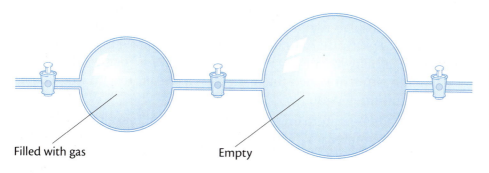

Figure 4.18 *Two glass bulbs are separated by a closed valve. The bulb on the left is filled with gas; the bulb on the right is empty. What would happen to the gas pressure if the middle valve were opened?*

Filled with gas   Empty

A.10 EXPLORING TEMPERATURE–VOLUME RELATIONSHIPS

Lab Video: Temperature–Volume Relationships

Introduction

Most matter is observed to expand when heated and contract when cooled. As you know, gas samples expand and shrink to a much greater extent than either solids or liquids. In this investigation, you will study how temperature changes influence the volume of a gas sample, assuming that pressure and the amount of gas remain unchanged. To do this, you will heat a thin glass tube containing a trapped air sample and record the changes in volume as the air sample cools.

Before starting, read the procedure to learn what you will need to do, note safety precautions, and plan necessary data collecting and observations.

Procedure 👓

1. Using two small rubber bands, fasten a capillary tube to the lower end of a thermometer. (See Figure 4.19.) Place the open end of the tube close to the thermometer bulb; about 5 to 7 mm above the bulb's tip.

2. Immerse the tube and thermometer in a hot oil bath that has been prepared by your teacher. Be sure the entire capillary tube is immersed in oil. Wait for your tube and thermometer to reach the temperature of the oil (approximately 100 °C). Record the temperature of the bath.

3. After your tube and thermometer have reached a steady temperature, lift them up until only about one-quarter of the capillary tube is still in the oil bath. Pause for about three seconds to allow some oil to rise into the tube. Then quickly place the tube and thermometer on a paper towel (to avoid dripping) and carry them back to your desk. (**Caution:** *Be careful not to touch the hot end of the thermometer or drips of hot oil.*)

4. Lay the tube and thermometer on a clean piece of paper towel on the desk. Make a reference line on the paper at the sealed end of the capillary tube. Also mark the upper end of the oil plug, as shown in Figure 4.19. Alongside this mark, write the temperature at which the mark was made, corresponding to that air-column length.

5. As the temperature of the gas sample drops, make at least six marks to represent the length of the air column trapped above the oil plug at various temperatures. Write the corresponding temperature next to each mark. Allow enough time so that the temperature drops by 50 to 60 °C.

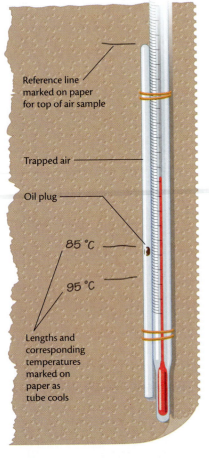

Figure 4.19 *Apparatus for studying how gas volume changes with temperature.*

6. When the thermometer shows a steady temperature (near room temperature), make a final observation of length and temperature. Discard the tube and the rubber bands according to your teacher's instructions. Wipe the thermometer clean.

7. Measure the length (in millimeters) from each marked line to the mark for the sealed end of the tube. Record each length of the gas sample. Have your teacher check your data before you discard your paper towel. Your teacher may ask you to submit your paper towel along with your report, so be sure to get specific instructions before disposing of the towel.

8. Wash your hands thoroughly before leaving the laboratory.

Data Analysis

1. Plot the length–temperature data for your air sample, with length on the vertical axis and temperature on the horizontal axis. The y-axis (length) should range from 0 to 10 cm; the x-axis (temperature) should include values from −350 to 150 °C. Label your axes, and arrange the scales so that the graph nearly fills the space available.

2. Draw the best straight line through your plotted points with a ruler. Using a dashed line, extend this straight graph line so that it intersects the x-axis. Use your completed graph to help you answer the following questions.

Questions

1. At what temperature does your extended graph line intersect the x-axis?

2. a. What would be the volume of your gas sample at that temperature?
   b. Why is that predicted volume only a theoretical value?

3. Renumber the temperature scale on your graph, assigning the value "zero" to the temperature at which your plotted graph line intersects the x-axis. The new scale now expresses temperature in *kelvins* (K), the **kelvin temperature scale.** One kelvin is the same size as one degree Celsius. However, unlike zero degrees Celsius, zero kelvins is the lowest temperature theoretically possible. It is called **absolute zero.**

4. Based on your graph,
   a. what temperature in kelvins (K) would correspond to 0 °C, the freezing point of water?
   b. what kelvin temperature would correspond to 100 °C, the normal boiling point of water?

A.11 TEMPERATURE–VOLUME BEHAVIOR OF GASES

Your plotted data from the preceding laboratory investigation should indicate the relationship between volume and temperature for a sample of gas at constant pressure. The volume–temperature relationship for gases was also demonstrated in Activity 2 (page 306), when you placed the two balloons of equal size into water baths of differing temperatures. As you observed, increasing or decreasing the temperature of a gas sample produces changes in its volume (Figure 4.20).

In the 1780s, French chemists (and hot-air balloonists) Jacques Charles and Joseph Gay-Lussac studied the changes in gas volume caused by temperature changes at constant pressure. Data for oxygen gas and nitrogen gas are shown in Figure 4.21. The plots for different gases and different sample sizes have different appearances. However, if all the graph lines are smoothly extended down to the x-axis (an extrapolation), the lines meet at the same low temperature. Lord Kelvin (an English scientist) used the work of Charles and Gay-Lussac to establish a simple mathematical temperature–volume relationship for gases, known as *Charles' law*. Lord Kelvin based his own new temperature scale on this relationship.

Doubling the kelvin temperature of a gas sample doubles its volume at constant pressure and amount of gas. Lowering the kelvin temperature in half causes the gas sample to decrease to one-half of its original volume, and so on. These relationships are summarized in **Charles' law**. At constant pressure and amount of gas, the volume (*V*) of a gas sample divided by its temperature in kelvins (*T*) is always a constant value:

$$\frac{V}{T} = k$$

where *k* is a constant. This is true even when the temperature and volume of that same sample of gas are changed, as long as pressure and amount of gas are kept constant:

$$\frac{V_2}{T_2} = k$$

Figure 4.20 *The gas held in hot-air balloons expands when heated.*

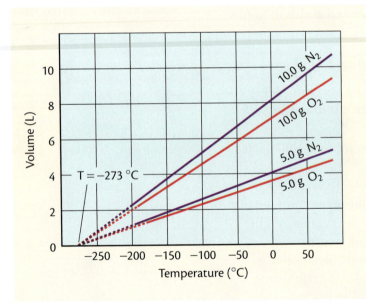

Figure 4.21
Temperature–volume measurements of various gas samples at 1.0 atm pressure. Extrapolation has been made for temperatures below liquefaction.

Because both expressions are equal to the same value (*k*), a temperature–volume relationship can be expressed between the initial and final state of the gas sample:

$$\frac{V_1}{T_1} = \frac{V_2}{T_2}$$

This expression of Charles' law is useful in predicting changes in gas volume and temperature at constant pressure and amount of gas.

A.12 PREDICTING GAS BEHAVIOR: TEMPERATURE–VOLUME

Developing Skills

Apply Charles' law to answer the following questions. For some questions, it may be necessary to convert temperatures to kelvins.

$$K = {}^{\circ}C + 273$$

1. What would happen to a balloon's volume, originally at 20 °C, if you took the balloon outdoors in a temperature of 40 °C? Assume that pressure is constant and the balloon does not allow any gas to escape.

2. In planning to administer a gaseous anesthetic to a patient,

   a. why must the anesthesiologist take into account the fact that during surgery the gaseous anesthetic is used both at room temperature (18 °C) and at the patient's body temperature (37 °C)?

   b. what problems might arise if the anesthesiologist did not allow for the patient's higher body temperature?

3. An air bubble trapped in bread dough at room temperature (291 K) has a volume of 1.0 mL. The bread bakes in the oven at 623 K (350 °C):

   a. Predict whether the air-bubble volume will increase or decrease as the bread bakes. Explain your prediction, using kinetic molecular theory.

   b. Calculate the new volume of the air bubble, using Charles' law. (*Hint:* Remember that Charles' law only applies to temperatures expressed in kelvins.)

4. You buy a 3.0-L helium balloon in a mall and place it in a car sitting in hot summer sunlight. The temperature in the air-conditioned mall was 22 °C, and the temperature inside the closed car is 45 °C:

   a. What will you observe happening to the balloon as it sits in the warm car?

   b. What will be the new volume of the balloon?

   c. Sketch two illustrations that depict the helium atoms in the balloon, one when the balloon was in the mall and one when it is in the car.

   d. Explain your illustrations, using kinetic molecular theory.

A.13 TEMPERATURE–PRESSURE BEHAVIOR OF GASES

Picture a closed cylinder of gas, such as a deep-sea scuba tank where volume is constant. See Figure 4.22. What would happen to the average kinetic energy of gas molecules in the cylinder if you were to heat the tank? How would this affect the gas pressure?

If you concluded that raising the kelvin temperature of the gas at constant volume should cause an increase in gas pressure, you are correct. Increasing temperature increases the average kinetic energy (and thus average velocity) of gas molecules. Because the molecules, on average, are traveling faster, the number of molecular collisions with the container walls increases, and the energy involved in each collision also increases. These effects cause a corresponding increase in gas pressure. The mathematical expression for this relationship is

$$\frac{P}{T} = k$$

Figure 4.22 *Scuba divers in the Arctic ocean (top) and in the tropics (bottom). Assuming the cylinders are of equal volume and contain the same amount of gas, which of the cylinder pressure-gauge readings will be higher?*

where k is a constant. Particular changes in gas pressure and temperature at constant volume can be found, using this equation:

$$\frac{P_1}{T_1} = \frac{P_2}{T_2}$$

You can also reason out an answer using your understanding of the temperature–pressure relationship. For calculations, remember that all gas temperatures must be expressed in kelvins.

Developing Skills

A.14 USING GAS RELATIONSHIPS

Solve the following problems, using appropriate gas relationships:

1. a. If the kelvin temperature of a gas sample held in a steel tank increases to three times its original value, predict what will happen to the pressure of the gas. Will it increase or decrease? By what factor do you expect the gas pressure to change?
    b. Draw two molecular models that represent the movement of gas molecules inside the tank at the two temperatures.
    c. Explain your models, using kinetic molecular theory.

2. A gas sample at a constant pressure shrinks to one-fourth its initial volume. What must have happened to its temperature? Did it increase or decrease? By what factor did the kelvin temperature of the gas change?

3. Explain why automobile owners in severe northern climates often add air to their tires in the wintertime and release some air from the same tires in the summertime.

4. Use kinetic molecular theory and gas laws to explain why a weather balloon expands in size as it rises from Earth's surface.

5. Why does the label on an aerosol container caution you not to dispose of the container in a fire?

A.15 NONIDEAL GAS BEHAVIOR

All gas relationships considered up to now have related to *ideal gases.* That is, a gas sample that behaves under all conditions as the kinetic molecular theory predicts is called an **ideal gas.** Most gas behavior approximates that of an ideal gas. Thus, such gas behavior is satisfactorily explained by the kinetic molecular theory. However, at very high gas pressures or at very low gas temperatures, real gases do not behave ideally. That is, the gas laws you have considered do not accurately describe gas behavior under such extreme conditions.

On average, gas molecules move slowly at very low temperatures. As their average kinetic energy decreases, the weak intermolecular attractive forces among molecules may become such a significant factor (when compared to their relative motions) that the gas condenses to a liquid, such as when air is liquefied. At very high gas pressures, if the temperature is not too high, gas molecules become so close together that these same weak forces of attraction may also cause the gas to condense to a liquid. These extreme temperature and pressure conditions are well beyond normal values for atmospheric gases.

A.16 UNDERSTANDING KINETIC MOLECULAR THEORY

Previously, you modeled the structure of matter and changes in matter using pictures and symbols. Another way that matter can be modeled is through the use of *analogies.* An analogy can help you relate certain features of an abstract idea or theory to a situation that is familiar to you. Read the analogy provided in the next paragraph and answer the questions that follow concerning the kinetic molecular theory of gases.

Imagine that you cause several highly elastic, small "super-bounce balls" to bounce around inside a box that you steadily shake; this serves as an analogy for gas molecules randomly bouncing around inside a sealed container. The balls bounce randomly around inside the box. See Figure 4.23.

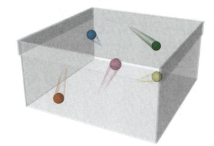

Figure 4.23 *Super-bounce balls moving about randomly inside a box.*

1. Decide which of these four gas variables—*volume, temperature, pressure,* or *number of molecules*—best matches each of the following factors, and explain each choice:

   a. the number of super-bounce balls inside the box
   b. the size of the box
   c. the vigor with which you shake the box
   d. the number and force of collisions with the box walls of the randomly moving super-bounce balls

2. How does each of the following changes relate to what you have learned about gases and the kinetic molecular theory?

   a. The vigor of shaking and the number of super-bounce balls remain the same, but the size of the box is decreased.
   b. The size of the box and the number of super-bounce balls remain the same, but the shaking becomes more vigorous.
   c. The size of the container and the vigor of shaking are kept the same, but the number of super-bounce balls is increased.

3. Suggest another situation similar to those in Question 2 that can serve as an analogy for the behavior of gases. Explain.

4. All analogies have limitations. For example, the super-bounce ball analogy fails to represent certain characteristics of gases. Gas molecules travel at *very* high velocities (on the order of 6000 km/h), while the super-bounce balls move much more slowly (on the order of 1 km/h). Suggest two other characteristics of actual gases that are not properly represented by this super-bounce ball analogy.

6000 km/h nearly equals 10 000 mph; 1 km/h is about 2 mph.

5. Describe your own analogy that might be useful for modeling gas behavior.

   a. Identify features of your analogy that relate to features of the kinetic molecular theory and *T*–*V*–*P* relationships for gases.
   b. Point out some key limitations of your analogy.

A.17 IDEAL GASES AND MOLAR VOLUME

Thus far, you have considered gas behavior under differing conditions of volume, pressure, and temperature. One variable that you have not yet fully considered is the actual *amount* of the gas sample; that is, the number of gas molecules contained in a particular sample. If you have the same volumes of oxygen gas, nitrogen gas, and carbon dioxide gas in three different balloons at the same temperature and pressure, how do the numbers of gas molecules compare? See Figure 4.24.

That same question was investigated in the early 1800s by the Italian lawyer and mathematical physics professor Amedeo Avogadro. By making careful observations of gas samples

Figure 4.24 *These three equally sized balloons contain three different gases at the same temperature and pressure. Are the total gas molecules in each balloon the same or different?*

like those you have worked with, he proposed: *Equal volumes of all gases at the same temperature and pressure contain the same number of molecules.* This important statement is commonly known as **Avogadro's law.**

One useful consequence of Avogadro's law is that *all* gases have equal molar volumes if they are measured at the same temperature and pressure. The **molar volume** is the volume occupied by one mole of a substance. At conditions of 0 °C and 1 atm, the molar volume of *any* gas sample is 22.4 L. (See Figure 4.25.) There is no similar simple relationship between moles of various solids or liquids and their corresponding volumes.

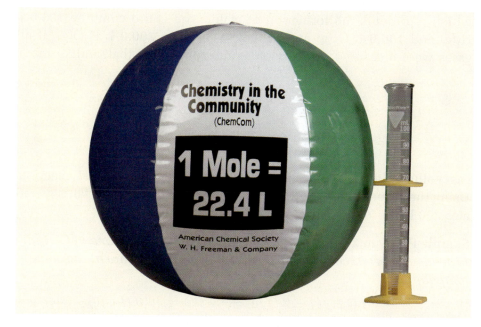

Figure 4.25 *This beach ball contains one mole of gas.*

The realization that all gases have the same molar volume under the same conditions greatly simplifies our thinking about chemical reactions involving gases. For example, consider the following chemical equations that involve gaseous reactions:

$$N_2(g) \ + \ O_2(g) \longrightarrow 2\ NO(g)$$

$$2\ H_2(g) \ + \ O_2(g) \longrightarrow 2\ H_2O(g)$$

You have learned that coefficients in chemical equations indicate the relative numbers of molecules or moles of reactants and products. Based on Avogadro's law, similar calculations can also involve *volumes* of gaseous reactants and products.

Hence, the first equation above can now be interpreted: 1 volume of $N_2(g)$ and 1 volume of $O_2(g)$ combine to form 2 volumes of $NO(g)$, assuming all gases were measured at the same conditions of temperature and pressure. Based on this information, consider the sample problem at the top of the next page.

Sample Problem 1: *What volume (in liters) of NO(g) is produced as 3 L N$_2$(g) and 3 L O$_2$(g) react? See equation, page 327.*

Because the amounts of both the reactants have tripled, the amount of product will also be tripled: 6 L NO(g) will be produced.

Similar relationships hold for the volumes of all gaseous reactants and products. Using Avogadro's law, you can interpret the coefficients in terms of gas volumes:

$$1 \text{ volume } N_2(g) + 1 \text{ volume } O_2(g) \longrightarrow 2 \text{ volumes NO}(g)$$

$$2 \text{ volumes } H_2(g) + 1 \text{ volume } O_2(g) \longrightarrow 2 \text{ volumes } H_2O(g)$$

The measured volumes could be expressed in any convenient units, such as liters (L) or cubic centimeters (cm^3). In the second equation, which represents the formation of water, you could combine 200.0 L H$_2$(g) and 100.0 L O$_2$(g) and expect to produce 200.0 L H$_2$O(g), if all gases were measured at the same conditions. Note that, unlike mass,

Figure 4.26 *Two volumes H$_2$(g) combine with one volume O$_2$(g) to form two volumes H$_2$O(g).*

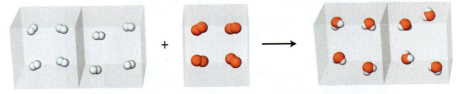

gas volumes are *not* necessarily conserved in a chemical reaction. Using the second equation, note that 200 L + 100 L of reactants generates 200 L of product, and *not* 300 L. See Figure 4.26.

Sample Problem 2: *In the following equation, what volume of C$_2$H$_6$(g) is needed to produce 12 L CO$_2$(g)? All measurements are at the same temperature.*

$$2 \, C_2H_6(g) + 7 \, O_2(g) \longrightarrow 4 \, CO_2(g) + 6 \, H_2O(g)$$

From the chemical equation, we know that 2 L C$_2$H$_6$(g) will produce 4 L CO$_2$(g). If we want to produce 12 L of CO$_2$(g), which is three times more than is represented in the equation, then we must start with *three times* as much C$_2$H$_6$(g); that is, 6 L C$_2$H$_6$(g).

Sample Problem 3: *Using this same scenario, how many liters of O$_2$(g) would be needed to produce 12 L CO$_2$(g)? Think through this question and make a prediction.*

Problems such as this can also be solved using ratios made from coefficients from the balanced equation. If we know from the previous equation that

$$7 \text{ L } O_2(g) \longrightarrow 4 \text{ L } CO_2(g)$$

and we want to know how many liters O$_2$(g) are needed to produce 12 L CO$_2$, we can set up a conversion factor to solve for the answer:

$$? \text{ L } O_2 = 12 \text{ L } CO_2 \times \frac{7 \text{ L } O_2}{4 \text{ L } CO_2} = 21 \text{ L } O_2$$

According to this calculation, 21 L O$_2$(g) are needed to produce 12 L CO$_2$(g). Was your prediction correct?

These volume relationships allow chemists to monitor gaseous reactions by measuring the gas volumes involved. In the activity that follows, you have an opportunity to check your understanding of this concept.

A.18 MOLAR VOLUME AND REACTIONS OF GASES

Developing Skills

1. What volume would be occupied by 3 mol $CO_2(g)$ at 0 °C and 1 atm?

2. In a certain gaseous reaction, 2 mol NO react with 1 mol O_2:

$$2 NO(g) + O_2(g) \longrightarrow 2 NO_2(g)$$

> 22.4 L = 1 mol of any gas at 0 °C and 1 atm (see Figure 4.25, page 327).

   a. Given the same conditions of temperature and pressure, what volume of $O_2(g)$ would react with 4 L NO gas?
   b. How might chemists use their knowledge of molar volumes to monitor the progress of this reaction?

3. Toxic carbon monoxide (CO) gas is produced when fossil fuels, such as gasoline, burn without sufficient oxygen gas. The CO can eventually be converted to CO_2 in the atmosphere. Automobile catalytic converters are designed to speed up this conversion:

   Carbon monoxide gas + Oxygen gas ⟶ Carbon dioxide gas

   a. Write the balanced equation for this conversion.
   b. How many moles of oxygen gas would be needed to convert 50.0 mol carbon monoxide to carbon dioxide?
   c. What volume of oxygen gas would be needed to react with 968 L CO? (Assume both gases are at the same temperature and pressure.)

A.19 THE IDEAL GAS LAW

Starting on page 316, you learned about the relationship between the pressure and volume of an ideal gas sample when its temperature is held constant (Boyle's law). Similarly, you are now familiar with the relationship between the temperature and volume of an ideal gas sample (Charles' law) and the relationship between the amount of gas (in moles) and the volume of the gas sample (Avogadro's law).

In 1834, a more general relationship—the **ideal gas law**—was derived from these experimentally based laws of researchers such as Boyle and Charles. All of the variables represented in the laws of Boyle, Charles, and Avogadro are included in the ideal gas law:

$$P \times V = n \times R \times T \quad \text{or} \quad PV = nRT$$

where P = pressure, V = volume, n = moles (amount) of gas, R = a constant, and T = temperature.

Figure 4.27 *What would you need to know to estimate the number of molecules of gas inside this blimp?*

As you probably noticed, the ideal gas law contains a constant (R). Before using this equation on page 329, we must find the value of the constant. Recall (see page 327) that the volume of one mole of any ideal gas at 1.00 atm and 0 °C is equal to 22.4 L. Substituting this information into the ideal gas law will allow us to find the value of R:

$$1.00 \text{ atm} \times 22.4 \text{ L} = 1.00 \text{ mol} \times R \times 273 \text{ K}$$

$$R = 1.00 \text{ atm} \times \frac{22.4 \text{ L}}{1.00 \text{ mol}} \times \left(\frac{1}{273 \text{ K}} \right)$$

$$R = 0.0821 \frac{\text{atm} \times \text{L}}{\text{mol} \times \text{K}}$$

The value of the constant, R, depends on the units in which pressure, volume, temperature, and amount of gas are expressed. The value of R given in the previous calculations is used when pressure is expressed in atmospheres, volume is measured in liters, temperature is expressed in kelvins, and amount of gas is expressed in moles.

Under particular circumstances, the ideal gas law can be reduced to each of the gas laws upon which it is based. Boyle discovered that for a gas sample at constant temperature, the product of its measured pressure (P) and volume (V) remains the same, or

$$P \times V = k$$

In this case, a relationship between pressure and volume is apparent if both the temperature and the amount of gas are kept constant. Consider these variables as you reexamine the ideal gas law:

$$P \times V = n \times R \times T$$

If n and T are both held constant, then the product $n \times R \times T$ must also be equal to a constant value. In other words, the equation could be rewritten as

$$P \times V = k$$

Figure 4.28 *Incandescent light bulbs are filled with inert (unreactive) gas. When the light is turned on, what will happen to the gas pressure inside the bulb? Why?*

The ideal gas law can be similarly simplified to Charles' law (when P and n are held constant), and to Avogadro's law (when P and T are held constant). However, holding certain variables constant is not always possible when working in the laboratory with gas samples. The ideal gas law is useful if you know the values of any three of the gas variables (P, V, n, or T) and seek to find the fourth value. See Figures 4.27 and 4.28.

Developing Skills

A.20 USING THE IDEAL GAS LAW

Solve the following problems, using the ideal gas law equation.

1. A 0.50-L canister of hazardous sarin gas was discovered at an old, abandoned military installation. To properly dispose of this gas, technicians must know how much gas is held in the canister. The gas pressure is measured as 10.0 atm at room temperature (25 °C). How many moles of sarin are held inside the canister?

2. What volume will 2.0 mol H_2(g) occupy at 40.0 °C and 0.50 atm pressure?

3. When the volume of a gas sample is measured, its pressure and temperature must also be specified.

   a. Why?

   b. That practice is normally not necessary in measuring the volumes of liquids or solids. Why?

Sarin is considered a very toxic substance. It prevents chemical signals from being properly transmitted between nerve cells.

Making Decisions

A.21 PREPARING TO EXPLORE BUS-IDLING POLICY

At the start of this unit, you learned about the new school bus-idling policy at Riverwood High School (see Figure 4.29, page 332). You have been assigned to write an article detailing facts, benefits, and drawbacks of this school policy. Thus far, you have learned how gas behavior is influenced by various conditions related to gas temperature and pressure. Now you are prepared to investigate the school bus-idling policy and the concerns and opinions voiced by those quoted at the opening of this unit (see page 300).

In the following sections, you will learn about air pollutants and air-quality concerns. In preparing to write your article, it will be helpful to decide what aspects of this policy you intend to investigate and what questions you have *prior* to obtaining further information on the policy.

1. In your group, revisit the opening of this unit, and reread the statements of Principal Powers and others about the school bus-idling policy. Generate a table of *tentative* benefits and drawbacks of the new policy as claimed by these statements. Keep in mind that the information included in those statements has not yet been verified.

Figure 4.29 *What are some benefits and drawbacks of the new school bus-idling policy?*

2. Review the benefits and drawbacks summarized in your group's table, and identify questions you have about this table and the statements upon which it is based. As a group, list at least five factual questions you plan to explore to verify interview statements. Some typical questions might be: *What emissions are typically found in bus exhaust? How, if at all, does bus exhaust affect human health?*

3. Once you have completed Steps 1 and 2, select one group representative to share your list of questions with the class. Listen carefully to the questions generated by the other groups, and see if you detect similarities among them. Guided by your teacher, identify major categories and topics within which questions fall.

Later in this unit, you will learn about many of the topics and questions you have identified in this activity. It is likely that you will also consult sources other than this textbook as you seek answers to questions you have posed in preparing to write your article.

In the next section, you will learn how radiation from the Sun interacts with gases in the atmosphere to maintain a habitable climate on Earth. Learning about the natural processes involved in maintaining Earth's climate will provide you with further insight into benefits or drawbacks of the school bus-idling policy at Riverwood High School. It will also enable you to better evaluate the concerns and opinions expressed by individuals interviewed at the opening of this unit.

SECTION A SUMMARY
Reviewing the Concepts

Earth's atmosphere is composed of a mixture of gases.

1. List the percent values of the four most plentiful gases found in the atmosphere.

2. Is the mixture of gases in the atmosphere considered a solution, a suspension, or a colloid? (*Hint:* See Figure 1.24, page 30.) Explain your answer.

3. Sketch and label four layers of the atmosphere.

4. a. Has the atmosphere's composition changed significantly throughout human history?
   b. What evidence can you cite to support your answer?

5. List three changes in the atmosphere as altitude increases from sea level upward.

Pressure involves a force applied over a particular area. Air pressure is often expressed in units of atmospheres (atm), kilopascals (kPa), or millimeters of mercury (mmHg).

6. It is much easier to slice a piece of pie with the edge of a sharp knife than with the edge of a pencil. Explain this observation in terms of applied pressure.

7. U.S. weather reports generally express air pressure in units of inches of mercury. During a severe storm, the barometric pressure can drop to as low as 27.2 inches of mercury. Convert this air-pressure value to
   a. millimeters of mercury (mmHg).
      (*Hint:* 1 inch = 25.4 mm)
   b. atmospheres (atm).
   c. kilopascals (kPa).

8. Heavy vehicles that must move easily over loose sand are often equipped with special tires.
   a. Would you expect these tires to be wide or narrow?
   b. Explain your answer using the concept of pressure.

9. Which is more likely to cause damage to a wooden floor: A 1068-N (240-lb) basketball player in athletic shoes (floor-contact area = 420 cm²) or a 534-N (120-lb) sports reporter standing in high heels (floor-contact area = 24 cm²). Support your answer with suitable pressure calculations.

The volume of a sample of gas in a flexible, impermeable container will increase if external pressure is reduced and decrease if external pressure is increased.

10. Explain why a sealed bag of potato chips "puffs out" (increases in volume) when it is carried from sea level up to a high mountain pass.

11. A small quantity of the inert gas argon (Ar) is added to incandescent light bulbs to reduce vaporization of tungsten (W) atoms from the solid filament. What volume of argon gas measured at 760 mmHg is needed to fill a 0.21-L light bulb to a pressure of 1.30 mmHg? Assume the argon gas temperature remains constant.

12. If the same bag of potato chips (see Question 10) were carried under water by a scuba diver, what would happen to its volume as the sealed bag descended into deep water?

13. A party-supply store sells a helium-gas cylinder with a volume of 1.55×10^{-2} m³. If the cylinder provides 1.81 m³ of helium for balloon inflation (at STP), what must be the pressure inside the cylinder?

The volume of a sample of gas in a flexible, impermeable container at constant pressure will increase if its temperature is increased and decrease if its temperature is decreased.

14. A 2.50-L balloon at 25 °C is inflated with helium. At constant pressure, the temperature of the balloon is then lowered to −5 °C. What will be the balloon's volume at this new temperature?

15. What do you think a marshmallow, composed mainly of small pockets of air, would look like at temperatures close to absolute zero? Explain your answer.

The pressure exerted by a sample of gas in a rigid, impermeable container will increase if its temperature is increased and decrease if its temperature is decreased.

16. Experts recommend measuring automobile tire pressure when the tires are cold rather than just after driving for several hours. Why?

17. Soccer players often notice that kicking the ball feels different on cold days compared to warm days. Explain why this might be so.

18. A steel tank at 299 K contains 0.285 L of nitrogen gas at 1.92 atm pressure. The tank will burst if its internal pressure exceeds 7.34 atm. At what temperature should the tank burst?

19. An oxygen gas sample in a rigid container is heated from 15 °C to 30 °C. By what factor will the oxygen gas pressure increase?

The kinetic molecular theory states that gases are composed of particles of negligible size that are in constant random motion and undergo elastic collisions. The average kinetic energy of a gas sample is directly related to its kelvin temperature.

20. Use the kinetic molecular theory to explain each of the following observations:
    a. Decreasing the volume of a gas sample at constant temperature causes the gas pressure to increase.
    b. At constant volume and constant amount of gas, the pressure of a gas sample changes if its temperature is changed.

21. Sketch the gas particles inside an inflated balloon submerged in
    a. hot water. b. ice water.

22. Think of a pool-table set up as a model (analogy) for gas behavior:
    a. How would you use pool balls to demonstrate the effect of an increase in the temperature of a gas sample?
    b. How would you use this model to demonstrate the effect of adding more gas molecules to the sample?
    c. In what ways is the pool-table model unsuccessful and misleading in modeling gas behavior?

> Equal numbers of gas molecules at the same temperature and pressure occupy the same total volume. One mole of a gas sample at 0.0 °C and 1.0 atm occupies a volume of 22.4 L.

23. What volume would 8.0 g helium (He) gas occupy at 0.0 °C and 1.0 atm?

24. How many moles of gas molecules would you expect to find in a 2.0-L bottle of air at 0.0 °C and 1.0 atm?

25. What would be the mass of gas inside a 3.0-L balloon at 0 °C and 1.0 atm filled with
    a. He(g)? b. CO_2(g)? c. CH_4(g)?

26. a. How many moles of oxygen gas would be present in a 1.0-mL gas sample at 0.0 °C and 1.0 atm?

    b. How many *molecules* of oxygen gas does your answer to Question 26a represent?

> The coefficients in a balanced chemical equation that involves gases indicate the relative volumes of gaseous reactants or products.

27. The following equation represents the production of ammonia (NH_3) by the reaction of nitrogen gas with hydrogen gas

$$N_2(g) + 3 H_2(g) \rightleftharpoons 2 NH_3(g)$$

    a. If 1 mol N_2(g) reacts with 3 mol H_2(g) in a flexible container at constant temperature and pressure, would you expect the total gas volume to increase or decrease? Why?
    b. How many moles of NH_3 would form if 2.0 mol N_2 react completely with hydrogen gas?
    c. How many grams of hydrogen gas would be needed to react completely with 2.0 mol N_2?

28. In a chemical reaction, 1 L hydrogen gas (H_2) reacts with 1 L chlorine gas (Cl_2) to produce 2 L hydrogen chloride gas (HCl). All volumes are measured at the same temperature and pressure. Create a depiction of this chemical reaction, using

    a. a sketch involving molecular models.
    b. a balanced chemical equation.

> For a given gas sample, the ideal gas law expresses the relationship among its volume, pressure, temperature, and amount of gas molecules present.

29. Determine the pressure exerted by 0.122 mol oxygen gas in a 1.50-L container at room temperature (25 °C).

30. At what temperature will 2.5 mol nitrogen gas exert a pressure of 10.0 atm in a 2.0-L container?

Connecting the Concepts

31. If the temperature of one mole of gas increases while its volume decreases, would you expect the gas pressure to increase, decrease, or remain the same? Explain your reasoning in terms of the gas laws.

32. The volume of a hot-air balloon remains constant while the temperature of the gas in the balloon changes. How is this possible?

33. Based on your knowledge of gas properties, predict what will happen to the density of a 100.0-g helium gas sample if you heat it from room temperature to 62 °C at constant pressure.

34. If air at 25 °C has a density of 1.28 g/L and your classroom has a volume of 2.0×10^5 L,
    a. what mass of air is contained in your classroom?
    b. assume that the room temperature is increased. How would this affect the total mass of air contained in the room if the room were
       i. tightly sealed?
       ii. not sealed?

Extending the Concepts

35. a. In what ways is an analogy involving an "ocean of gases" useful in thinking about the atmosphere?
    b. In what ways does that analogy *not* apply to the atmosphere's observed behavior?

36. Some people suggest that they could dive to the bottom of a swimming pool and stay there, breathing through an empty garden hose that extends above the water surface. *(Caution: Do not try this!)* How would you argue against this plan?

37. The gas behavior described by Boyle's law is a matter of life and death to scuba divers. On the water's surface, the diver's lungs, tank, and body are at atmospheric pressure. However, under water, a diver's body experiences the combined pressures of the atmosphere and the water:
    a. Why do scuba divers need to use pressurized tanks?
    b. What would happen to the tank volume if it were not strong enough to withstand the pressure of the outside water?
    c. How do the problems of a diver compare with those of a pilot climbing to a higher altitude in an unpressurized airplane?

38. Consider the behavior of an ideal gas, as described by the postulates of the kinetic molecular theory (see page 315). For each postulate, predict how gas behavior might change if the postulate did not hold.

39. Oxygen gas is essential to life as we know it. Earth's atmosphere contains approximately 21% oxygen gas. Would a higher concentration of atmospheric oxygen gas be desirable? Explain.

40. The kelvin temperature scale is used extensively in theoretical and applied chemistry work.
    a. Find out how closely scientists have approached a temperature of zero kelvins (0 K) in a laboratory setting.
    b. Results from research in cryogenics have suggested intriguing applications of low-temperature chemistry and physics concepts. Report on activities in this field.

Radiation and Climate

Without Earth's atmosphere, the midday sunshine would heat a rock hot enough to fry an egg, and the Sun's ultraviolet rays would burn exposed skin quickly; nights would be so cold that carbon dioxide gas would freeze to a solid. Because Earth's moon lacks an atmosphere, these extreme conditions are not science fiction; they exist there now (Figure 4.30).

The Sun's radiant energy and the gases that make up Earth's atmosphere combine to maintain a hospitable climate for life on this planet. Along with its vital role as an abundant reservoir for both oxygen gas and carbon dioxide gas, Earth's atmosphere provides protection from some of the Sun's ultraviolet radiation. Additionally, the atmosphere helps moderate Earth's temperature by controlling how much solar radiation is trapped close to Earth's surface.

In this section, you will learn about the nature of solar radiation, how it affects humans directly, and how it interacts with gases in the atmosphere.

Figure 4.30 *Since the Moon lacks an atmosphere, it has very different temperatures on its "Sun-facing" side than it does on its other side.*

B.1 THE ELECTROMAGNETIC SPECTRUM AND SOLAR RADIATION

The enormous quantity of energy produced by the Sun, which is Earth's main external source of energy, is a result of the fusion of hydrogen nuclei to form helium. *Nuclear fusion* is the combining of two nuclei to form a new, heavier nucleus, with the accompanying release of energy. Nuclear fusion occurs at high temperature and pressure, powering the Sun and other stars (Figure 4.31).

Solid carbon dioxide, $CO_2(s)$, is often called, somewhat misleadingly, *dry ice*.

You can learn more about nuclear fusion in Unit 6.

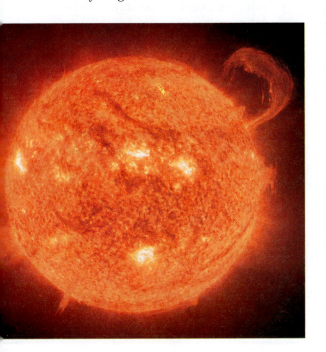

Figure 4.31 *The Sun is powered by nuclear fusion—a process that releases huge quantities of energy. This energy is used by growing plants, drives the hydrologic cycle, and warms the atmosphere on Earth.*

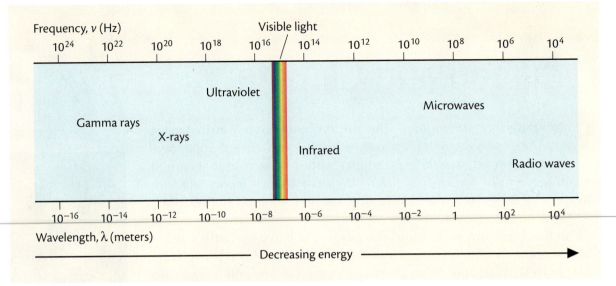

Figure 4.32
The electromagnetic spectrum.

Electromagnetic Radiation

Some energy liberated through nuclear fusion in the Sun is transmitted to Earth as **electromagnetic radiation,** which consists of a broad range of energetic emissions. It is useful to identify particular types of electromagnetic radiation, with each type representing a particular energy range, that together make up the **electromagnetic spectrum.** Figure 4.32 shows major types of electromagnetic radiation. Radio waves and microwaves are at the low-energy end of the spectrum; X-rays and gamma rays are at the high-energy end of the spectrum.

Between these extremes are infrared and ultraviolet radiation and the familiar visible spectrum, which is the only type of electromagnetic radiation that is visible to the unaided human eye.

Electromagnetic radiation is composed of **photons,** or bundles of energy. Photons travel as waves and, as you might expect, move at the speed of light. Unlike sound waves or ocean waves, electromagnetic waves do not require a medium, or substance, to support their movement. Electromagnetic radiation can move through a vacuum as well as through air and other media (Figure 4.33).

Photons of electromagnetic radiation have energy ranges characteristic of the type of radiation involved (such as visible, ultraviolet, and gamma radiation). All waves, including photon waves of electromagnetic radiation, involve oscillation. The rate of oscillation, or the number of waves that pass by a given reference point per second, is called **frequency.** The frequency of a wave is directly proportional to its energy; high-frequency radiation is also high-energy radiation.

Another characteristic of waves is *wavelength,* shown in Figure 4.34. The distance between the tops (or any corresponding part) of successive waves equals the **wavelength.** The wavelength and energy of a photon are inversely proportional; radiation with longer wavelengths is less energetic than is radiation with shorter wavelengths.

Photons can transfer their energy as they collide and interact with matter. The photon's energy, and thus either its wavelength or frequency, largely determines its effect on living things and other types of matter.

Figure 4.33 *Satellite dishes atop roofs in a French village receive electromagnetic signals from a satellite above Earth.*

Wavelength and frequency are inversely proportional.

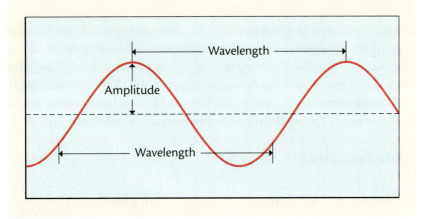

Figure 4.34 *Parts of a wave.*

Solar Radiation

The complete *solar spectrum* is shown in Figure 4.35. Most of the radiant energy emitted by the Sun is spread over a large portion of the electromagnetic spectrum: About 45% is in the infrared (IR) region, 46% is in the visible region, and 9% of this radiation is in the ultraviolet (UV) region. In the following paragraphs, you will be introduced to these three regions of the solar spectrum in more detail. Less than 1% of solar radiation falls outside of these three regions.

 Solar Radiation

Infrared Radiation

Electromagnetic radiation with frequencies slightly lower than that of red light is called **infrared (IR)** radiation. Infrared radiation causes certain bonded atoms to vibrate more energetically, which is observed as an increase in the material's temperature; infrared radiation is "heat" radiation. Most of the infrared radiation from the Sun cannot reach Earth's

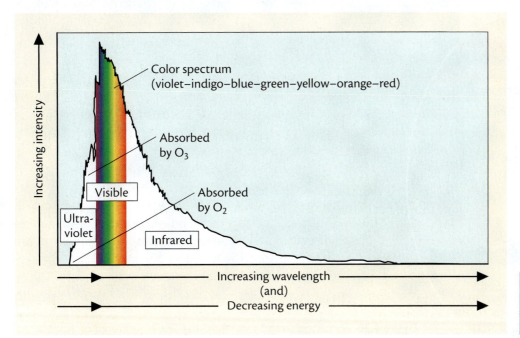

Figure 4.35 *The solar spectrum. Intensity, plotted on the y-axis, is an expression of the quantity of radiation at a given wavelength.*

In the 1860s, John Tyndall demonstrated the ability of CO_2 and gaseous H_2O to absorb radiant energy.

surface because it is absorbed by CO_2 and gaseous H_2O molecules in the atmosphere. However, some of the shorter wavelength (higher energy) solar radiation absorbed by Earth is transformed and reradiated as infrared energy. This reradiated energy is reflected back and retained by the atmosphere, a fact that will come into play later when we consider the role of CO_2 and certain other gases in the atmosphere.

Visible Radiation

On a clear day, more than 90% of the visible region of solar radiation directed toward Earth travels down to Earth's surface. The scattering of this portion of the Sun's radiation by water, air, and dust is the cause of red sunsets and blue skies, such as those depicted in Figure 4.36.

Photon–electron interactions also occur in the double bonds of certain molecules within your eyes, making it possible for you to see.

Visible radiation can energize electrons in some chemical bonds. An example of this is the interaction of visible light with electrons in chlorophyll molecules, which provides the energy needed for photosynthesis reactions.

Ultraviolet Radiation

There are three subcategories of **ultraviolet (UV)** radiation, all of which possess greater energy than does visible light. Of the three, *UV-A radiation* has the longest wavelengths and, thus, the lowest energy. *UV-B radiation* has more energy; it can cause sunburn; and, with long-term exposure, it is linked to skin cancer. *UV-C radiation,* the most energetic form of ultraviolet radiation, is useful for sterilizing materials because it can kill bacteria and destroy viruses. This is due to the fact that UV-C photons have enough energy to break covalent bonds. As a result,

Specially designed light bulbs are available that produce particular UV radiation suitable for specific applications, including use in tanning booths and sterilizing cabinets.

Figure 4.36 *Some visible radiation from the Sun is scattered by Earth's atmosphere. This scattering creates the colors we see in the sky.*

chemical changes can occur in materials exposed to some ultraviolet radiation, including damage to tissues of living organisms. UV-C consists of ultraviolet radiation with wavelengths shorter than 280 nm, UV-B wavelengths range from 280 to 320 nm, and UV-A radiation has wavelengths longer than 320 nm.

UV-C radiation is absorbed in the stratosphere before reaching Earth's surface. Most UV-B radiation, and much UV-A radiation, does not reach Earth's surface; it is absorbed by the stratospheric ozone layer, which you will learn more about later in this unit.

If all the UV radiation reaching the atmosphere actually reached the Earth's surface, it is likely that most life on Earth would be destroyed. Ultraviolet radiation, however, is not all bad. Humans and animals must have some exposure to it because vitamin D is produced when the skin receives moderate doses of ultraviolet radiation included in sunlight.

> Approximate Wavelengths
> Infrared: 10^3–10^4 nm
> Visible: 380–780 nm
> Ultraviolet: 180–380 nm
>   (1 nm = 10^{-9} m)

> Vitamin D is discussed further in Unit 7.

ChemQuandary 1

ALWAYS HARMFUL?

How accurate is the following statement: "All radiation is harmful and should be avoided"?

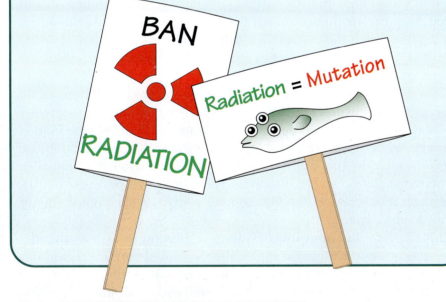

B.2 EARTH'S ENERGY BALANCE

The mild average temperature (15 °C, or 59 °F) at Earth's surface is determined partly by a balance between the inward flow to Earth of the Sun's energy and the outward flow into space of solar energy following its interaction with the Earth and its atmosphere. Certain properties of Earth's surface and atmosphere help determine how much thermal energy our planet can hold near its surface—where, of course, terrestrial life resides—and how much energy Earth radiates back into space. The combination of these two factors helps establish a balanced energy flow, leading to a hospitable climate here on Earth.

 Earth's Energy Balance

Figure 4.37 *One-third of the solar energy that reaches Earth powers the hydrologic cycle. Imagine how much thermal energy was generated as water vapor condensed in this developing thunderhead.*

Chlorofluorocarbons are discussed in more detail later in this unit.

About 30% of incoming solar radiation never reaches Earth's surface, but is reflected directly back into space by clouds and atmospheric particles. Solar radiation is also reflected when it strikes materials, such as snow, sand, or concrete on Earth's surface. In fact, visible light reflected in this way allows Earth's illuminated surface to be seen from space.

Of the remaining 70% of incoming solar radiation that actually reaches Earth's surface, about two-thirds is absorbed, warming the atmosphere, oceans, and continents. The other one-third of this energy powers the hydrologic cycle, which is the continuous cycling of water into and out of the atmosphere by evaporation and condensation. Solar energy causes water to evaporate from the oceans and land masses. The water condenses to form clouds, which then release water back to Earth as precipitation (see Figure 4.37).

Greenhouse Gases

All objects with temperatures above zero kelvins (0 K) radiate energy. The quantity of this radiated energy is directly related to an object's kelvin temperature. Specifically, Earth's surface reradiates most absorbed solar radiation, but usually at longer wavelengths (lower energy) than that of the original incoming radiation. This reradiated energy plays a major role in Earth's energy balance. Certain types of molecules in the air do not absorb the Sun's UV and visible radiation, allowing it to reach the surface of Earth, but absorb any infrared radiation that is reradiated from Earth's surface, thus holding warmth in the atmosphere.

Carbon dioxide and water readily absorb infrared radiation, as do methane (CH_4), nitrous oxide (N_2O), and halogenated hydrocarbons such as CF_3Cl and other chlorofluorocarbons (CFCs). Because clouds are composed of droplets of water or ice, they absorb infrared radiation. Energy absorbed by these molecules in the atmosphere is reradiated in all directions. Thus, energy can pass back and forth between Earth's surface and molecules in the atmosphere many times before it finally escapes to outer space.

This trapping and returning of infrared radiation by carbon dioxide, water, and other atmospheric molecules is known as the **greenhouse effect** because this process resembles, to some extent, the way thermal energy is held in a greenhouse (or in a closed car) on a sunny day. (See Figure 4.38.) Atmospheric gas molecules that effectively absorb infrared radiation are classified as **greenhouse gases.**

Without water and carbon dioxide molecules in the atmosphere to absorb and reradiate thermal energy back to Earth, scientists estimate that our planet would have an average temperature of a frigid −18 °C (0 °F). At the other thermal extreme is the planet Venus, demonstrating a runaway greenhouse effect. The Venusian atmosphere is composed of 96% carbon dioxide (and clouds made of sulfuric acid), which prevents the escape of most infrared radiation. As a result, the average surface temperature on Venus (450 °C) is much higher than that on Earth (15 °C).

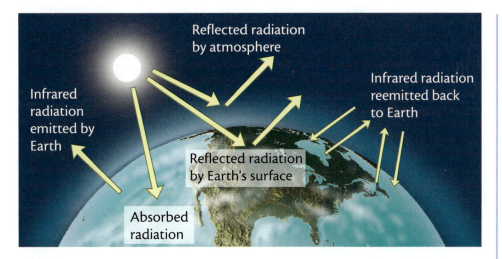

Reflected radiation by atmosphere

Infrared radiation emitted by Earth

Infrared radiation reemitted back to Earth

Reflected radiation by Earth's surface

Absorbed radiation

Figure 4.38 *When solar energy strikes Earth, clouds reflect about 30% back into space, and 70% moves through the atmosphere to reach Earth's surface. Energy that reaches the surface can be reflected or absorbed. Objects that absorb the Sun's UV and visible light reradiate it at lower-energy wavelengths such as infrared (IR) radiation, which is absorbed by atmospheric gases such as water, carbon dioxide, and methane (CH_4). IR radiation can be absorbed and reradiated by atmospheric gases and Earth's surface several times before escaping into space. This "greenhouse effect" helps maintain warm temperatures on Earth.*

Although some of this difference is due to their planetary positions relative to the Sun, Venus (the second planet from the Sun) is actually hotter than Mercury (the planet nearest to the Sun).

Climate Connections

In addition to maintaining a habitable average temperature on Earth, the interaction of solar radiation with Earth's atmosphere is a major factor in determining climates and weather. Radiant energy from the Sun warms Earth's land and water surfaces. Earth's warm surfaces, in turn, warm the air above them. As warmer air expands, its density decreases. This warmer air becomes displaced by colder, denser air, causing the warmer air to rise. These movements of warm and cold air masses help create continuous air currents that drive the world's weather (see Figure 4.39).

The differing thermal properties of materials on Earth's surface are another factor that influences climate. You will learn more about thermal properties, such as reflectivity and specific heat capacity, on pages 344–345.

Figure 4.39 *This hang glider depends on strong air currents as it is launched near San Francisco.*

The average or prevailing weather conditions in a region, commonly referred to as *climate,* are influenced by other factors. One factor is Earth's rotation, which causes day and night and also influences wind patterns. Other significant factors are Earth's yearly revolution around the Sun and Earth's tilt on its axis. The combination of these factors causes uneven distribution of solar radiation, which results in four distinct seasons in Earth's midlatitudes, a climatic pattern.

B.3 SOLAR RADIATION

Check your understanding of the interaction of radiation with matter and of radiation's role in controlling Earth's surface temperature by answering the following questions:

1. Why is human exposure to ultraviolet radiation potentially more harmful than exposure to infrared radiation?
2. Describe two essential roles played by visible solar radiation.
3. Explain why dry, arid regions in the United States, such as New Mexico and Arizona, experience wider air-temperature fluctuations from night to day than do states with more humid conditions, such as Florida.
4. Suppose Earth had a less dense atmosphere (fewer gas molecules) than it does now:
   a. How would average daytime temperatures be affected? Why?
   b. How would average nighttime temperatures be affected? Why?

If you had to walk barefoot across an asphalt parking lot on a hot, sunny day, you would probably find that carefully walking on the white, painted stripes is more comfortable than walking on the dark asphalt. Why?

B.4 THERMAL PROPERTIES AT EARTH'S SURFACE

If you live in the South or Southwest, or if you have visited these areas, you may have noticed that many cars are light colored, both inside and out. Do you know why? A particular property of materials helps keep these vehicles cooler in sunlight than those with darker surfaces.

Reflectivity

When solar radiation strikes a surface, some photons are absorbed, which increases the surface temperature, and other photons are reflected. The proportion of radiation that is reflected, expressed as the material's *reflectivity,* does not raise the object's temperature. Light-colored surfaces reflect more radiation, and therefore remain cooler, than dark-colored surfaces. For example, clean snow reflects almost 95% of solar radiation, whereas forests reflect very little. Variations in the reflectivity of materials at Earth's surface help determine local surface temperatures. See Figures 4.40 and 4.41. On a sunny, hot day, it is more comfortable to walk barefoot across a concrete sidewalk than across an asphalt parking lot. Walking across a lawn feels even more comfortable. The lawn reflects 15 to 30% of the Sun's rays, whereas black asphalt reflects almost no solar radiation. Instead, asphalt absorbs the radiation, which causes it to become hotter.

Figure 4.40 *Asphalt absorbs almost all solar radiation that strikes it, which causes asphalt roads to become quite hot and radiate heat.*

SPECIFIC HEAT CAPACITIES FOR COMMON MATERIALS AT 20 °C	
Material	**Specific Heat Capacity, J/(g·°C)**
air	1.00
aluminum	0.895
asphalt	0.92
brass	0.380
carbon dioxide	0.832
copper	0.387
ethyl alcohol	2.45
gold	0.129
granite	0.803
iron	0.448
lead	0.128
sand	0.29
silver	0.233
stainless steel	0.51
water (liquid)	4.18
zinc	0.386

Table 4.4

Specific Heat Capacity

Every material has a characteristic reflectivity and heat capacity, which together determine how much and how fast the material warms. As you already learned (refer to page 249), *specific heat capacity* is the quantity of thermal energy (heat) needed to raise the temperature of one gram of a particular material by one degree Celsius.

In effect, a specific heat capacity value for a material can be considered as its "storage capacity" for thermal energy. The lower a material's specific heat capacity, the higher its temperature will rise if a certain quantity of thermal energy is added. The higher the specific heat capacity of the material, the less its temperature will rise for a given quantity of added thermal energy.

Thus, materials with higher specific heat capacities can store more thermal energy. For instance, copper metal, with a specific heat capacity of 0.387 J/(g·°C), will have a larger temperature change in response to the same quantity of thermal energy added than will water, with a specific heat capacity of 4.18 J/(g·°C). In particular, 10.0 J of thermal energy will raise the temperature of 1.00 g of copper by 25.8 °C; however, it will increase the temperature of an equal mass of water by only 2.4 °C, which is less than 10% of copper's temperature increase. Table 4.4 summarizes the specific heat capacities of some common materials.

Water's unique properties make it influential in the world's climates. As you can see in Table 4.4, water's specific heat capacity is higher than that of most other common materials. Because of this, bodies of liquid water can store large quantities of thermal energy and are thus slow to heat up or cool down. By contrast, land surfaces have much lower specific heat capacities; they heat up and cool down much more readily.

B.5 THERMAL PROPERTIES OF MATERIALS

1. Explain each of the following observations made on a hot, sunny day:

   a. The chrome fender of a white convertible car with its top down is cooler to the touch than are its dark red plastic seats.
   b. An asphalt sidewalk is warmer to the touch on a sunny day than is a concrete sidewalk.

2. Use data in Table 4.4 (page 345) to explain why water is a better liquid to use in a hot pack than is ethyl alcohol.

3. a. The surface of beach sand feels hotter than does grass on a hot, sunny day. Which property is likely to be more responsible for this observation, *specific heat capacity* or *reflectivity*?
   b. Explain your choice.

4. a. On average, which would you expect to be hotter in the summer, a city far away from any large body of water or a similar city located near a large body of water?
   b. Explain your choice.

B.6 SPECIFIC HEAT CAPACITY

Lab Video: Specific Heat Capacity

Introduction

As you have learned, each material has a characteristic specific heat capacity. In Unit 3, Investigating Matter B.6, you used the specific heat capacity of water to determine the heat of combustion of candle wax. In this investigation, you will use this property to identify an unknown metal sample. To accomplish this, you will investigate the transfer of thermal energy from the hot metal sample to cool water. The quantity of thermal energy gained by the water is described by this formula:

Thermal energy gained (E) =

Mass of water (m) × Specific heat capacity of water (C) × Change in water temperature (ΔT)

or

$$E_{H_2O} = (m_{H_2O})(C_{H_2O})(\Delta T_{H_2O})$$

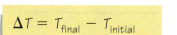

$$\Delta T = T_{final} - T_{initial}$$

The thermal energy gained by the water must equal the thermal energy lost by the metal. To indicate the loss of thermal energy, a negative sign is used:

$$E_{gained\ by\ water} = -E_{lost\ by\ metal}$$

or

$$(m_{H_2O})(C_{H_2O})(\Delta T_{H_2O}) = -(m_{metal})(C_{metal})(\Delta T_{metal})$$

Remember that you are trying to find the specific heat capacity of the metal, C_{metal}. All other values in the expression can be measured or calculated or they can be found in your textbook.

After reading the following procedure, construct a data table appropriate for recording the data you will collect in this investigation.

Procedure

1. Half-fill a 250- or 400-mL beaker with water. Place the beaker on a hot plate and start heating it to boiling.

2. Obtain a metal sample and record a description of its appearance.

3. Determine and record the mass of the metal sample.

4. Carefully place the metal sample into the water that is heating in the beaker. (*Caution: Do not drop the metal sample into the beaker.*) Allow the water to come to a boil, and keep boiling it for several minutes.

5. While the water is heating, obtain or assemble a *calorimeter*. A simple version of a calorimeter is shown in Figure 4.42.

6. Accurately measure a room-temperature water sample with a volume of about 60 to 80% of the capacity of the calorimeter. Record the exact volume of water you will add to the calorimeter. Carefully pour the water into the calorimeter.

7. Measure and record the temperature of the water in the calorimeter.

8. Measure and record the temperature of the water boiling in the beaker on the hot plate.

9. After the water has boiled for 2 to 3 min, use tongs to remove the metal sample from the water, and quickly lower the metal sample into the calorimeter. (*Caution: Do not drop the metal sample into the calorimeter.*)

10. Gently stir the water in the calorimeter, and record its temperature every 30 s until it reaches a maximum value and starts to drop.

11. Repeat Steps 3 through 10 using the same metal sample and average the recorded masses and temperature changes for the two trials.

12. Wash your hands thoroughly before leaving the laboratory.

Figure 4.42 *A Styrofoam-cup calorimeter consisting of two nested cups, a cardboard lid, and a thermometer.*

Data Analysis

1. Find the quantity of thermal energy absorbed by the water in the calorimeter.

2. Assume that the thermal energy absorbed by the water equals the thermal energy released by the metal sample. In addition, assume that the metal sample was initially at the temperature of the boiling water. Calculate the specific heat capacity of the metal.

Questions

1. Using your calculated specific heat capacity value, as well as the data provided in Table 4.4 (page 345), identify your metal sample.

2. The level of certainty you have about the actual identity of your metal sample depends in part on sources of error in the laboratory procedure. List and explain several potential sources of error in the procedure that you followed. Remember to distinguish between human error and experimental error.

3. Closely related to sources of error are any simplifying assumptions you made to complete your laboratory task and calculations.

   a. List some assumptions that you made in completing this investigation and its calculations.
   b. Evaluate each assumption in terms of how it could adversely affect your calculated result.

4. A recycling company decides to separate the various metals it receives by using the different specific heat capacities of the metals.

   a. Do you think that this is a reasonable plan? Explain.
   b. Propose an alternative method for separating the metals.

B.7 THE CARBON CYCLE

More than a century ago, it was suggested that a significant increase in burning fossil fuels might release enough carbon dioxide into the atmosphere to affect Earth's surface temperature. This suggestion was based on the idea that human activity can affect processes in natural ecosystems, producing changes that might not always be beneficial. In particular, burning fossil fuels might perturb the natural movement of carbon within Earth's systems—the global **carbon cycle.**

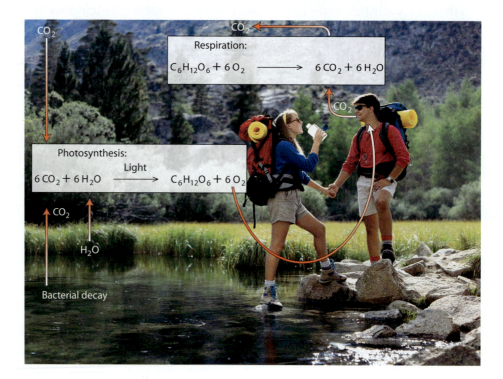

Figure 4.43 *The carbon cycle—major pathways within the biosphere.*

In the carbon cycle, which is illustrated in Figures 4.43 and 4.44, the different forms and compounds in which carbon atoms are found can be considered as "chemical reservoirs" of carbon atoms. These reservoirs include atmospheric CO_2 gas, solid calcium carbonate ($CaCO_3$) in limestone, natural gas (methane, CH_4), and organic molecules.

Each movement within the carbon cycle, and thus among these reservoirs, either requires energy or releases energy. For instance, plants use CO_2 and solar energy to form carbohydrates through photosynthesis. The carbohydrates are consumed by other organisms (or by the plant itself) and are eventually broken down or oxidized, releasing energy for use by organisms that consumed them.

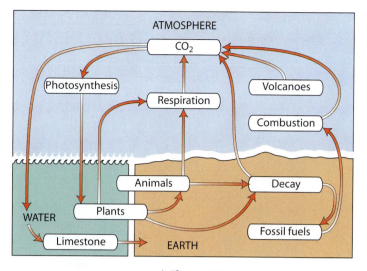

Figure 4.44 *The carbon cycle—relationships among major carbon reservoirs.*

The carbon atoms used and circulated in photosynthesis represent only a tiny portion of available global carbon. Gaseous CO_2 continually moves between the atmosphere and the oceans. In fact, 71% of Earth's carbon atoms, in the form of CO_2, are dissolved in the oceans. Another 22% are trapped in fossil fuels and in carbonate rocks formed when dissolved CO_2 reacted with water, which first produced carbonates, then sediments, then rocks. Dead organisms and terrestrial ecosystems (such as trees, crops, and other living matter) account for the remaining global carbon-atom inventory.

Without the influence of human activities, the distribution of carbon within various reservoirs would remain relatively unchanged over time. Atmospheric CO_2 levels, however, have increased by about 30% since 1800. This increase is the result of several processes. See Figure 4.45 (right). For example, clearing forests removes vegetation that would ordinarily consume CO_2 through photosynthesis; see Figure 4.45 (left). As cuttings and scrap timber are burned, they release CO_2 into the atmosphere. Most significantly, burning fossil fuel releases CO_2 into the air, as the following equations illustrate:

Burning coal: $$C(s) + O_2(g) \longrightarrow CO_2(g)$$

Burning natural gas: $$CH_4(g) + 2\,O_2(g) \longrightarrow CO_2(g) + 2\,H_2O(g)$$

Burning gasoline: $$2\,C_8H_{18}(g) + 25\,O_2(g) \longrightarrow 16\,CO_2(g) + 18\,H_2O(g)$$

As you continue to learn about the role of CO_2 in the atmosphere, recall the concerns raised at the start of this unit by individuals such as Principal Powers and Isaiah Martin about Riverwood High School's new school bus-idling policy.

Figure 4.45 *How does each of these scenes affect atmospheric carbon dioxide (CO_2) levels?*

B.8 CONSIDERING THE GLOBAL CARBON CYCLE

Idling a vehicle for 10 min typically consumes about 0.10 L of fuel and produces about 0.25 kg of carbon dioxide. On average, the ten Riverwood High School bus engines each idle 40 minutes daily.

1. What mass of carbon dioxide is produced daily by all 10 Riverwood High School buses?

2. How many liters of fuel are consumed daily by buses idling at Riverwood High School?

3. In a 180-day school year, what mass of carbon dioxide would be released into the atmosphere by these buses while idling? How many liters of fuel would be consumed?

4. Compare how long it takes for a Riverwood High School bus to convert a tank of fossil fuel to atmospheric carbon dioxide with the time required for that same quantity of fossil fuel to become originally formed by nature. Explain why atmospheric CO_2 levels have increased by about 30% since 1800 from the perspective of these two contrasting rates.

5. The values that you calculated in Questions 1 through 3 were for ten school buses over one school year. Do you think the school bus-idling policy at Riverwood High School will significantly affect the annual worldwide supply of fossil fuel that is converted to carbon dioxide gas? Explain.

6. In light of your answer to Question 5, why would Isaiah Martin, Principal Powers, and others believe that the school bus-idling policy is necessary?

B.9 CARBON DIOXIDE LEVELS

Introduction

Lab Video: Carbon Dioxide Levels

Air usually contains very low concentrations of CO_2. However, the CO_2 concentration in a small, enclosed airspace can be increased by burning coal or petroleum, by allowing organic matter to decompose, or by gathering a crowd of people or animals.

In this investigation, you will estimate and compare the amounts of CO_2 contained in several air samples. To accomplish this, air will bubble through water that contains an acid–base indicator, bromthymol blue. Carbon dioxide reacts with water to form carbonic acid, H_2CO_3(aq):

$$CO_2(g) + H_2O(l) \longrightarrow H_2CO_3(aq)$$

As the concentration of carbonic acid in the bromthymol blue solution increases (thus increasing acidity), the indicator color changes from blue to green and finally to yellow.

Before starting, read the procedure to learn what you will need to do, note safety precautions, and plan necessary data collecting and observations.

Procedure 👓

Part I: CO₂ in Normal Air

1. Pour 400 mL of distilled water into a 500-mL flask and add 1 mL bromthymol blue indicator solution. The solution color should appear blue. If it does not, add one drop of 0.1 M NaOH, and gently swirl the flask. Save this solution for later steps. ⚠ (**Caution:** *Sodium hydroxide solution is corrosive. If any splashes on your skin, wash it immediately with water and inform your teacher.*)

2. Pour 10 mL of the solution prepared in Step 1 into a test tube labeled "Control." Stopper the test tube. Set this control aside for later comparisons. Add 125 mL of the solution you prepared in Step 1 to a 250-mL filter flask.

3. Assemble the apparatus illustrated in Figure 4.46.

4. Record the starting time. Then gradually turn on the water tap until the aspirator pulls air through the flask. Mark or note the position of the faucet handle so that you can run the aspirator at the same flow rate later in the investigation.

5. Allow the aspirator to run until the indicator solution turns yellow. Record the total time needed to reach the yellow color. Detach the aspirator hose from the flask, then turn off the water. Remove the stopper from the flask.

6. Pour 10 mL of the used indicator solution from the flask into a second test tube labeled "Normal air" and stopper the tube. Compare this sample's color with the color of the control. Record your observations. Save the "Normal air" and "Control" test tubes.

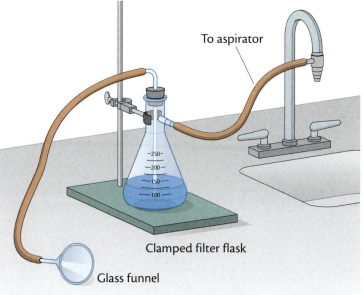

| **Figure 4.46** *Apparatus for collecting air.*

Part II: CO₂ from Hydrocarbon Combustion

7. Empty the filter flask and rinse it thoroughly with distilled water.

8. Pour 125 mL of indicator solution, prepared in Step 1, into the filter flask. Reassemble the apparatus as in Figure 4.46.

9. Light a candle. Hold the glass funnel above the flame so that the flame tip reaches just inside the base of the funnel.

10. Record the starting time. Then turn on the water tap to the same position you established in Step 4. Run the aspirator until the indicator solution color turns yellow. Record the total time needed to reach the yellow color. Detach the aspirator hose from the flask, then turn off the water.

11. Pour 10 mL of the solution into a clean test tube labeled "Combustion CO_2" and stopper the tube. Compare its color with that of the "Normal air" and the "Control" solutions. Record your observations.

Part III: CO_2 in Exhaled Air

12. Place 125 mL of indicator solution, prepared in Step 1, in a 250-mL Erlenmeyer flask.

13. Record the starting time. Then blow your breath through a clean straw into the solution until the indicator color changes to yellow. (*Caution: Do not draw any solution into your mouth.*) Record the total time it takes for the indicator color to change.

14. Pour 10 mL of the solution into a clean test tube labeled "Breath CO_2." Stopper the tube. Compare its color with those of the other three solutions. Record your observations.

15. Dispose of the waste solutions as directed by your teacher. Wash your hands thoroughly before leaving the laboratory.

Questions

1. Compare how long it took for the indicator solution to turn yellow in each test. Explain any differences you noted.

2. How accurate do you think this laboratory technique is for measuring the amount or concentration of CO_2 in gas samples? Explain.

3. Which sample contained a higher concentration of CO_2, the air surrounding the burning candle or exhaled air? Explain your answer.

4. If left exposed in a closed room indefinitely, the indicator solution would absorb CO_2 from the surroundings and could change color. Discuss the effect of each of the following conditions on the time required for the indicator color to change to green or yellow, or explain why you think the indicator color would not change.

   a. Many plants are growing in the room.
   b. Fifty people enter and remain in the room.
   c. Better ventilation is achieved in the room.
   d. Several people are using Bunsen burners in the room.

B.10 INCOMPLETE COMBUSTION

Modeling Matter

The products of hydrocarbon burning were shown as CO_2 and H_2O in the chemical equations summarized on page 349. As you probably know, however, these are not the only possible products of combustion. When hydrocarbons undergo *incomplete combustion*, carbon monoxide (CO) and soot (C) also form.

> CO is an air pollutant you will study further later in this unit.

Why might incomplete combustion occur? One possible reason is that there is not enough oxygen gas available for complete combustion; that is, there is not enough oxygen for all carbon atoms to form molecules of CO_2; CO forms instead. In such cases, oxygen would become the **limiting reactant** in the combustion process; it *limits* the amount of hydrocarbon that can be completely burned.

This is somewhat analogous to a situation you might encounter if you bought one package of hot-dog buns and one package of hot dogs for a picnic. Hot dogs are normally sold in packages of ten, but hot-dog buns are sold in packages of eight. Thus, hot-dog buns would become the *limiting reactant* if you had just one package of each. They limit the number of hot-dog sandwiches to eight sandwiches with two excess hot dogs.

1. Consider the combustion of propane, C_3H_8:

   a. Draw a molecular model of one propane molecule.
   b. Imagine taking that one propane molecule and allowing it to react with as much oxygen gas (O_2) as needed to get all the carbon atoms to form CO_2 and all the hydrogen atoms to form H_2O. Draw all the CO_2 and H_2O molecules formed by complete combustion of one propane molecule.
   c. How many oxygen molecules (O_2) are needed for the complete combustion of one molecule of propane?

2. Now consider the combustion of butane, C_4H_{10}:

   a. Draw models of one C_4H_{10} molecule and four O_2 molecules.
   b. Form as many molecules of CO_2 and H_2O as possible from the models you drew in answering Question 2a.
   c. Do four O_2 molecules provide sufficient oxygen atoms to support complete combustion of one C_4H_{10} molecule?
   d. Which is the limiting reactant, C_4H_{10} or O_2?

3. Assume that 1.0 mol C_4H_{10} is completely burned in excess oxygen gas to form carbon dioxide and water:

   a. How many moles of CO_2 would be produced?
   b. How many moles of H_2O would be produced?
   c. What is the minimum moles of O_2 required in that reaction?
   d. If it were possible to form CO instead of CO_2, how many moles of oxygen gas would be required?
   e. Consider your answers to Questions 3c and 3d. Under what conditions would CO formation be favored over CO_2 formation? Why?

B.11 GREENHOUSE GASES AND GLOBAL CHANGE

Greenhouse gases were first discussed on page 342.

As long as concentrations of carbon dioxide and other greenhouse gases in the atmosphere remain relatively constant, the greenhouse effect will comfortably maintain Earth's average temperature. The hydrologic cycle (page 87) and the carbon cycle (pages 348–349) maintain stable concentrations of water and carbon dioxide in their respective reservoirs, including the atmosphere. However, as you already realize, the effect of human activity must be considered.

"Spiraling-Up"

If more CO_2 is added to the atmosphere than can be removed by natural processes, its concentration will increase. Eventually, if sufficient CO_2 were added, the atmosphere could retain enough additional infrared radiation from the Sun to increase Earth's average surface temperature. With an increased surface temperature, CO_2 stored in ice, water, and the frozen floors of northern forests could also be released. A "spiraling-up" effect could occur, where warmer temperatures produce more carbon dioxide, which produces warmer temperatures, and so on.

The hydrologic cycle must also be considered. The atmosphere contains about 12 trillion metric tons of water vapor, a quantity so large that it might seem impossible that human activity could significantly affect it. However, if global temperatures increase, oceans and other bodies of water will also become warmer.

The amount of water vapor released increases as temperature increases, so more of this slightly warmer water will evaporate, increasing the atmospheric concentration of water vapor. As a greenhouse gas, increased water vapor may cause an even greater increase in global temperatures due to absorption and the release of infrared radiation, which causes another "upward spiral."

This spiraling-up effect is also commonly known as a *runaway greenhouse effect*. However, by contrast, increased water vapor concentration would also lead to increased cloud cover, which would reflect more solar radiation, thus counteracting some of the predicted temperature increase.

Two other naturally occurring greenhouse gases are also produced by human activity: nitrous oxide (N_2O) and methane (CH_4). Many agricultural and industrial activities, as well as the burning of solid waste and fossil fuels, contribute to the concentration buildup of N_2O in the atmosphere. CH_4 occurs naturally as a decomposition product of plant and animal wastes, but it also is produced from refining fossil fuels and raising livestock. See Figure 4.47. Finally, some gases that do not occur naturally can also contribute to the greenhouse effect. Of particular significance are fluorocarbons used in refrigeration and air conditioning. As you can see in Table 4.5, some of these gases are much more effective than CO_2 is in retaining heat at Earth's surface.

| Table 4.5

GREENHOUSE GAS EFFECTIVENESS

Greenhouse Gas	Relative Effectiveness	Percent Abundance in Troposphere
Carbon dioxide (CO_2)	1 (assigned value)	3.3×10^{-2}
Methane (CH_4)	30	1.7×10^{-4}
Nitrous oxide (N_2O)	160	3×10^{-4}
Water vapor (H_2O)	0.1	1
Ozone (O_3)	2 000	4×10^{-6}
Trichlorofluoromethane (CCl_3F)	21 000	2.8×10^{-8}
Dichlorodifluoromethane (CCl_2F_2)	25 000	4.8×10^{-8}

Figure 4.47 *In addition to carbon dioxide, atmospheric gases such as methane (CH_4) and nitrous oxide (N_2O) are considered greenhouse gases. Raising cattle (top) increases methane concentrations, while burning solid waste (right) adds nitrous oxide to the atmosphere.*

Temperature Increases

What are implications of increased atmospheric concentrations of greenhouse gases? Has any effect on Earth's climate been noted thus far? Examine Figure 4.48. The 20th century's 10 warmest years all occurred within the last 15 years of that century. Of these, 1998 was the warmest year on record.

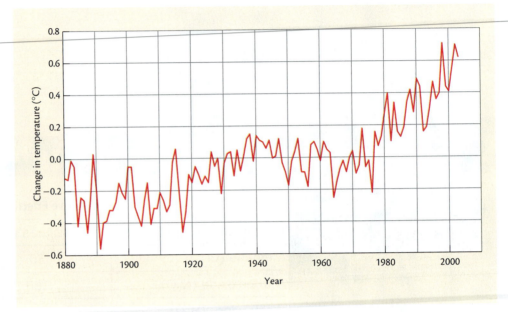

Figure 4.48 *Global annual mean surface air temperature trends since 1880. The 1951–1980 mean surface air temperature has been used as the zero-point base for comparison.*

These observed and predicted increases in average global surface temperatures are often referred to as global warming, which was a concern mentioned by Isaiah Martin at the beginning of this unit.

An international panel of climate scientists has predicted that under a business-as-usual scenario, in which no steps are taken to control the release of human-generated CO_2 and other greenhouse gases, the average global surface temperature could rise 0.8 °C to 2.9 °C (1.4 °F to 5.8 °F) in the next 50 years and 1.4 °C to 5.8 °C (2.2 °F to 10 °F) in the next 100 years, with significant regional variation. This prediction can be compared to evidence that Earth has warmed only 3 °C to 5 °C since the depths of the last Ice Age, some 20 000 years ago.

The 2001 report of the United Nations Intergovernmental Panel on Climate Change reported that Earth's average global surface air temperature increased by 0.4 °C to 0.8 °C over the past 100 years. The 2000-page report was prepared by 500 climate scientists and reviewed by another 500 climate experts.

Although most climate specialists agree that Earth's average temperature will increase, there is still some disagreement about causes of this predicted warming. New data indicate that the lower atmosphere may not be warming at the same rate as Earth's surface, which suggests that additional factors contribute to the buildup of greenhouse gases and to the warming trend. Among these additional factors is the circulation of ocean water. Ocean currents, such as El Niño and La Niña, transport thermal energy within the oceans and can affect climates when this energy is transferred to land and to the atmosphere.

Possible Impact

What are the projected effects of global warming? Based on the assumption that the average temperature increases 0.5 °C per decade, the oceans are predicted to rise about 5 cm each decade over this century. This is due to the major effect of melting polar ice caps and also to a very small contribution of expansion of ocean water with increasing temperature. Such changes could produce an approximately 15-cm increase in ocean levels by 2030, a situation that could cause flooding in coastal cities, such as Miami and New York. Possible regional climate changes, including reduced summer precipitation and soil moisture in North America, might be expected (see Figure 4.49). For example, northern regions might benefit from lengthened growing seasons, while growers in southern regions would probably shift to crops that can benefit from warmer winters.

What Can Be Done?

Scientists today understand the influence of CO_2 and other greenhouse gases on world climate better than they did even a decade ago. Sophisticated computer modeling has enhanced their research efforts. This new understanding supports the idea of a surface global warming trend. In 1992, representatives of more than 150 nations developed the Framework Convention on Climate Change, where they agreed to develop policies and procedures to reduce greenhouse gas emissions. In 1997, the third meeting of parties to this agreement, held in Kyoto, Japan, resulted in a protocol to address climate change. By late 2004, 126 nations had signed the protocol; thus, this international agreement entered into force in 2005, since, according to the protocol's stipulations, it had been ratified by nations producing at least 55% of 1990 emissions of six gases, including CO_2.

The *Kyoto Protocol* sets ambitious goals for reducing greenhouse gas emissions. For industrialized nations, this involves developing energy-efficient technologies, relying more heavily on renewable energy, and applying alternative processes that do not release greenhouse gases into the atmosphere. For instance, an alternative process has been developed for producing the polystyrene foam used to make such items as plastic egg cartons and meat trays. This process uses CO_2 that would normally be released during production of ammonia. The captured CO_2 replaces chlorofluorocarbons (CFCs) that had been used in this foam-producing process, thus reducing greenhouse gases in two ways. First, CO_2 that would normally be released by another industry is used. Second, CO_2 replaces CFCs, which have a global-warming potential much greater than that of CO_2. In an application of Green Chemistry principles, Hangers Cleaners, a chain of dry-cleaning stores, also uses liquid carbon dioxide (formed by compressing the gas under high pressure) and special detergents to replace potentially harmful dry-cleaning solvents.

Figure 4.49 *The glacier on Mt. Kilimanjaro as photographed in February 1993 (top) and in February 2000 (bottom).*

As of 2005, some industrialized nations, including the United States and Australia, had not ratified the Kyoto Protocol.

B.12 TRENDS IN ATMOSPHERIC CO₂ LEVELS

The CO_2-level data reported in Table 4.6 summarize average measurements of trapped gas bubbles in Antarctic ice or, more recently, of air at the Mauna Loa Observatory in Hawaii. These data are also available on the Internet. If possible, locate the most recent annual data before you start this activity. Graph the data summarized in Table 4.6. Prepare the *x*-axis scale to include the years 1800 to 2050 and the *y*-axis to represent CO_2 levels from 280 ppm to 600 ppm. Plot the data and draw a smooth curve to indicate the trend among plotted points.

| Table 4.6

APPROXIMATE CARBON DIOXIDE LEVELS IN THE ATMOSPHERE, 1800–2003

Year	Approximate CO₂ Level (ppm by volume)
1800	283
1820	284
1840	285
1860	286
1880	291
1900	297
1920	303
1940	309
1960	317
1965	320
1970	326
1975	331
1980	339
1985	346
1990	354
1995	361
2000	369
2001	371
2002	373
2003	376

Note: Data before 1960 are based on air bubbles trapped in ice-core samples, Law Dome, East Antarctica. Data since 1960 are based on air samples collected at Mauna Loa Observatory, Hawaii.

1. Assuming the trend in your smooth curve will continue, extrapolate your curve with a dashed line from the last year for which you have data to the year 2050. You can now make and evaluate some predictions, using the graph you have just completed.

2. What does your graph indicate about the general change in CO_2 levels since 1800?

3. Based on your extrapolation, predict CO_2 levels for
   a. next year.
   b. the year 2020.
   c. the year 2050.

4. a. Which of your predictions from Question 3 is likely to be the most accurate?
   b. Why?

5. a. Does your graph predict a doubling of the 1900 CO_2 level?
   b. If yes, in what year will this doubling presumably occur?

6. a. What assumptions must you make when extrapolating from known data trends?
   b. How do these assumptions affect the accuracy of your predictions?

7. Why might present CO_2 data for air samples collected at Mauna Loa Observatory be different from data collected at other locations around the planet?

8. At the beginning of this unit, Kayla Johnson was quoted as claiming that the carbon dioxide added to the atmosphere by bus exhaust should not be problematic because it is already a natural component of the atmosphere. Based on your graph and what you have learned about greenhouse gases, such as carbon dioxide, how would you respond to Kayla's assertion?

B.13 CONSIDERING GREENHOUSE GASES

Later in this unit, you will design and write a special newspaper feature on the school bus-idling policy that recently was approved at Riverwood High School. Consider the following questions as you prepare to inform *Riverwood High Gazette* readers about this new policy.

1. Do idling school buses generate any greenhouse gases? If yes, list the typical gases produced.

2. If the school bus-idling policy were implemented, would the policy resolve the issue of school buses and greenhouse gases? Why?

3. Assume that Riverwood High School buses were hybrid vehicles, as shown in Figure 4.50. How would this fact alter the benefits or drawbacks of the school bus-idling policy?

See Unit 3, pages 291 to 293, for background on hybrid vehicles.

4. List three other forms of transportation that students could use to get to and from school. For each transportation mode listed, discuss its potential for emitting greenhouse gases, such as CO_2. Consider whether strategies, such as the school bus-idling policy, could reduce CO_2 emissions from these transportation sources. Describe such strategies, if they are applicable.

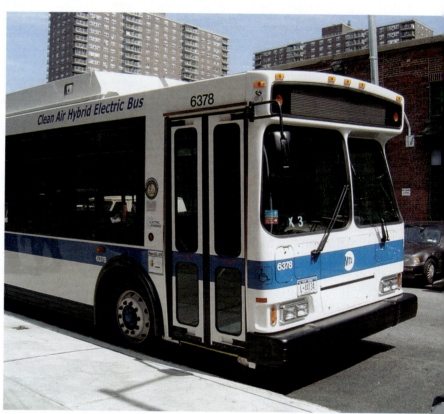

| Figure 4.50 *A hybrid electric bus.*

As you learned in Unit 1, water is naturally cleansed through processes within the hydrologic system. Similar cleansing of air takes place in the atmosphere; impurities are converted to more polar, water-soluble forms (hydrocarbons are converted to CO_2 and H_2O, for instance). These substances are then removed from the atmosphere through precipitation, which also is called a *raining-out process*. The consequences and effects of these natural processes, as well as the possible effects of human influences on these processes, are considered in Section C.

SECTION B SUMMARY
Reviewing the Concepts

Electromagnetic radiation includes X-rays; gamma rays; ultraviolet (UV), visible, and infrared (IR) radiation; radio waves; and microwaves. The energy transmitted by electromagnetic radiation varies according to its wavelength; the shorter the wavelength, the higher the energy.

1. Write an equation or a sentence that describes the relationship between the frequency of electromagnetic radiation and its energy.

2. Why is the word *spectrum* a good descriptor of the types of energy found in electromagnetic radiation?

3. Why is visible light useful in plant photosynthesis, while other forms of electromagnetic radiation are not?

4. a. List the main types of electromagnetic radiation in order of increasing energy.
   b. Describe how each type of radiation listed in your answer to Question 4a affects living things.

5. Ultraviolet light is often used to sterilize chemistry laboratory protective goggles. Why is ultraviolet light effective for this use, while visible light is not effective?

Earth's atmosphere protects living organisms by absorbing and distributing solar radiation.

6. Compare infrared, visible, and ultraviolet radiation in terms of how well they are absorbed by the atmosphere.

7. Describe two main effects of the solar radiation that reaches Earth's surface.

Electromagnetic radiation can interact with matter to transfer energy. The change in temperature of a material when thermal energy is absorbed depends on its specific heat capacity.

8. a. Compare lake water and asphalt in terms of how readily each warms up when exposed to sunlight.
   b. What properties of these two materials account for differences in their behavior?

9. From a scientific viewpoint, why do many desert dwellers elect to wear white or light-colored garments?

10. Given their respective functions, compare the specific heat capacities of the material from which the *handle* of a frying pan is made and the material from which the frying pan itself is made.

11. Given the same energy input, which would you expect to have the greater temperature increase, equal masses of aluminum or iron? Refer to Table 4.4 on page 345. Explain your answer in terms of the concept of specific heat capacity.

12. Lake Erie has a water volume of about 484 km³. How much energy would be required to raise its temperature 1 °C? Show your calculations.

13. Normal human body temperature is about 37 °C. Room temperature is usually about 25 °C:
    a. How much energy is required to raise the temperature of 1.0 L of air (density = 0.0012 g/mL) from room temperature to body temperature?
    b. How much energy is required to raise the temperature of 1.0 L of liquid water (density = 1.0 g/mL) from room temperature to body temperature?
    c. Using your answers to Questions 13a and 13b, explain why cold water, or even room-temperature water, feels colder on the skin than does air at the same temperature.

14. Describe how atmospheric CO_2 and water vapor help maintain moderate temperatures at Earth's surface.

15. List two natural processes and two human activities that can increase the amount of

    a. CO_2 in the atmosphere.
    b. CH_4 in the atmosphere.

16. What changes in the composition of the atmosphere would cause the average surface temperature of the Earth to

    a. increase?   b. decrease?

17. Explain why, on a sunny winter day, a greenhouse with transparent glass walls is much warmer than is a structure with opaque wooden walls.

18. Draw sketches to show how

    a. a greenhouse works.
    b. the global greenhouse effect works.

19. List three chemical reservoirs of carbon atoms.

20. Explain how, over time, a particular carbon atom can be part of the atmosphere, biosphere, lithosphere, and hydrosphere.

21. Write a chemical equation that depicts the transfer of a carbon atom between any two "spheres" listed in Question 20.

Connecting the Concepts

22. Many inexpensive sunglasses block visible light, but they do not block ultraviolet radiation. Explain the hazards of wearing such sunglasses at the beach or on a ski slope during a bright, sunny day.

23. If the air–fuel mixture is incorrect when using a Bunsen burner, the yellow flame will deposit black carbon on the bottom of a heating beaker of water. Explain why, in terms of limiting reactants.

24. Consider the following equation, which is part of a series of chemical changes used to recover sulfur from hydrogen sulfide (H_2S) produced during crude-oil processing and natural-gas processing:

$$2\ H_2S + SO_2 \longrightarrow 3\ S + 2\ H_2O + Heat$$

    a. If 4.0 mol H_2S and 3.0 mol SO_2 are the reactants, which is the limiting reactant?
    b. If 7.5 mol H_2S and 2.8 mol SO_2 are the reactants, which is the limiting reactant?

25. Consider the equation shown in Question 24:

    a. What mass of H_2S would be required to react completely with 205 g SO_2?
    b. If 45.4 kg SO_2 and 67.3 kg H_2S were present in the reaction chamber, which would become the limiting reactant?

26. If you see considerable black exhaust emitted by a bus, what might that indicate about the combustion processes in the engine?

27. a. What would be the effect on the average global temperature if significant atmospheric increases occurred in
      i. carbon dioxide?
      ii. methane?
      iii. water vapor?
    b. How would such global temperature changes, in turn, affect Earth's atmosphere? Explain.

Extending the Concepts

28. Investigate and describe how night-vision goggles allow objects to become visible in low-light conditions.

29. Some animals "see" a portion of the electromagnetic spectrum different from the wavelengths to which human eyes respond. Investigate some examples and speculate on the benefit of these abilities to the animals you research.

30. In terms of the kinetic molecular theory, describe changes in the behavior of gas molecules as they absorb electromagnetic radiation.

31. Investigate and describe the characteristics of molecules that allow them to absorb various types of radiation.

32. The presence of carbon dioxide in the atmosphere is only one aspect of the global carbon cycle. Assess current scientific knowledge of the global carbon cycle and evaluate the roles of one or more major carbon sinks (carbon reservoirs).

33. Scientists analyze deep-ice samples from the Antarctic to estimate the atmospheric carbon dioxide concentrations many thousands of years ago (see Figure 4.2, page 302). Research and prepare a report on how such ice samples are analyzed to obtain this information.

Acids in the Atmosphere

SECTION C

As you learned in Section B, the burning of fossil fuels has world-wide implications. Scientists predict that every region on Earth will be affected, in some way, by increasing concentrations of carbon dioxide in the atmosphere. Unfortunately, this is not the only wide-ranging concern associated with fuel burning. The effects of fuel combustion can be experienced hundreds of kilometers away from the original source, when by-products of combustion (along with other air pollutants) combine with water to produce *acid rain*.

C.1 ACID RAIN

What is acid rain? To answer that question, it is worth noting that "acid rain" does not have to be rain; it can be any form of atmospheric precipitation. **Acid rain** is defined as fog, sleet, snow, or rain with a pH lower than about 5.6, which is the average pH of natural precipitation in the absence of air pollution. You probably recall that pure water is neutral, with a pH of 7. (See Figure 1.44, page 70.) Why, then, does natural precipitation have a pH less than 7?

In Unit 1, you learned that water samples, although they may be clean and healthful, are rarely pure. Even water that evaporates from Earth's surface condenses and dissolves atmospheric gases, such as nitrogen, oxygen, and carbon dioxide. The pH of rainwater is normally slightly acidic, about 5.6, mainly due to the reaction of carbon dioxide with water, forming carbonic acid, H_2CO_3:

$$CO_2(g) + H_2O(l) \rightleftharpoons H_2CO_3(aq)$$

This dilute solution of carbonic acid then falls to Earth as rain, snow, or sleet.

Other natural events can contribute to the acidity of precipitation. Volcanic eruptions, forest fires, and lightning bolts produce sulfur dioxide (SO_2), sulfur trioxide (SO_3), and nitrogen dioxide (NO_2). These gases can then react with atmospheric water in much the same way that carbon dioxide does:

$$SO_2(g) + H_2O(l) \rightleftharpoons H_2SO_3(aq)$$
Sulfurous acid

$$SO_3(g) + H_2O(l) \rightleftharpoons H_2SO_4(aq)$$
Sulfuric acid

$$2\,NO_2(g) + H_2O(l) \rightleftharpoons HNO_3(aq) + HNO_2(aq)$$
Nitric acid   Nitrous acid

The result, as the previous equations suggest, is the formation of acidic precipitation. If rain customarily is acidic, why does acid rain

> The expression *acidic precipitation* is sometimes used instead of *acid rain* to indicate that any form of precipitation—fog, sleet, snow, or rain—can be acidic.

> Double arrows indicate that both forward and reverse reactions occur simultaneously. You will learn more about reversible reactions on page 375.

> The actual processes that generate acid rain are more complex than depicted here.

pose a problem? The fact that this precipitation is *more* acidic than normal leads to the problems associated with acid rain.

Impact of Acid Rain

Why would Principal Powers and others at Riverwood High School be concerned about acid rain? The reality is that sulfur oxides and nitrogen oxides emitted from power plants, various industries, and fossil-fuel-burning vehicles react with water vapor to form acids that lower the pH of rainwater—at times to 4.5 or lower in the northeastern United States. See Figure 4.51.

Figure 4.51 *pH of precipitation throughout the continental United States (National Atmospheric Deposition Program).*

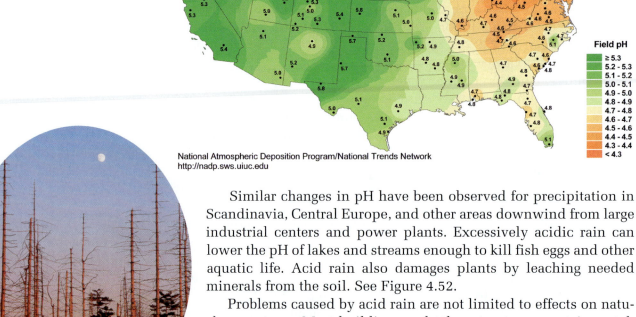

Hydrogen ion concentration as pH from measurements made at the field laboratories, 2003

National Atmospheric Deposition Program/National Trends Network
http://nadp.sws.uiuc.edu

Field pH

	≥ 5.3
	5.2 - 5.3
	5.1 - 5.2
	5.0 - 5.1
	4.9 - 5.0
	4.8 - 4.9
	4.7 - 4.8
	4.6 - 4.7
	4.5 - 4.6
	4.4 - 4.5
	4.3 - 4.4
	< 4.3

Figure 4.52 *This forest in the Czech Republic has been badly damaged by acid rain.*

Similar changes in pH have been observed for precipitation in Scandinavia, Central Europe, and other areas downwind from large industrial centers and power plants. Excessively acidic rain can lower the pH of lakes and streams enough to kill fish eggs and other aquatic life. Acid rain also damages plants by leaching needed minerals from the soil. See Figure 4.52.

Problems caused by acid rain are not limited to effects on natural ecosystems. Most buildings and other structures contain metal, limestone, or concrete, which are all materials susceptible to damage by acids. Statues and monuments (such as the Parthenon in Greece) that have stood for centuries show signs of significant surface damage, due, in part, to acid rain. Acid attacks calcium carbonate in limestone, marble, and cement according to this equation:

$$H_2SO_4(aq) + CaCO_3(s) \longrightarrow CaSO_4(s) + H_2O(l) + CO_2(g)$$

| Sulfuric acid (in acid rain) | Calcium carbonate | Calcium sulfate | Water | Carbon dioxide |

Calcium sulfate is much more soluble in water than is calcium carbonate. Thus, as calcium sulfate forms, it washes away, uncovering fresh solid calcium carbonate that reacts further with acid rain. See Figure 4.53.

Addressing Acid Rain

Control of acid rain is difficult because air pollutants do not respect political boundaries. Carried by air currents, acid rain often shows up hundreds of kilometers from the original sources of air contamination. For example, much of the acid rain that falls in Scandinavia has its origins in the industrial regions of Germany and the United Kingdom. Acid rain falling on New England and southern Canada originates largely from the industrial Ohio Valley in the United States.

Figure 4.53 *Acid rain has disfigured these limestone gargoyles on the Cathedral of Notre Dame in Paris, France.*

The Clean Air Act Amendments of 1990 were enacted, in part, to address issues of acid rain. The compliance plan created by these amendments imposed emissions restrictions on fossil-fueled power plants. Figure 4.54 (page 366) shows that fuel combustion is the largest contributor to SO_2 emissions. To achieve SO_2 reductions, the U.S. Environmental Protection Agency (EPA) issued permits to electric utility plants, particularly coal-burning power plants that have consistently been the largest sulfur dioxide emitters. The permits specify the maximum amount of SO_2 each plant can release annually. As a bonus for reducing emissions, plants with low emissions are allowed to auction or sell their excess permits to plants that produce more emissions. National SO_2 emissions have decreased steadily since the early 1980s.

Reductions in SO_2 emissions were accomplished primarily by using more expensive lower-sulfur coal. From 1988 to 1997, the percent of low- and medium-sulfur coal shipped by rail increased from 82% (221 million short tons) to 92% (337 million short tons), while the percent of high-sulfur coal fell from 18% (49 million short tons) to 8% (29 million short tons). Some power plants have installed scrubbers to remove SO_2 from smokestack emissions, but the cost of this option is currently higher than that associated with using low-sulfur coal.

Scrubbers are described in Section D.

Unlike SO_2 emissions, which are produced by burning fuels containing sulfur, nitrogen oxide (NO_x) emissions are formed by the reaction of atmospheric nitrogen gas with oxygen gas at the high temperatures associated with most forms of combustion—including internal combustion engines.

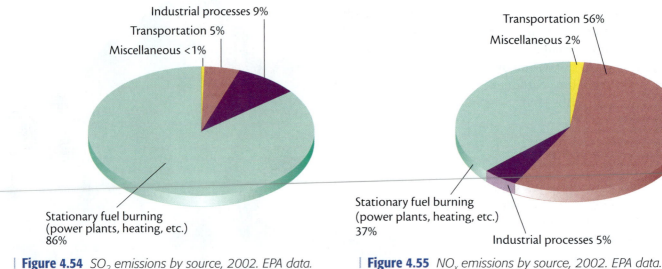

| **Figure 4.54** SO_2 emissions by source, 2002. EPA data. | **Figure 4.55** NO_x emissions by source, 2002. EPA data. |

Nitrogen monoxide (NO), which is also called nitric oxide, and nitrogen dioxide (NO_2) are sometimes referred to collectively as NO_x (pronounced "nocks").

Figure 4.55 indicates that fuel combustion by power plants is joined by transportation as major sources of nitrogen oxide emissions. Although the 2002 national SO_2 emissions were 39% lower than in 1993, total national NO_x emissions were only 12% lower. The EPA estimates that reducing NO_x emissions from fixed-location sources, such as power plants, is a cost-effective strategy. But the issue of NO_x emissions from mobile sources, such as automobiles and other modes of transportation, has yet to be addressed.

ChemQuandary 2

THE RAIN IN MAINE . . .

The state of Maine has no coal-fired power plants and relatively few residents. (The average population density of Maine is approximately 41 people per square mile; Maine's largest city, Portland, has about 64 000 residents). However, Figure 4.51 (page 364) indicates that the average pH of precipitation in the state is lower than that of rainfall free of air contaminants (pH ~ 5.6). Why might this be?

Investigating Matter

C.2 MAKING ACID RAIN

Introduction

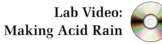

Lab Video: Making Acid Rain

Have you seen any evidence of the effects of acid rain near where you live? In this investigation, you will generate some laboratory-scale "acid rain." You then will observe the effects this acidic solution has on plant material (an apple skin), on a chemically reactive metal (magnesium, Mg), and on marble (calcium carbonate, $CaCO_3$). You will use acid–base indicators to estimate the acidity of the solution. After reading the following procedure, construct a data table appropriate for recording the data you will collect in this investigation.

Procedure 😎

1. Place a 1-g sample of sodium sulfite (Na_2SO_3) in a test tube. Place the test tube inside a one-pint reclosable zip-seal bag, holding it upright from outside the bag.

2. Carefully fill a Beral pipet with 6 M hydrochloric acid (HCl). (*Caution: 6 M hydrochloric acid is corrosive. If you spill it on yourself or others, wash it off thoroughly and inform your teacher. Avoid breathing any HCl fumes.*) Using a wash bottle, carefully rinse off any acid left on the outside of the pipet.

3. Without squeezing the Beral pipet bulb, place the filled pipet inside the test tube containing sodium sulfite. Then carefully add 10 mL of distilled water to the inside of the bag. Make sure that the distilled water does not come in contact with the hydrochloric acid or sodium sulfite. Carefully smooth the bag to force out most of the air. Close the bag with the zip-seal strip. See Figure 4.56.

4. Slowly squeeze the Beral pipet bulb through the outside of the bag so that hydrochloric acid drips onto the solid sodium sulfite. Keep the test tube upright, so its contents do not spill out. Keep the bag sealed.

5. Allow the reaction in the test tube to proceed for 1 to 2 min, gently tapping the test tube every few seconds. After the reaction in the test tube has stopped, gently swirl the water in the bottom of the bag for another 1 to 2 min. Do not swirl the water so vigorously that it mixes with the contents of the test tube.

6. (*Caution: Avoid inhaling the gas released when you open the bag.*) Carefully open a top corner of the bag and, using a clean, dry Beral pipet, transfer three pipets of water from the bottom of the bag to a clean, dry test tube. Reseal the bag. Label the test tube "A."

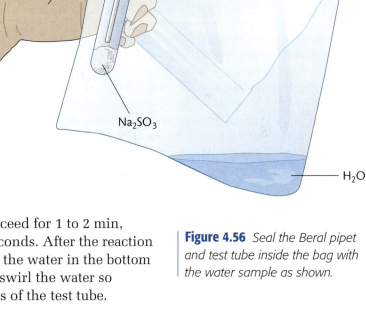

6 M HCl

Na_2SO_3

H_2O

Figure 4.56 *Seal the Beral pipet and test tube inside the bag with the water sample as shown.*

Recall from Unit 1 (page 68) that the vegetable dye litmus turns blue in a basic solution and red in an acidic solution.

7. Peel two pieces of fresh apple skin, approximately 2 cm long, and place them on a paper towel. Add 4 to 5 drops of liquid from test tube A to the outside of one piece of apple skin. Add 4 to 5 drops of distilled water to the second piece of apple skin. (The second apple-skin sample serves as a control. Why?) After 3 min, observe the two apple-skin samples; record your observations.

8. Place one drop of distilled water each on fresh strips of red litmus paper, blue litmus paper, and pH paper. Record your observations.

9. Repeat Step 8, except use solution from test tube A in place of the distilled water. Record your observations.

10. Place a 1-cm length of magnesium ribbon in a separate clean, dry test tube. Add one pipet of solution from test tube A. Observe the reaction for 3 min. Record your observations.

11. Add two small marble chips (calcium carbonate, $CaCO_3$) to the solution remaining in test tube A. Observe the marble chips for 3 min; record your observations.

12. Dispose of all remaining solutions as directed by your teacher.

13. Wash your hands thoroughly before leaving the laboratory.

Questions

1. The gas formed in the test tube by the reaction of hydrochloric acid with sodium sulfite was sulfur dioxide, SO_2. The other products were water and $NaCl(aq)$. Write a balanced equation for the reaction that produced SO_2 gas.

2. What effect did this gas have on the acidity of the distilled water placed inside the plastic bag? Support your answer by referring to specific observations and test results.

3. Write a chemical equation that shows how "acid rain" (H_2SO_3) was produced from SO_2 gas and water inside the zip-seal bag.

4. Explain how this laboratory investigation models the production of acid rain.

5. As you know, precipitation with a pH less than 5.6 is defined as acid rain. How does the pH of your solution compare to this value?

6. If a liquid similar to the solution in the plastic bag moistened a marble statue or steel girders, what effect might it have?

7. a. Write an equation for the reaction between your "acid rain" (H_2SO_3) and marble chips ($CaCO_3$). (*Hint:* Carbon dioxide gas and calcium sulfite solution were two of the three products formed.)

   b. Explain how your equation in Question 7a relates to acid rain's effects on marble statues and building materials.

Using Chemistry to Bring Criminals to Justice

For a moment, no one in the courtroom moved. Aware that everyone was staring at her, Susan could hear the sound of her own heart beating.

Then the prosecutor continued. "Ms. Ragudo, please tell this court," he said, holding up the small plastic bag of white powder, "the results of your analysis of the substance labeled Exhibit A."

For many people, the thought of testifying in a criminal court case is enough to make them break out in a sweat. But for Susan Ragudo, it's just another part of her job as a forensic scientist employed by the Commonwealth of Virginia. *Forensic science*—the use of science in criminal investigations—can describe many types of inquiry. Susan's specialized area involves analyzing crime scene material for evidence of illegal substances.

To identify the unknown white material, for instance, Susan begins by adding the material to a test reagent. Depending on the resulting color change, Susan knows what the substance most probably is. For example, the drug heroin will turn purple, cocaine will turn blue, and so on. Next, Susan uses thin-layer chromatography (TLC) to compare the substance to a known sample of the suspected drug. This narrows Susan's choices even more and also tells her whether one or more substances are present. Finally Susan tests her conclusion by using gas chromatography/mass spectroscopy. Then she returns the material—and her analysis data—to law enforcement officers.

Susan explains that in her field it is essential to know one's way around a laboratory. It's also important to have an analytical mind—to "zero in on" an unknown material logically, by conducting and interpreting tests. Susan adds that she must also be comfortable speaking in public, because she sometimes testifies in court about her findings.

Calling All Sleuths

Susan Ragudo uses chemistry to help determine whether materials are illegal drugs. In what other types of criminal investigation could chemistry be helpful? What types of tests could be used? Write a story about a crime in which everyone overlooks a small chemical clue—everyone but you, that is. Don't be bashful—make yourself the hero, and explain how you used a basic understanding of chemistry to unravel the crime and convict the criminal!

Truth vs. Evidence

1. Some chemical tests can conclusively prove what a substance is *not*; it is often much harder to establish what it *is*. Explain why one type of proof is more difficult than the other.

2. Most tests are quite reliable but still may only be 95% or 99% accurate. Given that no test is likely to be 100% accurate, how sure must evidence be for you to accept it as "the truth"? Explain your answer.

3. Juries sit in judgment regarding the facts in a criminal case. Who serves as the jury in deciding which scientific research results are regarded as true?

C.3 PREVENTING ACID RAIN

You have learned about the roles of sulfur oxides and nitrogen oxides in forming acid rain, as well as current efforts to control these emissions. Use this information to answer the following questions:

1. a. In what ways is the large-scale generation of sulfur oxides easier to control than the generation of nitrogen oxides?
   b. In what ways is it more difficult?

2. a. Describe incentives that you think could be offered by the government to encourage decreasing SO_x and NO_x emissions.
   b. Which emissions might be reduced more effectively through the use of penalties?

3. In terms of decreasing the total release of sulfur oxides and nitrogen oxides into the atmosphere, how could you decide whether an electric vehicle that uses electricity from a fossil-fuel-burning power plant is preferable to a gasoline-powered vehicle?

4. If students at Riverwood High School decided to reduce their personal contributions to acid rain, what changes could they make in their daily activities? List three reduction strategies they might employ, explaining how each would lead to reduced amounts of acid-rain-causing emissions. It may be helpful to refer to the source categories of acid rain in Figures 4.54 and 4.55 (page 366) and to think about your own daily activities.

C.4 STRUCTURE DETERMINES PROPERTIES

What Is an Acidic Solution?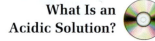

Arrhenius also proposed that increases in atmospheric carbon dioxide may increase global temperatures.

Swedish chemist Svante Arrhenius, in 1887, defined an acid as any substance that generates hydrogen ions (H^+) when dissolved in water. He defined a base as a substance that generates hydroxide ions (OH^-) in water. In aqueous solutions, the hydrogen ion released by the acid is bonded to water. Such a combination can be represented as $H^+(aq)$, $H(H_2O)^+$, or more commonly as $H_3O^+(aq)$, the *hydronium* ion:

$$H^+(aq) + H_2O(l) \rightleftharpoons H_3O^+(aq)$$

Consider the earlier example of carbon dioxide dissolving in water to produce carbonic acid, $H_2CO_3(aq)$:

$$H_2O(l) + CO_2(g) \rightleftharpoons H_2CO_3(aq)$$

The transfer of a hydrogen ion, H^+, from carbonic acid to a water molecule produces a hydrogen carbonate (bicarbonate) ion and a hydronium ion (H_3O^+), which is responsible for the solution's acidity:

$$H_2O(l) \;+\; H_2CO_3(aq) \rightleftharpoons H_3O^+(aq) \;+\; HCO_3^-(aq)$$

Water   Carbonic   Hydronium   Bicarbonate

           acid   ion   ion

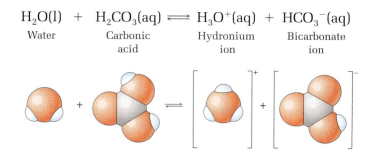

Most common acids are substances containing one or more hydrogen atoms that can be released as $H^+(aq)$ in water. The remainder of the original acid molecule, with hydrogen's single electron still attached to it, becomes an anion. As shown in the following equations for two common acids, the dissolved acid produces hydronium ion and an anion.

Nitric acid: $\quad H_2O(l) + HNO_3(aq) \longrightarrow H_3O^+(aq) + NO_3^-(aq)$

Sulfuric acid: $H_2O(l) + H_2SO_4(aq) \longrightarrow H_3O^+(aq) + HSO_4^-(aq)$

Bases are generally composed of a cation and the hydroxide anion OH^-. When such bases dissolve in water, the cations and hydroxide ions separate and disperse uniformly throughout the solution. The hydroxide ions (OH^-) give basic solutions, which also are referred to as *alkaline* solutions, their characteristic properties:

Potassium hydroxide: $\;KOH(s) + H_2O(l) \longrightarrow K^+(aq) + OH^-(aq)$

Barium hydroxide: $Ba(OH)_2(s) + H_2O(l) \longrightarrow Ba^{2+}(aq) + 2\,OH^-(aq)$

Acidic solutions have more $H_3O^+(aq)$ than $OH^-(aq)$, while alkaline solutions have more $OH^-(aq)$ than $H_3O^+(aq)$. Pure water is a neutral substance (pH = 7.0 at 25 °C); it contains equal amounts of $H_3O^+(aq)$ and $OH^-(aq)$, as is true of all neutral solutions.

Mixing equal amounts of $H_3O^+(aq)$ and $OH^-(aq)$ produces a **neutralization** reaction. When these two ions react, water is the only product formed:

$$H_3O^+(aq) + OH^-(aq) \rightleftharpoons 2\,H_2O(l)$$

Both acidic and basic characteristics are destroyed in this neutralization reaction; pure (neutral) water is produced, which contains small, equal amounts of $H_3O^+(aq)$ and $OH^-(aq)$.

Consider the neutralization of hydrochloric acid with sodium hydroxide, according to this equation:

$$HCl(aq) + NaOH(aq) \longrightarrow H_2O(l) + NaCl(aq)$$

As is true for all neutralization reactions between compounds, this reaction produces water and a salt (an ionic compound).

C.5 ACIDS AND BASES

1. Hydrochloric acid (HCl) is an acid found in gastric juice (stomach fluid):

   a. Write an equation for the formation of ions as HCl dissolves in water.

   b. Name each ion formed.

2. Separate equations for carbonic acid, nitric acid, and sulfuric acid reacting with water are shown on page 371. Note that the expression for carbonic acid's reaction also includes names and molecular models for each reactant and product. Complete the following steps for both nitric acid and sulfuric acid:

   a. Write the equation for the acid's reaction with water.

   b. Write the name of each reactant and product.

   c. Draw molecular models for each reactant and product. (*Hint:* Just as in carbonic acid, each hydrogen atom in nitric acid and sulfuric acid is connected directly to an oxygen atom.)

3. Each base in the following list has commercial and industrial uses. Name each base and write an equation showing the ions that are liberated when the base dissolves in water:

   a. $Mg(OH)_2$, the active ingredient in milk of magnesia, an antacid

   b. $Al(OH)_3$, a compound used to bind dyes to fabrics

   c. NaOH, a compound often found in drain and oven cleaners

4. A person consumes some antacid containing magnesium hydroxide, $Mg(OH)_2$, to relieve discomfort from excess stomach acid, HCl(aq):

   a. Write the overall equation for the neutralization of HCl(aq) with $Mg(OH)_2$(aq).

   b. Name each product shown in the equation you wrote in answer to Question 4a.

C.6 ACIDITY, MOLAR CONCENTRATION, AND pH

Water and its solutions contain both hydronium ions (H_3O^+) and hydroxide ions (OH^-).

▶ In pure water and in a **neutral solution,** the concentrations of these ions are equal, but very small.

▶ In an **acidic solution,** the hydronium-ion concentration is larger than the hydroxide-ion concentration. In *very* acidic solutions, the hydronium-ion concentration is much larger than the hydroxide-ion concentration.

▶ In a **basic solution,** the hydroxide-ion concentration is larger than the hydronium-ion concentration. In *very* basic solutions, the hydroxide-ion concentration is much larger than the hydronium-ion concentration.

Molar Concentration

Chemists often express a solution's concentration as the amount of solute dissolved in a particular volume of solution. When the amount of solute (in moles) is divided by the volume of solution (in liters), the resulting value is called the **molar concentration,** or **molarity (M).** Figure 4.57 illustrates how a solution of a particular molar concentration (molarity) can be prepared in the laboratory.

Expressed in symbolic terms,

$$M = \frac{\text{mol solute}}{\text{L solution}}$$

Thus, if 2.0 mol sodium chloride (NaCl) were dissolved in 1.0 L of total solution, the molar concentration of the NaCl solution would be

$$\frac{2.0 \text{ mol NaCl}}{1.0 \text{ L solution}} = 2.0 \text{ M NaCl}$$

A solution containing 1.0 mol NaCl in 0.50 L of solution is *also* a 2.0-M NaCl solution. Does this make sense? Take a moment to think about why these two solutions would have identical molar concentrations.

Sample Problem: *If 0.16 mol NaCl were completely dissolved in 2.0 L of solution, what would be the solution's molar concentration of NaCl?*

Applying the definition of molar concentration, the answer is

$$\frac{0.16 \text{ mol NaCl}}{2.0 \text{ L solution}} = 0.080 \text{ M NaCl}$$

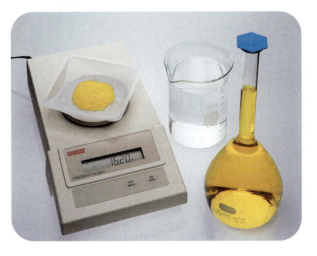

Figure 4.57 *Preparing a 0.1000-M solution of sodium chromate, Na_2CrO_4. The molar mass of Na_2CrO_4 is 162.0 g/mol. A 0.1000-mol sample of that solute (16.20 g) is dissolved in enough water to produce 1.000 L of final solution. A line etched into the neck of the volumetric flask shown in the photo precisely indicates that final volume measurement.*

The pH Scale

The pH value of an aqueous solution is related to its hydronium-ion concentration (H_3O^+). The symbol *pH* stands for "power of hydronium ion," where "power" refers to the mathematical power (exponent) of 10 that expresses the hydronium's ion's molar concentration.

 **What Is pH?**

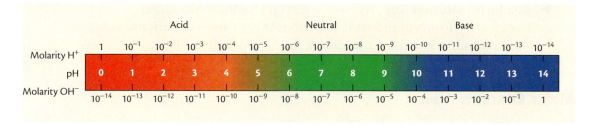

| Molarity H⁺ | 1 | 10⁻¹ | 10⁻² | 10⁻³ | 10⁻⁴ | 10⁻⁵ | 10⁻⁶ | 10⁻⁷ | 10⁻⁸ | 10⁻⁹ | 10⁻¹⁰ | 10⁻¹¹ | 10⁻¹² | 10⁻¹³ | 10⁻¹⁴ |

Figure 4.58 *The relationships among pH, molar concentration of H_3O^+, and molar concentration of OH^- at 25 °C.*

> Using similar reasoning, a solution with a pOH of 2 would contain 1×10^{-2} mol OH^- (0.01 mol OH^-) per liter.

For example, a solution containing 0.001 mol H_3O^+ (1×10^{-3} mol H_3O^+) per liter has an H_3O^+ concentration of 0.001 M (10^{-3} M), and, thus, a pH value of 3. The pH value can be interpreted as the exponent on the power of 10 with its sign changed. Thus, a solution with 0.000 000 01 mol H_3O^+ (1×10^{-8} mol H_3O^+) per liter (0.000 000 01 M) would have a pH of 8. Figure 4.58 shows how the pH of an aqueous solution is related to the molar concentration of H^+ (H_3O^+) and OH^-.

As you have just learned, the pH scale expresses acidity and alkalinity based on powers of 10. Thus, lemon juice, with pH 2.0, is *10 times* more acidic than is a soft drink at pH 3.0, which is *10 000 times* more acidic than is pure water, at pH 7.0 (four steps farther up the pH scale). Similarly, a solution with pH 11.0 is *100 times* more basic than is a solution with pH 9.0.

Developing Skills

C.7 INTERPRETING THE pH SCALE

1. Using the values shown in Figure 4.58, describe the mathematical relationship for aqueous solutions at 25 °C

   a. between pH values and H_3O^+ (H^+) concentrations.
   b. between H_3O^+ and OH^- concentrations.

2. Some common aqueous solutions and their typical pH values appear in the following list. Classify each solution as *acidic, basic,* or *neutral* and estimate the hydronium-ion and hydroxide-ion concentrations found in each:

   a. milk, pH = 6.0
   b. stomach fluid, pH = 1.0
   c. a solution of drain cleaner, pH = 13.0
   d. a cola drink, pH = 3.0
   e. sugar dissolved in pure water, pH = 7.0
   f. a solution of household ammonia, pH = 11.0

3. How many times more acidic is a cola drink than milk? (See the information provided in Question 2.)

4. Clouds over Clingmans Dome, a peak in Great Smoky Mountains National Park, have had pH levels as low as 2.0. Compared to rainfall at pH 5.5, estimate how many times more acidic the moisture has been in these clouds.

5. Products used to remove clogs in drains contain strong bases, such as sodium hydroxide (NaOH). If drain cleaner containing 10.0 g NaOH(s) is added to a clogged drain pipe filled with 2.5 L water,

a. how many moles of NaOH were added?
b. what is the molar concentration of NaOH in the resulting solution? (*Hint:* This is also equal to the molar concentration of OH^- in the solution.)
c. use Figure 4.58 to find the pH of this drain-cleaner solution.

C.8 STRENGTHS OF ACIDS AND BASES

Acid Concentration and Strength

In everyday discussions of solutions, *strong* tends to be associated with the idea of "concentrated," while *weak* usually means "diluted." (See Figure 4.59.)

However, to a chemist, *strong* and *weak*—in relation to acid and base solutions—have very different meanings. Acids and bases are classified as strong or weak according to the extent to which they **ionize,** or form ions in solution. An acid is considered a **strong acid** if it ionizes completely, meaning that every dissolved acid molecule or unit reacts to produce a hydronium ion and an anion. A base is considered a **strong base** if every dissolved molecule or unit produces an hydroxide ion and a cation. The total number of dissolved molecules or ions does not matter. *Strong* signifies *complete* ionization, while *weak* implies only *partial* ionization.

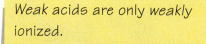

Strong acids are strongly ionized.

When a strong acid dissolves in water, none of the original acid molecules remain. Nitric acid (HNO_3), which is found in acid rain, is a strong acid. The complete reaction within a nitric acid solution is represented this way:

$$H_2O(l) + HNO_3(aq) \longrightarrow H_3O^+(aq) + NO_3^-(aq)$$

All of the original HNO_3 molecules have reacted with water, forming hydronium ions and nitrate ions.

In a **weak acid,** by contrast, only a relatively few dissolved molecules ionize to form hydronium ions and anions. That is, weak acids are only slightly ionized. Nitrous acid (HNO_2), sometimes also found in acid rain, is a typical weak acid. The partial ionization within aqueous nitrous acid solution is represented this way:

Weak acids are only weakly ionized.

$$H_2O(l) + HNO_2(aq) \rightleftharpoons H_3O^+(aq) + NO_2^-(aq)$$

Note that a double arrow is written in the equation. It is used to indicate that it is a **reversible reaction;** that is, it proceeds at equal rates in each direction, both to the right and to the left.

An analogy is useful to illustrate this idea of reversible reactions: At a basketball game, if the number of people walking out of the arena to buy snack foods and drinks equals the number of people returning to their seats after buying their snacks and drinks, the system is described as being in **dynamic equilibrium;** two offsetting processes occur at equal rates, producing a state of balance where no net change is observed.

Figure 4.59 *What terms would you use to express the relative concentrations of these two tea beverages?*

It is important to realize that the forward and reverse *rates*—not the *amounts* of products and reactants—are *equal* in a system at equilibrium. In the basketball-arena analogy, it is clear at any particular time that many more people remain seated watching the game than are moving to and from the snack counters. Likewise, at equilibrium, the concentration of HNO_2 is much larger than are the concentrations of H_3O^+ and NO_2^-. However, a state of balance (dynamic equilibrium) is attained in both cases.

In addition to nitrous acid, acid rain may contain other weak acids, commonly sulfurous acid (H_2SO_3) and carbonic acid (H_2CO_3). Strong acids, such as sulfuric acid (H_2SO_4), may also be present.

Sodium hydroxide (NaOH) and potassium hydroxide (KOH) are two strong bases. Their aqueous solutions completely consist of sodium or potassium cations and OH^- ions. Even though the OH^- concentration may be limited by the solubility of a particular solid base in water, such a substance is regarded as a strong base. Magnesium hydroxide, $Mg(OH)_2$, a commercial antacid ingredient, is an example of a strong (but sparingly soluble) base.

A weak base commonly found in the environment is the carbonate anion (CO_3^{2-}), which is a component of limestone and marble. This poses an apparent contradiction: Why are carbonate anions considered basic, even though they have no hydroxide ions (OH^-) in their formula? The answer is shown by the following equation, where water reacts with carbonate anions to *produce* basic OH^- ions in an aqueous solution:

$$H_2O(l) + CO_3^{2-}(aq) \rightleftharpoons OH^-(aq) + HCO_3^-(aq)$$

Other anions that are basic, such as phosphate (PO_4^{3-}) and sulfite (SO_3^{2-}), also produce OH^- ions in water.

> Only 0.01 g $Mg(OH)_2(s)$ can dissolve in 1 L of water at room temperature, but magnesium hydroxide is still regarded as a strong base.

Modeling Matter

C.9 STRONG VERSUS CONCENTRATED

In this activity, you will depict concentrated and dilute solutions of two imaginary acids—one strong (HSt) and the other weak (HWe).

In referring to an acidic solution, as you recall, *strong* indicates that all (or nearly all) acid molecules have reacted with the water to produce hydronium ions and anions. By contrast, *concentrated* indicates that there are a very large number of particles—molecules or ions—dissolved in a particular volume of solution.

Figure 4.60 depicts possible models for strong and weak acids. In this activity, you will illustrate the ideas you have just learned about.

Draw four equally sized rectangular boxes. Each box represents 1.0 L of solution. Use H^+ to represent hydrogen ions. Each H^+ symbol will represent 0.1 mol of dissolved hydrogen ions. Do *not* draw the solution's water molecules.

Inside each box, draw between 2 and 20 acid molecules, or the appropriate numbers of hydrogen ions and anions, to illustrate the following conditions:

- In the first box, depict a *concentrated strong acid* (HSt) solution.
- In the second box, depict a *dilute strong acid* (HSt) solution.
- In the third box, depict a *concentrated weak acid* (HWe) solution.
- In the fourth box, depict a *dilute weak acid* (HWe) solution.

Use your depictions to answer the following questions:

1. Explain the difference between each pair of words used to describe a solution of an acid or base:

   a. weak, dilute   b. strong, concentrated

2. Calculate the molar concentration (molarity) of hydrogen ions (H^+) depicted in each of your four drawings. Recall that each ion you drew represented 0.1 mol of particles.

3. How would the pH values for the four solutions that you depicted compare to one another?

4. a. Why is it easier to use a single symbol to represent 0.1 mol of particles, rather than trying to draw every particle in 0.1 mol of ions?

   b. How many total particles are actually in 0.1 mol of particles? (Refer to page 161 in Unit 2.)

5. How might your model drawings mislead a viewer about what is actually present in each solution?

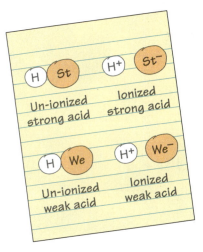

Figure 4.60 *Suggested models used to represent strong and weak acids in solution.*

C.10 ACIDS, BASES, AND BUFFERS

 **Buffer Action**

One early mystery associated with acid rain was the observation that the pH of some lakes and streams seemed unaffected by acidic precipitation. What protected these bodies of water from undergoing large changes in pH?

The key to solving this mystery came when researchers realized that bodies of water that suffered acidification due to acid rain (Figure 4.61) had two features in common. First, they were downwind from a dense array of power stations, metal-processing plants, or large cities, all of which produce nitrogen oxides and sulfur oxides. Second, these bodies of water were often surrounded by soils that are unable to neutralize acid carried by the precipitation. If the soil cannot neutralize the acidic precipitation, the lake or stream into which the precipitation drains then becomes acidified.

By contrast, some bodies of water benefit from surrounding and underlying rock and soils that can neutralize the acidic precipitation and, consequently, they are not seriously affected by acid rain. In particular, the effects of acidic precipitation can be greatly reduced by limestone ($CaCO_3$), which reacts with acids to produce soluble calcium hydrogen carbonate:

$$CaCO_3(s) + H_3O^+(aq) \longrightarrow Ca^{2+}(aq) + HCO_3^-(aq) + H_2O(l)$$

However, calcium carbonate can only prevent the pH from changing when an acid is added to the water. If a base is added, there is nothing available to neutralize it and the pH will increase. To keep the pH of a solution relatively steady—that is, neither increasing nor decreasing appreciably if small amounts of base or acid are added—a *buffer* is needed.

Figure 4.61 *pH sampling of lake water*

A **buffer** is a substance or a combination of substances capable of neutralizing limited quantities of either an acid (H_3O^+) or a base (OH^-) that are added to it without significantly altering its pH value by more than one pH unit. Thus, a buffer must contain two components: (1) an acid to neutralize added base and (2) a base to neutralize added acid. Although calcium carbonate does not act as a true buffer—it cannot prevent added base from increasing the pH—the bicarbonate (HCO_3^-) anion produced serves as a buffer capable of neutralizing *either* added acid or base:

Neutralizing added acid: $H_3O^+(aq) + HCO_3^-(aq) \rightleftharpoons 2\,H_2O(l) + CO_2(g)$

Neutralizing added base: $OH^-(aq) + HCO_3^-(aq) \rightleftharpoons H_2O(l) + CO_3^{2-}(aq)$

In the next investigation, you will observe the unique behavior of buffers and better understand how some bodies of water can withstand the effects of acid rain.

Investigating Matter

C.11 BUFFERS

Introduction

Lab Video: Buffers

In this investigation, you will test a buffer solution. In particular, the results of adding acid and base to water, which serves as the control, will be compared to the results of adding acid and base to a buffered solution. Read the entire procedure and construct a data table suitable for collecting all relevant data.

Procedure

1. Place a wellplate on top of a sheet of white paper. The white paper will help you to observe more easily any color changes during this investigation. (**Caution:** *The solutions in this activity are corrosive. If you splash anything on your skin, wash the affected area thoroughly with water and promptly inform your teacher.*)

2. Add 20 drops of distilled water to each of two wells. Add one drop of universal indicator to each well that contains water. Record the color of the resulting solutions in your data table. Retain the two solutions.

3. Add 20 drops of buffer solution to each of two other empty wells. This buffer solution contains 0.1 M sodium hydrogen phosphate (Na_2HPO_4) and 0.1 M sodium dihydrogen phosphate (NaH_2PO_4).

4. Add one drop of universal indicator to each well containing buffer solution. Record the resulting colors.

5. Add 5 drops 0.1 M NaOH to one of the previous wells containing water and universal indicator. Note and record the color produced.

6. Carefully counting each drop, slowly add 0.1 M NaOH to one well containing buffer solution and universal indicator until the color just matches the well color produced in Step 5. Record the number of drops required.

7. Add 5 drops of 0.1 M HCl to the second well that contains water and universal indicator. Record the color produced.

8. Carefully counting each drop, slowly add 0.1 M HCl to the second well that contains buffer solution and universal indicator until its color just matches the well color produced in Step 7. Record the number of drops required.

9. Retain the solutions until you have gathered all data required.

10. Report your group's data to the class as instructed.

11. Dispose of all solutions as directed by your teacher.

12. Wash your hands thoroughly before leaving the laboratory.

Questions

1. What observations suggest the solution you tested is a buffer?

2. a. How many drops of NaOH solution were needed to bring the buffer solution to the same color (or pH value) that was produced by adding 5 drops of NaOH to the distilled water? Explain any difference.

   b. Make the same comparison as in Question 2a for added HCl.

3. The buffer used in this investigation included hydrogen phosphate ions (HPO_4^{2-}). Write an equation that shows how hydrogen phosphate ions could prevent the pH of lake water from decreasing if limited quantities of acid (H_3O^+) were added.

Making Decisions

C.12 CONSIDERING ACID RAIN

As you plan your feature for the *Riverwood High Gazette,* it will be helpful to consider how school bus idling may or may not be linked to acid rain.

1. Isaiah Martin said he heard that bus air pollution creates acid rain (page 300). As you know, sulfur oxides and nitrogen oxides are the primary emissions that lead to acid rain. Does an idling school bus generate either of these gases? Is Mr. Martin correct? Explain.

2. Ms. Powers, principal of Riverwood High School, stated that reducing school bus idling is motivated by concern for students' well-being and for protecting the environment from damage (page 300). To what damage might she be referring?

3. If a "no bus-idling" policy is enforced, will issues of petroleum-burning vehicles and associated air pollution be resolved? Why?

4. At the larger scale of acid rain or global warming, Riverwood idling school bus emissions are relatively small. So why do you think Principal Powers and other citizens believe that a "no bus-idling" policy is useful?

SECTION C SUMMARY
Reviewing the Concepts

Rainwater is naturally acidic due to dissolved CO_2, but contaminants in the atmosphere can produce precipitation that is even more acidic than normal.

1. What is the pH range of
   a. typical rainwater?
   b. acid rain?

2. Provide an explanation (including a chemical equation) for why rain is naturally acidic.

Gaseous sulfur oxides and nitrogen oxides generated from natural and human sources contribute to acid rain.

3. List two natural sources of sulfur oxides and nitrogen oxides and two human-generated sources of these two pollutants.

4. What strategies can reduce emissions of sulfur oxides and nitrogen oxides to the atmosphere?

5. Identify the major acidic components of acid rain.

6. Write a balanced equation representing what happens when SO_3 gas dissolves in water.

7. What pattern would you expect to find among pH measurements of precipitation in the immediate vicinity of an acid-rain-producing smokestack and at regular distances down-wind and upwind from the site? Sketch a graph of what you would expect to find.

Acids produce hydrogen (or hydronium) ions in water, while bases produce hydroxide ions. Strong acids and bases ionize completely; weak acids and bases ionize only partially.

8. Which of the following substances are acids? Which are bases? Which are neutral substances?
   a. LiOH
   b. CH_3COOH
   c. $C_{12}H_{22}O_{11}$ (table sugar)
   d. H_2SO_3

9. Complete the following equations, depicting reactions that yield hydroxide ions:
   a. $??? + ??? \longrightarrow Ca^{2+}(aq) + 2\ OH^-(aq)$
   b. $NH_3(g) + H_2O(l) \rightleftharpoons ??? + ???$

10. Write an equation showing production of hydronium ions as the following acids are mixed with water:
    a. sulfuric (H_2SO_4) acid, a strong acid used in automotive batteries
    b. hydrofluoric (HF) acid, a weak acid some-times used to etch glass

11. What is the difference between a *strong* acid and a *concentrated* acid?

12. What is the difference between a hydrogen ion and a hydronium ion?

Acidic solutions contain a higher concentration of hydronium ions than hydroxide ions; basic solutions contain a higher concentration of hydroxide ions than hydronium ions.

13. Compare the concentration of hydroxide ions and hydronium ions in pure water.

14. What would happen to the concentration of hydronium ions in a solution of a strong acid if an equal number of hydroxide ions were added to the solution?

15. One liter of room-temperature solution contains 1×10^{-2} mol H_3O^+ and 1×10^{-12} mol OH^-. Is this solution acidic or basic? Explain.

pH is an expression of the molar concentration of hydronium ions in an aqueous solution. Water solutions with pH 7.0 at room temperature are neutral; those with lower pH are acidic and those with higher pH are basic.

16. Why do chemists often express the acidity of a solution as a pH value rather than as a hydrogen ion concentration?

17. Coffee has a pH of 4.0, and pure water has a pH of 7.0. How many more times acidic is coffee than pure water?

18. A sample of acid rain has a pH of 3.0:

   a. What is its hydronium ion concentration?
   b. What is its hydroxide ion concentration?

19. Consider the three solutions in the following list, and then answer the questions that follow:
   - lemon juice, containing 0.001 M H_3O^+
   - stomach fluid, containing 0.1 M H_3O^+
   - drain cleaner, NaOH, containing 0.1 M OH^-

   a. What is the pH of each solution? See Figure 4.58, page 374.
   b. Which solution is most acidic?
   c. Which solution is most basic?

Living organisms and the reactions that sustain them may be adversely affected by changes in pH.

20. How would a lake surrounded by limestone respond to acid rain compared to a lake that was surrounded by rock that does not react appreciably with acid rain?

21. Someone proposes that the acid rain problem could be fixed by dissolving large quantities of basic substances in affected lakes. Do you agree? Explain.

22. Cite evidence from Investigating Matter C.2, pages 366–368, that low pH has a harmful effect on plant material.

23. Very acidic solutions can be used to preserve food. Explain how low pH prevents food spoilage.

A buffered solution is capable of neutralizing limited amounts of either added acid or base, thus resisting changes in solution pH.

24. What are the general components of a buffer?

25. $H_2PO_4^-$ acts as a buffer in cellular systems. Write an equation showing the reaction of a $H_2PO_4^-$ ion with

   a. hydronium ion.
   b. hydroxide ion.

26. Is it better to describe a buffered solution as *resisting* or *preventing* all changes in pH? Explain your answer.

27. Why do some lakes appear unaffected by acid rain?

Connecting the Concepts

28. Explain why the substance that could be called *hydrogen hydroxide* has neither acidic nor basic properties. What is a more familiar name for this compound?

29. A methane molecule (CH_4) contains several hydrogen atoms. Describe a laboratory investigation that would allow you to decide whether or not methane behaves as an acid.

30. Explain how it might be possible for a strong acid solution and a weak acid solution to have identical pH values.

31. Why are shampoos and medicines often buffered?

32. In terms of their chemical behavior, compare antacids to buffers.

33. In Unit 1, you learned that most fish can live in water with pH values between 5.0 and 9.0:
    a. What range of hydronium-ion concentrations corresponds to this pH range?
    b. By what factor does the hydronium-ion concentration differ between these two pH values?
    c. Explain how acid rain might affect the distribution of fish species in an affected lake.

Extending the Concepts

34. What characteristics make the hydrogen ion uniquely reactive compared to a sodium or potassium cation with the same 1+ electrical charge?

35. Construct models and draw electron-dot structures of a water molecule and hydronium ion. Compare their shapes, bonding, and interactions.

36. Use the Internet and the library to research and report on serious acid rain effects over the past decade.

37. Examine the list of ingredients for several antacid products. Which ingredients are responsible for neutralizing acids?

38. Is it possible to have a solution with a pH value of *zero*? Does this mean that the solution has no measurable acidity? Explain your answers.

39. Sometimes the zone separating warring or hostile countries is called a "buffer zone." How does that particular use of the word *buffer* compare to a chemist's use of the term?

Air Pollution: Sources, Effects, and Solutions

Polluted air is so common in the United States (see Figure 4.62) that weather reports for some cities include the levels of certain contaminants in the air.

Most air pollutants result from fossil-fuel-based energy production, such as gasoline-powered vehicles or coal-based electricity generation. Depending on where you live, motor vehicles (such as school buses), power plants, or industries may contribute to local air pollution. See Figure 4.63.

Contaminated air often smells unpleasant and looks unsightly. But even when it cannot be smelled or seen, air pollution can be harmful. Air pollution causes billions of dollars of damage every year. It can corrode buildings and machines, and it can stunt the growth of agricultural crops and weaken livestock. Even more grave, air pollution has been associated with diseases such as bronchitis, asthma (Figure 4.64), emphysema, and lung cancer, adding to human suffering and hospital costs worldwide.

Have you had any experiences with air pollution? Take a moment to consider what you know about air quality. What do you think are common sources of contaminated air near where you live? What is the impact of air pollution on public health and on the environment? This section addresses these and other questions.

Figure 4.62 *The downtown Los Angeles skyline shrouded in smog.*

Figure 4.63 *Small-scale, individual activities (such as mowing a lawn) can lead to build-up of air pollution in densely populated cities and towns.*

Figure 4.64 *Air pollution can exacerbate asthma symptoms.*

D.1 SOURCES OF AIR POLLUTANTS

Many air contaminants such as carbon dioxide, which contribute to rainwater's natural acidity, are the result of natural processes. Natural sources produce more of almost every major air contaminant (except SO_2) than do human sources. In most cases, natural air contamination occurs over wide regions and is seldom noticed. Furthermore, the environment may dilute, transform, or disperse such naturally emitted substances before they accumulate to harmful levels in the air.

By contrast, air pollutants from human activities are usually generated within more localized areas, such as in the vicinity of smokestacks, exhaust pipes, or large population centers. When the amount of an air contaminant overwhelms the ability of natural processes to disperse or dispose of it, air quality can become a serious problem.

Many large cities are prone to high concentrations of air pollutants. If air pollutants generated by activity in large cities were evenly spread over the entire nation, they would have substantially lowered concentrations and be much less noticeable. Unfortunately, weather-related air movement does not produce such even spreading.

Several categories of air contaminants can be identified.

Primary air pollutants directly enter the atmosphere. For example, methane (CH_4) that enters the atmosphere is a by-product of fossil-fuel use and a component of natural gas. Methane is also produced naturally in large quantities by anaerobic bacteria as they break down organic matter, a process that occurs in wetlands, landfills, and even within the digestive systems of animals such as cows and sheep.

Another primary pollutant is actually a class of compounds known as *volatile organic compounds* (VOCs). The term *VOC* refers to a group of hydrocarbons that easily evaporate or that are gaseous at room temperature. Many VOCs released to the atmosphere are unburned fuel molecules emitted from automobiles, trucks, and buses. Release of these compounds has also been linked to natural processes (several types of trees are known to emit large amounts of VOCs) and to human activities that involve easily vaporized liquid hydrocarbons, such as those in some paints, dry-cleaning processes, and near gasoline stations. VOCs are also involved in smog formation, as you will later learn.

Secondary air pollutants are formed in the atmosphere by reactions between primary air pollutants and natural air components. For example, atmospheric sulfur dioxide (SO_2) and oxygen gas react to form sulfur trioxide (SO_3), a secondary air pollutant. Reactions with water in the atmosphere can convert sulfur trioxide to sulfates (SO_4^{2-}) or sulfuric acid (H_2SO_4), a secondary contaminant partly responsible for acid rain.

Particulate pollutants, as shown in Figure 4.65, include microscopic particles that enter the air from either human activities (such as power plants, waste burning, diesel combustion, road building, or mining) or natural processes (such as forest fires, wind erosion, or

> For safety, an odoriferous substance is added to household natural gas, which is mainly methane; the methane itself is odorless.

> As the name indicates, hydrocarbons are made up of C and H. VOCs are primarily composed of these elements, although they may also contain smaller amounts of O, N, and other elements.

> Particulate pollution is also called *particle pollution.*

volcanic eruptions). Common particle pollution includes emissions from smokestacks and vehicle tailpipes, which occasionally are observed as "smoke," but may not always be visible. In fact, virtually undetectable smaller particles that commonly are emitted from diesel-burning vehicles are known as the most dangerous to human health.

Synthetic substances are present in air solely as a result of human activity; if not for people, they would not be there at all. For example, fluorine is released in the stratosphere by high-energy ultraviolet photons that interact with a group of synthetic compounds known as **chlorofluorocarbons** (CFCs). These compounds will be discussed later in this section. Because no natural sources of stratospheric fluorine are known, the presence of fluorine has provided evidence that CFCs play a key role in stratospheric ozone depletion—a topic that you will consider soon.

Figure 4.65 *Particulate contaminants found in air may include dust, pollen, and soot, all shown here under magnification.*

Developing Skills

D.2 IDENTIFYING MAJOR AIR CONTAMINANTS

Table 4.7 (page 386) gives a detailed picture of the sources of some major U.S. air contaminants. Use the information provided in Table 4.7 to answer the following questions:

1. Overall, what is the main source of U.S. air contaminants?

2. For which contaminants is one-third or more contributed by
   a. industry?
   b. transportation?
   c. fuel combustion for electrical, industrial, home heating, and other so-called stationary fuel-burning applications?

3. Name one general type of volatile organic compound (VOC) that might be associated with transportation. Explain its source.

4. Based on data in Table 4.7, would the replacement of gasoline-fueled vehicles by gasoline–electric hybrid vehicles eliminate transportation as a source of sulfur dioxide (SO_2) emissions? Why?

What proportion of total air contamination does human activity produce? Automobile use alone contributes about half the total mass of human-generated air contaminants. However, air is also contaminated each time we heat or cool a building, when we use fossil-fuel-based electricity, or even when food items and other products are delivered to retail stores.

SELECTED U.S. AIR POLLUTANTS, 2000 (IN 10^3 METRIC TONS PER YEAR)							
Source	CO	Pb	NO_x	VOCs	PM_{10}	SO_2	Totals
Transportation	69 332	0.513	12 021	7 617	642	1 637	91 250
Fuel combustion							
Electric utilities	404	0.065	4 777	58	245	10 332	15 816
Industrial	1 108	0.015	2 923	168	221	2 625	7 045
Other	2 653	0.374	1 053	868	438	538	5 550
(residential, commercial, institutional)							
Industrial Processes	6 823	2.869	877	7 288	1 127	1 359	17 477
Miscellaneous (fires, agriculture and forestry, fugitive dust)	18 875	–	523	2 459	19 891	19	41 767
Totals	99 195	3.836	22 174	18 458	22 564	16 510	

Key:
CO   Carbon monoxide
Pb   Lead
NO_x Nitrogen oxides
VOCs Volatile organic compounds
PM_{10} Particulate matter ($<$10 μm diameter)
SO_2 Sulfur dioxide

Source: U.S. EPA, *National Air Quality and Emissions Trends Report, 2003 Special Studies Edition.*

| **Table 4.7**

| **Figure 4.66** *Donora, PA, at midday in late October, 1948.*

D.3 SMOG: HAZARDOUS TO YOUR HEALTH

Sometimes meteorological conditions interact with air contaminants to form a potentially hazardous condition called **smog.** (The word *smog* is a combination of *smoke* and *fog*). In 1952, a deadly smog over London, England, lasted five days and contributed to the deaths of nearly 4000 people. Four years earlier in the coal-mining town of Donora, Pennsylvania (see Figure 4.66), smog-laden air killed 20 people and hospitalized hundreds of others. Similar, though less deadly, episodes have occurred historically in other cities.

Because substances in smog can endanger health, their levels in the air are of major public interest; many weather forecasters report an *air quality rating*, along with humidity and temperature data. The U.S. Environmental Protection Agency (EPA) has devised the Air Quality Index (AQI) based on concentrations of pollutants that are major contributors to smog in metropolitan areas. Table 4.8 emphasizes that the combined health effects of these pollutants can become quite serious.

U.S. AIR QUALITY INDICES

Air Quality Index (AQI) Values	Air Quality Description	Particulate Matter (<10 μm diameter; 24 hour) μg/m³	Sulfur Dioxide (24 hour) ppm	Carbon Monoxide (8 hour) ppm	Ground-Level Ozone (1 hour) ppm	Nitrogen Dioxide (1 hour) ppm	Levels of Health Concern
0–50	Good	0–54	0.000–0.034	0.0–4.4	–	Not reported	Satisfactory; little or no risk.
51–100	Moderate	55–154	0.035–0.144	4.5–9.4	–	Not reported	Acceptable air quality; for some pollutants there may be a moderate health concern for a very small number of people.
101–150	Unhealthful for sensitive groups	155–254	0.145–0.224	9.5–12.4	0.125–0.164	Not reported	Members of sensitive groups* may experience health effects. General public is not likely to be affected.
151–200	Unhealthful	255–354	0.225–0.304	12.5–15.4	0.165–0.204	Not reported	Everyone may begin to experience health effects. Members of sensitive groups* may experience serious health effects.
201–300	Very unhealthful	355–424	0.305–0.604	15.5–30.4	0.205–0.404	0.65–1.24	Health alert; everyone may experience serious health effects.
301–500	Hazardous	425–604	0.605–1.004	30.5–50.4	0.405–0.604	1.25–2.04	Health warnings of emergency conditions. The entire population is likely to be affected.

*High-risk group includes elderly people, children, and those with heart or lung diseases.

Source: Adapted from U.S. EPA, *Air Quality Index: A Guide to Air Quality and Your Health,* 2004.

| Table 4.8

Fatality rates in severe smog episodes have been higher than predicted from known hazards of sulfur oxides or particulates alone. According to some researchers, this increase may be due to *synergistic interactions,* where the combined effect of several substances is greater than the sum of their separate effects alone.

A brownish haze that irritated the eyes, nose, and throat and that also damaged crops first appeared in the air above and around Los Angeles, California, in the 1940s. Researchers were puzzled for some time because Los Angeles had no significant industrial or heating activities. However, the city did have many motor vehicles and abundant sunshine. These factors contributed to the formation of *photochemical smog,* which, as you will soon learn, is different than the previously described industrial smog of London or Donora, Pennsylvania. Another factor was that Los Angeles is bordered by mountains on three sides. Although its valley location makes Los Angeles smog-prone, the smog experienced there was much worse than seemed reasonable. There had to be more to this story.

Normally, air at Earth's surface is warmed by solar radiation and by reradiation from surface materials. This warmer, less dense air rises, carrying pollutants away with it. Cooler, less polluted air then moves in below. In a *temperature inversion,* a cool air mass is trapped beneath a less dense warm air mass, often in a valley or over a city (see Figure 4.67).

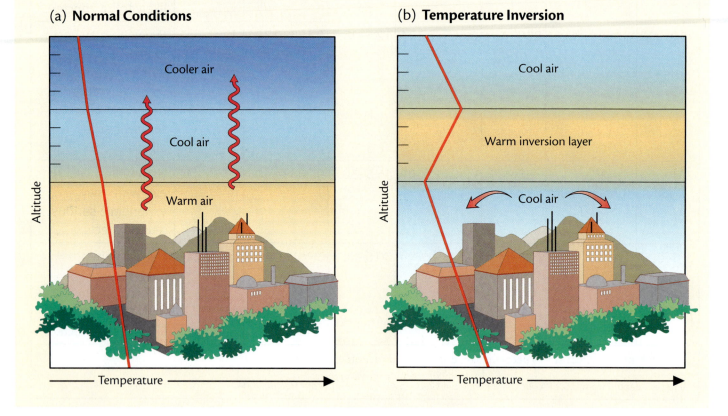

(a) Normal Conditions

Altitude

Cooler air

Cool air

Warm air

Temperature

(b) Temperature Inversion

Altitude

Cool air

Warm inversion layer

Cool air

Temperature

Figure 4.67 *A temperature inversion. Unlike normal conditions (a), in a temperature inversion (b), a layer of cool air is trapped close to Earth's surface below a warm layer, preventing air contaminants from escaping.*

In Los Angeles, the combination of sunny weather and mountains can produce temperature inversions about 320 days annually. During a temperature inversion, air pollutants cannot escape, so their concentrations may rise to dangerous levels. The production and severity of photochemical smog, which can occur even without a temperature inversion (particularly in regions congested with automobiles, such as Los Angeles), are amplified by that accumulation of pollutants.

The simplified equation that follows represents the key ingredients and products of **photochemical smog.** Hydrocarbons, carbon monoxide, and nitrogen oxides from vehicle exhausts are irradiated by sunlight in the presence of oxygen gas. The resulting reactions produce a potentially dangerous mixture, including other nitrogen oxides, ozone, and irritating organic compounds, as well as carbon dioxide and water vapor:

Any reaction initiated by light is a photochemical reaction.

$$\text{Hydrocarbons} + \text{Sunlight} + O_2(g) + CO(g) + NO_x(g) \rightarrow O_3(g) + NO_x(g) + \text{Organic compounds} + CO_2(g) + H_2O(g)$$

Auto exhaust   Oxidizing agents
                                                                                                           and irritants

Nitrogen oxides are essential ingredients of photochemical smog. As noted in Section C, the high temperatures and pressures of automotive combustion (nearly 1000 °C and about 10 atm) promote a reaction between nitrogen gas and oxygen gas in the vehicle's engine cylinders that produces the pollutant nitrogen monoxide (NO):

See Unit 2 (page 149) for background on oxidizing agents. Such materials are also known as oxidants.

$$N_2(g) + O_2(g) + \text{Energy} \longrightarrow 2\ NO(g)$$

Exposed to the atmosphere, nitrogen monoxide gas from the vehicle's exhaust is oxidized to reddish brown nitrogen dioxide gas (NO$_2$), which often is visible in polluted urban air:

$$2\ NO(g) + O_2(g) \longrightarrow 2\ NO_2(g)$$

The photochemical-smog cycle begins when photons from sunlight—the second essential ingredient—break N–O bonds in NO$_2$. This forms NO molecules and highly reactive oxygen (O) atoms, also known as *oxygen radicals.*

$$NO_2(g) + \text{Sunlight} \longrightarrow NO(g) + O(g)$$

Oxygen radicals then react with oxygen molecules (O$_2$) in the troposphere to produce ozone (O$_3$), just as in the stratosphere:

$$O(g) + O_2(g) \longrightarrow O_3(g)$$

Two harmful and unpleasant components of photochemical smog have now been identified—NO$_2$ and O$_3$. Nitrogen dioxide has a pungent, irritating odor at detectable concentrations. Even at relatively low concentrations (0.5 ppm), nitrogen dioxide can inhibit plant growth. At 3 to 5 ppm, this pollutant can cause human respiratory distress after an hour of exposure.

Ozone is a very powerful oxidizing agent (oxidant). At concentrations as low as 0.1 ppm, ozone can crack rubber, corrode metals, and

Radicals are molecules or atoms that contain at least one unpaired electron, causing them to be extremely unstable. Radicals are stabilized by taking electrons from nearby molecules, which can lead to bodily cell damage and also to destruction of the protective ozone layer in the upper atmosphere.

Ozone Formation

damage plant and animal tissues. In terms of human health, relatively low ozone concentrations can cause chest pain, shortness of breath, throat irritation, and coughing. Ozone may also worsen chronic respiratory diseases such as asthma, as well as compromise the body's ability to fight respiratory infections.

Ozone undergoes a complex series of reactions with hydrocarbons, principally VOCs, which are the third essential ingredient of photochemical smog. Hydrocarbons escape from vehicle gasoline tanks and also are emitted due to incomplete gasoline combustion.

The concentrations in air of substances associated with photochemical smog customarily vary over a 24-hour period. In the following activity, you will consider factors that may influence such concentration changes.

Making Decisions

D.4 VEHICLES AND SMOG

Use the data in Figure 4.68 to answer the following questions:

1. a. Between what hours do the concentrations of nitrogen oxides and hydrocarbons peak?
   b. Account for this fact in terms of automobile traffic patterns.

2. Give two reasons why a given pollutant may decrease in concentration over several hours.

3. The concentration maximum for NO_2 occurs at the same time as the concentration minimum for NO. Explain this phenomenon.

4. Although ozone is necessary in the stratosphere to protect Earth from excessive ultraviolet light, at Earth's surface, it is a major component of photochemical smog.

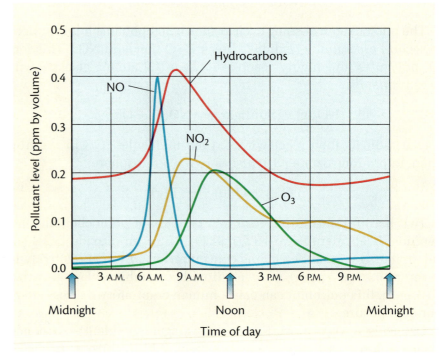

Figure 4.68 *Photochemical smog formation.*

a. Using Figure 4.68, determine which substances are at their minimum concentrations when O_3 is at its maximum concentration.

b. What does this suggest about the production of O_3 in polluted tropospheric air?

5. None of the individuals interviewed at the beginning of this unit (see page 300) mentioned smog in conjunction with the school-bus-idling policy. Are the two related? Explain your answer, and be sure to include what you have learned from Figure 4.68.

Many nations, including the United States, have made considerable progress in smog control. Many cities have cleaner air today than they did decades ago. As you read on, you will explore how the reduction and control of air pollutants formerly produced by motor vehicles and other sources have been successfully addressed.

D.5 STATIONARY-SOURCE AIR POLLUTION CONTROL AND PREVENTION

Several policy options can limit and prevent air pollution from stationary sources:

▶ Energy technologies that cause air pollution can be replaced by technologies that do not involve combustion, such as solar power, wind power, or nuclear power.

▶ Pollution from combustion can be reduced by energy-conservation measures, such as extracting more energy from fuels and thereby burning less of them.

▶ Potentially harmful substances can be removed from fuel before burning; for example, most sulfur can be removed from coal.

▶ The combustion process can be modified so that the fuel is more completely burned (oxidized).

▶ Pollutants can be removed after combustion.

All pollution reduction options involve new costs. When evaluating a pollution prevention or control measure, decision makers must answer three key questions: (1) *What would the prevention or control measure cost?* (2) *What benefits would it offer?* (3) *What costs or risks would be involved in* not *using the prevention or control measure?*

Power plants and smelters generate more than half of all particulate matter emitted by human activity in the United States. However, notable progress has been made in cleaning up the air pollution from these sources (and others) in recent years. Since the 1970s, there has been a significant decrease in the amount of most air pollutants in the United States. See Figure 4.69 (page 392).

Manufacturers have used large-scale techniques to reduce particulates emerging from industrial plants. Several cost-effective methods, such as those described next, are used for controlling particulates and other emissions.

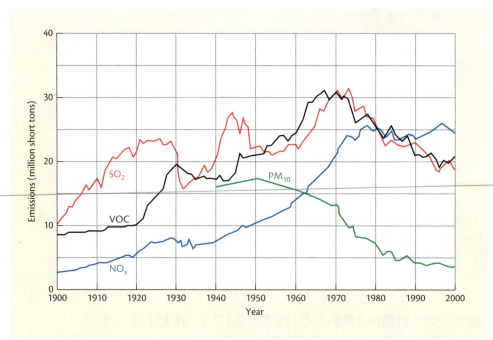

Figure 4.69 *Air pollutants in the United States, 1900–2000. One short ton equals 2000 pounds.*

Electrostatic precipitation is currently the most common technique used to control particulate pollutants. Combustion waste products pass through a strong electrical field, where they become electrically charged. The charged particles then collect on plates of opposite charge. This technique can remove 99% (or more) of particulates, leaving only particles with diameters smaller than about one-tenth of a micrometer (0.1 μm, where 1 μm = 1 × 10^{-6} m). Dust and pollen collectors installed in home ventilation systems are often based on this technique.

Mechanical filtering works much like a vacuum cleaner. Combustion waste products pass through a cleaning room (bag house) where huge filters trap up to 99% of the particles.

Scrubbing is a method that controls particles and also sulfur oxides. In the example of *wet scrubbing* shown in Figure 4.70, sulfur dioxide gas is removed by an aqueous solution of calcium hydroxide, $Ca(OH)_2$(aq), also known as *limewater*. The sulfur dioxide reacts to form solid calcium sulfite, $CaSO_3$:

$$SO_2(g) + Ca(OH)_2(aq) \longrightarrow CaSO_3(s) + H_2O(l)$$

Scrubbers can remove up to 95% of sulfur oxides. Unfortunately, because of higher operating costs resulting from maintenance and disposal expenses, combined with lower net power production, scrubbers add significantly to the cost of electrical-power generation.

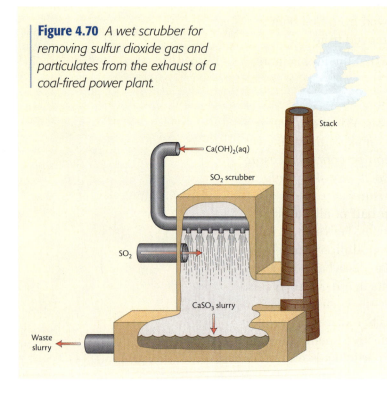

Figure 4.70 *A wet scrubber for removing sulfur dioxide gas and particulates from the exhaust of a coal-fired power plant.*

LIMITING REACTANTS AND SCRUBBERS

In Section B (page 353), you were introduced to the concept of limiting reactants. Consider the stoichiometry involved in removing sulfur dioxide emissions with a wet scrubber (see the equation on page 392). In properly designing this scrubber, which reactant should be designated as the limiting reactant? Why? What would be the consequences if the other reactant were the limiting reactant?

Investigating Matter

D.6 CLEANSING AIR

Introduction

Your teacher will demonstrate two control methods for air pollutants: *electrostatic precipitation* and *wet scrubbing*. Prepare a sketch of each laboratory setup as part of your note-taking for this demonstration.

 **Lab Video: Cleansing Air**

Part I: Electrostatic Precipitation

1. Observe the generation of smoke and what happens to it. Record your observations.
2. Observe the chemical reaction that takes place on the copper rod. Record your observations.

Part II: Wet Scrubbing

3. As the reaction proceeds, observe the color of the universal indicator in the liquid and the color of the pH paper in each flask. Compare those colors with the color-key information provided with the pH paper. Record your observations.

Questions

1. Write an equation representing the smoke-generating reaction that occurred in Part II between HCl and NH_3.
2. What information did the universal indicator provide about the contents of each flask in Part II?
3. What information did the pH paper in each flask provide?
4. What was the overall effect of the scrubbing, as shown by the acid-base indicators?
5. List two ways in which the quality of the air in the reaction vessel was improved by wet scrubbing.
6. a. What advantages do electrostatic precipitators have over wet scrubbers?
   b. What are their disadvantages?

D.7 CONTROLLING AUTOMOBILE EMISSIONS

The Clean Air Act of 1970 authorized the Environmental Protection Agency to set emissions standards for new automobiles. Maximum allowable limits were set for hydrocarbon, nitrogen oxide, and carbon monoxide emissions. One major contribution toward meeting those standards was the development of the *catalytic converter,* which is a reaction chamber built into the exhaust system of motor vehicles, as shown in Figure 4.71.

As you may recall from Unit 3 (page 254), a catalyst is a material that speeds up a chemical reaction that would proceed far slower without the presence of that catalyst. Enzymes that aid in digesting food and in promoting other bodily functions are organic catalysts. Although a catalyst participates in the chemical reaction, it is not considered a reactant because it emerges unchanged when the reaction is completed.

How can a catalyst speed up a reaction and still remain unchanged? **Collision theory** states that reactions can occur only if molecules collide with sufficient energy and with suitable orientation to disrupt chemical bonds. The minimum energy required for such effective collisions is called the **activation energy.** You can think of the activation energy as an energy barrier that stands between the reactants and products. See Figure 4.72. Reactants must have enough energy to get over the barrier before a reaction can occur. That is, the reactants must have sufficient kinetic energy to participate in effective

Figure 4.71 *A catalytic converter (top) is installed in an automobile's exhaust system (bottom).*

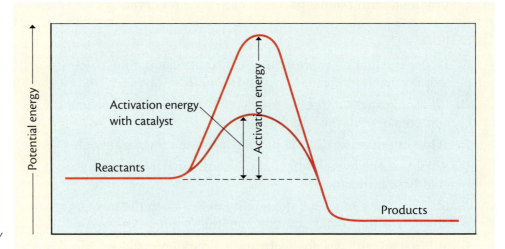

Figure 4.72 *Activation-energy diagram for a typical chemical reaction. The activation energy represents the minimum energy needed to initiate the reaction.*

molecular collisions. The higher the energy barrier, the fewer the molecules with sufficient energy to get over it and the slower the overall reaction proceeds.

A catalyst works by providing a different reaction pathway, one with a lower activation energy. In effect, the catalyst lowers the energy barrier. The result is that more molecules now have sufficient energy to react and to form products within a given period of time. In automotive catalytic converters, a few grams of platinum, palladium, or rhodium embedded in a noncatalytic material such as aluminum oxide (Al_2O_3) act as the catalyst.

In the **catalytic converter,** exhaust gases and outside air pass over several solid catalysts that help speed the conversion of potentially harmful gases to harmless products—nitrogen oxides to nitrogen gas, carbon monoxide to carbon dioxide, and hydrocarbons to carbon dioxide and water. Exhaust gases enter the catalytic converter where, for example, nitrogen oxides and carbon monoxide are removed as shown in the following equation.

Figure 4.73 *Many cities and towns, and some states, require annual testing of exhaust emissions.*

$$2\ NO(g)\ +\ 2\ CO(g)\ \xrightarrow{\text{Catalyst}}\ N_2(g)\ +\ 2\ CO_2(g)$$

The Clean Air Act amendments of 1990 set new tailpipe emission standards for U.S. automobiles (see Figure 4.73) and also established emission standards for vehicles other than automobiles. Those national tailpipe-emission levels set achievable goals that significantly helped reduce photochemical smog. Between 1993 and 2002, total CO tailpipe emissions decreased 21%. NO_x emissions from transportation decreased 5% within the same period.

Transportation is still a major air-pollution source in many states despite improvements made by catalytic converters.

ChemQuandary 4

CONTROLLING AIR POLLUTION

From a practical point of view, why do you think that controlling air pollution from all U.S. automobiles, recreational vehicles, buses, and trucks would be more difficult than controlling air pollution from U.S. power plants and industries?

D.8 OZONE AND CFCs: A CONTINUED SUCCESS STORY

Although small doses of ultraviolet A and B radiation are necessary for health, too much is dangerous. In fact, if all ultraviolet radiation in sunlight reached Earth's surface, many forms of life would be seriously damaged. Ultraviolet B and C photons, you learned in Section B, have enough energy to break covalent bonds, even within biological structures such as DNA. The resulting chemical changes can cause sunburn and cancer in humans and can damage many biological systems (Figures 4.74 and 4.75).

Earth's Ozone Shield

Fortunately, Earth has an ultraviolet shield in the stratosphere, near the troposphere. The shield consists of a thin region of gaseous ozone (O_3) that absorbs ultraviolet radiation. However, this stratospheric **ozone shield** is a fragile system that contains a remarkably low concentration of ozone. If all ozone molecules in the shield were moved to Earth's surface at atmospheric pressure, they would form a 3-mm layer (about the thickness of this textbook's cover).

This ozone layer is vital to life on Earth. A National Research Council study suggests that for each 1% decrease in stratospheric ozone, a 2% to 5% increase in various forms of skin cancer will occur. Such ozone depletion could also lower yields of some food crops, sharply increase sunburn and eye cataracts, and damage some aquatic plant species.

As sunlight penetrates the stratosphere, high-energy ultraviolet photons react with oxygen gas molecules, splitting them into individual oxygen atoms (the dot in the following equation represents an unpaired electron):

$$O_2 + \text{High-energy UV photon} \longrightarrow \cdot O + \cdot O$$

These reactive oxygen atoms are an example of a **free radical.** Radicals contain unstable arrangements and an odd number of electrons. Radicals quickly enter into chemical reactions that allow them to attain stable arrangements of electrons. For example, oxygen free radicals in the stratosphere can combine with oxygen molecules to form ozone, as the following equation shows. A third molecule (typically N_2 or O_2, represented by M in the equation) carries away excess energy from the reaction, but remains unchanged:

$$O_2 + \cdot O + M \longrightarrow O_3 + M$$

Each ozone molecule formed in the stratosphere can absorb an ultraviolet B photon with a wavelength of less than 320 nm. In this energy absorption, which keeps potentially harmful ultraviolet radiation from reaching Earth's surface, ozone decomposes, producing an oxygen molecule and an oxygen free radical:

$$O_3 + \text{Medium-energy UV photon} \longrightarrow O_2 + \cdot O$$

Figure 4.74 *Sunscreen products protect skin from the harmful effects of the Sun's ultraviolet rays. This is particularly necessary now that scientists have determined that stratospheric ozone has been significantly depleted.*

Figure 4.75 *Ultraviolet rays can damage eyes as well as skin. Many sunglasses are made of materials that block UV rays from reaching one's eyes.*

These products can then continue the reaction cycle by replacing ozone in the protective stratospheric layer:

$$O_2 + \cdot O + M \longrightarrow O_3 + M$$

CFCs and Ozone Depletion

 **Ozone Depletion**

Some human activities have already affected Earth's stratospheric ozone layer. Riverwood's Isaiah Martin speculated that one possible cause might be bus exhaust.

Actually, the major culprit has been identified as chlorine atoms from chlorinated hydrocarbon molecules known as *chlorofluorocarbons* (CFCs). The Freons, CCl_3F (CFC-11) and CCl_2F_2 (CFC-12), are the most familiar CFCs. Chlorofluorocarbons are synthetic substances that were once widely used (and still are in some countries) as propellants in aerosol cans (see Figure 4.76), as cooling fluids in air conditioners

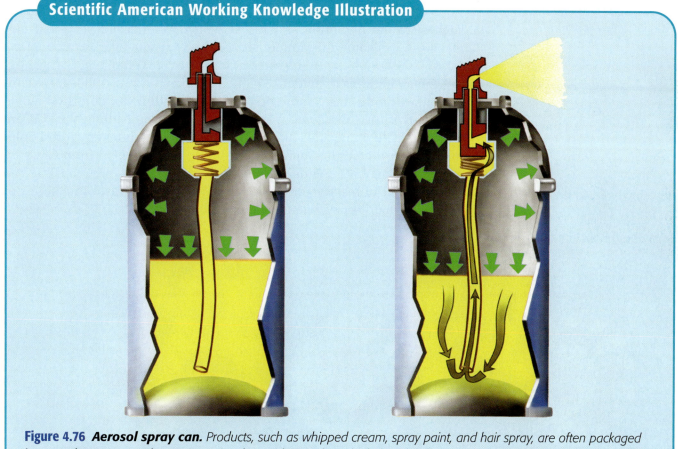

Scientific American Working Knowledge Illustration

Figure 4.76 *Aerosol spray can.* Products, such as whipped cream, spray paint, and hair spray, are often packaged in aerosol spray cans. The can contains the product and a volatile liquid and gas, called the propellant. The propellant gas exerts high pressure above the liquid product. (As you know, gaseous particles are constantly in motion and exert pressure as they collide with surfaces.) Left: Propellant gas inside the can normally exerts pressure (green arrows) on can walls and liquid (yellow). Right: As the spray can nozzle is pushed down, a passage opens to the outside atmosphere. Because gas pressure exerted on the liquid is greater than air pressure outside the can, the liquid is propelled up toward the nozzle and exits as sprayed droplets.

Figure 4.77 *Nobel Prize Winners: F. Sherwood Rowland (top) and Mario Molina (bottom).*

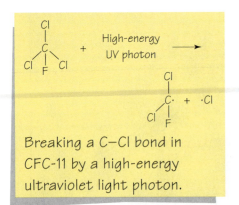

Breaking a C–Cl bond in CFC-11 by a high-energy ultraviolet light photon.

and refrigerators, and as cleaning solvents for computer chips. Developed in the 1930s, nontoxic and nonflammable CFCs quickly replaced toxic ammonia or sulfur dioxide as coolants of choice in refrigerators and air conditioners. For almost half a century, CFCs were well accepted as refrigerants, propellants, and cleaning solvents.

However in the 1970s, two U.S. chemists—F. Sherwood Rowland and Mario Molina (Figure 4.77)—proposed that chlorofluorocarbons (CFCs) posed a threat to the world's protective ozone layer. Since then, extensive research has developed a more complete picture of that threat, as you will soon learn.

A satellite launched later in the 1970s carried instruments to monitor changes in worldwide stratospheric ozone concentrations. By 1986, satellite data showed seasonal ozone reduction, or the formation of a so-called **ozone hole,** in the stratosphere above Antarctica, as shown in Figure 4.78. The data indicated an annual variation in ozone concentration of 1% to 2.5%, well beyond normal variation noted before the production of CFCs. That depletion cycle was observed to increase annually over subsequent years. In fact, by the late 1990s, nearly 70% of stratospheric ozone disappeared for a time over Antarctica, creating an area of ozone depletion as large as North America. More recent evidence shows some thinning of stratospheric ozone over Arctic regions as well, although not to the same extent as over Antarctica.

Several bodies of evidence gathered since the mid-1980s have converged, linking the Antarctic ozone depletion to chlorine atoms obtained from CFCs, not from natural sources. Although CFCs are highly stable molecules in the troposphere, high-energy ultraviolet photons in the stratosphere split chlorine radicals (·Cl) from CFCs by breaking their C–Cl bonds.

The highly reactive chlorine radicals can participate in a series of reactions that destroy ozone by converting it to O_2:

$$2 \cdot Cl + 2 O_3 \longrightarrow 2 \cdot ClO + 2 O_2$$

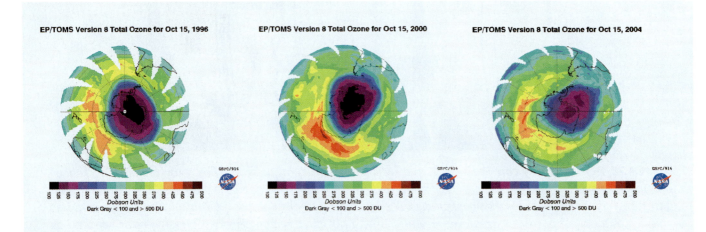

Figure 4.78 *Total October ozone concentrations in the stratosphere over Antarctica, 1996–2004. Reds and greens indicate high ozone concentrations; blues and purples indicate low ozone concentrations. Lower Dobson units indicate lower ozone concentration.*

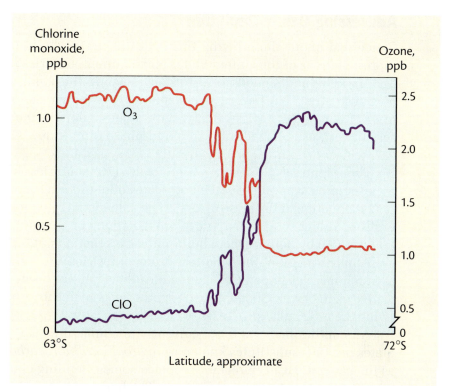

Figure 4.79 *Stratospheric ozone (red) and chlorine monoxide (purple) concentrations over Antarctica.*

The reactive chlorine monoxide radical (·ClO) produced by that ozone-destroying reaction can combine with itself:

$$·ClO + ·ClO \longrightarrow ClOOCl$$

The resulting ClOOCl then can decompose in the presence of solar radiation through a two-step process:

$$ClOOCl + UV \text{ photon} \longrightarrow ·ClOO + ·Cl$$

$$·ClOO + UV \text{ photon} \longrightarrow ·Cl + O_2$$

Note that these last two reactions replace the "2 ·Cl" that was consumed in the first ozone-destroying step. That is, the chlorine free radical (·Cl) emerges unchanged in this series of reactions, even though it participated in them. Thus, the ·Cl serves as a *catalyst* for the destruction of stratospheric O_3. The net result (the arithmetic sum) of all of these reactions can be written this way:

$$2 O_3 \longrightarrow 3 O_2$$

Each chlorine radical participates in not just one O_3-destroying reaction, but in an average of 100 000 of these reactions. In so doing, it speeds up ozone destruction, but the chlorine radical remains ultimately unchanged. The net effect is depletion of ozone molecules (O_3) by their conversion into oxygen molecules (O_2), a chemical reaction catalyzed by chlorine radicals released from CFC molecules. See Figure 4.79.

> The "dot" in the symbols ·Cl and ·ClO signifies a single, unpaired electron; the resulting symbol represents a highly reactive free radical.

> A particular chlorine radical is eventually removed from this stratospheric cycle due to other chemical changes, such as its reaction with methane (CH_4) to produce hydrogen chloride, HCl.

Addressing Ozone Depletion

Concerned about the growing threat to the stratospheric ozone layer, representatives of the United States and 55 other countries met in Montreal in 1987 and signed an historic agreement, the Montreal Protocol on Substances That Deplete the Ozone Layer, known also as the *Montreal Protocol*. This initial treaty, with amendments in the 1990s (and again later), established a timetable to phase out all new CFC production within developed nations by 1996.

The Montreal Protocol, which over 170 countries have joined, appears to be working. The United States and 140 other nations halted CFC production at the close of 1995. The amended treaty includes a multimillion dollar fund to help developing nations phase out CFC production by 2010 by helping them build industries that do not rely on CFCs.

Another result of the Montreal Protocol restrictions is that chemical research has developed CFC substitutes that will not harm Earth's protective ozone layer. Chemists have synthesized such substitutes by modifying the molecular structures of CFCs.

Air-conditioned vehicles currently manufactured in the United States use CFC substitutes known as *hydrochlorofluorocarbons* (HCFCs) as the coolant. HCFC molecules replace some offending chlorine atoms with atoms of hydrogen and fluorine. The hydrogen atoms allow HCFCs to decompose in the troposphere and reduce the amount of catalytic chlorine free radicals released in the stratosphere. For example, the hydrochlorofluorocarbon CHF_2Cl has only about 5% of the ozone-depleting capacity of CFC-12.

Although some HCFC decomposition occurs in the troposphere, appreciable numbers of undecomposed HCFC molecules still rise into the stratosphere and decompose under exposure to UV radiation, releasing chlorine free radicals. And, like CFCs, HCFCs are greenhouse gases; thus, there is some concern that they may contribute to global warming.

Because of these potential problems, HCFCs are considered as only transitional replacements for CFCs. Montreal Protocol amendments require that companies accelerate phaseout of some HCFCs before 2030 and complete the phaseout of all HCFCs by then.

Thus, chemists continue research for even better CFC substitutes. This research has already produced a group of synthetic compounds known as *hydrofluorocarbons,* HFCs. Molecules of HFCs, such as CH_2FCF_3, contain only carbon, fluorine, and hydrogen atoms. They contain no ozone-depleting chlorine atoms, and the included hydrogen atoms promote HFC decomposition.

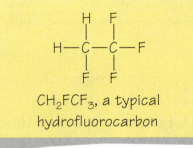

CH_2FCF_3, a typical hydrofluorocarbon

D.9 AIR-QUALITY CONCLUSIONS

As you have learned, human activity can have large-scale impact on gases surrounding our planet. Such impact may influence the world's average temperature by altering normal atmospheric concentrations of substances such as carbon dioxide. They increase potential exposure of living systems to ultraviolet radiation by thinning the protective stratospheric ozone layer. Air pollution resulting from human activity can also lead to regional effects, such as acid rain and smog, or even to localized problems, such as the potential health hazards associated with elevated levels of various air pollutants.

Atmospheric chemistry is quite complex and difficult to study. However, increased knowledge of chemical and physical processes involved in the atmosphere, all gained by scientific research, already has helped guide decision making to protect and enhance the vital envelope of gases encircling Earth.

However, what about the decision to prohibit school bus idling at Riverwood High School? Based on what you have learned in this unit, do benefits of that decision outweigh its drawbacks? By now, you have probably formed some thoughts on these issues, as you prepare to write your feature article for the *Riverwood High Gazette.*

Concerns regarding Earth's air quality will continue into the future; you have gained useful background in this unit to comprehend these issues.

SECTION D SUMMARY
Reviewing the Concepts

Air pollution results from contributions of both primary and secondary pollutants.

1. What distinguishes a primary air pollutant from a secondary air pollutant?

2. Write and balance the equation for formation of SO_3 from SO_2 and oxygen gas.

3. Characterize each of the following as a primary or secondary pollutant, and identify its major source:
   a. CO_2
   b. CH_4
   c. NO_x
   d. H_2S

Photochemical smog can intensify due to temperature inversions and adverse wind patterns.

4. What ingredients are needed to create photochemical smog?

5. What conditions are shared by cities that experience photochemical smog?

6. Are all cities subject to temperature inversions? Explain your answer.

7. What do oxygen free radicals contribute to photochemical smog?

8. What are some ways that photochemical smog can be reduced?

9. What does the formula NO_x represent?

10. Why is the general expression NO_x used, rather than a specific molecular formula?

Air pollution can be affected by everyday activities and by changes in chemical technologies.

11. List five ways that air pollutants from motor vehicles and industries can be reduced.

12. Describe how a catalyst can speed up a chemical reaction.

13. Consider these three air-pollution control technologies: *scrubbers, catalytic converters,* and *electrostatic precipitators.* Which technique is most effective in combating each of the following types of air pollutants?
    a. sulfur oxides
    b. particulates
    c. hydrocarbons

14. Explain how electrostatic precipitators can reduce air pollution.

15. Some Green Chemistry programs seek to prevent pollution before it starts. What are advantages of this strategy over combating existing air pollution?

16. The following list provides some examples of decisions that one might make to reduce personal contributions to air pollution. For each example, identify particular air pollutants involved, and explain how their release will potentially be reduced:
    a. ordering a fast-food lunch inside a restaurant, rather than waiting in a line of cars at the drive-up window
    b. hanging wet laundry outside to dry
    c. avoiding "topping off" a car's gasoline tank by adding more gasoline after the pump stops automatically
    d. buying groceries and other locally produced items
    e. turning off lights, the TV, and other electrical devices when exiting rooms

17. Draw Lewis dot structures for CFC-11 and CFC-12.

18. What properties of CFCs made them useful in industry?

19. What are some alternatives for industrial uses of CFCs?

20. Why are problems associated with CFCs regarded as *global* rather than *national*?

21. What is a free radical, and why is it so reactive?

22. Explain how a chlorine free radical can act as a catalyst in the stratosphere.

Connecting the Concepts

23. Explain the pattern in the data shown in Figure 4.79, page 399. If comparable data were obtained in the Northern Hemisphere, would you expect to observe the same pattern?

24. What are advantages and disadvantages of scrubbing as a strategy for air-pollution reduction?

25. Why is the *presence* of ozone regarded as a problem at ground level, while the *absence* of ozone is regarded as a problem in the stratosphere?

26. Find Internet sites that report current patterns of ozone distribution in the stratosphere. Write a report to summarize this current information.

27. How might monitoring practices be different for primary and secondary air pollutants?

28. Which primary and secondary air pollutants are typically produced by diesel engines, such as those that power buses?

Extending the Concepts

29. Why would scrubbing probably *not* be a good strategy for removing NO_x from vehicle exhaust gases?

30. Major automobile manufacturers have abandoned their research and development of pure electric-powered vehicles. Investigate some factors, in terms of total pollution, that led to this decision.

31. Investigate the historical use of refrigerants. What did CFCs replace? What current products have replaced CFCs?

32. Investigate reported progress of phasing out CFCs and HCFCs. What has been accomplished and what challenges remain?

33. What advances recently have been accomplished in the design of internal combustion engines to minimize pollution?

34. Investigate the topic of diesel engines. Are catalytic converters used with diesel engines? If so, do they differ from catalytic converters used with gasoline engines?

35. Investigate differences in the atmospheric movement of CFCs and ozone. How does this movement affect stratospheric conditions?

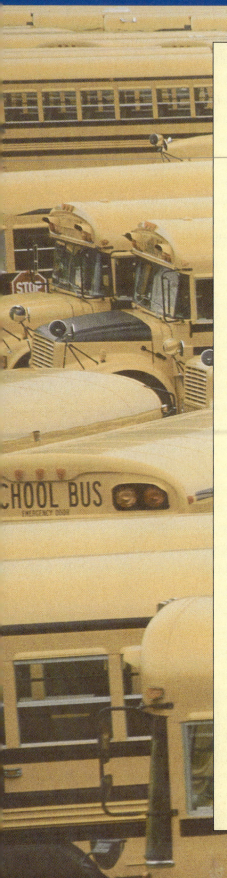

BUS-IDLING POLICY AND AIR QUALITY

As you learned in this unit, Earth's atmosphere is composed of various gases and particulates, some naturally occurring and others produced by human activities. Health and environmental concerns sometimes arise when these atmospheric components interact with one another and with solar radiation. Which interactions will be affected by the school bus-idling policy recently approved at Riverwood High School?

In your role as a *Riverwood High Gazette* reporter, you are expected to provide up-to-date, accurate information about this issue clearly and objectively. Good reporters confirm new information with multiple valid sources, and they judge the credibility of each source. Keep this in mind as you begin writing your feature article on Riverwood High School's bus-idling policy.

1. Each reporter in your class will design and draft a feature article to inform readers about implications of the school bus-idling policy. The following questions (at minimum) should be addressed:

   ▶ What emissions are produced when a bus is idling?
   ▶ What air-quality concerns are associated with a no-idling policy?
   ▶ What benefits are gained from minimizing school bus-idling time?
   ▶ What problems arise with minimizing bus-idling time?
   ▶ How valid or accurate are the opinions and concerns expressed by those interviewed at the start of this unit? (See page 300.) Remember, the people interviewed represent some typical views of the readers that you intend to inform about this issue.

2. In some situations, diagrams or illustrations can convey key concepts or ideas more concisely than words alone. Include at least two attractively designed and informative diagrams or illustrations in your article.

3. Before an article is published in a newspaper, it is usually proofread by one or more editors. Thus, after each reporter has completed an initial draft, they should trade their draft articles with at least two classmates for further assistance. The "editors," or proofreaders, will then provide constructive, useful feedback.

The following questions will guide your proofreading efforts:

▶ Is the science correct?
▶ Are the text and illustrations informative and educational to the reader?
▶ Does the article address any common misconceptions or incorrect ideas mentioned in this unit's opening interviews?
▶ Does the article address both benefits and drawbacks of the new policy? Are benefits and drawbacks summarized in an objective, unbiased style?
▶ Is the information in the article confirmed by multiple credible sources? Are references cited for all sources used?
▶ Is the article well organized, clearly written, free of grammatical and spelling errors, and visually appealing?

4. After receiving feedback from their proofreading partners, reporters will revise their articles (if necessary) and each reporter will create a final "ready for publication" draft, which will be submitted to your teacher.

LOOKING BACK AND LOOKING AHEAD

This completes the first four units of *Chemistry in the Community*. Many chemistry topics remain to be explored further, such as chemical equilibrium, biochemistry, and nuclear chemistry. The last three *ChemCom* units also focus on important chemistry-related issues; they can be studied in any order desired. These three units build upon and apply the chemistry knowledge and skills you acquired through your study of the first four units.

UNIT 5

Industry: Applying Chemical Reactions

HOW do chemical industries convert substances into other useful materials by large-scale manufacturing?

SECTION A
Some Chemistry of Nitrogen (page 410)

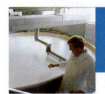

WHAT chemical principles are involved in converting nitrogen gas to useful nitrogen-based compounds?

SECTION B
Nitrogen and Industry (page 433)

HOW does electrical energy cause chemical changes?

SECTION C
Metal Processing and Electrochemistry (page 452)

The town of Riverwood seeks to generate new employment opportunities. Two large chemical companies are interested in locating in Riverwood. In this unit, you will learn about issues Riverwood must address before the town invites either company to build a chemical plant in the community.

EKS or WYE?
More Job Options for Riverwood

A *RIVERWOOD NEWS* COMMENTARY BY GRETA B. LEDERMAN
Riverwood Resident

Several weeks ago, the town council announced that two industrial firms were interested in building chemical manufacturing plants in Riverwood: EKS Nitrogen Products Company and WYE Metals Corporation. Because of a lack of employment opportunities in Riverwood, many of your friends and neighbors have expressed relief at this news.

Greta Lederman

When the Riverwood Corporation declared bankruptcy last year, many residents lost their jobs. Although some have found new jobs, many have not. The unemployment rate in Riverwood has hovered near 15% for most of the past year.

Can anything be done to boost our local economy? Yes. We can allow one of these eager companies to build a chemical plant in our town. New jobs would be created for chemists, chemical technicians, and office personnel. Each company claims it would need to hire about 200 new employees.

In addition, both companies confirm that they plan to occupy the former Riverwood Corporation site. Recently, the Riverwood Corporation parking lot seems to attract only weekend flea-market vendors and people selling used cars.

We need more job opportunities; no one denies that. But some people may wonder whether a chemical plant is the best option for Riverwood. Why not a furniture manufacturer or a book warehouse? Aren't chemical plants dangerous?

The reality is that no other companies have expressed an interest in locating in Riverwood. EKS and WYE are attracted by our well-educated workforce and by our access to abundant resources and electrical energy, among other factors. Both companies are well suited to our community.

Both companies have strong safety and environmental records. They are members of the American Chemistry Council's Responsible Care initiative. They both advocate green technologies.

We have a unique opportunity to choose between two companies that are competing to locate here. Usually it's the reverse—two or more cities competing for one company. Learn more about each company. Visit the Web site of each company and attend the scheduled citywide discussions.

If you still have doubts, notice how this news has affected your friends and neighbors. A new sense of hope has emerged over the recent weeks. Let's turn weeks of hope into years of Riverwood prosperity.

EKS or WYE?
New Jobs May Become too Costly

A *RIVERWOOD NEWS* COMMENTARY BY PAK JIN-WOO
Riverwood Resident

Although no one can deny that our Riverwood community needs a source of new jobs, it's foolish to invite either EKS Nitrogen Products Company or WYE Metals Corporation to locate here to manufacture, respectively, ammonia or aluminum, without first considering all the consequences.

Pak Jin-Woo

The promise of 200 new jobs is alluring. However, all of these jobs would be in only one company. What will happen if there's a decline in the market for ammonia or aluminum?

Either of the proposed Riverwood plants would be small compared to other plants operated by EKS and WYE in other cities. Thus, the Riverwood plant would probably be among the first to be shut down if an economic slowdown occurs. If that happens, the Riverwood area would quickly return to the same economic situation it is in now.

Wouldn't it be more prudent to distribute those 200 jobs among several different companies? Let's learn from our recent experience with the city's former principal employer, the now-bankrupt Riverwood Corporation.

Although long-term economic health is important, we must question the potential safety and environmental risks that each of these two chemical plants would pose to our community.

For instance, ammonia production requires very high pressures and temperatures. Although accidents are uncommon, the potential consequences of an explosion or spill are still great. Several illnesses and even deaths of workers have been documented at ammonia plants.

Aluminum production also involves risks to its workers. For every 100 employees, job-related illnesses and injuries involve, on average, about 20 workers annually. Which 40 Riverwood residents would be included in this statistic if we allow an aluminum plant to be established here?

Perhaps the greatest long-term concern is the potential for environmental harm. Ammonia-based contamination of the Snake River could, for example, cause another fish-kill crisis.

Along the same lines, manufacturing aluminum metal produces, among other things, toxic carbon monoxide gas. What might happen if the company mishandles the containment of this gas?

I realize that not all of these scenarios are likely to occur. However, we must be absolutely willing to accept the consequences should something negative happen. With what level of risk are you and your friends willing to live?

As you can infer from the newspaper commentaries, two companies want to establish a chemical plant in Riverwood. EKS Nitrogen Products Company wants to establish a plant that produces ammonia. WYE Metals Corporation would build an aluminum-production plant. As both commentators acknowledge, either company would offer at least 200 new job opportunities to the Riverwood community. However, as the commentaries suggest, job creation is not the only factor to consider.

Later in this unit, you will help to decide whether to invite EKS or WYE (or neither company) to operate a plant in Riverwood. The chemistry you learn in this unit will prepare you to make informed decisions about the risks and benefits of operating such plants. Keep in mind the concerns in the two commentaries as you learn about the chemistry involved in manufacturing ammonia and aluminum.

Some Chemistry of Nitrogen

In the first four units, you learned that chemistry is concerned with the composition and properties of matter, changes in matter, and the energy involved in those changes. In this unit you will explore how chemical industries use chemical knowledge and reactions to produce a wide range of useful material goods on a large scale.

In particular, this unit offers you an opportunity to evaluate the chemical operations of EKS and WYE. This knowledge will help you later as you debate whether or not a new chemical plant should locate in Riverwood, and, if so, which one.

Making Decisions

A.1 CHEMICAL PROCESSING IN YOUR LIFE

To become more aware of how pervasive the products of chemical processing are in everyday life, list five items or materials around you that have *not* been manufactured, processed, or altered from their natural form. Start by considering everyday items, such as clothes, objects in your home, modes of transportation, books, foods, communication devices such as phones and computers, and sports and recreation equipment—all things you routinely encounter.

Now answer the following questions. Be prepared to share and discuss your answers in class.

Figure 5.1 *The chemical industry manufactures many products, some of which are used in shipping and supplying food products.*

1. a. Which items on your list were wrapped, boxed, or shipped in materials that had been manufactured (Figure 5.1)? Explain.
    b. Is the packaging or shipping material essential or simply a convenience? Why?

2. In what ways might each item or material on your list be better than or inferior to a manufactured, processed, or synthetic alternative? Consider factors such as cost, convenience, availability, and quality.

3. If a product is "100% natural," does that necessarily mean it was not involved in any processing or chemical or physical changes? Why? Support your answer with at least one example.

A.2 CHEMICAL PRODUCTS

Consider the items you listed in Making Decisions A.1. There are very few things that we use regularly that have not been modified in some way. This is the chemical industry's focus—transforming natural resources into useful products to meet a wide variety of needs and purposes. The chemical industry also creates new substances and materials as replacements for natural ones. For example, plastics often replace wood or metals, and synthetic fibers often replace cotton or wool.

Even though the chemical industry is a worldwide, multibillion-dollar enterprise that affects everyone's daily life through its products and economic impact, most people are not aware of what happens when new materials are produced. This raises questions about how chemical industries operate, how they manufacture new products, and what those products contain.

The modern chemical industry employs well over a million people worldwide. Over the past 80 years, it has grown through mergers of smaller companies and creation of new companies. During that time, the industry's focus has expanded from a limited range of basic products to more than 70 000 products. Hundreds of chemical companies form the third largest manufacturing group in the United States; only industries that produce machinery and electrical equipment are larger. Indeed, if we include the food and petroleum industries in the chemical-industry category, this represents the world's *largest* industry.

Most chemical products reach the public indirectly because they are used to produce other consumer materials. For instance, the automobile and home-construction industries use enormous supplies of industrial chemicals. They use paints and plastics for automobile body parts such as bumpers, dashboard panels, upholstery and carpeting, and synthetic rubber in tires. Home construction involves large quantities of plastics for carpeting and flooring, insulation, siding, window frames, piping, and appliances. It also involves using paints, metals, and air-conditioning coolants. Figure 5.2 shows a range of products from various chemical industries.

Figure 5.2 *The chemical industry produces many materials that have useful properties.*

You will learn about aluminum production in Section C.

Fertilizer Components

In Riverwood, the two industrial companies under consideration manufacture products involving nitrogen and aluminum. Among the products manufactured by EKS Nitrogen Products Company are nitric acid and ammonia, which are often used in chemical reactions that produce other materials. By contrast, the sheet aluminum produced by WYE Metals Corporation is most often used directly in its manufactured form.

EKS is committed to producing high-quality nitrogen-based fertilizer in Riverwood at a reasonable cost, using the best available technologies. Fertilizer may sound unappealing as a product, but its manufacture and sale represent a multi-million-dollar business that employs thousands of people worldwide and affects the lives of nearly everyone, from farmers and gardeners to food producers and consumers.

Figure 5.3 *This label indicates that this fertilizer contains 4% nitrogen, 5% P_2O_5, 2% K_2O, and 1% iron.*

Ideal for roses, azaleas, camellias, ferns, fuchsias, begonias, and other acid or shade loving plants.

GUARANTEED ANALYSIS

Total Nitrogen (N)..4.00%
Available Phosphoric Acid (P_2O_5) 5.00%
Potash (K_2O)...2.00%
Iron (Fe)... 1.00%

Derived from processed organic materials: Ammonium Nitrate, Ammonium Phosphate, Sulfate of Potash, and Iron Sulfate.

DIRECTIONS

Apply evenly by hand around base of plant out to drip line and water in well. A 1/8" deep layer around plant three times annually. For potted plants apply 2 tablespoons full per 6" pot.

LAWN & DICHONDRA — Apply by hand or spreader 2-1/2 lbs. per 100 sq. ft. (10 x 10). A 1 lb. coffee can holds approximately 2-1/2 lbs. of 4-5-2.

How can you decide whether a particular fertilizer is best for a specific application, such as on houseplants, lawns, or cornfields? One way is to find out if the fertilizer contains the proper ingredients. Complete fertilizers, such as those manufactured by EKS, contain the three main elements growing plants need—*nitrogen, phosphorus,* and *potassium*—as well as trace ions and filler material.

Figure 5.3 shows the label of a typical commercial fertilizer bag. The numbers on the label indicate the percent values of key ingredients contained in the fertilizer: nitrogen, N; phosphorus, P (expressed as percent P_2O_5); and potassium, K (expressed as percent K_2O). The proportion of each component varies according to crop needs. Most lawn grasses need nitrogen; a 20–10–10 fertilizer is a good choice for lawns. Phosphorus is especially useful in promoting fruit and vegetable growth, so some gardeners prefer a 10–30–10 mixture over a balanced (10–10–10) fertilizer.

Many plant nutrients provided in fertilizers are in the form of cations and anions. Cations are likely to include potassium (K^+), ammonium (NH_4^+), and iron(II) (Fe^{2+}) or iron(III) (Fe^{3+}); anions include nitrate (NO_3^-), phosphate (PO_4^{3-}), and sulfate (SO_4^{2-}).

In the following laboratory investigation, you will use confirming tests to determine whether a fertilizer solution contains specified nutrients.

Reporting P and K as P_2O_5 and K_2O (as in Figure 5.3) is traditional and originates from early research on fertilizers. When chemists burned fertilized plants to ash and then analyzed the ash, they obtained the masses of P_2O_5 and K_2O.

You used confirming tests in water testing in Unit 1 (see page 42).

A.3 FERTILIZER COMPONENTS

Introduction

You will test a fertilizer solution for six particular ions (three anions and three cations). In Part I you will perform tests on known solutions of those ions to become familiar with each confirming test. In Part II you will decide which ions are present in an unknown fertilizer solution. Read the complete procedure and prepare a suitable data table to record your observations.

 **Lab Video: Fertilizer Components**

Procedure

Part I. Ion Tests

1. Prepare a warm-water bath by adding about 30 mL of water to a 100-mL beaker. Place the beaker on a hot plate. The water must be warm, but it should not boil. Control the hot plate accordingly. You will use this bath in Step 6d.

2. Obtain a Beral pipet set containing test solutions of each of six known ions: nitrate (NO_3^-), phosphate (PO_4^{3-}), sulfate (SO_4^{2-}), ammonium (NH_4^+), iron(III) (Fe^{3+}), and potassium (K^+). Observe and record the colors of all six solutions.

BaCl₂ Test

Several of the ions you are studying in this investigation can be identified first by their reaction with barium cations (Ba^{2+}) and then their behavior in the presence of an acid.

3. a. Add 2 to 3 drops of each test solution to six separate, clean test tubes. Add 2 to 3 drops of distilled water to a seventh clean test tube to serve as a blank.

   b. Test each solution individually by adding 1 to 2 drops of 0.1 M barium chloride ($BaCl_2$) solution (see Figure 5.4). Record your observations.

   A blank was explained and used in a Unit 1 investigation (see page 42).

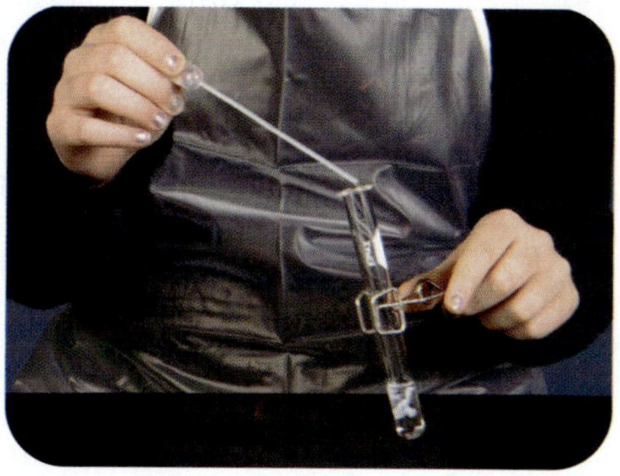

   c. Add 3 drops of 6 M hydrochloric acid (HCl) to each of the seven test tubes. *Caution: 6 M HCl is very corrosive. If any splashes on your skin, wash it off thoroughly with water and inform your teacher. Do not inhale any HCl fumes.* Record your observations.

   d. Dispose of the solutions as instructed, then clean and rinse the test tubes.

Figure 5.4 *Adding drops to a test tube.*

Brown-Ring Test

In the presence of nitrate ions (NO_3^-), mixing iron(II) ions (Fe^{2+}) with sulfuric acid (H_2SO_4) produces a distinctive result—a brown ring. You can use this *brown-ring test* to detect nitrate ions in a solution.

4. a. Add eight drops of sodium nitrate ($NaNO_3$) solution to a small, clean test tube. Place eight drops of distilled water in a second small, clean test tube, which will serve as a blank.

   b. Add about one milliliter of iron(II) sulfate ($FeSO_4$) solution to each of the two test tubes. Gently swirl to mix the contents of each tube.

   c. Have your teacher carefully pour about one milliliter of concentrated sulfuric acid (H_2SO_4) along the inside of each test tube wall so the acid forms a second layer under the undisturbed solution already in the tube (Figure 5.5).

   ⚠ *Caution: Concentrated H_2SO_4 is a very strong, corrosive acid. If any contacts your skin, immediately wash affected areas with abundant running tap water and inform your teacher.*

   d. Allow the test tubes to stand, undisturbed, for 1 to 2 min.

   e. Observe any change that occurs at the interface of the two liquid layers. Record your observations.

5. Dispose of the solutions as instructed, then clean and rinse the test tubes.

Figure 5.5 *To minimize mixing, concentrated sulfuric acid is allowed to run down the inside surface of the test tube. Possessing a larger density than water, concentrated sulfuric acid forms a layer beneath the water.*

NaOH and Litmus Test

Now you will investigate which of the three cations (NH_4^+, Fe^{3+}, or K^+) you can identify by observing their characteristic behavior in the presence of a strong base.

6. a. Add four drops of each cation test solution to three separate, clean test tubes. Add four drops of distilled water to a fourth clean test tube, which will serve as a blank.

   b. Moisten four strips of red litmus paper with distilled water; place them on a watch glass.

   c. Add 10 drops of 3 M sodium hydroxide (NaOH) directly to the solution in one of the four test tubes. *Caution: 3 M NaOH is corrosive. If any splashes on your skin, wash it off thoroughly with water and inform your teacher. Do not allow any NaOH solution to contact the test tube lip or inner wall.* Immediately stick one of the four premoistened red litmus-paper strips from Step 6b across the top of the test tube. The strip must not contact the solution.

   d. Warm the test tube gently in a hot water bath for 1 min. Wait 30 s and then record your observations.

   e. Repeat Steps 6c and 6d for each of the other three test tubes.

Flame Test

You can identify many metal ions by the characteristic color they emit when heated in a burner flame. For example, it is common to use a flame test to identify potassium ions.

7. a. Obtain a platinum or nichrome wire inserted into glass tubing or a cork stopper, which serves as its handle.

b. Set up and light a bunsen burner. Adjust the flame to produce a light blue, steady inner cone and a luminous, pale blue outer cone.

c. To clean the wire, place about 10 drops of 2 M hydrochloric acid (HCl) in a small test tube. ***Caution:*** *2 M HCl is corrosive. If any splashes on your skin, wash it off thoroughly with water and inform your teacher. Do not inhale any HCl fumes.* Dip the wire into the hydrochloric acid and then heat the wire tip in the burner flame. Position the wire at the intersection of the two parts of the flame, not in the center cone. As the wire heats bright red, the burner flame may become colored, as illustrated in Figure 5.6. The characteristic colors are due to metallic cations held on the wire's surface.

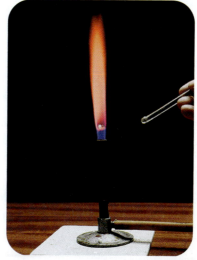

Figure 5.6 *Completing a flame test. Metallic ions emit characteristic colors when heated*

d. Continue dipping the wire into the acid solution and inserting the wire into the flame until there is little or no change in flame color as the wire heats to redness.

e. Place 7 drops of potassium ion solution into a clean test tube. Dip the cool, cleaned wire into this solution. Then insert the wire into the flame. Note any change in the flame color, the intensity of the color, and the estimated time (in seconds) that the color was visible.

> The lack of color change in the flame indicates that the wire is clean.

f. Repeat the potassium-ion flame test, this time observing the burner flame through cobalt-blue or didymium glass (Figure 5.7). Again, note the color, intensity, and duration of the color. Your partner can hold the wire in the flame while you observe through the colored glass. Then exchange roles. Record all observations.

> The colored glass may make the characteristic potassium flame color easier to see against the color of the burner flame.

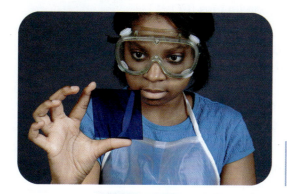

Figure 5.7 *The emission produced by excited potassium ions is best observed through cobalt glass.*

KSCN Test

In Unit 1 (page 44), you learned that when potassium thiocyanate (KSCN) is added to an aqueous solution containing iron(III) ions (Fe^{3+}), a deep red color appears due to the formation of $[FeSCN]^{2+}$ cations. Appearance of this characteristic color confirms the presence of iron(III) in a solution.

8. a. Place 3 drops of iron(III)-containing solution into a clean test tube. Place 3 drops of water into a second clean test tube.

   b. Add 1 drop of 0.1 M KSCN solution to each test tube. Record your observations.

   c. Dispose of the solutions as instructed, then clean and rinse the test tubes.

Part II. Tests on an Unknown Fertilizer Solution

1. Obtain a Beral pipet containing an unknown fertilizer solution. Record its code number. Your unknown solution contains one of the cations and one of the anions that you tested in Part I. Observe and record the color (if any) of the unknown solution.

2. Complete suitable laboratory tests of the fertilizer solution to identify the two unknown ions it contains. Record all observations and conclusions. Repeat a particular test, if you wish, to confirm your initial observations.

3. Dispose of all solutions used in this investigation as directed by your teacher.

4. Wash your hands thoroughly before leaving the laboratory.

Questions

1. Identify the ions that were present in your unknown fertilizer sample, citing evidence from your investigation.

2. Look up the components of common commercial fertilizers and provide the name and formula for two of the ingredients that contain one or both of the ions in your fertilizer sample. For example, sodium chloride (NaCl) would supply sodium ions (Na^+) and chloride ions (Cl^-) to a solution, while potassium carbonate (K_2CO_3) would furnish potassium ions (K^+) and carbonate ions (CO_3^{2-}).

3. Using evidence collected in your investigation, describe a test you could complete to verify whether a fertilizer sample contains phosphate ions (PO_4^{3-}).

4. Is barium nitrate, $Ba(NO_3)_2$, soluble or insoluble in water? Cite specific evidence from your investigation to support your answer.

5. Explain why the kind of data gathered in this investigation about a specific fertilizer solution is inadequate for you to judge whether the fertilizer is suitable for a particular use.

A.4 FERTILIZER AND THE NITROGEN CYCLE

Each year EKS manufactures about 3 million tons of ammonia and more than 1.5 million tons of nitric acid. Most is used to manufacture fertilizers sold to farmers and gardeners.

The purpose of all fertilizers is to add nutrients to soil so growing plants have adequate supplies. The raw materials that growing crops use are mainly carbon dioxide from the atmosphere and water and nutrients from the soil. Plant roots absorb water and nutrients such as phosphate, magnesium, potassium, and nitrate ions.

Phosphate becomes part of the energy-storage molecule ATP (adenosine triphosphate) and the nucleic acids RNA and DNA. Magnesium ions are a key component of chlorophyll, which is essential for photosynthesis. Potassium ions—found in the fluids and cells of most living things—help maintain a growing plant's ability to convert carbohydrates from one form to another and to synthesize proteins. Nitrate ions aid vigorous plant growth and supply nitrogen atoms that are incorporated into proteins. Figure 5.8 illustrates how plants use these nutrients.

> Liquified ammonia can also be directly applied to soil as a fertilizer.

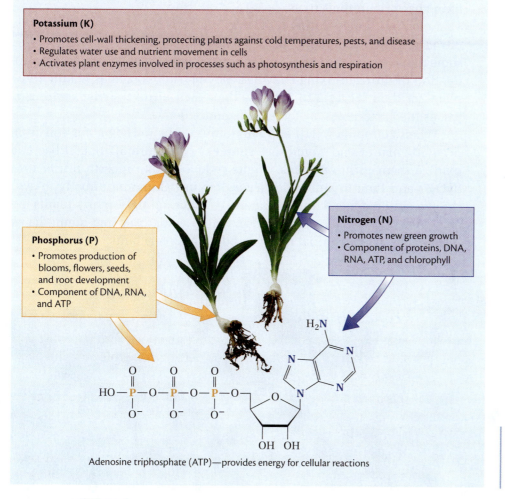

Potassium (K)
• Promotes cell-wall thickening, protecting plants against cold temperatures, pests, and disease
• Regulates water use and nutrient movement in cells
• Activates plant enzymes involved in processes such as photosynthesis and respiration

Phosphorus (P)
• Promotes production of blooms, flowers, seeds, and root development
• Component of DNA, RNA, and ATP

Nitrogen (N)
• Promotes new green growth
• Component of proteins, DNA, RNA, ATP, and chlorophyll

Adenosine triphosphate (ATP)—provides energy for cellular reactions

Figure 5.8 *All plants depend on certain nutrients participating as reactants in vital chemical reactions.*

Figure 5.9 *Energy supplied by lightning facilitates reactions between atmospheric nitrogen and oxygen. This is one way to "fix" nitrogen, that is, to incorporate it into nitrogen-based compounds.*

Proteins are large molecules made up of various combinations of 22 different amino acids.

Proteins are a major constituent of all living organisms, including plants. In plants, proteins usually account for 5–20% of the plant's mass. Nitrogen makes up about 16% of the mass of those protein molecules.

Although nitrogen gas (N_2) is abundant in the atmosphere, it is so chemically stable (unreactive) that plants cannot use it directly. However, nitrogen gas can be *fixed*—that is, combined with other elements to produce nitrogen-containing compounds that plants can use chemically. Lightning or combustion can fix atmospheric nitrogen (see Figure 5.9) by causing it to combine with other elements, especially oxygen. Some plants called *legumes,* such as clover and alfalfa, have nitrogen-fixing bacteria in their roots.

Scientists are exploring biological methods for making atmospheric nitrogen more available to plants. These methods include engineering some microorganisms and plants to contain genes that will direct the production of nitrogen-fixing enzymes. Having such enzymes would allow plants or their bacteria to produce their own nitrogen-based fertilizers, just as the bacteria associated with legumes do.

When ammonia (NH_3) and ammonium ions (NH_4^+) enter soil from decaying matter and from other sources (see Figure 5.10), soil bacteria oxidize them to nitrate ions. Plants reduce nitrate ions to nitrite ions (NO_2^-) and then to ammonia. They then use ammonia directly to synthesize amino acids. Unlike humans or other animals, many plants are able to synthesize all their needed amino acids by using ammonia or nitrate ions as their initial nitrogen-containing reactants.

Figure 5.10 *Many U.S. farmers routinely apply anhydrous liquid ammonia fertilizer to their fields.*

When organic matter decays, much of the released nitrogen recycles among plants and animals, and some returns to the atmosphere. Thus, some nitrogen gas removed from the atmosphere through nitrogen fixation eventually cycles back to the atmosphere.

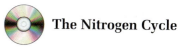

The Nitrogen Cycle

The **nitrogen cycle** consists of the following steps:

1. Nitrogen-fixing bacteria that live in legume root nodules or in soil convert atmospheric nitrogen (N_2) to ammonia molecules or ammonium ions. Alternatively, lightning converts N_2 to nitrogen oxides (NO_x).

2. Various soil bacteria oxidize ammonia and ammonium ions to nitrite ions and then to nitrate ions.

3. Plants take in nitrate ions from the soil, and subsequently incorporate nitrogen atoms in synthesizing proteins and other nitrogen-containing compounds.

4. The nitrogen passes along the food chain to animals that feed on these plants and to animals that feed on other animals.

5. When those plants and animals die, bacteria and fungi take up and use some of the nitrogen from plant/animal proteins and other nitrogen-containing molecules. The remaining nitrogen atoms are released from the decaying matter as ammonium ions and ammonia gas.

6. Denitrifying bacteria convert some ammonia, nitrite ions, and nitrate ions back to nitrogen gas, which returns to the atmosphere.

> Nitrifying bacteria are found in many plants, blue-green algae, some marine algae, and lichens.

> Some nitrogen atoms recycle through the living world without returning to the atmosphere.

A.5 PLANT NUTRIENTS

Making Decisions

Consider and answer the following questions:

1. Why do some farmers alternate plantings of legumes and grain crops over several growing seasons (Figure 5.11)?

2. Why is it beneficial to return unused parts of harvested crops to the soil?

3. How might research on new ways to fix nitrogen help lower farmers' operating costs?

4. What consequences might result from over-fertilizing?

Figure 5.11 *A contour strip farm in Wisconsin. Legumes (green) and grain (gold) are rotated over several growing seasons. How might this practice be beneficial?*

A.6 THE NITROGEN CYCLE

In Unit 1, you learned how the hydrologic cycle can purify water (pages 86–87). You also learned in Unit 4 that carbon-containing molecules change as carbon atoms cycle among living and nonliving components of Earth (page 348–349). For instance, plants transform carbon found as CO_2 into complex molecules, such as carbohydrates. Carbon is also chemically bound as dissolved CO_2 in the oceans and as carbonate rocks in Earth's crust (Figure 5.12).

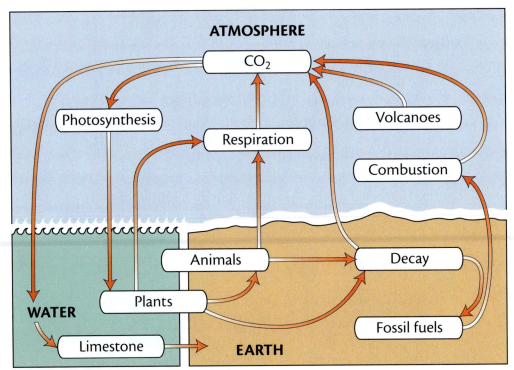

| **Figure 5.12** *The carbon cycle.*

In addition to studying the hydrologic and carbon cycles, you examined diagrams illustrating those processes. Such visual models help you organize related information, allowing you to easily trace interactions and connections.

You have just learned that nitrogen atoms also cycle among the atmosphere, soil, and organisms. However, a visual model depicting this nitrogen-cycle information has not been presented. In this activity you will create that missing diagram.

Look at Figure 5.12 and also figures on pages 87 and 348. Notice how they depict the hydrologic and carbon cycles. Then review Fertilizer and the Nitrogen Cycle A.4 (pages 417–419) in this unit (and earlier textbook material if needed) to guide your completion of the following steps.

1. Construct your own diagram of the nitrogen cycle.
   a. Use arrows to show the direction of flow as nitrogen atoms cycle among the atmosphere, soil, and living organisms.
   b. Include symbols and names for key molecules and ions at each cycle stage.
   c. Use pictures and color as needed to clarify details in your model.
   d. Make your model easy to follow. A classmate should be able to summarize the steps correctly by referring to your diagram.

2. Exchange your model with a classmate's model.

3. Select an appropriate starting point on your classmate's diagram and trace nitrogen through its cycle.

4. Repeat Step 3, but use your classmate's diagram to write a description of the key steps in the nitrogen cycle. Your written description should be limited to information contained in your classmate's diagram, even if some features are different from those in your diagram.

5. Exchange diagrams and written descriptions with your classmate. You now have the nitrogen-cycle diagram you originally drew and your classmate's written description based on it.

6. Compare the difficulty you experienced in completing these two tasks:
   a. transforming the book's description of the nitrogen cycle into your diagram
   b. transforming your classmate's diagram of the nitrogen cycle into a written description

7. How closely did your classmate's written description reflect the actual structure and details of your diagram? Explain.

8. Compare the description that your classmate wrote about your diagram with the description of cycle steps on page 419.
   a. Compared to the textbook description, did your classmate's description omit or add any details or steps? Explain.
   b. Which description is more detailed? Explain.

9. Based on your classmate's description,
   a. how easy and convenient was your diagram to interpret and follow? Explain.
   b. how could you modify your diagram to improve its accuracy and clarity? Explain.

10. Considering your answers to Question 9, make any needed changes to your diagram so it more clearly illustrates the nitrogen cycle.

In what ways is this step similar to drawing a map for someone who asks you to provide directions to travel to a particular location?

This is similar to using a drawn map to navigate unfamiliar territory. What are the consequences of missing details or inadequate labeling in such a map?

In Investigating Matter A.3 (page 413), you identified some major ions present in fertilizer. In the next investigation you will complete a more detailed study of one of these ions—the phosphate anion (PO_4^{3-}).

A.7 PHOSPHATES

Introduction

We can partially evaluate the quality of fertilizers by examining the percent (by mass) of essential nutrients they contain. In this investigation, you will analyze a fertilizer solution to determine the mass and percent phosphate (PO_4^{3-}) contained in it.

Phosphate is a component of several critically important biomolecules, including DNA, RNA, ATP, and proteins. In DNA and RNA, phosphate links other molecular units together in long chains that store genetic information. In ATP, phosphate ions help store chemical energy. Changes in phosphate-ion concentration can also affect an enzyme's effectiveness in catalyzing a particular reaction. Thus, phosphate serves as a "chemical switch," turning particular chemical reactions on or off.

Colorimetry, the method you will use to determine a fertilizer sample's phosphate content, is based on the fact that the intensity of a solution's color is directly related to the concentration of the colored substance. To analyze colorless phosphate ions this way, you will first convert all phosphate ions to a new, colored ion. Then you will compare the color intensity of the unknown solution to the intensities of solutions with known phosphate concentrations—that is, to a set of phosphate-based *color standards* (see Figure 5.13).

If the color intensities of the sample and standard appear identical, then the phosphate concentrations must also be the same. Likewise, if the color intensity of the unknown sample is lower than that of the standard, the unknown sample has the lower of the two phosphate concentrations. One strategy of this laboratory procedure is to reduce the unknown phosphate solution concentration sufficiently through dilution so its color intensity will be within the range of the prepared color standards.

See page 412 for background on the components of commercial fertilizers, including phosphorus.

**Lab Video:
Phosphates**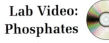

Figure 5.13 *The intensity of colors in several different solutions of known concentration can be used to estimate an unknown solution concentration.*

Procedure 🥽

1. Label five clean test tubes: *10 ppm, 7.5 ppm, 5.0 ppm, 2.5 ppm,* and *x ppm.*

2. Prepare a water solution of the unknown sample by placing 0.50 g of solid fertilizer in a 400-mL beaker. Add 250 mL of distilled water. Stir until the sample completely dissolves.

As prepared in Step 2, your fertilizer solution is too concentrated to analyze by colorimetry. In other words, milliliter for milliliter, your solution contains considerably more phosphate ion than you can evaluate by the known color standards you will use later. Therefore, you will have to prepare a more dilute fertilizer solution for analysis, one that is only 1/50 as concentrated.

Here is a very impractical way to accomplish that dilution goal of 1/50 concentration: Imagine adding distilled water (with stirring) to your 250-mL fertilizer solution until its volume becomes 50 times larger. The phosphate concentration would, indeed, be 1/50 of its initial value. However, you would probably be dismayed to discover that the diluted volume would be almost 13 L (0.250 L × 50 = 12.5 L)—over three gallons!

Fortunately, there is a more sensible dilution strategy. As you complete Steps 3 and 4, think about how this procedure accomplishes the same dilution goal more successfully.

3. Measure 5.0 mL of your unknown fertilizer solution. Retain the remaining solution (in case you need to start over).

4. Pour this 5.0-mL sample of fertilizer solution into a clean 400-mL beaker. Then add 245 mL of distilled water, measuring that volume with a graduated cylinder. Stir thoroughly.

5. Pour 20 mL of this diluted fertilizer solution into the test tube labeled *x ppm*.

6. Your teacher has already prepared a supply of 10-ppm phosphate-ion standard solutions. Pour 20 mL of that standard solution in the test tube labeled *10 ppm*.

7. Given supplies of 10-ppm solution and distilled water, decide what volumes of 10-ppm solution and distilled water you should measure and mix to prepare 20.0-mL samples, respectively, of 7.5-ppm, 5.0-ppm, and 2.5-ppm phosphate solutions. Write out your plan.

8. Ask your teacher to check your written plan. After receiving your teacher's approval, prepare the three solutions. Pour each standard solution into its appropriately labeled test tube.

9. Place a 400-mL beaker half-full of water on a hot plate. Start heating the water but do not allow the water to boil. You will use this warm-water bath in Step 12.

10. Add 2.0 mL of ammonium molybdate–sulfuric acid reagent to each of the four test tubes containing phosphate standards and to the diluted unknown solution.

11. Add a few crystals of ascorbic acid (no more than the size of a pencil eraser) to each tube. Using a rinsed and dried stirring rod each time, stir to dissolve.

> Ascorbic acid is another name for Vitamin C.

12. Carefully place your five test tubes into the warm-water bath held in the beaker on the hot plate. Continue heating the tubes in the water bath until a blue color develops in the 2.5-ppm solution. Do not allow the water to boil. Turn off the hot plate.

13. Using a test-tube holder, remove the test tubes from the water bath and place them in numerical order in a test-tube rack.

14. Compare the color intensity of the unknown solution ('*x ppm*') with the intensities of the four color-standard solutions. Place the unknown-solution test tube between the two tubes containing standard solutions that you judge to have the closest-matching color intensities to the unknown.

> Recall that *ppm* represents parts per million, or, as here, g PO_4^{3-} per 10^6 g solution.

15. Estimate the concentration (ppm) of your unknown phosphate solution from the known color standards. For example, if the unknown solution color falls between the 7.5-ppm and 5.0-ppm color standards, you might decide to call it *6 ppm* (or 6 g PO_4^{3-} per 10^6 g solution). Record the estimated phosphate ion concentration in ppm.

16. Discard your solutions as directed by your teacher.

17. Wash your hands thoroughly before leaving the laboratory.

Data Analysis

1. Calculate the mass (in grams) of phosphate present in your original 250-mL fertilizer solution (or in the original 0.50-g solid fertilizer sample). Note that you need to use a multiplication factor of 50 to compensate for the fact that you used only 1/50 of the original fertilizer sample (see Step 3). In other words, the phosphate concentration in the original fertilizer must have been 50 times greater than the value found for the solution in the *x ppm* test tube. Thus the calculation becomes

$$\text{mass of phosphate (in grams)} = \frac{?? \text{ g phosphate}}{10^6 \text{ g solution}} \times 250 \text{ g solution} \times 50$$

In place of *??*, substitute the value of the unknown solution concentration (in ppm) you obtained in Step 15. Record your calculated mass of phosphate.

2. Based on your calculated mass, find and record the percent phosphate (by mass) in your original 0.50-g fertilizer sample.

Questions

1. Name two common household solutions for which you can estimate their relative concentrations just by observing their color intensities.

2. Chemists often use an instrument called a *colorimeter* to determine solution concentrations. A **colorimeter** measures the quantity of light that can pass through an unknown sample and compares it to the quantity of light that can pass through a known standard solution. What advantages does a colorimeter offer over use of the human eye?

3. Explain this statement: "The accuracy of results of a colorimetric analysis depends, in part, on the care taken in preparing the standards used in the analysis."

4. In Investigating Matter A.3 (page 413), the tests were *qualitative* and were used only to indicate whether particular ions were present. How could you modify an ion test involving the formation of a colored solution so it would be *quantitative—* indicating the concentration of ion present?

5. In Step 3, you measured 1/50 of the total fertilizer solution volume and set aside the rest. Consider this seemingly simpler alternative: In Step 2 you could have taken 1/50 of the starting mass of fertilizer (0.50 g × 1/50 = 0.010 g) and then omitted Steps 3 and 4 completely. What are possible disadvantages of this shorter, more direct procedure?

A.8 FIXING NITROGEN BY OXIDATION–REDUCTION

Fertilizer Sources

Using commercial ammonia for fertilizer has had a huge impact on agriculture and world food supplies. World ammonia production has increased dramatically over the last 50 years, as farmers have increased their use of fertilizer to meet the food needs of growing populations. The U.S. chemical industry annually produces about 40 billion pounds of ammonia, mostly dedicated to making fertilizer.

> Liquified ammonia can also be directly applied to soil as a fertilizer. See Figure 5.10, page 418.

Before modern methods of manufacturing ammonia were developed, nitrogen-containing fertilizers came from either animal waste or nitrate compounds. Large quantities of such compounds came from Chilean guano beds (Figure 5.14). At the turn of the 20th century, speculation arose that guano beds would be depleted by about 1930, raising fears of an agricultural crisis and the spectre of world famine. Actually, the advent of commercial ammonia production largely eliminated dependence on all natural nitrate sources.

In seeking ways to fix nitrogen gas artificially, scientists in 1780 first combined atmospheric nitrogen and oxygen by exposing them to an electric spark. However, the cost of electricity made this too expensive for any commercial use. A less expensive method, the *Haber-Bosch process* replaced it. From 1912 to 1916, Fritz Haber and Karl Bosch in Germany developed the technique for making ammonia directly from hydrogen gas and nitrogen gas, according to this equation:

$$N_2(g) + 3\,H_2(g) \rightleftharpoons 2\,NH_3(g)$$

Figure 5.14 *Guano (seabird dung) deposits were one of the first commercial sources of fertilizer.*

Oxidation–Reduction

In the Haber-Bosch reaction of nitrogen gas with hydrogen gas, electrons are involved in an oxidation–reduction reaction. As you know, atoms that apparently lose their share of electrons are involved in a process called *oxidation* (see page 147). For example, the conversion of metallic sodium atoms (Na) into sodium ions (Na^+) is oxidation because electrically neutral sodium atoms are oxidized to +1 sodium ions by the loss of one electron per atom. Recall that the opposite process—the apparent gaining of electrons—is called *reduction.* The formation of chloride ions (Cl^-) from electrically neutral chlorine atoms is an example of reduction. Electrons can be transferred to or from particular atoms, molecules, or ions. The products of such oxidation–reduction reactions also include atoms, molecules, or ions.

You can judge the relative tendency of a covalently bonded atom to attract electrons within compounds based on that element's *electronegativity* (see page 71). Nonmetallic elements typically have higher electronegativities than do metallic elements. Figure 5.15 shows the electronegativity values for some common elements.

> Regardless of how the electron-transfer occurs, the same oxidation-reduction principles apply.

Increasing Electronegativity →

Increasing Electronegativity (vertical)

H 2.1																
Li 1.0	Be 1.5											B 2.0	C 2.5	N 3.0	O 3.5	F 4.0
Na 0.9	Mg 1.2											Al 1.5	Si 1.8	P 2.1	S 2.5	Cl 3.0
K 0.8	Ca 1.0	Sc 1.3	Ti 1.5	V 1.6	Cr 1.6	Mn 1.5	Fe 1.8	Co 1.8	Ni 1.8	Cu 1.9	Zn 1.6	Ga 1.6	Ge 1.8	As 2.0	Se 2.4	Br 2.8
Rb 0.8	Sr 1.0	Y 1.2	Zr 1.4	Nb 1.6	Mo 1.8	Tc 1.9	Ru 2.2	Rh 2.2	Pd 2.2	Ag 1.9	Cd 1.7	In 1.7	Sn 1.8	Sb 1.9	Te 2.1	I 2.5
Cs 0.7	Ba 0.9	Lu 1.2	Hf 1.3	Ta 1.5	W 1.7	Re 1.9	Os 2.2	Ir 2.2	Pt 2.2	Au 2.4	Hg 1.9	Ti 1.8	Pb 1.8	Bi 1.9	Po 2.0	At 2.2
Fr 0.7	Ra 0.9															

Figure 5.15 *Electronegativity values of selected elements.*

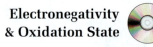

Electronegativity & Oxidation State

Oxidation Number

A convenient, yet arbitrary, way to express the degree of oxidation or reduction of particular atoms is by assigning each atom an **oxidation state.** The higher (more positive) the oxidation state becomes, the more an atom has become oxidized. The lower (less positive) the atom's oxidation state becomes, the more the atom has become reduced.

In binary compounds (compounds composed of two elements), we assign atoms of the element with the lower electronegativity a **positive oxidation state,** corresponding to an apparent loss of electrons. Likewise, we assign atoms of the more electronegative element a **negative oxidation state,** corresponding to an apparent gain of electrons.

Consider the key chemical change in the Haber-Bosch process, as depicted with electron-dot formulas and space-filling models:

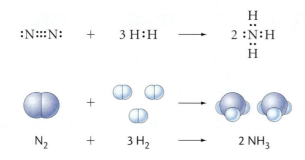

Note that each nitrogen atom in N_2 shares six electrons with another nitrogen atom, resulting in a triple covalent bond. Both nitrogen atoms exert equal attraction for these shared electrons. Each nitrogen atom in N_2 has a **zero oxidation state.** (This is true of any atom of an element that is not bonded to any other element.) The two nitrogen atoms share their bonding electrons equally; there is no separation of electrical charge. As the reaction progresses, each nitrogen atom becomes covalently bonded to three hydrogen atoms. Each bonded nitrogen and hydrogen atom shares an electron pair, but they do not share the pair equally. Nitrogen atoms (electronegativity value = 3.0) have a greater attraction for these shared electrons than do hydrogen atoms (electronegativity value = 2.1). Thus, the nitrogen atom in each NH_3 molecule acquires a greater portion of shared electrons than does each hydrogen atom.

Consequently, the nitrogen atom in NH_3 is assigned a negative oxidation state. Likewise, each hydrogen atom in NH_3 has lost some share of bonding electrons due to the reaction; hydrogen is assigned a positive oxidation state. Thus, the oxidation state of an atom in a particular substance depends on the identity of neighboring atoms to which it is covalently bonded.

The Haber-Bosch process converts nitrogen gas molecules into more useful ammonia molecules, a form of *fixed* nitrogen. Once nitrogen is chemically combined with another element, it can be readily converted to other nitrogen-containing compounds.

For example, under proper conditions, ammonia will react with oxygen gas, forming nitrogen dioxide:

$$4 \ NH_3(g) + 7 \ O_2(g) \longrightarrow 4 \ NO_2(g) + 6 \ H_2O(g)$$

This is an oxidation–reduction reaction. In forming NO_2, the nitrogen atom in ammonia has apparently been oxidized; it has lost part of its share of electrons. Why does this occur? Because oxygen is more electronegative than is nitrogen (O = 3.5; N = 3.0), oxygen attracts bonding electrons more strongly than does nitrogen. Each oxygen atom is considered to be reduced; each oxygen atom has gained more control of its bonding electrons than it originally had in O_2.

A.9 DETERMINING OXIDATION STATES

Sample Problem 1: *Which element in sulfur dioxide (SO_2) has a positive oxidation state?*

Figure 5.15 (page 426) indicates that sulfur's electronegativity value is 2.5 and oxygen's is 3.5. In sulfur–oxygen covalent bonding, sulfur has the weaker electron-attracting ability. Therefore, sulfur is assigned a positive oxidation state in SO_2.

Sample Problem 2: *The following equation represents the reaction of sulfur with oxygen gas to produce sulfur dioxide gas:*

$$S_8(s) + 8\ O_2(g) \longrightarrow 8\ SO_2(g)$$

Why do chemists consider this reaction as an oxidation–reduction reaction in which sulfur is oxidized?

The element sulfur is found in eight-atom rings, S_8.

The oxidation state of sulfur changes from zero (in the pure element) to a positive value in the product, sulfur dioxide (see Sample Problem 1)—sulfur has been oxidized. By contrast, oxygen gas becomes reduced; it has become reduced from a zero oxidation state to a negative oxidation state.

Now answer the following questions:

1. Consider each of the following covalent compounds. Using the electronegativity values from Figure 5.15 (page 426), decide which element in each compound has (i) a positive oxidation state and which has (ii) a negative oxidation state:

   a. ammonia, NH_3
   b. hydrogen chloride, HCl
   c. hydrogen fluoride, HF
   d. oxygen difluoride, OF_2
   e. iodine trifluoride, IF_3
   f. phosphorus trifluoride, PF_3

2. Each of the following compounds includes a metallic element and a nonmetallic element. Select the element in each compound possessing (i) a positive oxidation state and (ii) a negative oxidation state.

a. sodium iodide, NaI
b. lead(II) fluoride, PbF_2
c. lead(II) sulfide, PbS
d. potassium oxide, K_2O
e. iron(III) chloride, $FeCl_3$
f. sodium phosphide, Na_3P
g. sodium chloride, NaCl

3. Consider your answers to Questions 1 and 2. What conclusions can you draw about the oxidation states of metals and nonmetals in binary compounds?

4. Consider the chemical equation $8\,Ni + S_8 \longrightarrow 8\,NiS$.

a. Does this equation represent an oxidation–reduction reaction?
b. If so, identify the element oxidized and the element reduced. If not, explain why.

5. The element iron is part of an essential system of energy transfer within human cells. In that system, Fe^{2+} ions are converted to Fe^{3+} ions. Does that change represent an oxidation or a reduction?

6. Within the nitrogen cycle (see page 419), nitrogen gas (N_2) undergoes particular chemical reactions in which it is oxidized and other reactions in which it is reduced. Identify, by name and formula, a nitrogen-cycle product that forms by nitrogen gas being

a. oxidized.
b. reduced.

SECTION A SUMMARY
Reviewing the Concepts

Fertilizers contain many essential plant nutrients.

1. List the three main elements in fertilizer.

2. What would the expression *20–10–15* mean if found on a fertilizer label?

3. Why is potassium content expressed as percent K_2O in fertilizer?

4. Describe the role in plant growth of each of the three essential elements in a typical fertilizer.

5. Why is it useful for fertilizers, such as 7–7–7 and 20–10–10, to be available with different compositions?

6. List each test completed in Investigating Matter A.3 (page 413), along with the ions that the test identified.

Nitrogen is transformed chemically as it cycles through living systems and the physical environment.

7. Why do plants need nitrogen?

8. Given the fact that nitrogen is abundant in the atmosphere, why is it included in fertilizers?

9. What does it mean to *fix* nitrogen gas?

10. List three ways in which atmospheric nitrogen can be fixed.

11. List two nitrogen-containing ions that are useful to plant growth.

12. How do plants and animals differ in the ways they obtain
    a. nitrogen?
    b. amino acids?

13. What is one role of denitrifying bacteria in the environment?

14. Summarize the steps of the nitrogen cycle.

The relative tendency of an atom to attract electrons within a covalent chemical bond can be estimated by the electronegativity of that element.

15. Referring to Figure 5.15 (page 426), identify the element that is *most* electronegative.
    a. Identify the element's symbol and name.
    b. Is this element a metal or nonmetal?

16. Referring to Figure 5.15, identify the element that is *least* electronegative.
    a. Identify the element's symbol and name.
    b. Is this element a metal or nonmetal?

17. Describe how electronegativity values for elements change as one moves
    a. from left to right across any period of the periodic table.
    b. down a group of the periodic table.

18. Arrange each of the following sets of elements in order of their increasing attraction for electrons within a bond:
    a. silicon, sodium, and sulfur
    b. nitrogen, phosphorus, and potassium
    c. bromine, fluorine, lithium, and potassium

19. What is the oxidation state of an atom in its elemental form (that is, not combined with an atom of another element)?

20. How does the oxidation state of an atom change when the atom is

a. oxidized?
b. reduced?

21. How is it possible for the same element to be oxidized in one reaction and reduced in another?

22. What type of element (metal or nonmetal) is more often found in negative oxidation states when bonded to atoms of other elements?

23. For each of the following compounds, write its chemical formula and identify which element has a positive oxidation state and which has a negative oxidation state.

a. water
b. ammonia
c. carbon dioxide
d. magnesium chloride

24. List four different ways that the chemical industry is involved in the production of a box of breakfast cereal.

25. Look around wherever you are sitting right now and identify four things you see that are products of the chemical industry.

26. What is the ultimate use of most of the ammonia produced in the United States?

27. What was the most common source of ammonia before the 20th century?

28. Write the chemical equation that represents the main reaction in the Haber-Bosch process.

Connecting the Concepts

29. Describe one advantage and one disadvantage of using commercial fertilizer instead of manure to fertilize crops.

30. The quantity of phosphorus in fertilizer is reported as *percent P_2O_5*.

a. What is the percent phosphorus in P_2O_5?
b. Fertilizers are burned (oxidized) for analysis, and the burning produces P_2O_5. In what form is the phosphorus actually found in the original fertilizer?

31. Describe a procedure for determining whether a soil sample contains any fixed nitrogen.

32. Compare the fertilizer tests you completed in this unit to the ion tests in Unit 1 (page 42). In what ways are these tests similar and in what ways are they different?

33. A magazine article claims that "oxygen is needed for all oxidation reactions." Do you agree or disagree with that statement? Support your answer with evidence.

34. What does the electronegativity of an electrically neutral atom indicate about its tendency to become oxidized?

35. In general, how do the electronegativities of metallic and nonmetallic elements compare?

36. Consider the Haber-Bosch process.

a. Write the Lewis-dot structure for each reactant and product involved.
b. Using the concept of electronegativity, determine which atoms in that reaction have a positive oxidation state and which have a negative oxidation state.

37. How does the concept of a limiting reactant apply to the use of fertilizers?

38. Why is colorimetry effective in measuring the concentration of only certain kinds of solutions?

Extending the Concepts

39. You have considered three major natural cycles: water, carbon, and nitrogen. Compare these cycles with respect to

    a. conservation of mass,
    b. types of chemical change, and
    c. participating organisms.

40. Why do some vegetarians claim that their diets make more economical use of world food resources than do the diets of nonvegetarians?

41. Some historians claim that development of the Haber-Bosch process prolonged World War I. Explain.

42. Research and draw a diagram of the basic parts of a colorimeter, which is a simple version of a *spectrophotometer*.

43. a. Magnesium is a key component of chlorophyll in green plants. Explain why magnesium is generally not included in commercial fertilizers.
    b. Identify some other substances required by growing plants that are not included in fertilizers.

44. Review the list of ingredients in a multipurpose vitamin capsule for humans and compare this to the ingredients in a typical commercial fertilizer. Suggest reasons for the similarities and differences you find.

45. How does lightning fix nitrogen?

46. Research and report on denitrifying bacteria. Include their typical habitats and any unusual characteristics.

Nitrogen and Industry

As you have learned, many industrial raw materials are extracted from Earth's crust (such as minerals, precious metals, sulfur, and petroleum), oceans (for example, magnesium and bromine), and atmosphere. Nitrogen gas and oxygen gas, both obtained by low-temperature distillation of liquefied air, are valuable starting materials in the production of ammonia and nitric acid. As you will soon learn, producing ammonia also depends on understanding implications of reversible reactions and chemical equilibrium.

B.1 KINETICS AND EQUILIBRIUM

Producing ammonia from nitrogen gas and hydrogen gas is a chemical challenge. As you learned in Section A, molecular nitrogen (N_2) is very stable. This means that nitrogen fixation, the chemical combination of nitrogen gas with other elements, has a substantial activation-energy barrier. As you learned in Unit 4 (page 394), a reaction with a large energy barrier requires either that the reactant particles have substantial kinetic energy or that a catalyst reduces the energy required to initiate the reaction.

The reaction of nitrogen gas with hydrogen gas is also difficult because some ammonia molecules tend to decompose back to nitrogen gas and hydrogen gas during the synthesis reaction. As you learned in Unit 4 (page 375), this kind of reaction—one in which products re-form reactants at the same time that reactants form products—is known as a *reversible reaction.* The double arrows used in the following equation indicate that both forward and reverse reactions occur simultaneously:

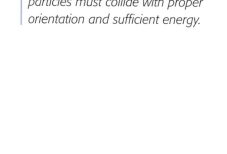

Figure 5.16 *For a chemical reaction to occur, reactant particles must collide with proper orientation and sufficient energy.*

$$N_2(g) + 3\,H_2(g) \rightleftharpoons 2\,NH_3(g)$$

How do chemists and chemical engineers, whether at EKS Nitrogen Products Company or elsewhere, overcome these obstacles to produce ammonia?

Kinetics: Producing More Ammonia in Less Time

The rate at which nitrogen fixation occurs determines the time required for a certain amount of ammonia to form. The **reaction rate** expresses how fast a particular chemical change occurs. The study of reaction rates is often referred to as chemical **kinetics.**

For chemical reactions to occur, reactant molecules, atoms, or ions must collide with one another. According to collision theory, the reaction rate depends on the collision frequency and the energy involved in each collision (Figure 5.16).

You first learned about collision theory in Unit 4 (page 394). According to this theory, the reaction rate also depends on the orientation of the colliding molecules.

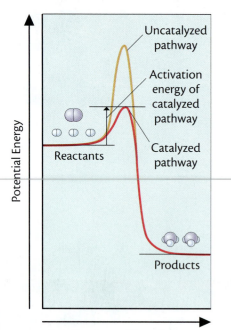

Figure 5.17 *A catalyst reduces the size of the activation-energy barrier involved in synthesizing ammonia from its elements.*

Increasing concentrations of reactants speeds up reaction rates by increasing the number of collisions. High temperatures increase the reaction rate by providing more reacting molecules with sufficient energy to overcome the activation-energy barrier. Catalysts, on the other hand, increase the reaction rate by lowering the activation-energy barrier required for the reaction to occur, as shown in Figure 5.17.

Although a higher reaction temperature increases the average kinetic energies of the nitrogen and hydrogen molecules that react to form ammonia, ammonia itself becomes increasingly unstable at higher temperatures. The result is that ammonia decomposes back to nitrogen gas and hydrogen gas. If the reaction takes place at lower temperatures, fewer nitrogen and hydrogen molecules have enough energy to overcome the activation energy barrier, thus slowing the net rate of ammonia formation (even though less ammonia decomposes at that lower temperature). What can we do to increase the rate and yield of ammonia production?

The major breakthrough that led to profitable ammonia production was the discovery of a suitable catalyst. Catalysts made it possible to produce ammonia at lower temperatures (450–500 °C), thus slowing the rate of ammonia decomposition. Fritz Haber and his colleagues spent a great deal of time and effort in the early 1900s systematically searching for good catalysts. Today, the ammonia industry commonly employs iron as an ammonia-synthesis catalyst.

Equilibrium: Favoring the Forward Reaction

Any reversible reaction appears to stop when the rate at which product forms equals the rate at which product reverts back to reactants—that is, when reactants and products attain *dynamic equilibrium.* At equilibrium, both the forward and the reverse reactions continue, but there is no further change in the amounts of reactants or products. At the point of dynamic equilibrium, the two opposing changes are in exact balance, as modeled in Figures 5.18 and 5.19.

> All reversible reactions in closed containers eventually reach equilibrium if conditions such as temperature remain constant.

Visualizing an Iconic Equilibrium

Figure 5.18 *A system at dynamic equilibrium involves two ongoing processes acting in opposition to one another.*

Scientific American Conceptual Illustration

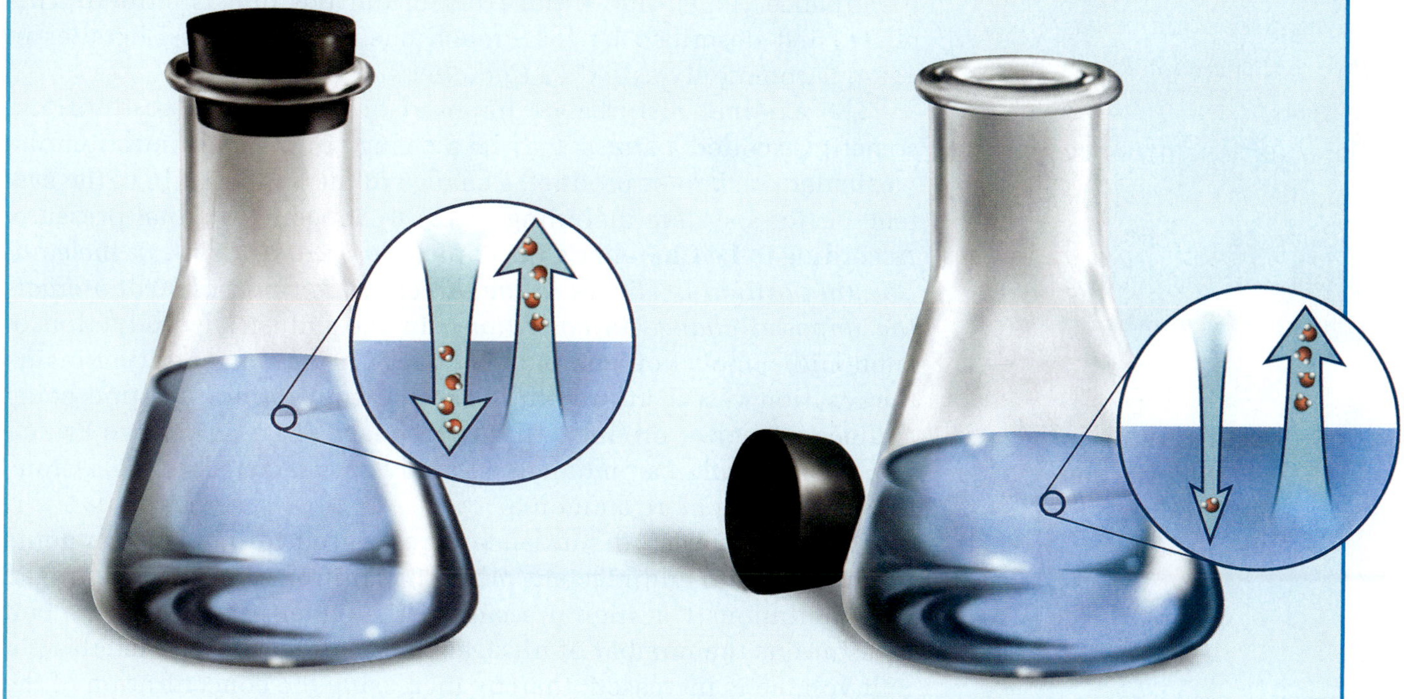

Figure 5.19 Dynamic equilibrium. *In the stoppered flask (left), we observe no overall change in the water level because its evaporation rate equals its rate of condensation. Water contained in the open flask (right) slowly escapes from the flask as water vapor. This happens because the rate of evaporation of water molecules is greater than the rate of condensation—this system is not in equilibrium. The closed system in the stoppered flask is, by contrast, an example of dynamic equilibrium.*

The net amount of ammonia formed from a certain amount of nitrogen gas and hydrogen gas at a fixed temperature is limited. One way to increase the amount of ammonia produced is to cool the ammonia as soon as it forms until it turns to a liquid and remove it from the reaction chamber. This prevents ammonia from decomposing back into nitrogen gas and hydrogen gas. If ammonia is continuously removed, the rate of the reverse reaction (decomposition of ammonia) is significantly decreased because there is less gaseous ammonia available to decompose. This causes the overall reaction (see equation, page 433) to favor the production of more ammonia.

Disturbing an Equilibrium

This example illustrates that a system at equilibrium can often be disturbed by changing the concentration of either reactants or products or by changing the temperature of the system. When one of these disturbances occurs, it causes either the forward or the reverse reaction rate to become larger than the other and thus to be favored over the other—a change that *partially counteracts* the initial effect of the disturbance. Thus, the initial equilibrium position is shifted. This effect, first described by the French chemist Henri Le Châtelier in 1884, is commonly called *Le Châtelier's Principle.*

The external disturbance imposed on a system at equilibrium, sometimes called a *stress,* may be a change in the concentration of a particular reactant or product, a change in the temperature of the system, or (for a system including gases) a change in the total pressure. According to **Le Châtelier's Principle,** *the predicted shift in the equilibrium position is always in the direction that partially counteracts the imposed change in conditions.* In the industrial production of ammonia, removal of ammonia (a change in its concentration inside the reaction vessel) results in the initial equilibrium position being shifted in favor of products. In the language of Le Châtelier's Principle, the removal of ammonia is partially counteracted by the system, thus producing *more* ammonia.

Another external disturbance (stress) used to increase ammonia production is to add reactant molecules (nitrogen gas and hydrogen gas) continuously at high pressure. This higher pressure of reactant gases means the number of nitrogen and hydrogen gas molecules per unit volume is increased, thereby increasing the concentration of the reactants. The frequency of molecular collisions increases, which favors the forward reaction, and thus increases the amount of ammonia formed. This change can be viewed as partially counteracting the initial increased pressure because the total number of gas molecules have been decreased. (Four molecules of gas—three molecules H_2 and one molecule N_2—are replaced by two molecules of NH_3 gas.)

In many cases, changing the system's temperature can also cause an equilibrium system to shift. The direction of that effect can be predicted based on whether the forward reaction is exothermic or endothermic. For example, the synthesis of ammonia is exothermic:

$$N_2(g) + 3\ H_2(g) \rightleftharpoons 2\ NH_3(g) + Heat$$

Heat (thermal energy) can be regarded as a product of the forward (left to right) reaction. Raising the temperature would tend to favor the reverse (right to left) reaction, a chemical change that absorbs heat. Consequently, less ammonia is formed at conditions of higher temperatures. Remember, though, that the temperature must be high enough to provide the nitrogen and hydrogen molecules with adequate kinetic energy to react. A delicate balance is needed; the temperature must be high enough to produce significant amounts of ammonia but not so high that it promotes an excessive rate of ammonia decomposition.

B.2 CHEMICAL SYSTEMS AT EQUILIBRIUM

Sample Problem: *For the following equilibrium system, describe three changes you could make to increase the formation of nitric oxide gas, NO(g), at equilibrium.*

$$\text{Heat} + 2\ NO_2(g) \rightleftharpoons 2\ NO(g) + O_2(g)$$

We need to consider factors that will cause the equilibrium to shift to the right, favoring NO production. Based on Le Châtelier's Principle, these factors apply: (1) The system could be heated to a higher temperature. (2) The concentration of NO_2 could be increased. (3) The concentration of O_2 or NO could be decreased.

Now answer the following questions:

1. For each equilibrium system, describe three changes you could make to favor the forward reaction.

   a. $2\ SO_2(g) + O_2(g) \rightleftharpoons 2\ SO_3(g) + \text{Heat}$
   b. $H_2(g) + Cl_2(g) \rightleftharpoons 2\ HCl(g) + \text{Heat}$

2. Examine the graph in Figure 5.20:

   a. What generalization can you make about the effect of temperature on the yield of ammonia?
   b. Do you think your generalization would remain valid for temperatures much lower than 400 °C? Explain your answer.
   c. What generalization can you make about the effect of total pressure on the yield of ammonia?
   d. What combination of temperature and pressure results in the highest ammonia yield?

3. For Questions 1a and 1b, describe three changes you could make to favor each reverse reaction.

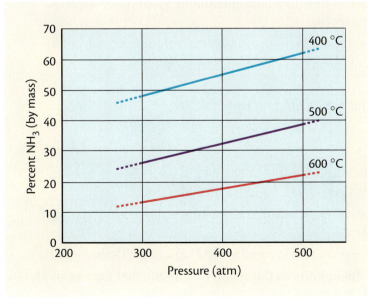

Figure 5.20 *Ammonia production in the system* $N_2(g) + 3\ H_2(g) \rightleftharpoons 2\ NH_3(g) + \text{Heat}.$ *The graph depicts the effect of pressure and temperature changes on the percent NH_3 present at equilibrium.*

B.3 LE CHÂTELIER'S PRINCIPLE

Introduction

Lab Video:
Le Châtelier's
Principle

In this investigation of a system at equilibrium, you will use what you have learned about Le Châtelier's Principle to explore the effects of changing concentration and temperature on the position of an equilibrium system. The chemical system that you will investigate is described by the following equilibrium equation involving *complex ions:*

$$\text{Heat} + [Co(H_2O)_6]^{2+}(aq) + 4\,Cl^-(aq) \rightleftharpoons [CoCl_4]^{2-}(aq) + 6\,H_2O(l)$$

A **complex ion** is composed of a single central atom or ion, usually a metal ion, to which other atoms, molecules, or ions are attached. Part of your task will be to decide which complex ion above—$[CoCl_4]^{2-}$ or $[Co(H_2O)_6]^{2+}$—is blue and which is pink.

Procedure 👓

1. Prepare a hot-water bath (60–70 °C) to use in Step 4.

2. Add 20 drops of 0.1 M cobalt(II) chloride ($CoCl_2$) solution to a clean, dry test tube. Record the color.

3. Add 7 drops of 0.1 M silver nitrate ($AgNO_3$) solution to the same test tube. **Caution:** *AgNO₃ solution can stain your skin or clothing. Handle with care.* Gently swirl the tube to ensure good mixing. Record the color.

4. Heat the tube in a hot-water bath for 30 s. Record the color.

5. Remove the tube from the hot-water bath. Add approximately 0.3 g sodium chloride (NaCl). Gently agitate the tube. Heat the solution for 30 s in the hot-water bath. Record the color.

6. Place the test tube in a beaker containing ice water for 30 s. Record the color.

7. Reheat the test tube in the hot-water bath. Record the color.

8. Dispose of the mixture in the test tube as directed by your teacher.

9. Wash your hands thoroughly before leaving the laboratory.

Questions

In answering the following questions, refer to your observations and to the equilibrium equation contained in the Introduction.

1. Which reaction (forward or reverse) was favored by cooling the solution?

2. Which reaction was favored by adding more chloride ions?

3. What is the identity of the white precipitate that formed in Step 3? (*Hint:* Refer to Unit 1, page 44.)

4. Why did adding $AgNO_3$ solution affect the equilibrium, even though neither Ag^+ ions nor NO_3^- ions appear in the equilibrium equation? Write an equation to support your answer.

5. Why did the solution's color change after heating in Step 5 but not in Step 4?

6. a. Which complex ion is pink: $[CoCl_4]^{2-}$ or $[Co(H_2O)_6]^{2+}$?
   b. Which complex ion is blue?

   Provide evidence that supports both of your answers.

B.4 INDUSTRIAL SYNTHESIS OF AMMONIA

Large-scale ammonia production involves much more than allowing nitrogen gas and hydrogen gas to react in the presence of a catalyst. First, of course, the plant must obtain a continuous supply of the reactants. Nitrogen gas, which represents 78% of Earth's atmosphere, is liquefied from air through a series of steps involving cooling and compression. As you will soon learn, hydrogen gas can be obtained chemically from natural gas (mainly methane, CH_4). Opening an ammonia plant in Riverwood would mean building a pipeline from a natural gas source (or distribution center) to the chemical plant.

To produce hydrogen gas, chemical engineers first treat natural gas to remove sulfur compounds; then they allow methane to react with steam:

$$\text{Heat} + CH_4(g) + H_2O(g) \rightleftharpoons 3\,H_2(g) + CO(g)$$

In modern ammonia plants, this endothermic reaction takes place at 200–600 °C and at pressures of 200–900 atm. Technicians, such as the one shown in Figure 5.21, must carefully control the ratio of methane to steam to prevent the formation of carbon compounds that would lower the yield of hydrogen gas.

Carbon monoxide, a product of the hydrogen-generating reaction shown above, is converted to carbon dioxide, which is accompanied by the production of additional hydrogen gas:

$$CO(g) + H_2O(g) \rightleftharpoons H_2(g) + CO_2(g)$$

All the hydrogen gas produced is then separated from carbon dioxide and from any unreacted methane.

In the Haber-Bosch process (see page 425), the reactants (hydrogen gas and nitrogen gas) are first compressed to high pressures (150–300 atm). Ammonia forms as the hot gases (at about 500 °C) flow over an iron catalyst. Ammonia gas is removed by converting it, by means of cooling and added pressure, to liquid ammonia, which is then removed from the reaction chamber, thus reducing the rate of ammonia decomposition. Unreacted nitrogen gas and hydrogen gas are recycled, mixed with additional supplies of reactants, and passed through the reaction chamber again.

Figure 5.21 *This technician checks conditions within an industrial chemical storage system. What evidence suggests that safety is a priority?*

The CO_2 formed in this reaction can be removed in several ways, including allowing it to react with calcium oxide (CaO or lime), which forms solid calcium carbonate ($CaCO_3$), or by dissolving the CO_2 gas at high pressure in water.

B.5 NITROGEN'S OTHER FACE

About 80% of NH₃ produced is used in fertilizer and 5% is used in explosives. What other uses for NH₃ might account for the other 15%?

The Haber-Bosch process has provided relatively inexpensive ammonia for a variety of applications. For example, ammonia reacts directly with nitric acid to produce ammonium nitrate, a substitute for natural nitrates traditionally used as fertilizers.

$$NH_3(g) + HNO_3(aq) \longrightarrow NH_4NO_3(aq)$$

Ammonia   Nitric   Ammonium
             acid   nitrate

The widespread availability of ammonia and nitrates has changed the course of warfare as well as of agriculture. Ammonia is a reactant in the production of explosives, most of which are nitrogen-containing compounds (see Figures 5.22 and 5.23). The development of the Haber-Bosch process provided a convenient source of ammonia for making both fertilizers and military munitions. This allowed Germany to continue fighting in World War I even after its shipping connections to Chilean nitrate deposits were cut off by the British Navy.

Nitrogen-based explosives also have non-hostile, and even life-saving, uses. Air bags in automobiles are one such modern application. An air bag quickly inflates like a big pillow during a collision to reduce injuries to the driver and passengers. The uninflated air-bag assembly contains solid sodium azide, NaN₃. In a collision, sensors initiate a sequence of events that rapidly decompose the sodium azide to form a large volume of nitrogen gas:

$$3\ NaN_3(s) \longrightarrow Na_3N(s) + 4\ N_2(g)$$

Figure 5.22 *Explosive substances, such as those used here, are used in many mining and construction operations.*

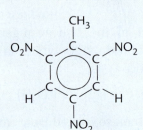

2,4,6-Trinitrotoluene (TNT)

NH₄NO₃
Ammonium nitrate

Nitroglycerin

Pb(—N=N=N)₂
Lead azide

Hexahydro-1,3,5-trinitro-1,3,5-triazine (RDX)

Figure 5.23 *Structural formulas of some common substances used as explosives. What similarities are shared by these chemical structures?*

Nitrogen gas quickly inflates the air bag (to about 50 L) within 50 ms (0.05 s) after the collision starts (see Figure 5.24).

The forces released by nitrogen-based explosives also blast roadbeds through solid rock during highway construction. To cut through the stone faces of hills and mountains, road crews drill holes, drop in explosive canisters, and then detonate the explosives.

Explosions in general result from the rapid formation of gaseous products from liquid or solid reactants. Detonating an explosive such as sodium azide, dynamite, or nitroglycerin produces gases that occupy more than a thousand times the volume of the original solid or liquid.

Many compounds used as explosives involve nitrogen atoms in a positive oxidation state and carbon in a negative oxidation state within the same reactant molecule. This creates conditions for the release of vast quantities of energy. The energy released in this type of explosive reaction is due in part to the formation of N_2, a highly stable molecule.

The powerful explosive nitroglycerin (see Figure 5.23) was invented in 1846. However, it was too sensitive (unstable) to be useful; one never knew when it was going to explode. The Nobel family built a laboratory in Stockholm to explore ways to control this unstable substance (Figure 5.25). Although the father and his four sons were interested in explosives, one son, Alfred, was the most persistent experimenter.

Carelessness led to accidental explosions. One killed Alfred's brother Emil. The city of Stockholm finally insisted that Alfred move his experimenting elsewhere. Determined to continue research to make nitroglycerin less unpredictable and dangerous, Alfred rented a barge and performed experiments in the middle of a lake.

Alfred finally discovered that mixing oily nitroglycerin with a finely divided solid (diatomaceous earth) caused nitroglycerin to become stable enough for safe transportation and storage. However, the nitroglycerin would still explode if a blasting cap activated it. This new, more stable form of nitroglycerin carried a new name—*dynamite.*

A new era in explosives had begun. At first, dynamite served peaceful uses in mining and in road and tunnel construction (see Figure 5.22). By the late 1800s, dynamite also found destructive use in warfare.

Figure 5.24 *Automobile air bags deploy due to nitrogen gas released from a nitrogen-based explosive compound.*

Chemical explosions are often rapid, exothermic oxidation–reduction reactions that produce large volumes of gaseous product.

Figure 5.25 *The production of nitroglycerin, devised by Alfred Nobel, was a dangerous process that needed careful monitoring. Note the one-legged stool that ensured that the attendant didn't fall asleep on the job.*

Military use of his invention caused Alfred Nobel considerable anguish and motivated him to use his fortune to benefit humanity. His will specified that his money be dedicated to annual international prizes for advances in physics, chemistry, physiology and medicine, literature, and peace. (The Swedish parliament later added economics as an award category.) Nobel prizes, first awarded in 1901, are still regarded as the highest honors individuals can receive in these fields. Table 5.1 lists recent Nobel Prize recipients in chemistry, together with their contributions to chemical science.

Table 5.1 *Afred Nobel's will established several annual awards, one of which is bestowed upon those who "shall have made the most important chemical discovery or improvement."*

NOBEL LAUREATES IN CHEMISTRY, 2001–2005

Year	Awardees	Contributions
2005	Yves Chauvin, France, Institut Français du Pétrole; Robert H. Grubbs, United States, California Institute of Technology; and Richard R. Schrock, United States, Massachusetts Institute of Technology	Development of the metathesis method in organic chemistry, reducing potentially hazardous waste through "Green-Chemistry" synthesis methods
2004	Aaron Ciechanover and Avram Hershko, Israel, Technion-Israel Institute of Technology; and Irwin Rose, United States, University of California, Irvine	Discovery of ubiquitin-mediated protein degradation
2003	Peter Agre, United States, Johns Hopkins University School of Medicine; and Roderick MacKinnon, United States, Rockefeller University	Discoveries about channels in cell membranes
2002	John B. Fenn, United States, Virginia Commonwealth University; Koichi Tanaka, Japan, Shimadzu Corporation; and Kurt Wüthrich, Switzerland, Swiss Federal Institute of Technology	Development of analytical methods for identification and structure analyses of biological macromolecules
2001	William S. Knowles, United States, Monsanto Company; and Ryoji Noyori, Japan, Nagoya University	Work on chirally catalyzed hydrogenation reactions
	K. Barry Sharpless, United States, The Scripps Research Institute	Work on chirally catalyzed oxidation reactions

B.6 EXPLOSIVE NITROGEN CHEMISTRY

The following equation describes the explosion of nitroglycerin:

$$4\ C_3H_5(NO_3)_3(l) \longrightarrow 12\ CO_2(g) + 6\ N_2(g) + 10\ H_2O(g) + O_2(g) + Energy$$
Nitroglycerin

1. How many total moles of gaseous products form in the explosion of one mole of nitroglycerin?

2. One mole of gas at standard temperature and pressure occupies 22.4 L. One mole of liquid nitroglycerin occupies approximately 0.1 L. By what factor does the volume increase when one mole of nitroglycerin explodes? (Assume temperature remains constant.)

3. When nitroglycerin explodes, the actual temperature increase causes gas volume to increase eight times more than the factor you just calculated in Question 2. By what combined factor does the total volume suddenly increase during an actual nitroglycerin explosion?

4. How do your answers to Questions 2 and 3 help explain the destructive power of a nitroglycerin explosion?

B.7 FROM RAW MATERIALS TO PRODUCTS

Some chemical reactions you have observed in this chemistry class are essentially the same as reactions used in industry to synthesize chemical products. However, chemical reactions in industry must be scaled up to produce very large quantities of high-quality products at low cost.

Four considerations become crucial in attempting to scale up chemical reactions: *engineering, profitability, waste,* and *safety.* Sometimes an industrial reaction is conducted as a *batch process,* a single run of converting reactants to products (see Figure 5.26, top). You completed such reactions to produce esters in Unit 3 (page 275). Additional batch runs can produce more product. The early industrial production of nylon was also a batch process.

More commonly and less expensively, industrial reactions can involve a *continuous process:* Reactants flow steadily into the reaction chamber, and products continuously flow out (see Figure 5.26, bottom). Engineers carefully control the rate of flow, time, temperature, and catalyst composition to ensure a successful continuous-production operation.

Figure 5.26 *On an industrial scale, materials can be made through a batch process (top) or a continuous process (bottom). What benefits might each production method offer?*

Chemical engineers face many challenges in designing manufacturing systems for industry. In your classroom laboratory, the small quantity of thermal energy generated by a reaction in a test tube or wellplate may seem insignificant. However, in exothermic industrial processes, thousands of liters may react in huge vats. Engineers have to anticipate and manage the release of large quantities of heat. Otherwise, reaction temperatures may rise, creating potentially dangerous, costly, and destructive situations.

The chemical industry also faces challenges in dealing with unwanted materials that result from chemical processes. When reactions occur on an industrial scale, large quantities of waste can accumulate. A major responsibility of the Environmental Protection Agency (EPA) is regulating the cleanup of hundreds of chemical-waste sites in the United States (Figure 5.27). Those sites are legacies of earlier times when a common attitude was "out of sight, out of mind."

When the EPA put an end to such releases, many chemical industries discovered that, with some additional processing, they could turn some previously unwanted materials or products into valuable commodities. Former "waste" compounds can often become intermediates in producing useful substances. Thus, instead of contaminating the environment, such wastes-turned-resources offer new sources of income. Additionally, recently developed catalysts and new or modified processes have allowed manufacturers to increase their efficiency and decrease the amounts of starting materials (reactants) they need.

Chemical manufacturers have learned that pollution prevention pays off. For example, 3M Corporation, a major producer and user of chemical materials, has maintained a pollution prevention program for more than 25 years. During that time, 3M has saved nearly a billion dollars by eliminating more than 1.5 billion pounds of pollutants. 3M's goal now is to approach zero pollution. Most of these pollution-prevention suggestions have come directly from 3M employees.

Figure 5.27 *The Environmental Protection Agency (EPA) regulates and monitors pollutants, including those associated with industrial processes.*

ChemQuandary 1

THE TOP CHEMICALS

Table 5.2 lists the top-produced chemical substances in the United States. Notice that production is reported in billions of pounds. What quantities other than mass or weight could you use to compare the relative production levels of chemical materials? How would the relative rankings be affected— if at all—if you used each quantity?

TOP 25 CHEMICAL SUBSTANCES PRODUCED IN THE UNITED STATES

Rank	Name	Formula	Billions of Pounds
1	sulfuric acid	H_2SO_4	95.2
2	nitrogen	N_2	75.7
3	oxygen	O_2	57.7
4	ethene (ethylene)	C_2H_4	51.7
5	calcium oxide (lime)	CaO	45.0
6	ammonia	NH_3	39.5
7	phosphoric acid	H_3PO_4	28.8
8	propene (propylene)	C_3H_6	28.7
9	chlorine	Cl_2	25.7
10	sodium hydroxide	$NaOH$	23.0
11	sodium carbonate	Na_2CO_3	21.4
12	1,2-dichloroethane (ethylene dichloride)	$C_2H_4Cl_2$	19.5
13	methyl *tert*-butyl ether (MTBE)	$C_5H_{12}O$	18.9
14	nitric acid	HNO_3	18.7
15	urea	$(NH_2)_2CO$	17.6
16	ammonium nitrate	NH_4NO_3	17.2
17	vinyl chloride	C_2H_3Cl	17.0
18	benzene	C_6H_6	16.3
19	ethylbenzene	C_8H_{10}	13.0
20	methanol	CH_3OH	12.5
21	styrene	C_8H_8	11.4
22	terephthalic acid	$C_8H_6O_4$	8.7
23	formaldehyde	H_2CO	8.6
24	hydrochloric acid	HCl	8.6
25	toluene	$C_6H_5CH_3$	8.1

Based on data for 1998, *U.S. Chemical Industry Handbook,* Chemical Manufacturers Association, 1999 (Table 2.17, page 44).

Table 5.2 *The U.S. chemical industry produces billions of pounds of chemical substances annually.*

B.8 GREEN CHEMISTRY AND RESPONSIBLE CARE

On learning that EKS and WYE may want to locate plants in Riverwood, the town council created a draft of general criteria that either company must meet. These criteria are based, in part, on two national initiatives regarding the chemical industry's social and technical responsibilities.

The first initiative, Green Chemistry, addresses the chemistry involved in an industrial process and its effects on humans and the environment. The second initiative, Responsible Care, considers how industries can safely and productively contribute to communities.

The impact of these initiatives can be partly monitored by the EPA's Toxics Release Inventory (TRI). The TRI is a state-by-state database of annually reported releases of about 600 toxic materials into the air, water, or land by more than 21 000 facilities. Anyone, including citizens of Riverwood, interested in monitoring local releases of toxic materials can consult this TRI database.

Green Chemistry encourages monitoring all industrial practices (Figure 5.28). The goal is to make chemical production less hazardous to human health and to the environment. This initiative is sometimes termed *Benign by Design*. In meeting the objectives of Green Chemistry, industries also try to make their processes more efficient and profitable.

Principles guiding Green Chemistry include these general points:

▶ It is better to prevent waste than treat it or clean it up after it forms.

▶ Synthetic methods should be designed so that the desired product contains as much of the material used in the process as possible.

▶ Whenever feasible, reactants used and waste generated should be as benign as possible.

▶ Solvents and other potentially hazardous substances should be made unnecessary wherever possible and innocuous where not possible.

▶ Energy requirements should be recognized for their environmental and economic impact and should be minimized wherever possible.

▶ Catalysts should be used whenever appropriate.

▶ Raw materials (reactants) should be obtained from renewable resources wherever possible.

▶ Chemical products should be designed so that if they decompose, the resulting products are not harmful.

▶ Chemical processes should be designed to minimize the chance for chemical accidents, including accidental releases, explosions, and fires.

Figure 5.28 *Industrial processes must be carefully monitored.*

The American Chemical Society's Green Chemistry Institute promotes education and research that encourages Green-Chemistry practices.

Green Chemistry

These principles encourage work that has already devised new industrial techniques, such as bleaching paper without using chlorine, recycling cellulose-based wastes into fuels, synthesizing drugs and polymers more efficiently, and developing reduced-risk pesticides highly specific for certain pests.

Responsible Care, initiated in 1988, is an international program in which chemical manufacturing companies voluntarily agree to public scrutiny and evaluation according to specific criteria. In the United States, member companies and partners of the American Chemistry Council (ACC) pledge to follow these principles:

▶ To seek and incorporate public input regarding products and operations.

▶ To provide chemicals that can be manufactured, transported, used, and disposed of safely.

▶ To make health, safety, the environment, and resource conservation critical considerations for all new and existing products and processes.

▶ To provide information on health or environmental risks and pursue protective measures for employees, the public, and other key stakeholders.

▶ To work with customers, carriers, suppliers, distributors, and contractors to foster the safe use, transport, and disposal of chemicals.

▶ To operate facilities in a manner that protects the environment and the health and safety of employees and the public.

▶ To support education and research on the health, safety, and environmental effects of products and processes.

▶ To work with others to resolve problems associated with past handling and disposal practices.

▶ To lead in the development of responsible laws, regulations, and standards that safeguard the community, workplace, and environment.

▶ To practice Responsible Care by encouraging and assisting others to adhere to these principles and practices.

As with Green Chemistry, adhering to the Responsible-Care pledge has encouraged many companies to increase safety for their workers and their surrounding communities, reduce the quantity of waste they generate, and achieve greater overall efficiency.

The next activity will help you develop criteria that any chemical industry interested in establishing a plant in Riverwood should meet.

B.9 WHAT DOES RIVERWOOD WANT?

The first of several town meetings to discuss the possibility of a chemical plant in Riverwood is coming up soon. Representatives from EKS and WYE and town council members will attend. All interested local citizens are also encouraged to attend.

1. Examine and compare the principles of Green Chemistry with those of the Responsible Care program (pages 446–447).

   a. Identify at least six expectations that the town council should clearly communicate to any chemical company that wants to locate in Riverwood.

   b. Classify your expectations as either 'mandatory' or simply 'desirable.'

2. Related to the expectations that you identified in Question 1, identify at least two questions that you would like to ask company representatives.

SECTION B SUMMARY
Reviewing the Concepts

The rate of a particular reaction depends on temperature, reactant concentration(s), and the influence of a catalyst.

1. What is meant by the *rate* of a reaction?

2. What does *chemical kinetics* mean?

3. Explain why reactions tend to speed up
   a. with increased temperature.
   b. with increased concentration.
   c. when a suitable catalyst is added.

When a system is in dynamic equilibrium, the rate of the forward reaction equals, and is thus balanced by, the rate of the reverse reaction.

4. What is equal about equilibrium?

5. What is dynamic about dynamic equilibrium?

6. How can you tell if chemical equilibrium is represented in a chemical equation?

7. What do *forward* and *reverse* mean when speaking about an equilibrium system?

Le Châtelier's Principle can be used to predict a shift in the equilibrium position of a reversible reaction.

8. Summarize Le Châtelier's Principle.

9. List three types of stress that can be applied to an equilibrium system.

10. Consider the following system at equilibrium:

$$PCl_3(g) + Cl_2(g) \rightleftharpoons PCl_5(g) + Heat$$

What effect (if any) will each of the following changes have on the position of that equilibrium system?

a. adding more Cl_2
b. lowering the temperature
c. removing some PCl_5 as it forms
d. decreasing the total pressure

11. Consider the following system at equilibrium:

$$C(s) + H_2O(g) + Heat \rightleftharpoons CO(g) + H_2(g)$$

What effect (if any) will each of the following changes have on the position of that equilibrium system?

a. lowering the temperature
b. adding steam at constant volume to the equilibrium system
c. adding a catalyst
d. increasing the total pressure

12. What is the effect of removing some heat from an exothermic reaction that is at equilibrium?

13. What are sources of the reactants for Haber-Bosch process?

14. The Haber-Bosch process works most effectively under particular conditions. What are optimal conditions for

a. pressure?
b. temperature?

15. What is the advantage of removing ammonia from the reaction mixture as it forms in the Haber-Bosch process?

16. Why does the yield of ammonia decrease if the Haber-Bosch process is conducted at too high a temperature?

17. What are the characteristics of a chemical explosion?

18. What two characteristics of nitrogen are responsible for the particular effectiveness of nitrogen-based explosives? Explain.

19. What were Alfred Nobel's main contributions to chemistry?

20. Consider the equation for the principal reaction that quickly causes an automobile air bag to inflate:

$$3\ NaN_3(s) \longrightarrow Na_3N(s) + 4\ N_2(g)$$

What volume of nitrogen gas (at STP) would an air-bag reaction involving 1.0 g NaN_3 produce?

21. Identify four principles of Responsible Care that differ from Green-Chemistry principles.

22. "An ounce of prevention is worth a pound of cure." Explain how this saying relates to the Green-Chemistry initiative.

23. Provide an example of how following Green-Chemistry principles can help a chemical-manufacturing company become more profitable.

Connecting the Concepts

24. Consider the following equilibrium system:

$$2\ SO_2(g) + O_2(g) \rightleftharpoons 2\ SO_3(g) + Heat$$

Explain how this equilibrium system would be affected by

a. adding more molecules of oxygen gas at constant volume.
b. increasing the temperature.
c. increasing the volume of the reaction vessel.
d. increasing the total pressure on the system.

25. Scaling up a reaction to production levels involves many challenges not necessarily apparent at the lab scale. Discuss three such challenges faced by chemical engineers at EKS Nitrogen Products Company.

26. Refrigerated food lasts longer than food stored at room temperature. Explain.

27. Many explosive compounds contain nitrogen atoms in a positive oxidation state and carbon atoms in a negative oxidation state. What happens to the oxidation states of each type of atom if these compounds explode?

Extending the Concepts

28. Compare the nitrogen-supply crisis of the early 20th century to current concerns about petroleum supplies.

29. Before chemists can convert methane to hydrogen gas as a reactant in the synthesis of ammonia, they must remove sulfur-containing compounds. How is sulfur removed from natural gas? Why is this necessary? Include and explain relevant chemical equations as part of your answer.

30. In several situations, an industrial by-product has become a valuable commodity. Discuss, in detail, one example.

31. Select two common substances that the chemical industry produces (Table 5.2, page 445) and investigate uses of these substances.

32. Describe some examples or analogies, other than chemical reactions, of dynamic equilibria.

33. The Toxics Release Inventory (TRI) is available from the EPA. Use the TRI to judge how much your state or metropolitan area has accomplished in reducing emissions from manufacturing facilities.

34. A *pressure cooker* reduces the total time needed to cook foods. Investigate the design of a pressure cooker and explain why it speeds up cooking times.

35. Select three Green-Chemistry or Responsible-Care principles and, for each, describe how you could apply and follow a comparable principle in your daily activities.

SECTION C

Metal Processing and Electrochemistry

You have learned how EKS Nitrogen Products Company produces ammonia using the Haber-Bosch process. Such a chemical plant could affect Riverwood both positively and negatively. You will now learn more about WYE Metals Corporation. Then you will be able to determine whether EKS or WYE (or neither company) should be invited to build a chemical plant near Riverwood.

In producing sheet aluminum, WYE specializes in **electrochemistry,** chemical changes that produce or are caused by electrical energy. The following discussions and laboratory investigation provide background on electrochemistry principles. This information will help you understand how WYE's proposed new plant would operate.

C.1 ELECTROLYSIS AND VOLTAIC CELLS

WYE Metals Corporation proposes to produce aluminum metal using oxidation–reduction methods. Because aluminum is the most abundant metallic element in Earth's crust, such production might sound easy. However, aluminum in Earth's crust is not found in its elemental form. Instead it is found in clay soils and is industrially extracted from the ore bauxite (Figure 5.29), in which aluminum ions (Al^{3+}) are strongly bonded to other atoms. Extraction of aluminum metal involves **electrolysis,** in which passing an electrical current through a solution of ions (an **electrolyte**) causes a chemical reaction. In this case, the flow of electrons reduces Al^{3+} ions in the electrolyte to aluminum metal, Al(s) (Figure 5.30).

Electrolysis requires considerable electrical energy, so the cost of electricity is a factor in plant location. The hydroelectric plant at the Snake River Dam generates considerably more electricity than needed by Riverwood and its surrounding communities. To encourage WYE Corporation to consider locating in Riverwood, power company officials have offered WYE large quantities of electrical power at very competitive rates.

In Unit 2, Investigating Matter B.5 (page 142), you learned that some metals lose electrons (become oxidized) more readily than others; that is, some metals are more chemically active than others. The relative tendencies of metals to release electrons can be summarized in the *activity series* of metals (see Table 5.3). A metal higher in the activity series will give up electrons more readily than will a metal that is lower. For example, according to Table 5.3, aluminum atoms are oxidized (lose electrons) more easily than are iron atoms.

Connecting two metals from different positions on the activity series in a voltaic cell creates an electrical potential between the

Figure 5.29 *Aluminum metal can be obtained from bauxite, an aluminum-containing ore. Bauxite is a mixture mainly of aluminum oxides and aluminum hydroxides.*

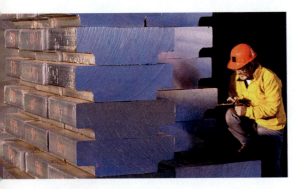

Figure 5.30 *Aluminum ingots are produced by an oxidation–reduction process.*

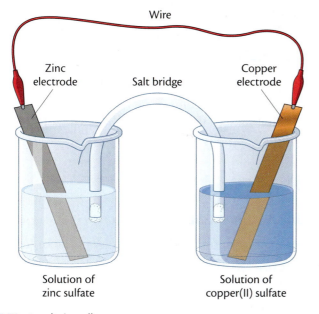

Wire

Zinc electrode   Salt bridge   Copper electrode

Solution of zinc sulfate   Solution of copper(II) sulfate

Figure 5.31 *A voltaic cell*

ACTIVITY SERIES OF COMMON METALS

Metal	Products of Metal Reactivity		
Li(s)	$\longrightarrow$ Li$^+$(aq)	+	e^-
Na(s)	$\longrightarrow$ Na$^+$(aq)	+	e^-
Mg(s)	$\longrightarrow$ Mg^{2+}(aq)	+	$2\,e^-$
Al(s)	$\longrightarrow$ Al^{3+}(aq)	+	$3\,e^-$
Mn(s)	$\longrightarrow$ Mn^{2+}(aq)	+	$2\,e^-$
Zn(s)	$\longrightarrow$ Zn^{2+}(aq)	+	$2\,e^-$
Cr(s)	$\longrightarrow$ Cr^{3+}(aq)	+	$3\,e^-$
Fe(s)	$\longrightarrow$ Fe^{2+}(aq)	+	$2\,e^-$
Ni(s)	$\longrightarrow$ Ni^{2+}(aq)	+	$2\,e^-$
Sn(s)	$\longrightarrow$ Sn^{2+}(aq)	+	$2\,e^-$
Pb(s)	$\longrightarrow$ Pb^{2+}(aq)	+	$2\,e^-$
Cu(s)	$\longrightarrow$ Cu^{2+}(aq)	+	$2\,e^-$
Ag(s)	$\longrightarrow$ Ag$^+$(aq)	+	e^-
Au(s)	$\longrightarrow$ Au^{3+}(aq)	+	$3\,e^-$

Table 5.3 *The higher a metal is positioned in the activity series, the more readily it can give up one or more electrons, as shown.*

 Using the Metal Activity Series

Constructing a Voltaic Cell

Electrolysis uses electrical energy to cause a nonspontaneous reaction. A voltaic cell generates electrical energy through a spontaneous reaction.

metals. Electrical potential (volts) is somewhat like water pressure in a pipe. Just as pressure causes water to flow in the pipe, **electrical potential** "pushes" electrons through the wire connecting the two metals. The greater the difference between the chemical activities of the two metals, the greater the electrical potential that the cell generates.

The differing tendency of metals to lose electrons allows an oxidation–reduction reaction to generate electrical energy. You can make a simple device called a **voltaic cell** from two half-cells connected in a circuit (see Figure 5.31). Each **half-cell** contains a metal partially immersed in a solution of ions of that metal. For example, one half-cell could contain copper metal immersed in a solution of Cu^{2+} ions. Another could contain a piece of zinc metal immersed in a solution of Zn^{2+} ions. You know from Unit 2, page 142, that the oxidation–reduction reaction between Cu^{2+} and Zn readily occurs. By separating these reactants into half-cells, the electrons are forced to flow through a wire and provide electrical energy to an external circuit.

A barrier prevents the solutions in the two half-cells from mixing. A wire connects the two metals, which act as **electrodes,** and allows electrons to flow between them, as shown in Figure 5.31. Such an electron flow constitutes an **electric current.** To complete the circuit and maintain a balance of electrical charges within the system, the two half-cells must be connected by a **salt bridge.** Dissolved ions in the salt bridge complete the internal electrical circuit by allowing ions to move freely between the half-cells, preventing build up of electrical charge near the electrodes. Without the flow of ions, a positive electrical charge would build up in one half-cell and a negative electrical charge would build up in the other half-cell. That situation would prevent any further flow of electrons in the wire.

C.2 VOLTAIC CELLS

In this investigation, you will construct several voltaic cells and measure and compare the electrical potentials that they generate. You will also explore factors that may help determine the actual electrical potential a particular voltaic cell generates.

Lab Video:
Voltaic Cells

Procedure

Part I: Constructing a Voltaic Cell

(Figure 5.32 shows a completed cell.)

1. Add 1 mL of 0.1 M $Cu(NO_3)_2$ to one well in a wellplate.
2. Add 1 mL of 0.1 M $Zn(NO_3)_2$ to an adjacent well.
3. Clean strips of Cu and Zn with steel wool.
4. Add a Cu strip (electrode) to the well containing $Cu(NO_3)_2$ solution.
5. Add a Zn strip (electrode) to the well containing $Zn(NO_3)_2$ solution.
6. Drape a small strip of filter paper saturated with KNO_3 solution between the wells containing the two solutions. Ensure that each end of the filter paper strip is immersed in one of the two solutions. Do not allow the metal strips to touch each other or the filter paper.

Figure 5.32 *The voltaic cells you will complete resemble the one depicted here. Voltaic cells result in spontaneous flow of electrons from the metal of higher activity to ions of the less-active metal.*

7. Obtain a voltmeter and two insulated wires with alligator clips connected at each end. Attach one end of each wire to a separate voltmeter terminal.
8. Attach one wire from the voltmeter to the copper electrode. Lightly touch the second wire to the zinc electrode. If the needle deflects in the direction of a positive potential or if the digital readout is positive, attach the second clip to the zinc metal. If the needle deflects in a negative direction (or the readout value is negative), reverse the clip connections to the electrodes.
9. Record the electrical potential indicated by the voltmeter.

Part II: Measuring Electrical Potentials

10. Using what you learned in Part I about building a voltaic cell, construct additional cells and measure electrical potentials generated by voltaic cells composed of all possible combinations of the half-cells listed below. (See Figure 5.33.) Record your data.

a. 0.1 M $Cu(NO_3)_2$ and Cu strip
b. 0.1 M $Zn(NO_3)_2$ and Zn strip
c. 0.1 M $Mg(NO_3)_2$ and Mg strip
d. 0.1 M $Fe(NO_3)_2$ and Fe sample (a nail)

Figure 5.33 *Representative metal electrodes used in this investigation.*

Part III: Investigating Electrode Size Effects

11. Based on Parts I and II, design a laboratory investigation, accompanied by an appropriate data table, to explore whether an electrode's size influences the total electrical potential (volts) generated by a cell and, if so, how you can describe that effect. Base your design on copper and zinc metal electrodes of different widths: 0.25 cm (narrow), 0.50 cm (medium), and 1.0 cm (wide).

 Maximizing the Electrical Potenial

12. Set up and conduct your investigation. Record your data.

13. Dispose of your experimental materials as directed by your teacher.

14. Wash your hands thoroughly before leaving the laboratory.

Questions

1. a. In Part II, you constructed voltaic cells using copper and three other metals: zinc, magnesium, and iron. List these cells in order of decreasing electrical potential (highest cell potential first).

b. Explain the resulting list in terms of the relative activities of the metals involved.

2. Would the electrical potential generated by cells composed of each of the following pairs of metals be larger or smaller than that of the Zn–Cu cell? Refer to Table 5.3 (page 453).

a. Zn and Cr   b. Zn and Ag   c. Sn and Cu

3. How did changing the size of the zinc and copper electrodes affect the measured electrical potential? Explain and provide evidence supporting your answer.

4. Could an Ag–Au cell serve as a commercially feasible voltaic cell? Explain and provide evidence for your answer.

C.3 VOLTAIC CELLS AND HALF-REACTIONS

In the voltaic cells you constructed in the laboratory, each metal immersed in a solution of its ions represented a half-cell. The activity series predicts that zinc is more likely to be oxidized (lose electrons) than copper. Thus, in the zinc–copper cell you investigated, oxidation (electron loss) occurred in the half-cell with zinc metal immersed in zinc nitrate solution. Reduction (electron gain) took place in the half-cell of copper metal in copper(II) nitrate solution. The half-reactions (individual electron-transfer steps) for that cell are:

$$\text{Oxidation half-reaction: } Zn(s) \longrightarrow Zn^{2+}(aq) + 2\ e^-$$

$$\text{Reduction half-reaction: } Cu^{2+}(aq) + 2\ e^- \longrightarrow Cu(s)$$

The electrode at which oxidation takes place is the *anode.* Reduction occurs at the *cathode.* (See Figure 5.34.)

The overall reaction in the zinc–copper voltaic cell is the sum of the two half-reactions, added so that the electrical charges balance—the total electrons lost and gained are the same—resulting in a net electrical charge of zero.

$$\begin{aligned} Zn(s) &\longrightarrow Zn^{2+}(aq) + 2\ e^- \\ Cu^{2+}(aq) + 2\ e^- &\longrightarrow Cu(s) \\ \hline Zn(s) + Cu^{2+}(aq) &\longrightarrow Zn^{2+}(aq) + Cu(s) \end{aligned}$$

Because a barrier separates the two reactants (Zn and Cu^{2+}) in the cell, the electrons released by zinc must travel through the external wire to reach (and reduce) the copper ions.

The greater the difference in the reactivity of the two metals in a voltaic cell, the greater the tendency for electron transfer to occur, and the greater the electrical potential (volts) of the cell. A zinc–gold voltaic cell, therefore, would generate a larger electrical potential than a zinc–copper cell. (See Table 5.3, page 453, to compare the placement of these metals in the activity series.)

> One way to remember these electrode processes: Note that **a**node and its associated process (**o**xidation) both begin with vowels, while **c**athode and its process (**r**eduction) both start with consonants.

Figure 5.34 REDuction always occurs at the **CAT**hode.

Developing Skills

C.4 GETTING CHARGED BY ELECTROCHEMISTRY

Each of the following questions deals with voltaic cells, their properties, and equations to describe them.

Sample Problem: *Consider a voltaic cell containing lead metal (Pb) immersed in lead(II) nitrate solution, $Pb(NO_3)_2$, and a half-cell containing silver metal (Ag) in silver nitrate solution, $AgNO_3$.*

> a. *Predict the direction of electron flow in the wire connecting the two metals.*
>
> b. *Write equations for the two half-reactions and the overall reaction.*
>
> c. *Which metal is the anode and which is the cathode?*

The answers are as follows:

   a. Table 5.3 (page 453) shows that lead is a more active metal than is silver. Therefore, lead will be oxidized, and electrons will flow from lead to silver.

   b. One half-reaction involves forming Pb^{2+} from Pb, as shown in Table 5.3. The other half-reaction produces Ag from Ag^+, which can be written by reversing the equation in Table 5.3 and doubling it so electrons lost and gained are the same:

$$Pb(s) \longrightarrow Pb^{2+}(aq) + 2e^-$$
$$2\,Ag^+(aq) + 2e^- \longrightarrow 2\,Ag(s)$$
$$\overline{Pb(s) + 2\,Ag^+(aq) \longrightarrow Pb^{2+}(aq) + 2\,Ag(s)}$$

   c. In the cell reaction, each Pb atom loses two electrons. Pb is therefore oxidized, making it the anode. Each Ag^+ ion gains one electron; a reduction reaction. Because reduction takes place at the cathode, Ag must be the cathode.

> The total electrons consumed in reduction equals the total electrons liberated in oxidation. Overall, the voltaic cell based on these two half-reactions involves twice as many silver ions reduced as lead atoms oxidized.

Now answer the following questions.

1. Predict the direction of electron flow in a voltaic cell made from each specified pair of metals partially immersed in solutions of their ions.

   a. Al and Sn
   b. Pb and Mg
   c. Cu and Fe

2. A voltaic cell uses tin (Sn) and cadmium (Cd) as the electrodes. The overall equation for the cell reaction is as follows:

$$Sn^{2+}(aq) + Cd(s) \longrightarrow Cd^{2+}(aq) + Sn(s)$$

   a. Write the two half-reaction equations for this cell.
   b. Which metal, Sn or Cd, loses electrons more readily?

3. Sketch a voltaic cell composed of a $Ni–Ni(NO_3)_2$ half-cell linked to a $Cu–Cu(NO_3)_2$ half-cell.

4. For each voltaic cell designated below, identify the anode and the cathode. Assume that each voltaic cell uses appropriate ionic solutions.

   a. Cu–Zn cell
   b. Al–Zn cell
   c. Mg–Mn cell
   d. Au–Ni cell

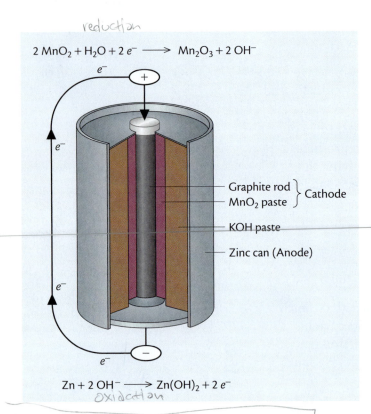

reduction

$$2 \, MnO_2 + H_2O + 2 \, e^- \longrightarrow Mn_2O_3 + 2 \, OH^-$$

Graphite rod
MnO$_2$ paste } Cathode

KOH paste

Zinc can (Anode)

$$Zn + 2 \, OH^- \longrightarrow Zn(OH)_2 + 2 \, e^-$$
oxidation

Figure 5.35 *An alkaline battery. What substances are contained in this fresh alkaline battery? How will those substances change as the battery is used?*

C.5 BATTERIES

Voltaic cells are a convenient way to convert chemical energy to electrical energy. The cells can fit in small, portable containers. We can use various combinations of metals and ions to make commercial voltaic cells. In ordinary *dry cells*—often called *batteries*—zinc is the anode, and a graphite rod surrounded by a water-based paste of manganese(IV) oxide (MnO_2) and graphite serves as the cathode. A mixture of ammonium chloride and zinc chloride in an aqueous paste serves as the electrolyte. Alkaline batteries, common in portable music players, have similar zinc and graphite–MnO_2 electrodes, but the electrolyte is an alkaline (basic) aqueous potassium hydroxide (KOH) electrolyte paste. See Figure 5.35.

Both zinc–MnO_2 dry cells and alkaline batteries generate an electrical potential of 1.54 V. The following oxidation–reduction equations describe the chemical changes involved.

Dry Cell

Oxidation: $\quad\quad\quad\quad\quad\quad\quad Zn(s) \longrightarrow Zn^{2+}(aq) + 2e^-$

Reduction: $\quad 2 \, MnO_2(s) + 2 \, NH_4^+(aq) + 2e^- \longrightarrow 2 \, MnO(OH)(s) + 2 \, NH_3(g)$

Overall: $\quad\;\; Zn(s) + 2 \, MnO_2(s) + 2 \, NH_4^+(aq) \longrightarrow Zn^{2+}(aq) + 2 \, MnO(OH)(s) + 2 \, NH_3(g)$

Alkaline Battery

Oxidation: $\quad\quad\quad Zn(s) + 2 \, OH^-(aq) \longrightarrow Zn(OH)_2(s) + 2e^-$

Reduction: $\quad 2 \, MnO_2(s) + H_2O(l) + 2e^- \longrightarrow Mn_2O_3(s) + 2 \, OH^-(aq)$

Overall: $\quad\;\; Zn(s) + 2 \, MnO_2(s) + H_2O(l) \longrightarrow Zn(OH)_2(s) + Mn_2O_3(s)$

ChemQuandary 2

BATTERY SIZES

Each battery in this photo generates the same electrical potential, 1.5 V. So why does anyone need 1.5-V batteries larger than the smallest one? Wouldn't it save space, weight, and perhaps even resources to restrict consumer use to the smallest batteries?

Dry cells and alkaline batteries are called *primary batteries.* They generate electrical potential only as long as all starting materials (reactants) remain. When the limiting reactant (see Unit 4, page 353) is depleted, primary batteries cannot be recharged.

By contrast, nickel–cadmium (NiCad) and automobile batteries are rechargeable; we can return their systems to their original states and thus reuse the batteries. The common NiCad rechargeable battery is based on a nickel–iron battery that Thomas Edison developed in the early 1900s. The NiCad battery anode is cadmium, and the cathode is nickel oxide. These electrodes have a rolled design (Figure 5.36) that increases surface area and, thus, increases the generated current (electron flow).

When this battery operates, Cd and NiO_2 are converted to $Cd(OH)_2(s)$ and $Ni(OH)_2(s)$. These solid products cling to the electrodes, allowing them to convert back to reactants when you connect the used battery in a recharging circuit. In recharging, an external electrical potential causes electrons to flow in the opposite direction, thus reversing the reaction (see Figure 5.37). Although you can recharge NiCad batteries many times, some of the same processes that prevent recharging of primary batteries can take their toll and eventually reduce the efficiency of recharging.

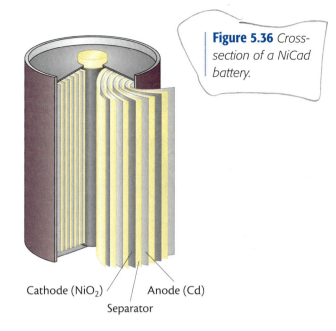

Figure 5.36 *Cross-section of a NiCad battery.*

Cathode (NiO_2)   Anode (Cd)

Separator

> NiCad batteries should be recycled to prevent the cadmium, which is toxic, from leaching into groundwater.

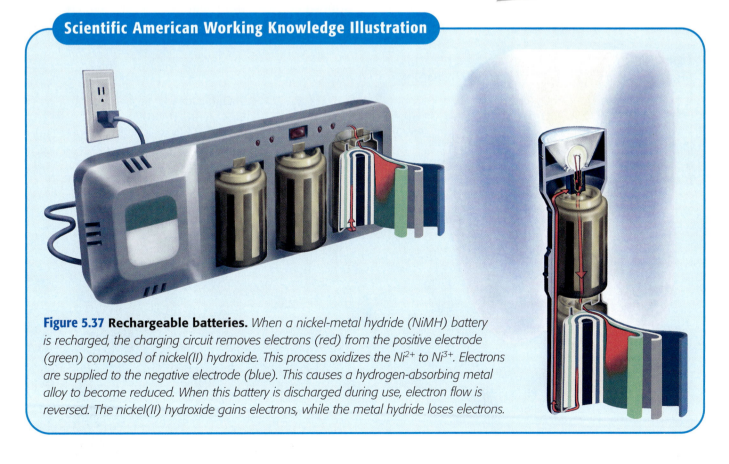

Scientific American Working Knowledge Illustration

Figure 5.37 Rechargeable batteries. *When a nickel-metal hydride (NiMH) battery is recharged, the charging circuit removes electrons (red) from the positive electrode (green) composed of nickel(II) hydroxide. This process oxidizes the Ni^{2+} to Ni^{3+}. Electrons are supplied to the negative electrode (blue). This causes a hydrogen-absorbing metal alloy to become reduced. When this battery is discharged during use, electron flow is reversed. The nickel(II) hydroxide gains electrons, while the metal hydride loses electrons.*

A 12-V automobile battery, known as a *lead-acid battery,* is composed of a series of six electrochemical cells. As illustrated in Figure 5.38, each cell consists of an anode of uncoated lead (Pb) plates and a cathode of lead plates coated with lead(IV) oxide (PbO_2). The electrodes are immersed in a dilute solution of sulfuric acid (H_2SO_4), the electrolyte in this system.

When the vehicle's ignition is turned on, the electrical circuit is completed. As electrons travel from one electrode to the other, they provide energy for the car's electrical systems. Metallic lead at the anode is oxidized to Pb^{2+}. The freed electrons travel through the wire to the lead(IV) oxide cathode, reducing Pb^{4+} in PbO_2 to Pb^{2+}. Lead ions (Pb^{2+}) produced at both electrodes then form $PbSO_4$ by reacting with the electrolyte, as shown in the following equations.

Oxidation: $$Pb(s) + SO_4{}^{2-}(aq) \longrightarrow PbSO_4(s) + 2e^-$$

Reduction: $$PbO_2(s) + SO_4{}^{2-}(aq) + 4\,H^+(aq) + 2e^- \longrightarrow PbSO_4(s) + 2\,H_2O(l)$$

Overall: $$PbO_2(s) + Pb(s) + 4\,H^+(aq) + 2\,SO_4{}^{2-}(aq) \longrightarrow 2\,PbSO_4(s) + 2\,H_2O(l)$$

If an automobile battery is used too long without being recharged, it runs down; that is, the redox reaction stops. Lead(II) sulfate eventually coats the electrodes, which reduces their ability to react and produce a current and electrical potential. In an automobile, recharging is accomplished by an alternator or generator, which converts some mechanical energy from the vehicle's engine into electrical energy that forces electrons to move in the opposite direction through the battery. This reverses the direction of the battery's chemical reactions (see Figure 5.38).

Figure 5.38

Discharging and recharging an automobile battery. What furnishes the energy to recharge this battery?

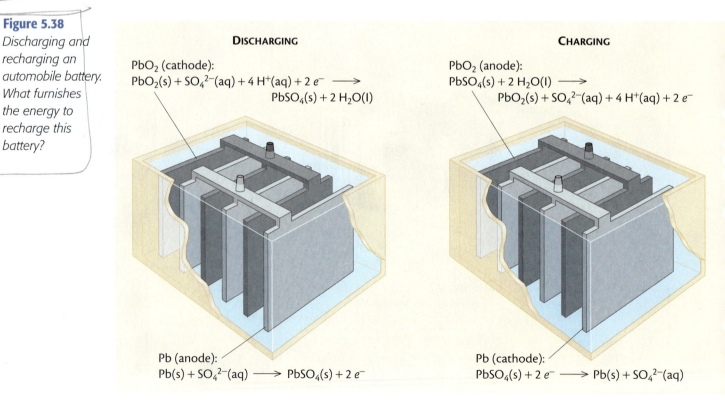

DISCHARGING

PbO_2 (cathode):
$$PbO_2(s) + SO_4{}^{2-}(aq) + 4\,H^+(aq) + 2\,e^- \longrightarrow PbSO_4(s) + 2\,H_2O(l)$$

Pb (anode):
$$Pb(s) + SO_4{}^{2-}(aq) \longrightarrow PbSO_4(s) + 2\,e^-$$

CHARGING

PbO_2 (anode):
$$PbSO_4(s) + 2\,H_2O(l) \longrightarrow PbO_2(s) + SO_4{}^{2-}(aq) + 4\,H^+(aq) + 2\,e^-$$

Pb (cathode):
$$PbSO_4(s) + 2\,e^- \longrightarrow Pb(s) + SO_4{}^{2-}(aq)$$

Lead-acid batteries can pose dangers if rapidly charged by an out-side source of electrical energy. Hydrogen gas, formed by the reduction of H^+ in the acidic electrolyte, is released at the lead electrode. A spark or flame can ignite the hydrogen gas, causing an explosion. This is one reason you must be careful to avoid sparks when using jumper cables to start a vehicle with a dead battery.

C.6 INDUSTRIAL ELECTROCHEMISTRY

Chlorine and Sodium Hydroxide Production

One economically significant application of electrolysis in the chemical industry is the *chlor-alkali process.* In this process, sodium chloride dissolved in water (a solution known as *brine*) is electrolyzed, producing chlorine gas (Cl_2) at the anode and hydrogen gas (H_2) at the cathode (Figure 5.39). Positive sodium ions and negative hydroxide ions remain in solution. The hydroxide ions replace the chloride ions lost from the brine as chlorine gas, thus maintaining the charge balance, as shown by the following equation:

$$\text{Electrical energy} + 2\,Na^+(aq) + 2\,Cl^-(aq) + 2\,H_2O(l) \longrightarrow 2\,Na^+(aq) + 2\,OH^-(aq) + H_2(g) + Cl_2(g)$$

This electrolysis reaction generates three widely produced industrial substances: hydrogen gas, chlorine gas, and sodium hydroxide.

We can use the hydrogen gas and chlorine gas produced in the chlor-alkali process without additional purification because no other gases are released by this reaction. We must purify the sodium hydroxide because it remains in solution with unreacted sodium chloride.

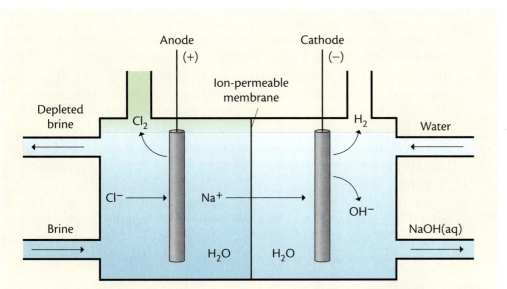

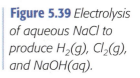
Figure 5.39 *Electrolysis of aqueous NaCl to produce $H_2(g)$, $Cl_2(g)$, and NaOH(aq).*

From: Kotz, J.C. and Treichel, P.: *Chemistry & Chemical Reactivity*, Saunders College Publishing, Fort Worth, TX, 1999.

Originally, the large industrial electrolysis cells for the chlor-alkali process included mercury cathodes. Before people understood the environmental hazards of metallic mercury, they dumped it into rivers and lakes while renewing the cells. Mercury is no longer used in chlor-alkali cells. Research chemists and chemical engineers modified the chlor-alkali process considerably, replacing both electrodes, and developed new permeable polymer membranes that separate the anode and cathode compartments (Figure 5.39, page 461).

Aluminum Production

The industrial production of aluminum, the focus of WYE Metals Corporation, is another major process that uses electrical energy to produce chemical change. It is very difficult to reduce Al^{3+} held by ore into pure metal, Al(s). Aluminum was first isolated as a metal in the 1820s by an expensive and potentially dangerous process that used highly reactive sodium or potassium metal as the reducing agent. As a result, for more than 60 years, aluminum was very expensive, despite the fact that aluminum is the most plentiful metal ion in Earth's crust. Aluminum was even used in jewelry, including the French crown jewels and the Danish crown. In 1884, a 2.8-kg (6-lb) aluminum cap was installed on top of the Washington Monument in Washington, D.C., as ornamentation and as the tip of a lightning-rod system. At that time, the aluminum cap cost considerably more than did the same mass of silver.

The challenge for industrial users was how to produce aluminum at a much lower cost. No common substance gives up electrons readily enough to reduce aluminum cations to aluminum metal. For example, carbon, an excellent reducing agent for metal compounds such as iron oxide or copper sulfide, cannot reduce aluminum compounds.

A young chemist named Charles Martin Hall solved the problem. In 1886, within a year after graduating from Oberlin College (Ohio), Hall devised a method for reducing aluminum using electricity (Figure 5.40). Hall's breakthrough was in discovering that aluminum oxide (Al_2O_3), which melts at 2000 °C, dissolved in molten cryolite (Na_3AlF_6) at 950 °C. The lower temperature meant that he had found an easier way to get aluminum ions from aluminum ore into solution for electrolysis. Hall's discovery became the basis of a rapidly growing aluminum industry. He founded the Aluminum Company of America (Alcoa) and was a multimillionaire when he died in 1914.

In the Hall-Héroult process, aluminum oxide (bauxite) is dissolved in molten cryolite in a large steel tank lined with carbon. The carbon tank lining is given a negative electrical charge by a source of electrical current. This carbon cathode transfers electrons to

Within two months after Hall developed his process, Paul-Louis Héroult, a young French scientist, independently developed the same process. These two workers, linked through their common discovery, also shared the same birth and death years (1863–1914).

aluminum ions, reducing them to molten metal. The molten aluminum sinks to the bottom, where it is periodically drawn off.

The anode, also made of carbon, is oxidized during the reaction. As the reaction consumes the tips of the carbon-rod anodes, the rods are gradually lowered deeper into the molten cryolite bath.

The half-reactions for producing aluminum metal by the Hall-Héroult process are as follows:

$$4\ Al^{3+}(melt) + 12\ e^- \longrightarrow 4\ Al(l)$$

$$3\ C(s) + 6\ O^{2-}(melt) \longrightarrow 3\ CO_2(g) + 12\ e^-$$

Ions from cryolite carry the electric current in the molten mixture.

Before that industrial process was devised, only about 1000 kg of aluminum was produced annually worldwide. In 1884, one pound of aluminum metal sold for about $12, a significant sum in those days. Within three years of their discovery, Hall and Héroult each started commercial operations, creating a worldwide aluminum production rate of one million tons annually by the turn of the 20th century (see Figure 5.41, page 464). The price of aluminum plummeted. Five years after large-scale production started, aluminum sold for 70 cents per pound. Thus, this process opened doors for wide-ranging uses of aluminum: from beverage cans and stepladders to foil wrap and aircraft components.

> The designation "(melt)" indicates that at very high cryolite temperatures, ions in the bauxite can dissolve and thus are free to move within the molten material.

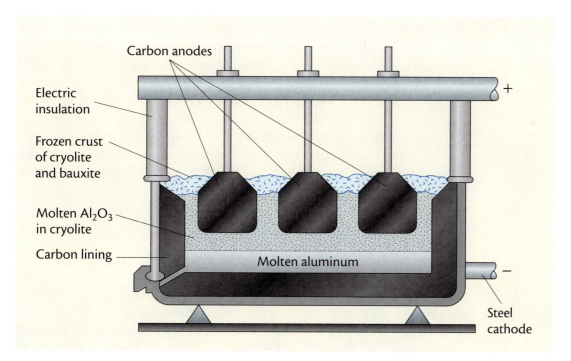

| Figure 5.40 *The Hall-Héroult process for producing aluminum.*

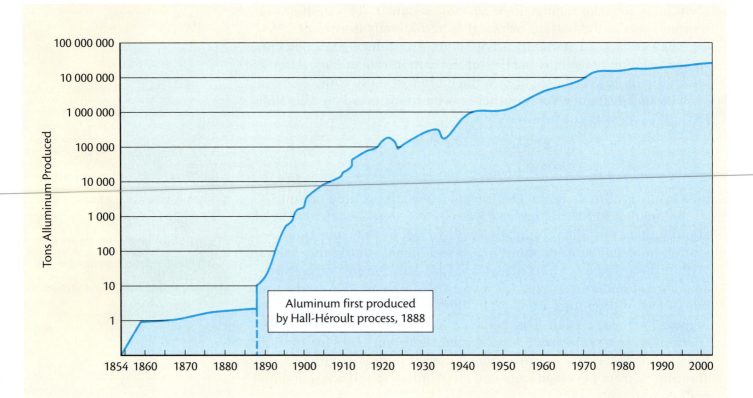

Figure 5.41 *Worldwide aluminum production.*

More than 10 billion kilograms (over 22 billion pounds) of aluminum are produced in the United States each year, requiring about $2 billion worth of electrical energy (about 20 cents per kilogram of aluminum metal produced). Because aluminum plants use such large quantities of electricity, they are often located near sources of hydroelectric power. Electricity from hydropower generally costs less than that generated from the combustion of fossil fuels. Many U.S. aluminum plants are located in the Pacific Northwest because of the hydroelectric power sources there. WYE Metals Corporation is interested in Riverwood due to the availability of hydroelectric power and the town's proximity to aluminum-ore sources.

C.7 MOVING TOWARD GREENER METHODS

As you learned earlier in this unit, industries, governments, and citizens in the United States and elsewhere have recognized their joint responsibility to ensure that chemical products are manufactured with maximum benefit and minimum risk to society. The EPA's Design for Environment program, which includes Green Chemistry (pages 446–447), has already influenced the way chemical-processing, printing, and dry-cleaning industries conduct business.

The chemical industry has developed new synthesis methods based on using safe starting materials to replace toxic or environmentally unsafe substances. Green Chemistry focuses on preventing environmental pollution directly at the point of manufacturing. In this approach, and in others such as Responsible Care (page 447), the chemical industry works as a social partner to sustain development and international trade without damaging the environment.

New Green Chemistry-based processes use more environmentally benign reactants than do traditional processes and create waste products that are less damaging to the air and water. For example, chemical processes can now use D-glucose, ordinary table sugar, to replace benzene, a known carcinogen. The D-glucose serves as a raw material for synthesizing other reactants that are used to produce nylon and various medicinal drugs. Also, in some processes, chemists can substitute nontoxic food dyes for catalysts composed of toxic metals such as lead, chromium, and cadmium.

Isocyanates are substances used to make polyurethanes, which in turn are used to produce seat cushions, insulation, and contact lenses. Carbon dioxide can replace phosgene, a toxic gas, in manufacturing isocyanates. Green Chemistry has led to methods for synthesizing industrial chemicals using water as a solvent instead of toxic alternatives, whenever possible, and for employing materials that can be recycled and reused, thus significantly reducing waste-disposal problems.

The chemical industry has the responsibility to make products in ways that are as hazard-free as possible. The chemical industry must deal honestly with the public to ensure that people clearly understand risks and benefits of chemical operations. Chemistry companies also must assure consumers that their products are safe when used as intended.

Chemical industries must comply with relevant laws and regulations, as well as with voluntary standards set by many manufacturers themselves. Independent, outside organizations, such as the International Organization for Standardization (ISO) in Geneva, Switzerland, encourage worldwide compliance.

No initiatives of the chemical industry or the government can eliminate all risks involved in manufacturing chemical substances, any more than we can completely eliminate the risks of automobile travel. However, knowing the risks, continuing to explore the sources of and alternatives to those risks, and prudent decision making all remain essential.

As users of the industry's chemical products, all consumers share these responsibilities. Having studied some basic concepts about the manufacturing of chemical products, you can now turn to using such knowledge to weigh risks and benefits of particular decisions—to you, to the community, and to the environment.

In 1998, the occupational injury and illnesses rate of the U.S. chemical industry was less than half that of U.S. manufacturing industries as a whole.

C.8 ASSET OR LIABILITY?

If you were a Riverwood citizen, what would be your view? Should the town invite a chemical plant to locate in Riverwood? On what basis should such a decision be made?

To clarify this challenge, it is helpful to consider and evaluate both benefits and burdens—positive and negative factors—associated with any choice. Then you will be ready to address key questions that confront Riverwood citizens as they decide on their choices.

Positive Factors

▶ The Riverwood economy would improve. Either plant would employ about 200 local residents. This would add about $8 million to Riverwood's economy annually. In addition, each plant employee would indirectly provide jobs for another four people in local businesses. This is very desirable because 15% of Riverwood's labor force of 21 000 is currently unemployed.

▶ Farming costs might go down. Each year farmers in the Riverwood area spread 700 tons of fertilizer (see Figure 5.42). The fertilizer comes from a fertilizer plant 200 miles away. Transport costs increase farmers' expenses by $15 for each ton of ammonia-based fertilizer. Local farmers thus stand to save about $10 000 annually in transportation costs.

Figure 5.42 *Fertilizers used to increase yields of farm crops are cost-effective due to large-scale fertilizer production by chemical industries.*

▶ New local industries could result. With direct access to aluminum locally, entrepreneurs might start a factory to make aluminum window frames and siding for home construction. Without significant transportation costs for raw materials, such a company could establish itself in the market by selling its products at lower prices. The factory would probably hire 30–45 employees, mainly shop workers and drivers, adding $700 000 annually to the local economy.

▶ As a commercial refrigerant, ammonia is used to produce large quantities of ice. Ready access to ammonia supplies could support a commercial ice-making plant in Riverwood. Such a company might employ 25–35 individuals, principally drivers and some plant workers, which could add $500 000 annually to Riverwood's economy.

▶ Riverwood air quality might improve. A Riverwood ammonia plant would require natural gas, and a natural-gas transmission company could build a pipeline to deliver the gas (mainly methane, CH_4). Residents have traditionally burned fuel oil in their home furnaces, but they could convert to natural gas. If all 11 000 Riverwood homes and businesses burned natural gas rather than oil, emissions of sulfur dioxide and particulate matter would decrease significantly.

▶ The tax base would improve. Although tax incentives are part of the appeal of a Riverwood site, the aluminum or ammonia plant will still contribute to the town's tax base. This will provide a large increase in revenues for the community.

Negative Factors

▶ Rates of worker-related injuries and accidents could increase. Ammonia is manufactured at high pressures and high temperatures. At ordinary temperature and pressure, ammonia gas is extremely toxic at high concentrations. Large amounts of ammonia released due to an accident on the road or at the plant could injure or kill workers and other community members within the vicinity. Several cases of work-related injury or illness or even death in ammonia-based fertilizer plants have been reported annually.

▶ Aluminum itself is nontoxic, but its production has its own dangers. Molten aluminum is drawn from large electrolytic cells. As the carbon anodes burn off in cryolite bath, toxic carbon monoxide gas may be produced as a by-product and must be handled properly. Mining and transporting aluminum ore may also involve risk of injury. Aluminum-production illness and injury rates annually are approximately 20 per 100 full-time employees. (The injury and accident rates in aluminum- and ammonia-based fertilizer production, by contrast, are both lower than rates found in motor-vehicle manufacturing, meat packing, or steel foundries.)

- Water quality might suffer. If ammonia, ammonia-bearing wastewater, or wastes from aluminum production leaked into the Snake River, the resulting water contaminants could threaten aquatic life.

- Aluminum or ammonia markets might decline. Aluminum is commercially valuable because it is inexpensive, resists corrosion, and has low density. The market for aluminum is driven by applications such as those in the building trades (including insulation, window casings, reinforcement strips, and ventilation devices), transportation (including sheeting for truck trailers, campers, and mobile homes), and packaging. If aluminum substitutes emerged (from new alloys, plastics, or ceramics) in any of these major areas, the demand for aluminum would weaken.

- The fertilizer industry is among the largest consumers of ammonia. Current fertilizer-intensive agricultural methods have created controversy. In some cases, crop yields have declined despite use of increased quantities of synthetic fertilizer. Some farmers have elected to use less synthetic fertilizer, so ammonia demand may decline in coming years.

Considering Burdens and Benefits

The poet William Wordsworth used the phrase *Weighing the mischief with the promised gain . . .* in judging technological advancements (railroads, new in his time). Since the dawn of civilization, people have often accepted the burdens of new technologies to gain technology's benefits. Fire, one of civilization's earliest useful devices, gave people the ability to cook, warm themselves, and forge tools from metals. Yet fire out of control can destroy property and life. Every technology offers its benefits at a price.

One way to identify an acceptable new technology or venture is to evaluate it, finding an option that has a relatively low probability of producing harm—that is, it delivers benefits that far outweigh the burdens. Unfortunately, benefit–burden analysis—weighing what Wordsworth called *mischief* against *promised gain*—is not an exact science.

For instance, some technologies may present high burdens immediately, while others may be associated with chronic, low-level risks for years or even decades. Many burdens are impossible to predict or assess with certainty. Individuals can control some potential burdens, but others must be addressed and controlled at regional or national levels. In short, it is quite difficult to conduct a thorough burden–benefit analysis. However, complete these activities:

1. Based on what you have learned in this unit, work in groups to create two lists for each chemical plant; one summarizing benefits and one summarizing burdens/risks.

> In many cases, electing not to make a decision is, in fact, also a decision—one accompanied by its own burdens and benefits.

2. Review your summaries of the burdens/risks. Are any negative factors completely unacceptable? If so, the plant associated with that risk or burden is probably not a viable option for Riverwood.

3. Within each list, mark the most valuable benefits with a plus symbol and the most serious burdens with a minus symbol.

4. Also consider the likelihood of occurrence of each benefit and burden. A burden that is fairly minor but almost certain to occur might merit more consideration than a burden that is more serious but extremely unlikely. Using a scale from *1* (highly unlikely) to *5* (extremely likely), rate each item on your lists in terms of its likelihood of occurrence.

5. Based on your responses to Questions 3 and 4, decide your personal position on the question of building a chemical plant near Riverwood. Discuss your view with your group.

6. As a class, discuss whether Riverwood should invite a chemical plant to the community. Be sure that the concerns of students who do not share the opinion of the majority are also heard and considered.

CHEMISTRY at Work

Searching for Solutions in Chemistry

Todd Blumenkopf has applied his chemistry knowledge in many different ways, in and out of the laboratory, during his 20-year career in the pharmaceutical industry.

Someday, when you reach for a pill to relieve an aching muscle or a headache, you might have Todd Blumenkopf to thank. Todd, who has used a wheelchair throughout his life, is a research chemist at Pfizer, Inc. Todd and his laboratory staff have worked on medications that reduce the swelling, or inflammation, of joints and muscles.

Todd and his colleagues in the Research Division worked to conceive of and synthesize new compounds that target the enzymes and receptors that contribute to certain disease states in the body. Each compound is tested *in vitro* (in an artificial environment outside the body) and *in vivo* (in the living body of an animal) to determine whether the compound produces desirable results without unwanted side effects.

After a drug has passed the required laboratory testing, researchers collect additional data about the drug to file with the Food and Drug Administration (FDA).

More recently, Todd took a position in Pfizer to identify and establish research collaborations with universities and biotechnology companies. Pfizer is thus able to expand opportunities to discover, develop, and market new drugs, while smaller companies and university research groups partner with Pfizer to complete the expensive clinical trials required to determine that a potential drug is both efficacious and safe.

As a child, Todd had a strong interest in science. His parents encouraged him to pursue a career in the profession, in part because they felt his disability might be met with less resistance in science than in other fields. As he progressed in his studies, he chose a career in pharmaceuticals. After receiving his Ph.D. and completing some postdoctoral work, Todd began his career as a research chemist. He now works at Pfizer's pharmaceutical research and development headquarters in New London, Connecticut.

Todd also works on the American Chemical Society's Committee on Chemists with Disabilities, which sponsors the development of materials and other efforts to improve opportunities in the chemical sciences for individuals with disabilities.

Todd believes the qualities needed for success in chemistry research are curiosity, creativity, and motivation. He also stresses the importance of patience and perseverance, since the process of developing a new product from an initial idea can take many years.

SECTION C SUMMARY
Reviewing the Concepts

Electrochemistry involves chemical changes that produce or are caused by electrical energy.

1. What is electrolysis?

2. What is a voltaic cell?

3. What is a half-cell?

4. Voltaic cells require a salt bridge.

   a. What is a salt bridge?
   b. Why is a salt bridge necessary for the operation of a voltaic cell?
   c. Describe one way to make a salt bridge.

5. How are half-cells linked to produce electricity?

6. Diagram a simple Ag–Cu voltaic cell. Label electrodes, solutions, and salt bridge.

7. Considering your results in Investigating Matter C.2 (page 454), does the electrical potential produced by a voltaic cell depend on the

   a. size of the electrodes? Explain.
   b. specific metals used? Explain.

The activity series of metals can be used to predict the direction of electron flow within a particular voltaic cell.

8. In a voltaic cell, what process takes place at the

   a. anode?   b. cathode?

9. Sketch a voltaic cell made from Ni and Zn in solutions of their ions. Label anode and cathode (identifying each metal) and show the direction of electron flow.

10. Predict the direction of electron flow in a voltaic cell made from each of these metal pairs in solutions of their ions. See Table 5.3 (page 453).

    a. Ag and Sn   c. Cu and Pb
    b. Cr and Ag

Batteries, which consist of one or more voltaic cells, provide convenient, portable ways to energize many common electrical devices.

11. What is the source of electrical energy in a battery?

12. Two types of voltaic cells are dry cells and alkaline batteries.

    a. List three similarities of these two cell types.
    b. How do these cells differ?

13. Write half-reaction equations for each of these oxidation–reduction processes. For each, identify (i) what is oxidized and (ii) what is reduced.

    a. $Pb(s) + Cu^{2+}(aq) \longrightarrow Pb^{2+}(aq) + Cu(s)$
    b. $Cr(s) + 3 Ag^+(aq) \longrightarrow Cr^{3+}(aq) + 3 Ag(s)$

14. Consider the following equation:

$$PbO_2(s) + Pb(s) + 4 H^+(aq) + 2 SO_4^{2-}(aq) \longrightarrow 2 PbSO_4(s) + 2 H_2O(l)$$

   a. In what type of battery does this reaction occur?
   b. What is (i) oxidized and (ii) reduced?
   c. Does this equation represent the charging or discharging of the battery?
   d. Identify the substance that you might observe as a white coating on battery electrodes.
   e. Under what conditions could this battery produce hydrogen gas?

Electrolysis is a useful application of electrochemistry.

15. What three products does the chlor-alkali industrial process generate?

16. How was the chlor-alkali process changed to address environmental concerns?

17. Why was aluminum so expensive in the early 19th century, despite its abundance as an element in Earth's crust?

18. Why are aluminum-producing plants generally located in areas with inexpensive electrical power?

19. Consider the Hall-Héroult process.

    a. Write the half-reaction equations.

    b. Label each equation as an oxidation or reduction.

    c. Write the overall redox equation.

    d. What component is missing from this equation?

    e. What is the role of that component?

    f. How many moles of electrons would you need to reduce enough Al^{3+} ions to produce a 378-g roll of aluminum foil?

Burden–benefit analysis is useful in weighing both positive and negative consequences when making decisions.

20. List two positive and two negative aspects of the technologies involved in producing

    a. ammonia.   **b.** aluminum.

21. Describe a specific change that the chemical industry has implemented in response to the principles of Green Chemistry.

Connecting the Concepts

22. Consider voltaic cells and electrolytic cells (cells where electrolysis occurs).

    a. How are they similar?

    b. How do they differ?

    c. Sketch a diagram of each type of cell and highlight the differences. (*Hint:* Refer to Investigating Matter, pages 202 and 454)

    d. Are these processes opposites? Explain.

23. Why is reducing aluminum oxide more difficult than reducing copper(II) oxide?

24. Describe the economic and societal impact of the Hall-Héroult process.

25. Why is recycling aluminum more cost-effective than processing aluminum ore by the Hall-Héroult method?

26. Explain how you can test the condition of a lead storage battery with a *hydrometer*, which measures liquid density. (*Hint:* Sulfuric acid solutions are more dense than liquid water.)

27. Why do batteries eventually stop operating? Explain in terms of limiting reactants.

28. Why are some batteries rechargeable and some not?

29. Identify one benefit and one burden or risk associated with each of the following:

    a. Playing high-school basketball

    b. Driving a car

    c. Jogging

    d. Receiving a dental X-ray

    e. Applying pesticides to garden plants

    f. Using food preservatives

Extending the Concepts

30. Identify the top four substances produced by the chemical industry in the United States. List two uses of each of those substances.

31. Identify and report on three commercial uses of chlorine gas. Are there alternatives to these uses that do not involve chlorine?

32. Find out how aluminum metal is welded. What techniques involved in welding aluminum differ from those used in welding other metals?

33. Investigate how the printing and dry-cleaning industries are changing their operations to become more environmentally sensitive and responsible.

34. Identify several technologies in your community. Use the risk-assessment table below to assign an appropriate letter to each activity. Discuss your decisions and compare your rankings with those made by others.

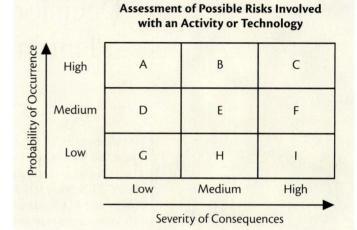

Assessment of Possible Risks Involved with an Activity or Technology

Probability of Occurrence	Low	Medium	High
High	A	B	C
Medium	D	E	F
Low	G	H	I

Severity of Consequences

A CHEMICAL PLANT FOR RIVERWOOD?

Decision Nears: Is a Chemical Plant in Riverwood's Future?

BY GARY FRANZEN
Riverwood News **Staff Reporter**

After months of study and discussion, the Riverwood Town Council is prepared to act on separate proposals from EKS Nitrogen Products Company and WYE Metals Corporation to locate a plant near Riverwood. At tonight's special meeting, the council will decide which, if either, plant to approve. The meeting starts at 7:30 p.m. in the town hall.

Mayor Cisko remarked, "I'm very pleased with the turnout we've had for the town meetings held on this issue. Tonight's council meeting is open to the public. I encourage all community members to attend and express their views about a chemical plant for Riverwood."

At the request of the Riverwood Town Council, both companies have prepared comprehensive summaries to inform the citizens of their plans for a Riverwood plant. These summaries were circulated at previous town meetings and have been widely distributed in the community through newspapers, pamphlets, and on the town's Web site. Additional summaries can be obtained from the mayor's office.

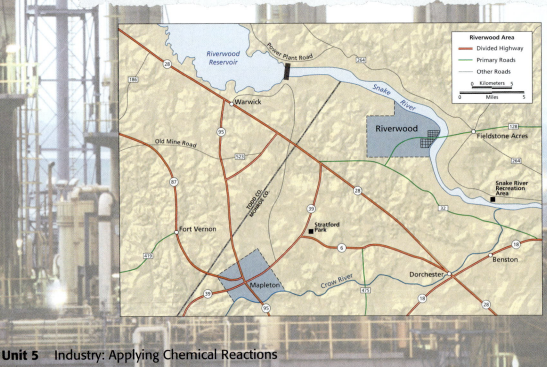

TOWN COUNCIL MEETING

Your teacher will assign you to one of the following groups to participate in the special town council meeting and will provide some background information concerning your group's viewpoints on both plant's proposals:

- ▶ Town Council Members
- ▶ EKS Nitrogen Products Company Officials
- ▶ WYE Metals Corporation Officials
- ▶ Riverwood Industrial Development Authority Members
- ▶ Riverwood Environmental League Members
- ▶ Riverwood Taxpayer Association Members

Meeting Rules and Penalties for Rule Violations

1. Council members decide and announce the presentation order at the start of the meeting.

2. Each group will be allowed a specified time for its presentation. Time cards will notify each speaker of the time remaining.

3. If a member of one group interrupts the presentation of another group, the offending group will be penalized 30 seconds for each interruption, to a maximum of two minutes. If the offending group has already made its presentation, it will forfeit its rebuttal time. (If interruptions persist, your teacher will regard this as a classroom disruption and deal with it accordingly.)

LOOKING BACK

Whether or not you decided to allow a chemical plant to locate near Riverwood, you have learned some valuable chemistry in the process of making that decision. You learned how key substances, such as ammonia, are produced and used, and how nitrogen cycles among air, soil, and living organisms.

You also learned how chemists use electrochemistry principles to harness chemical energy from spontaneous chemical reactions and also to provide energy enabling other reactions to occur. In addition, your acquired chemical knowledge and skills in analyzing burdens and benefits can help you to deal more effectively with future decision-making challenges.

UNIT 6

Atoms: Nuclear Interactions

WHAT discoveries led to a modern understanding of the composition of atoms?

WHY does human exposure to some types of radiation cause health problems?

HOW do the rates of radioactive decay influence decisions about using nuclear radiation?

WHAT risks and benefits accompany uses of nuclear energy?

A citizen's group wants to ban all nuclear radiation from Riverwood and its surrounding area. Turn the page to learn what you might contribute to the community's discussions of the group's proposals.

Citizens Against Nuclear Technology (CANT)
invite you to an informative seminar on banning the use of all
nuclear energy and materials in the Riverwood area.
See you Friday, 7:00 p.m., Town Council Hall

Do you know whether each statement below is true or false? If not, you could be in danger of serious nuclear exposure!

1. Home smoke detectors contain radioactive materials.
2. Radioactive materials and radiation are unnatural—they did not exist on Earth until created by scientists.
3. All radiation causes cancer.
4. Human senses can detect radioactivity.
5. Individuals vary widely concerning how they are affected by exposure to radiation.
6. Small amounts of matter change to immense quantities of energy released by nuclear weapons.
7. Physicians can distinguish cancer caused by radiation exposure from cancer resulting from other causes.
8. Medical X-rays are dangerous.
9. Nuclear power plants create serious hazards to public health and to the environment.
10. An improperly operated nuclear power plant can explode like a nuclear weapon.
11. Some nuclear wastes must be stored for centuries to prevent dangerous radioactivity from escaping.
12. New, dangerous elements are being invented every day.
13. Nuclear power plants produce material that could be converted into nuclear weapons.
14. All nuclear medical techniques are highly dangerous.

A local Riverwood organization, the Citizens Against Nuclear Technology (CANT), has organized to prevent all uses of nuclear power, the disposal of nuclear waste, food irradiation, and nuclear medicine in the Riverwood area. The flyer reproduced on the opposite page is a sample of the organization's effort to communicate with the residents of the Riverwood area.

Some Riverwood citizens are wary about the restrictions that CANT proposes because they fear that such restrictions would hinder access to a full range of medical diagnosis and treatment options. The grandmother of Ms. Lynn Paulson, a Riverwood High School chemistry teacher, is among these citizens. She asked Ms. Paulson whether any of her chemistry students could provide some background about nuclear science and technology to the senior-citizen community. Several senior citizens are planning to attend the announced CANT meeting, but they would like to acquire some background knowledge and be ready to question CANT representatives about their proposal. Ms. Paulson has agreed to help. In this unit, you and your classmates will assume the role of Riverwood High School chemistry students preparing a community presentation.

The chemistry you have learned thus far involves chemical changes due to sharing or transferring outer-shell electrons among atoms. You will encounter very different *changes* in this unit—changes associated with nuclei (rather than electrons) of atoms. This unit examines nuclear radiation, radioactivity, and nuclear energy, plus implications of their use and development.

As you progress through this unit, record the ideas and applications you decide to share in your presentation. In particular, focus on the assertions contained in the CANT flyer. You may also focus on issues that are of interest to your audience.

The Nature of Atoms

In this section, you will learn how both well-planned experiments and chance observations led to the current model of the atom's structure and to the discovery of radioactivity. Determining and describing the structure of atoms rank among the greatest scientific accomplishments of the past centuries. The stories of these discoveries will help you understand atomic and nuclear chemistry and how researchers have used methods of science in their investigations.

A.1 A GREAT DISCOVERY

The history of modern atomic investigation began with the study of **radiation,** which constitutes the emission of energetic particles or electromagnetic waves (such as visible light, X-rays, and microwaves). Scientists have long been interested in light—because of its essential role in life—and other types of radiation. By the end of the 19th century, scientists had already studied many types of radiation from a variety of sources.

In 1895, a series of observations significantly broadened scientific understanding of radiation. The German physicist W. K. Roentgen was studying **fluorescence,** a phenomenon where certain materials emit light when struck by radiant energy, such as ultraviolet rays (Figures 6.1 and 6.2). Roentgen found that certain materials fluoresced when exposed to beams of **cathode rays** emitted from the cathode when electricity passed through an evacuated glass tube (see

> You learned about electromagnetic radiation in Unit 4 (see page 337).

Figure 6.1 *French physicist Henri Becquerel's investigations of fluorescence led to the discovery of radioactivity (see pages 482–483).*

> Older computer monitors and TV screens were based on a version of the cathode-ray tube.

Figure 6.2 *Ultraviolet light shining on particular objects, such as certain fabrics, produces fluorescence—visible light emitted by material exposed to such electromagnetic radiation.*

Figure 6.3). A few years after Roentgen's work, cathode rays were identified as beams of electrons.

Roentgen was working with a cathode-ray tube covered by black cardboard. He observed an unexpected glow of light on a piece of paper across the room. The paper was coated with a fluorescent material, and Roentgen expected it to glow when exposed to radiation. However, visible radiation could not pass through the black cardboard covering the cathode-ray tube, and the fluorescent paper was not in the path of electrons in the tube. Roentgen hypothesized that some other radiation passing through the black cardboard had been emitted by the cathode-ray tube. He named the mysterious radiation *X-rays*, where *X* represented the unknown radiation. Scientists now know that **X-rays** are a form of high-energy electromagnetic radiation.

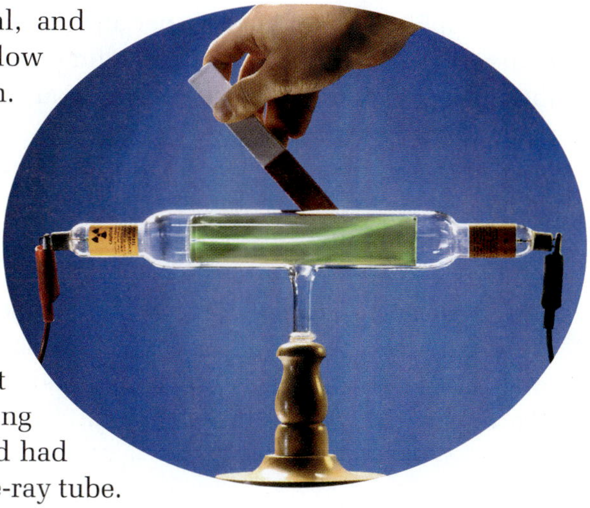

Further experiments revealed that these X-rays could penetrate many materials but could not easily pass through dense materials such as lead or bone. Scientists soon realized how useful these X-rays could be in medicine. In fact, one early X-ray image Roentgen obtained was of his wife's hand. Figures 6.4 and 6.5 show some modern X-ray images.

Figure 6.3 *A beam of electrons moves from the cathode (left) to the anode (right). The visible light emitted results from collisions of electrons with the fluorescent screen inside the tube. The deflection of the beam by a magnet indicates that the beam particles have a negative electrical charge. Collision of the electron beam with the glass or anode produces X-rays.*

In modern X-ray devices, the X-rays are generated when an electron beam strikes a metal target, which is often made of tungsten.

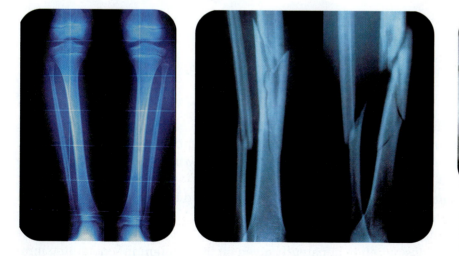

Figure 6.4 *These X-ray images reveal a pair of normal human legs (left) and a pair of broken legs (right). Development of this useful medical diagnostic tool emerged from the study of fluorescence.*

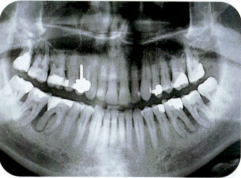

Figure 6.5 *An X-ray image of a human jaw. Such images help dentists detect cavities and other dental problems.*

Roentgen's discovery intrigued other scientists, including the French physicist Henri Becquerel. Because X-rays could produce fluorescence, Becquerel wondered if fluorescent minerals might give off X-rays as they fluoresce. In 1896, Becquerel placed in sunlight some crystals of a fluorescent mineral that contained uranium. He then wrapped an unexposed photographic plate in black paper and placed the mineral crystals on top of the wrapped plate. If the mineral did emit X-rays, they would penetrate the black paper and the exposed film would darken, even though the film was shielded from light.

Cloudy weather prevented Becquerel from completing his experiments. He stored the wrapped photographic plates in a drawer with the uranium-containing mineral. After several days, he decided to develop some of the stored plates, thinking that perhaps some fluorescence might have persisted, causing some fogging of the photographic plates. When Becquerel developed the plates, he was astounded. Instead of faint fogging, the plates had been strongly exposed. Figure 6.6 illustrates the chain of events in Becquerel's investigation.

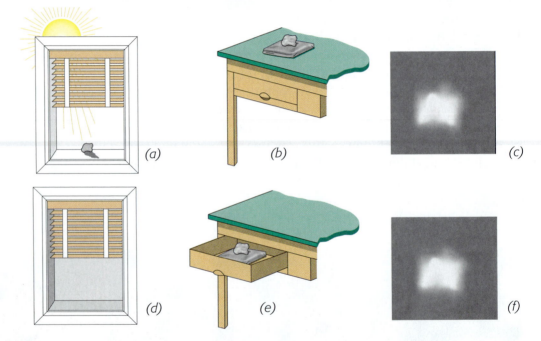

Figure 6.6 *Becquerel's investigation. Becquerel placed a fluorescent mineral in direct sunlight (a), then put it on an unexposed, wrapped photographic plate (b). Radiation exposed the plate (c). On a cloudy day (d), the wrapped plate was placed in a drawer with the mineral and kept from light (e), however, the mineral sample still exposed the photographic plate (f).*

Figure 6.7 *Many minerals exhibit fluorescence when exposed to ultraviolet light.*

Fluorescence stops as soon as the external radiation (in this case, from the Sun) is removed from the object. Thus, a fluorescent mineral in a desk drawer should not cause such an intense exposure. Scientists at that time could not offer a satisfactory explanation for Becquerel's observations. Becquerel suspected that the rays that exposed the photographic plates in the drawer were more energetic and had much greater penetrating ability than did X-rays. Thus he interrupted his study of X-rays to investigate the mysterious radiation apparently given off by the uranium-containing mineral. Although he was unable to explain it, Becquerel had discovered **radioactivity,** which is now known to involve the spontaneous emission of particles and energy from atomic nuclei. This phenomenon is distinctly different from X-ray production or fluorescence (Figure 6.7).

Becquerel suggested that Marie Curie (Figure 6.8), a graduate student working with him, attempt to isolate the radioactive component of pitchblende, a uranium ore, for her PhD research. Her preliminary work was successful; her physicist husband, Pierre Curie, changed his research focus to join her on the pitchblende project. Working together, Marie and Pierre Curie discovered that the level of radioactivity in pitchblende was four to five times greater than that expected from its known uranium content. The Curies suspected the presence of another radioactive element. After processing more than a thousand kilograms of pitchblende, they isolated tiny quantities (measured in milligrams) of two previously unknown radioactive elements. These elements later became known as polonium (Po) and radium (Ra).

Marie Curie first proposed the term "radioactivity."

Figure 6.8 *Marie Curie discovered two highly radioactive elements— radium (named for the radiation it emitted) and polonium (named for her native Poland). Her work earned her two Nobel Prizes.*

ChemQuandary 1

SCIENTIFIC DISCOVERIES
What do the following events have in common with Becquerel's discovery of radioactivity?

1. As Charles Goodyear experimented with natural rubber (a sticky material that melts when heated and cracks when cold), a mixture of the rubber and sulfur came in contact with a hot stove top. He noted that the rubber and sulfur mixture did not melt. *Vulcanization,* a process which makes rubber more durable, resulted from this observation.

2. Roy Plunkett, a research chemist, used gaseous tetrafluoroethene ($F_2C{=}CF_2$) from a storage cylinder, but the gas flow stopped long before the cylinder should have completely emptied. He cut open the cylinder and discovered a new, white solid that is now known as polytetrafluroroethene or *Teflon.*

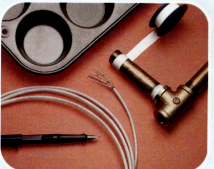

Products made with Teflon

3. James Schlatter, a research chemist trying to produce an antiulcer drug, accidentally got some of the substance on his fingers. When he later licked his fingers to pick up a piece of paper, his fingers tasted very sweet, and he correctly linked the sweetness to the antiulcer drug. Instead of finding an antiulcer drug, he had discovered *aspartame,* an artificial sweetener.

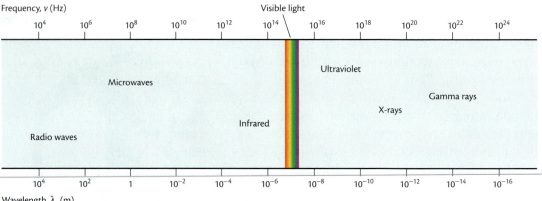

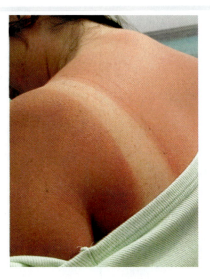

Figure 6.9 *The electromagnetic spectrum. Microwaves, infrared radiation (from heat lamps), and X-rays are all examples of electromagnetic radiation.*

A.2 NUCLEAR RADIATION

Many people respond with alarm and even panic when they hear the word *nuclear.* In addition, the general term *radiation,* which is sometimes used to refer particularly to *nuclear radiation* can also cause anxiety. In reality, radiation falls into two general types: *ionizing radiation* and *nonionizing radiation.*

Electromagnetic radiation in the visible and lower-energy regions of the spectrum (see Figure 6.9) is **nonionizing** (long-wavelength) **radiation.** Nonionizing radiation transfers its energy to matter, causing atoms or molecules to vibrate (infrared radiation), move their electrons to higher energy levels (visible radiation), or heat up (such as in microwave ovens). Although this radiation is generally considered safe, excessive exposure can be harmful. Sunburn, for example, results from an overexposure to nonionizing radiation from the Sun (Figure 6.10). In fact, intense microwave and infrared radiation can cause lethal burns.

Ionizing radiation, which includes high-energy electromagnetic radiation (short-wavelength ultraviolet radiation, X-rays, and gamma rays) and all nuclear radiation, carries more energy and potential for harm than nonionizing radiation. Energy from ionizing radiation can eject electrons from atoms and molecules, forming molecular fragments and ions. These fragments and ions can be highly reactive. If formed within a living system, they can disrupt normal cellular chemistry, causing serious cell damage.

Figure 6.10 *Sunburn often results from overexposure of skin to ultraviolet radiation.*

Nuclear radiation is a form of ionizing radiation that is caused by changes in the nuclei of atoms. In chemical reactions, the atomic number (number of protons) does not change; an atom of aluminum (13 protons) always remains an aluminum atom, and an iron atom (26 protons) always remains an iron atom. However, atoms with unstable nuclei—radioactive atoms—can spontaneously change their identities. A radioactive atom changes spontaneously through disintegration of its nucleus, which results in the emission of high-speed particles and energy. When this happens, the identity of the radioactive atom often changes; an atom of a different element forms. This process is **radioactive decay.** The emitted particles and energy make up **nuclear radiation.**

Ernest Rutherford showed in 1899 that nuclear radiation included at least two different types of emissions, which he named *alpha rays* and *beta rays*. Shortly afterward, scientists discovered a third kind of nuclear radiation: *gamma rays*.

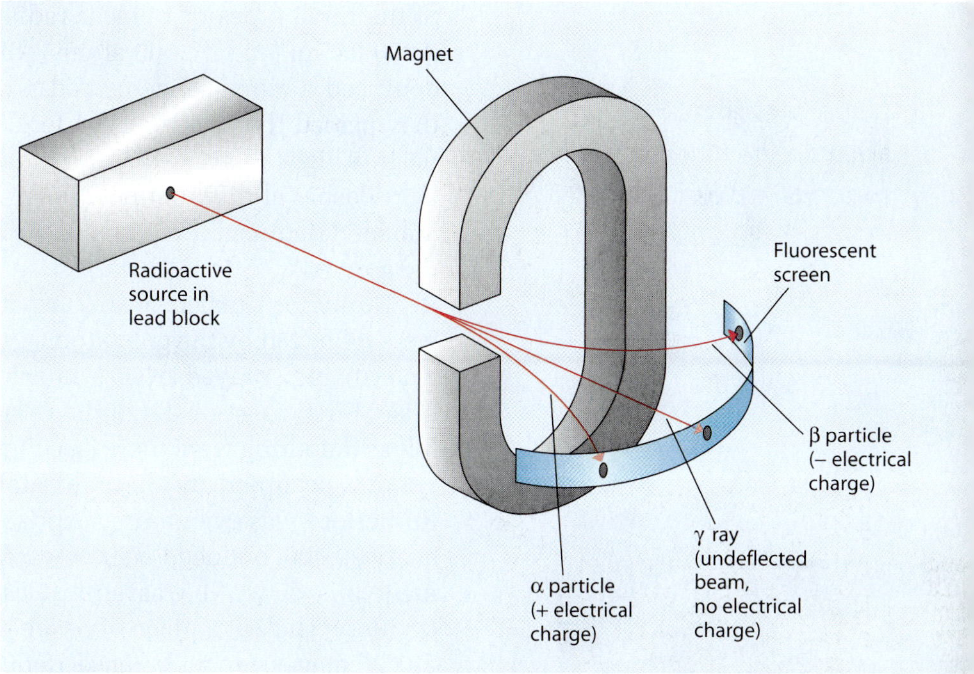

Researchers allowed the three types of nuclear radiation to pass through magnetic fields to investigate their electrical properties. Scientists already knew that when electrically charged particles move through a magnetic field, the magnetic force deflects them. See Figure 6.11. Scientists also knew that positively charged particles are deflected in one direction, negatively charged particles are deflected in the opposite direction, and electrically neutral particles and electromagnetic radiation are not deflected. Further experiments revealed that alpha emissions were composed of positively charged particles, and that beta emissions were composed of negatively charged particles. Thus these two types of emissions are referred to as **alpha particles** and **beta particles** (not *rays* as originally named). **Gamma rays,** not deflected by a magnetic field, have no electric charge; they are high-energy electromagnetic radiation similar to X-rays.

By thus describing the nature of radioactivity, scientists toppled an old theory (a common event in scientific progress—new knowledge replaces old knowledge). Once scientists knew about alpha particles, beta particles, and gamma rays, they became convinced that atoms, which were originally thought to be the smallest, most fundamental units of matter, must be composed of even *smaller* particles.

From the results of another investigation, the *gold-foil experiment,* Rutherford proposed a fundamental model of the atom that is still useful today. To do so, he developed an ingenious, indirect way to "look" at the structure of atoms.

Radioactivity

Magnet

Fluorescent screen

Radioactive source in lead block

β particle (– electrical charge)

γ ray (undeflected beam, no electrical charge)

α particle (+ electrical charge)

Figure 6.11 *Behavior of alpha (α) particles, beta (β) particles, and gamma (γ) rays passing through a magnetic field.*

A.3 THE GOLD-FOIL EXPERIMENT

The Gold Foil Experiment

Prior to Rutherford's research, scientists had tried to explain the arrangement of electrons and positively charged particles within atoms in several ways. In the most widely accepted model, the atom was viewed as a volume of positive electrical charge, with the negatively charged electrons embedded within, like peanuts in a candy bar. In the late 1800s this model was known as the "plum pudding" model because it resembled the distribution of raisins within that traditional English dessert.

About 1910, Rutherford decided to test the plum pudding model. Working in Rutherford's laboratory in Manchester, England, Hans Geiger and Ernest Marsden focused a beam of alpha particles—the most massive of the three types of nuclear radiation—at a thin sheet of gold foil only 0.000 04 cm (about 2000 atoms) thick (see Figure 6.12). Geiger and Marsden used a zinc sulfide-coated screen to detect the alpha particles after they passed through the gold foil (see Figure 6.13). The screen emitted a flash of light where each alpha particle struck it. By observing the tiny light flashes at different positions with respect to the gold foil, Geiger and Marsden deduced the paths of the alpha particles as they interacted with the gold foil.

> Alpha rays and beta rays are more commonly referred to as *particles* because they possess measurable mass.

Rutherford expected alpha particles to be scattered slightly as they were deflected by the gold atoms in the foil, producing a pattern similar to water being sprayed from a nozzle. However, he was in for quite a surprise. First, most of the alpha particles passed straight through the gold foil as if nothing were there (see Figure 6.14). This implied that most of the volume occupied by the gold atoms was essentially empty space. But Rutherford was even more surprised that a few alpha particles, about 1 in every 20 000, bounced *back* toward the source. He described his astonishment this way: "It was almost as incredible as if you fired a 15-inch [artillery] shell at a piece of tissue paper and it came back and hit you."

Whatever repelled these deflected alpha particles must have been extremely small because most of the alpha particles went straight through the foil. Yet the particles encountered by alpha particles must also have been very massive (as compared to the alpha particles) and most likely possessed a positive electrical charge because they tended to repel positively charged alpha particles.

Figure 6.12 *Gold foil, similar to gold leaf being applied here, was used in Rutherford's early investigations of atomic structure.*

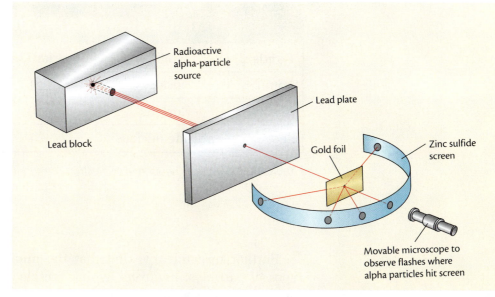

Figure 6.13 *The alpha-particle scattering experiment led Rutherford to conclude that the atom is largely empty space with an incredibly dense, positively charged nucleus at its center.*

From these results, Rutherford developed the modern model of the *nuclear atom.* He named the tiny, dense, positively charged region at the center of the atom the *nucleus.* He envisioned that electrons orbited the nucleus, somewhat like planets orbit the Sun. Figure 6.14 illustrates how the nuclear model explained the results of the gold-foil experiment, and also features of the nuclear model of the atom.

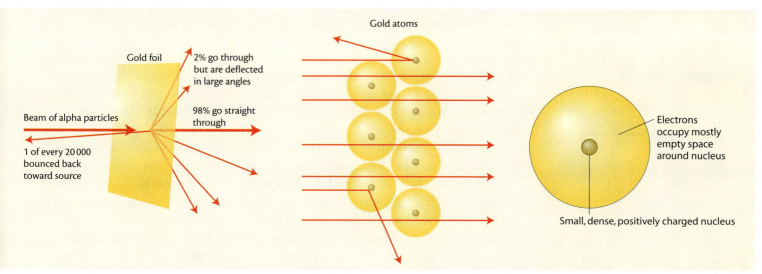

A.4 ARCHITECTURE OF ATOMS

Since Rutherford's time, scientific understanding of atomic structure has expanded and changed somewhat. Rutherford's model of a central, dense nucleus surrounded mostly by empty space is still valid. Subsequent research has shown that the idea of orbiting electrons is incorrect. The general regions in which electrons are most likely to be found can be described, but the particular movements or locations of electrons cannot be described.

Figure 6.14 *Result of the alpha-particle scattering experiment. Most alpha particles passed through the foil but a few were deflected, some at very large angles (left). Proposed model (middle) to account for the results. Nuclear model of the atom (right).*

THREE COMPONENTS OF ATOMS

Particle	Location	Electrical Charge	Molar Mass (g/mol)
Proton	Nucleus	+1	1
Neutron	Nucleus	0	1
Electron	Outside nucleus	−1	0.0005

| **Table 6.1** *The proton, neutron, and electron are subatomic particles.*

Neutrons are actually slightly more massive than are protons.

Protons: positive electrical charge
Neutrons: no electrical charge
Electrons: negative electrical charge

In most periodic tables, the atomic number is written above the symbol of each element.

Further research revealed that the nucleus is composed of two types of particles: *neutrons,* which are electrically neutral, and *protons,* which possess a positive charge. These particles, as well as *electrons,* are called **subatomic particles.** Protons and neutrons have about the same mass, 1.7×10^{-24} g. Although this mass is incredibly small, it is much greater than the electron's mass, which is 9.1×10^{-28} g. As shown in Table 6.1, one mole of electrons (6.02×10^{23} electrons) has a mass of only 0.0005 g. The same number of protons or neutrons would have a mass of about 1 g. In other words, a proton or a neutron is about 2000 times more massive than is an electron. Thus protons and neutrons account for nearly all the mass of every atom and also for nearly all of the total mass of every object that you encounter.

The diameter of a typical atom is about 10^{-10} m, and an average nuclear diameter is 10^{-14} m, which is only one ten-thousandth ($10^{-14}/10^{-10}$) of the entire atom's diameter. Looking at this another way, the nucleus occupies only about one trillionth (10^{-12}) of an atom's total volume.

Imagine that a billiard ball represents the diameter of an atom's nucleus. On that scale, electrons surrounding this billiard-ball nucleus would extend out in space more than one-half kilometer (about a third of a mile) away in all directions. This is consistent with the observation that most alpha particles, each the size of a helium nucleus, passed right through Rutherford's sheet of gold foil.

As you learned in Unit 2 (page 120), each atom of an element has the same number of protons in its nucleus, and each element has a unique number of protons. This number, called the *atomic number,* identifies the element. For example, each carbon atom nucleus contains six protons; therefore, the atomic number of carbon is 6.

All atoms of a given element do not necessarily have the same number of neutrons in their respective nuclei. Recall that atoms of the same element with different numbers of neutrons are called *isotopes* of that element. Naturally occurring carbon atoms, each containing six protons, may have six, seven, or even eight neutrons. The composition of these three carbon isotopes is summarized in Table 6.2. Figure 6.15 shows their nuclei.

Table 6.2 Composition of three
carbon isotopes.

THREE CARBON ISOTOPES

Name	Total Protons (Atomic Number)	Total Neutrons	Total Protons & Neutrons (Mass Number)	Total Electrons
Carbon-12	6	6	12	6
Carbon-13	6	7	13	6
Carbon-14	6	8	14	6

Isotopes are distinguished by their different mass numbers. The *mass number,* as you learned in Unit 2 (page 120), represents the total number of protons and neutrons in an atom. The three carbon isotopes in Table 6.2 have mass numbers 12, 13, and 14, respectively. To specify a particular isotope, the atomic number and the mass number are written in front of the element's symbol in a particular way. For example, an isotope of carbon (C) with an atomic number 6 and a mass number 12 is written as follows:

$$^{12}_{6}C$$

An atom of strontium with an atomic number 38 and mass number 90 would be symbolized as follows:

$$^{90}_{38}Sr$$

Another way to identify a particular isotope is to write the name or symbol of the element followed by a hyphen and its mass number. For example, an isotope for carbon may be called carbon-12, C-12, or ^{12}C. The symbols, names, and nuclear composition of some isotopes are summarized in Table 6.3.

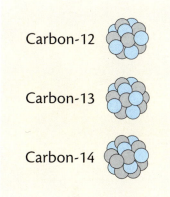

Carbon-12

Carbon-13

Carbon-14

Figure 6.15 *Nuclei of three carbon isotopes, containing, respectively, 6, 7, and 8 neutrons (gray) and 6 protons (blue).*

Table 6.3 *Names, symbols, and nuclear composition of some isotopes.*

SOME COMMON ISOTOPES

Symbol	Name	Total Protons (Atomic Number)	Total Neutrons	Total Protons + Neutrons (Mass Number)	Total Electrons
$^{7}_{3}Li$	Lithium-7	3	4	7	3
$^{16}_{8}O$	Oxygen-16	8	8	16	8
$^{20}_{10}Ne$	Neon-20	10	10	20	10
$^{67}_{31}Ga$	Gallium-67	31	36	67	31
$^{138}_{56}Ba$	Barium-138	56	82	138	56
$^{195}_{78}Pt$	Platinum-195	78	117	195	78
$^{201}_{81}Tl$	Thallium-201	81	120	201	81
$^{208}_{82}Pb$	Lead-208	82	126	208	82

A.5 INTERPRETING ISOTOPIC NOTATION

Sample Problem: *Suppose you know that one product of a certain nuclear reaction is an isotope containing 85 protons and 120 neutrons. The atom therefore has a mass number of 205 (85 p + 120 n). What is the symbol of this element?*

$$^{205}_{85}?$$

Consulting the periodic table, you find that the atomic number 85 represents the element astatine (At).

$$^{205}_{85}At$$

1. Prepare a summary chart similar to Table 6.3 (page 489) for the following six isotopes. (Consult the periodic table for any needed information.)

   a. $^{12}_{?}C$ d. $^{24}_{12}?$
   b. $^{14}_{7}?$ e. $^{202}_{?}Hg$
   c. $^{16}_{?}O$ f. $^{238}_{92}?$

2. Using Table 6.3 as a source of information, what general relationship do you note between the total number of protons and total number of neutrons for atoms of

   a. lighter elements with atomic numbers less than 20?
   b. heavier elements with atomic numbers greater than 50?

A.6 ISOTOPIC PENNIES

You learned earlier (Unit 2, page 113) that pre-1982 and post-1982 pennies have different compositions. As you would probably expect, these pennies also have different masses. In this activity, a mixture of pre- and post-1982 pennies will model or represent atoms of a naturally occurring mixture of two isotopes of the imaginary element *coinium.* Using the pennies, you will simulate one way scientists determine the relative amounts of different isotopes in a sample of an element.

You will receive a sealed container of 10 pennies that contains a mixture of pre-1982 and post-1982 pennies (Figure 6.16). Your container might hold any particular atomic mixture of these two isotopes. Your task is to determine the isotopic composition of *coinium—without opening the container.*

Figure 6.16 *How many pre- and post-1982 pennies are in your 10-coin sample of coinium?*

▶ Your teacher will give you some pre-1982 and post-1982 pennies, and a sealed container with a mixture of 10 pre- and post-1982 pennies, and will tell you the mass of the empty container. Record this information and the code number of your sealed container.

▶ Determine the isotopic composition of the element *coinium*. That is, find the percent pre-1982 and percent post-1982 pennies in your container. There is more than one way to find these answers.

Now answer these questions:

1. Describe the procedure that you followed to find the percent composition of *coinium*.

2. What property of *coinium* is different in its pre- and post-1982 forms?

3. Name at least one other familiar item that could serve as a model for isotopes.

A.7 ISOTOPES IN NATURE

Most elements in nature are mixtures of isotopes. Some isotopes of an element may be radioactive, while others are not. All isotopes of an element behave virtually the same way chemically, because they have the same electron distribution and differ only slightly in mass. If one considers chemical changes only, knowledge of isotopes is not particularly helpful. The atomic weight of an element, as shown on the periodic table, represents averages based on the relative natural abundances of the isotopes of that element (see Figure 6.17 for lithium).

Marie Curie originally thought that only heavy elements were radioactive. It is true that naturally occurring **radioisotopes** (that is, radioactive isotopes) are more common among the heavy elements. In fact, all naturally occurring isotopes of elements with atomic numbers greater than 83 (bismuth) are radioactive. However, many natural radioisotopes are also found among lighter elements. Modern technology has made it possible to create a radioisotope of any element. Table 6.4 lists some naturally occurring radioisotopes and their isotopic abundances.

What is the relationship between an element's molar mass and the percent abundance of the element's isotopes? To calculate the molar mass of an element, it is helpful to use the concept of a *weighted average,* as illustrated in Developing Skills A.8.

SOME NATURAL RADIOISOTOPES

Isotope	Abundance (%)
Hydrogen-3	0.000 13
Carbon-14	Trace
Potassium-40	0.0012
Rubidium-87	27.8
Indium-115	95.8
Lanthanum-138	0.089
Neodymium-144	23.9
Samarium-147	15.1
Lutetium-176	2.60
Rhenium-187	62.9
Platinum-190	0.012
Thorium-232	100

| Table 6.4

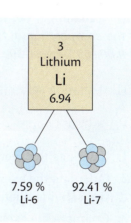

3
Lithium
Li
6.94

7.59 %   92.41 %
Li-6   Li-7

Figure 6.17 *Lithium's atomic weight is a weighted average of the molar masses of its two isotopes, Li-6 and Li-7. Most lithium atoms (92.41%) are Li-7, thus lithium's atomic weight is closer to 7 than to 6.*

No stable (nonradioactive) isotopes have yet been found for elements with atomic numbers of 83 or greater.

A.8 MOLAR MASS AND ISOTOPIC ABUNDANCE

You can calculate the average molar mass—and of the coins in the element *coinium* from the Modeling Matter activity on page 490—by using the mathematical relationship introduced in the sample problem below.

Sample Problem 1: *Consider an isotopic mixture of copper. Naturally occurring copper (Cu) consists of 69.1% copper-63 atoms and 30.9% copper-65 atoms. The molar masses of these two isotopes are:*

$$Copper\text{-}63 = 62.93 \text{ g/mol}$$

$$Copper\text{-}65 = 64.93 \text{ g/mol}$$

What is the average molar mass of naturally occurring copper?

The equation for finding average molar masses is as follows:

$$\text{Molar mass} = \begin{pmatrix} \text{Fractional} \\ \text{abundance} \\ \text{of isotope 1} \end{pmatrix} \times \begin{pmatrix} \text{Molar mass} \\ \text{of isotope 1} \end{pmatrix} + \begin{pmatrix} \text{Fractional} \\ \text{abundance} \\ \text{of isotope 2} \end{pmatrix} \times \begin{pmatrix} \text{Molar mass} \\ \text{of isotope 2} \end{pmatrix} + \dots \text{(for each isotope involved)}$$

Since there are two naturally occurring copper isotopes, the average molar mass of copper can be calculated as follows:

$$\text{Molar mass of Cu} = (0.691)(62.93 \text{ g/mol}) + (0.309)(64.93 \text{ g/mol}) = 63.5 \text{ g/mol}$$

> The decimal fractions must add up to 1. Why?

Sample Problem 2: *For the coinium example, suppose that you found that the composition of the mixture was 0.40 (40%) pre-1982 pennies and 0.60 (60%) post-1982 pennies. What is the average mass of a penny?*

The equation setup is shown here, using the 40/60 coin mixture:

$$\text{Average penny mass} = (0.40) \times (\text{Mass of pre-1982 penny}) + (0.60) \times (\text{Mass of post-1982 penny})$$

1. Calculate the average mass of a penny in your *coinium* mixture.

2. Calculate the average mass of a penny in your mixture another way: Divide the total mass of your entire penny sample by 10.

3. Compare the average masses that you calculated in Questions 1 and 2. These results should convince you that either calculation leads to the same result. If not, consult your teacher.

4. Naturally occurring boron (B) is a mixture of two isotopes. (See the table at the top of the next page.)

   a. Do you expect the molar mass of naturally occurring boron to be closer to 10 or to 11? Why?

   b. Calculate the molar mass of naturally occurring boron.

5. Naturally occurring uranium (U) is a mixture of three isotopes.

a. Do you expect the molar mass of naturally occurring uranium will be closest to 238, 235, or 234? Why?

b. Calculate the molar mass of naturally occurring uranium.

ISOTOPE MOLAR MASS AND ABUNDANCE		
Isotope	Molar Mass (g/mol)	% Natural Abundance
Boron-10	10.0	19.90%
Boron-11	11.0	80.10%
Uranium-234	234.0	0.0054%
Uranium-235	235.0	0.71%
Uranium-238	238.1	99.28%

Making Decisions

A.9 FACT OR FICTION?

Look again at the statements at the start of this unit (page 478). Answer the following questions about the CANT flyer. This will help you start preparing your presentation for Riverwood senior citizens (see Figure 6.18).

1. Identify the specific statements on the flyer that you can now conclusively identify as either true or false. For each statement,

a. list two pieces of evidence that helped you make your decision.

b. list two public concerns about the statement that you plan to address in your presentation.

2. a. Choose one statement from the flyer that you understand more completely now, but are still unable to confirm or deny.

b. What else do you need to know before you can make a decision about that statement?

3. How helpful do you think it will be to discuss the history of some discoveries that you studied in this section when you talk to the Riverwood senior citizens? Explain your answer.

4. a. Which new terms introduced in this section will you explain as part of your presentation?

b. Select two terms from your answer to Question 4a. Prepare an explanation of each term that your audience will understand. Describe examples or real-world applications that you may use in your explanation.

Figure 6.18 *What does your research lead you to conclude about the accuracy of statements made in the flyer? (See page 478.)*

The history of science is full of discoveries that build on earlier discoveries. The discovery of radioactivity was such an event. The investigations of Roentgen, Becquerel, the Curies, and Rutherford led to new knowledge and a better understanding of atomic structure. As you start to consider some current applications of nuclear radiation, think about the evidence and reasoning that supported these scientific advancements.

SECTION A SUMMARY
Reviewing the Concepts

Radioactive nuclei are unstable and undergo spontaneous changes in their structure.

1. Describe the sequence of events that led to Becquerel's discovery of radioactivity.

2. What three radioactive elements did the Curies find in pitchblende?

3. Define *radioactivity*.

4. List three types of nuclear radiation.

5. How did scientists determine that alpha particles have a positive electrical charge?

6. In what way is gamma radiation different from alpha and beta radiation?

7. How did the idea that the atom is the smallest particle of matter change after the discovery of radioactivity?

Radiation can be classified as either *ionizing* or *nonionizing*, depending on the type of energy it transmits.

8. Define and give an example of
   a. *ionizing radiation.*
   b. *nonionizing radiation.*

9. Why is ionizing radiation regarded as more dangerous than nonionizing radiation?

10. Classify each of the following as ionizing or nonionizing radiation:
    a. visible light   c. gamma rays
    b. X-rays   d. radio waves

11. How does ionizing radiation damage living cells?

Rutherford's gold-foil experiment results led to a new model of the atom.

12. Describe Rutherford's gold-foil experiment.

13. a. What happened to most of the alpha particles observed in the gold-foil experiment?
    b. What did Rutherford conclude from this observation?

14. a. What happened to about 1 in every 20 000 alpha particles in the gold-foil experiment?
    b. What did Rutherford conclude from this observation?

15. What was the general structure of the atom that Rutherford proposed?

16. Sketch models to show the concept of an atom before and after Rutherford's gold-foil experiment.

17. What characteristic of alpha particles made them desirable as the beam in Rutherford's gold-foil experiment?

18. Give the correct isotopic notation for copper-65.

19. Calculate the total neutrons present in an atom of sulfur-34.

20. Copy the following table and determine the value of each coded letter, *a* through *p*:

Symbol	Total Protons	Total Neutrons	Mass Number
$_1^2H$	*a*	1	*b*
$_c^{37}Cl$	*d*	*e*	*f*
$_h^gTc$	43	56	*i*
$_j^{137}Cs$	*k*	*l*	*m*
$_o^nAg$	47	60	*p*

21. How do the nuclei of carbon-12, carbon-13, and carbon-14 differ?

22. Consider the symbol $_{78}^{190}Pt$.

   a. What does the superscript 190 indicate?
   b. What does the subscript 78 indicate?
   c. How many neutrons are in a Pt-190 nucleus?

23. Neon (Ne) is composed of three isotopes with the following molar masses and relative abundances: Ne-20 (19.99 g/mol), 90.51%; Ne-21 (20.99 g/mol), 0.27%; and Ne-22 (21.99 g/mol), 9.22%.

   a. Based on these data, should neon's atomic weight be closest to 20, 21, or 22? Why?
   b. Calculate the actual atomic weight of neon. Show your calculations.

Connecting the Concepts

24. A local politician proposes to ban all radiation in your community. Explain why this proposal has little chance of success.

25. Why is it possible to receive a suntan from ultraviolet radiation but not from radio waves?

26. How does the gold-foil experiment demonstrate the importance of evidence regarding events that cannot be directly observed?

27. Describe how the development of atomic theory illustrates the way scientific discoveries build on previous scientific knowledge and experiments.

28. In what way is fluorescence different from radioactivity?

29. In what way is a cathode-ray tube similar to a

   a. modern X-ray device?
   b. television tube?

Extending the Concepts

30. Investigate the properties of gold and discuss why Rutherford probably selected that metal as an alpha-particle target.

31. Imagine that Rutherford proposed using beta particles rather than alpha particles in the gold-foil experiment. What result would you predict?

32. The neutron was discovered decades after the proton and electron were discovered. What made discovering the neutron such a challenge?

33. Compare earlier models of the atom with
   a. Rutherford's model.
   b. the currently accepted model of the atom.

Nuclear Radiation

Of nearly 2000 known isotopes, there are more radioactive (unstable) isotopes than there are nonradioactive (stable) isotopes. Actually, most isotopes you encounter are not radioactive. However, naturally occurring radioisotopes expose everyone to low levels of radiation. This radiation is from radioisotopes in building materials (such as brick and stone) in schools and homes (see Figure 6.19); in air, land, and sea; in foods you eat; and even within your own body. Because the human senses cannot detect nuclear radiation, various devices to detect and measure its intensity have been developed.

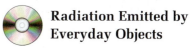

Radiation Emitted by Everyday Objects

Figure 6.19 *Bricks are a common source of background ionizing radiation.*

B.1 EXPOSURE TO IONIZING RADIATION

When radioisotopes spontaneously decay, they usually emit alpha, beta, and/or gamma radiation. The type and intensity of the radiation emitted helps to determine possible medical and industrial applications of particular radioisotopes. Each type of nuclear radiation also poses distinct hazards to human health.

A relatively constant level of radioactivity, called **background radiation,** is always present around and within you. Everyone receives background radiation at low levels from natural sources and from sources related to human activity. You will always experience at least some exposure to ionizing radiation.

Natural sources of background ionizing radiation include:

▶ High-energy particles from outer space that bombard Earth.

▶ Radioisotopes in rocks, soil, and groundwater: uranium (U-238 and U-235), thorium (Th-232), and the radioactive isotopes that form as they decay.

▶ Radioisotopes in the atmosphere: radon (Rn-222) and its decay products, including polonium (Po-210).

▶ Naturally occurring radioisotopes in foods and in the environment, such as potassium-40 and carbon-14.

Figure 6.20 *Air travel increases human exposure to ionizing radiation.*

Advances in science and technology have created additional sources of background radiation, such as:

▶ Residual radioactive fallout from aboveground nuclear-weapon testing.

▶ Increased exposure to radiation during high-altitude airplane flights (see Figure 6.20, page 497).

▶ Radioisotopes released into the environment from both fossil fuel and nuclear power generation as well as other nuclear technologies.

▶ Radioisotopes released through the disturbance and use of rocks in mining and in making cement, concrete, and sheet rock.

Because of its effect on living tissue, it is important to monitor the quantity of ionizing radiation to which people are exposed over time. The **gray** (Gy) is the SI unit that expresses the quantity of ionizing radiation absorbed by a particular sample, typically human tissue. An absorbed dose of one gray is defined as one joule of energy absorbed per kilogram of body tissue.

Not all forms of ionizing radiation, however, produce the same effect on living organisms. For example, alpha radiation will cause more harm internally to living organisms than will the same quantity of gamma radiation. The **sievert** (Sv) is the SI unit that expresses the ability of radiation—regardless of type or activity—to cause ionization in human tissue. Any exposure to radiation that produces the same detrimental effects as one gray of gamma rays represents one *sievert* of exposure. It is usually most convenient to express exposure in sieverts; this unit facilitates direct comparisons across different types of ionizing radiation.

While the SI units for radiation exposure are the gray and the sievert, two other units have traditionally been used in the U.S.: the *rad* and the *rem* (see Table 6.5). The **rad** (like the gray) expresses the absorbed dose of radiation, and the **rem** (like the sievert) indicates ionizing effects on living organisms. Both the rad and the rem are one-hundredth of their corresponding SI units.

> The total radioisotopes released to the environment from fossil-fuel power plants is greater than the total released from nuclear power plants.

> The units rad and rem are abbreviations for roentgen absorbed dose and roentgen equivalent man.

| Table 6.5

UNITS OF RADIATION DOSAGE			
	Unit Expresses:		
Unit	**Absorbed Dose**	**Ionizing Effects**	**Definition**
sievert (Sv)		X	1 Sv = dose equivalent of absorbed radiation that causes same biological effects as 1 Gy of gamma rays
rem		X	1 rem = 10^{-2} Sv
millirem (mrem)		X	1 mrem = 10^{-3} rem = 10^{-5} Sv
gray (Gy)	X		1 Gy = 1 J of energy absorbed per kilogram of body tissue
rad	X		1 rad = 10^{-2} Gy

Figure 6.21 *All living things contain some radioactive isotopes, including these that are located in particular parts of the human body.*

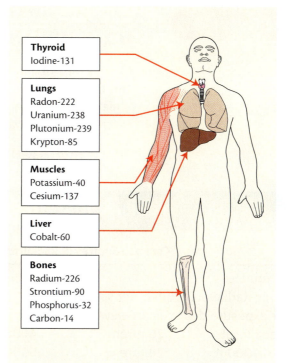

Thyroid	Iodine-131
Lungs	Radon-222, Uranium-238, Plutonium-239, Krypton-85
Muscles	Potassium-40, Cesium-137
Liver	Cobalt-60
Bones	Radium-226, Strontium-90, Phosphorus-32, Carbon-14

Even though one rem is only one-hundredth as large as one sievert, it is still much larger than typical radiation exposures. Normal human exposures are so small that doses are expressed in units of *millirem* (mrem), where 1 mrem = 0.001 rem. One millirem of any type of radiation produces essentially the same biological effects, whether the radiation is composed of alpha particles, beta particles, or gamma rays.

Some ionizing radiation comes from within your own body, as depicted in Figure 6.21. On average, people living in the United States receive about 360 mrem per person annually; about 300 mrem (83%) come from natural sources of radiation. Figure 6.22 shows the approximate proportion from each source.

It is in your best interest to avoid unnecessary ionizing radiation exposure. What ionizing radiation level is considered reasonably safe? The U.S. government's background radiation limit for the general public is 500 mrem (0.5 rem) per year for any individual. The U.S. average exposure value of 360 mrem falls below this. The established U.S. limit for an individual's annual maximum safe workplace exposure is 5000 mrem (5 rem).

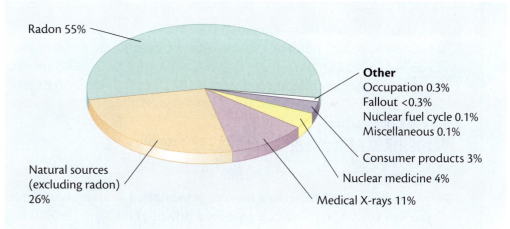

Radon 55%

Natural sources (excluding radon) 26%

Other
Occupation 0.3%
Fallout <0.3%
Nuclear fuel cycle 0.1%
Miscellaneous 0.1%

Consumer products 3%
Nuclear medicine 4%
Medical X-rays 11%

Figure 6.22 *Sources of ionizing radiation in the United States.*

ChemQuandary 2

RADIATION EXPOSURE STANDARDS

Why might radiation exposure standards for some individuals be different from those for the general public? Why do the standards differ for those who are exposed in their workplace?

Radiation badge

B.2 YOUR ANNUAL IONIZING-RADIATION DOSE

On a sheet of paper, list the numbers and letters of each category in the table on the following two pages. Then fill in the blanks on your sheet with appropriate quantities. Add all these quantities to estimate your annual ionizing-radiation dose.

YOUR ANNUAL IONIZING-RADIATION DOSE

Common Sources of Radiation	Your Annual Dose (mrem)
1. Where You Live	
a. Cosmic Radiation (from outer space)	
Your exposure depends on elevation. These are annual doses.	
sea level 26 mrem 4000–5000 ft 47 mrem	
0–1000 ft 28 mrem 5000–6000 ft 52 mrem	
1000–2000 ft 31 mrem 6000–7000 ft 66 mrem	
2000–3000 ft 35 mrem 7000–8000 ft 79 mrem	
3000–4000 ft 41 mrem 8000–9000 ft 96 mrem	_____ mrem
b. Terrestrial Radiation (from the ground)	
If you live in a state bordering Gulf or Atlantic coasts 16 mrem	
If you live in AZ, CO, NM, or UT 63 mrem	
If you live anywhere else in continental U.S. 30 mrem	_____ mrem
c. House Construction	
If you live in a stone, adobe, brick, or concrete building 7 mrem	_____ mrem
d. Power Plants	
If you live within 50 miles of a nuclear power plant 0.009 mrem	
If you live within 50 miles of a coal-fired power plant 0.03 mrem	_____ mrem
2. Food, Water, Air	
Internal Radiation (based on average values)	
a. from food (C-14, K-40) and water (radon dissolved in water) 40 mrem	_____ mrem
b. from air (radon) 200 mrem	_____ mrem

YOUR ANNUAL IONIZING-RADIATION DOSE (Continued)

Common Sources of Radiation		Your Annual Dose (mrem)
3. How You Live		
Weapons test fallout*	1 mrem	_____ mrem
Travel by jet aircraft (per hour of flight)	0.5 mrem	_____ mrem
If you have porcelain crowns or false teeth	0.07 mrem	_____ mrem
If you wear a luminous wristwatch	0.06 mrem	_____ mrem
If you go through airport security (each time)	0.002 mrem	_____ mrem
If you watch TV*	1 mrem	_____ mrem
If you use a video display (computer screen)*	1 mrem	_____ mrem
If you live in a dwelling with a smoke detector	0.008 mrem	_____ mrem
If you use a gas camping lantern with an old mantle	0.2 mrem	_____ mrem
If you wear a plutonium-powered pacemaker	100 mrem	_____ mrem
4. Medical Uses (radiation dose per procedure)		
X-rays: Extremity (arm, hand, foot, or leg)	1 mrem	_____ mrem
Dental	1 mrem	_____ mrem
Chest	6 mrem	_____ mrem
Pelvis/hip	65 mrem	_____ mrem
Skull/neck	20 mrem	_____ mrem
Barium enema	405 mrem	_____ mrem
Upper GI	245 mrem	_____ mrem
CT scan (head and body)	110 mrem	_____ mrem
Nuclear medicine (e.g., thyroid scan)	14 mrem	_____ mrem
Your Estimated Annual Radiation Dose		_____ *mrem*

*The value is less than 1 mrem but adding that value would be reasonable.
Adapted from American Nuclear Society (2000). Copyright 2000 American Nuclear Society.

1. Compare your annual ionizing-radiation dose to the
   a. U.S. limit of 500 mrem per person annually.
   b. average background radiation value (360 mrem).

2. a. Why is it useful to keep track of how many X-rays you receive annually?
   b. What geographic factors might decrease your annual ionizing-radiation dose?

3. a. What lifestyle changes could reduce a person's exposure to ionizing radiation?
   b. Would you decide to make those changes? Explain.

B.3 IONIZING RADIATION: HOW MUCH IS SAFE?

The two main factors that determine tissue damage due to ionizing radiation are *radiation density* (the number of ionizations within a given volume) and *dose* (the quantity of radiation received).

Gamma rays and X-rays are ionizing forms of electromagnetic radiation that penetrate deeply into human tissue. Ionizing radiation causes tissue damage by breaking bonds in molecules. At low levels of ionizing radiation, only a few molecules are damaged. In most low-dose cases, a body's systems can repair the damage. As the dose received increases, the total number of molecules affected by the radiation also increases. Generally, the damage to proteins and nucleic acids is of greatest concern because of their role in body structures and functions. Proteins form much of the body's soft tissue structure and compose enzymes, molecules that control the rates of cellular chemical reactions. If a large number of protein molecules are destroyed within a small region, too few functioning molecules may remain to enable the body to heal itself in a reasonable time (see Figure 6.23).

See Unit 7 for more information on proteins and enzymes.

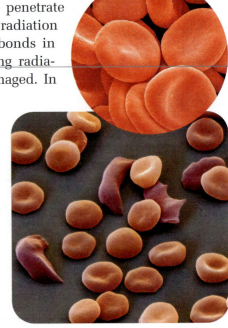

Figure 6.23 *Mutations are caused by damage to DNA which can result from exposure to radiation. Here, blood cells are deformed (normal cells are shown above) as a result of sickle-cell anemia, an inherited disease.*

Table 6.6 *These four factors determine the actual effects of particular ionizing-radiation exposure.*

BIOLOGICAL DAMAGE FROM IONIZING RADIATION	
Factor	**Effect**
Dose	Most scientists assume that an increase in radiation dose produces a proportional increase in risk.
Exposure time	The more a given dose is spread out over time, the less harm it does.
Area exposed	The larger the body area exposed to a given radiation dose, the greater the damage.
Tissue type	Rapidly dividing cells, such as blood cells and germ cells, are more susceptible to radiation damage than are slowly dividing or nondividing cells, such as nerve cells. Fetuses and children are more susceptible to radiation damage than are adults.

Nucleic acids in DNA can be damaged by ionizing radiation. Minor damage causes **mutations,** which are changes in the structure of DNA that may result in the production of altered proteins. Mutations often kill the cell in which they occur. If the cell is a sperm or ovum, a mutation may lead to birth defects in offspring. Some mutations can lead to *cancer,* a disease in which cell growth and metabolism are out of control. When

DNA molecules control cell reproduction and the synthesis of proteins.

the DNA in many body cells is severely damaged, cells cannot synthesize new proteins to replace the damaged ones, and the organism or person dies.

Table 6.6 lists factors determining the extent of biological damage from ionizing radiation. Table 6.7 summarizes the biological effects of large doses of ionizing radiation. Because the values in Table 6.7 are so large, they are reported in rems, not millirems.

Large ionizing-radiation doses can have drastic effects on humans. Conclusive evidence that such doses produce increased cancer rates has been gathered from uranium

IONIZING-RADIATION EFFECTS	
Dose (rem)	**Effect**
0–25	No immediate observable effects.
25–50	Small decreases in white blood cell count, causing lowered resistance to infections.
50–100	Marked decrease in white blood cell count; development of lesions.
100–200	Radiation sickness: nausea, vomiting, and hair loss; blood cells die.
200–300	Hemorrhaging, ulcers, death.
300–500	Acute radiation sickness; 50% of those exposed die within a few weeks.
>700	100% die.

miners and nuclear-accident victims. Some of the first cases of exposure to large doses of ionizing radiation occurred among workers who used radium compounds to paint numbers on watch dials that would glow in the dark (see Figure 6.24). The workers used their tongues to smooth the tips of their paintbrushes and unknowingly ingested small amounts of radioactive compounds. Later, these workers began to lose hair and became quite weak. Sometimes this exposure even led to death.

Table 6.7 *You can see how the consequences of radiation exposure change as dose increases.*

Becquerel discovered a red spot on his chest after carrying a radium sample in his breast pocket. The dangers of ionizing radiation were unknown at that time.

Figure 6.24 *These women painted radioactive radium onto watch dials so the watches could be read in the dark. Their exposure to ionizing radiation from radium often resulted in illness. Modern glow-in-the-dark watches do not contain radium.*

Leukemia, a rapidly developing cancer of white blood cells, is commonly associated with exposure to high doses of ionizing radiation. Exposure also promotes other forms of cancer, anemia, heart problems, and cataracts (opaque spots on an eye lens).

Considerable controversy continues regarding whether very low doses of ionizing radiation, such as those from typical background sources, can cause cancer. Most of the data on cancer incidence have been based on human exposure to high doses of radiation; these data are extrapolated to much lower doses. Few studies have directly linked low radiation doses with cancer development. Most scientists agree that typical background levels of ionizing radiation are safe for most people. Some authorities argue that any increase above normal background levels increases the probability of developing cancer.

Investigating Matter

B.4 ALPHA, BETA, AND GAMMA RADIATION

Introduction

Laboratory Video: Alpha, Beta, and Gamma Radiation

An early device used to detect radioactivity still used today is the Geiger-Mueller counter (Figure 6.25). As illustrated in Figure 6.26, ionizing radiation that strikes the detector produces electrical signals. In this investigation, a counter (or similar device) will detect the penetrating abilities of different types of ionizing radiation through cardboard, glass, and lead.

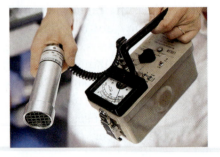

Figure 6.25 *Using a Geiger-Mueller counter to measure radiation dose.*

Ionizing radiation enters the counter's detecting tube (or probe), producing an electrical signal. Most radiation counters register these signals as both audible clicks and meter readings. The intensity of the radiation is indicated by the number of electronic signals or counts per minute (cpm).

You must first take a reading of background radiation to establish a *baseline* before taking readings from a known radioactive source. You then subtract the background count from each radioactive-source reading to find the actual radiation level emitted by that source.

With proper handling, radioactive materials in this investigation pose no danger to you. Nuclear materials are strictly regulated by state and federal laws. The radioactive sources you will use emit only very small quantities of radiation; using them requires no special license.

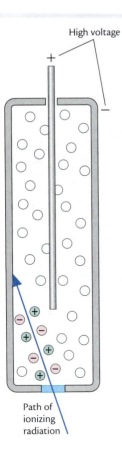

High voltage

Path of ionizing radiation

Figure 6.26 *As ionizing radiation passes into a Geiger-Mueller detector tube, ions form in the gas inside. Positive ions are attracted to the tube's negatively charged outer wall, while negative ions are attracted to the positively charged center. These electrically charged particle movements constitute a pulse of electrical current. Each pulse is detected and counted.*

Nevertheless, you should handle all radioactive samples with great care, including wearing protective gloves. Do not allow the radiation counter to come in direct contact with any radioactive material. Check your hands with a radiation monitor before you leave the laboratory.

Procedure

Part 1: Penetrating Ability

1. Read the following procedure and construct a data table suitable for recording all relevant data.

2. Set up the apparatus shown in Figure 6.27. There should be space between the source and the detector for several sheets of glass or metal.

3. Turn on the counter; allow it to warm up for at least 3 min. Determine the intensity of background radiation by counting the clicks for one minute without any radioactive sources present. Record this background-radiation value in counts per minute (cpm) in your data table.

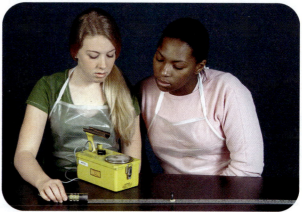

Figure 6.27 *You will need to determine the level of radiation detected in the absence of shielding.*

4. Put on protective gloves. Using forceps, place a gamma-ray source on the ruler at a point where it produces nearly a full-scale reading. Observe the meter for 30 s and estimate the number of counts per minute detected over this period. Record this approximate value. Then subtract the background reading from that value and record the corrected results.

5. Without moving the radiation source, place a piece of thin cardboard (or an index card) between the detector and the source, as shown in Figure 6.28.

6. Observe the meter for 30 s. Record the typical reading. Then correct the reading for background radiation and record the corrected result in your data table.

7. Repeat Steps 5 and 6, replacing the cardboard with a glass or plastic sheet.

8. Repeat Steps 5 and 6, replacing the cardboard with a lead sheet.

Figure 6.28 *How will the type of shielding affect the level of detected radiation?*

9. Repeat Steps 4 through 8, using a beta-particle source.

10. Repeat Steps 4 through 8, using an alpha-particle source.

Part 2: Effect of Distance on Intensity

11. Read the following procedure and construct a data table suitable for recording all relevant data.

12. Using forceps, place a radioactive source designated by your teacher at a ruler distance (usually about 5 cm) that produces nearly a full-scale reading.

13. Observe the reading over 30 s. Determine the average number of counts per minute (cpm) and record that value. Then correct this reading by subtracting the background value. Record your corrected values.

14. Move the source so its distance from the detector is doubled.

15. Observe the reading over 30 s. Record the average value and the corrected value.

16. Move the source two more times, so the original distance is first tripled, then quadrupled, recording the initial and corrected readings after each move. (For example, if your first reading was at 2 cm, you would also take readings at 4, 6, and 8 cm.)

17. Graph your data, plotting corrected cpm values on the y-axis and distances from the source to the detector (in cm) on the x-axis.

Figure 6.29 *Relationship between distance from the source (left) and radiation intensity. Intensity is expressed as counts per minute within a given area. Note how the same quantity of radiation spreads over a larger area (B vs. A) as distance from the source increases.*

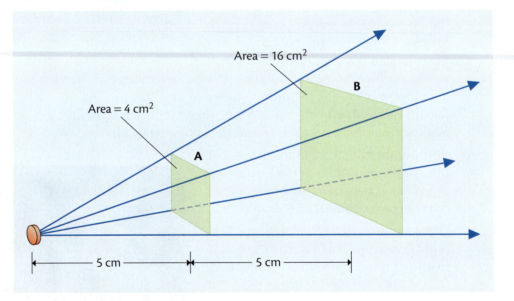

Area = 16 cm²

Area = 4 cm²

B

A

5 cm   5 cm

Part 3: Shielding Effects

18. Read the following procedure and construct a data table suitable for recording all relevant data.

19. Using forceps, place a radioactive source designated by your teacher on the ruler to give nearly a full-scale reading.

20. Take a typical reading over 30 s. Determine the number of counts per minute (cpm), correcting for background radiation. Record your actual and corrected values.

21. Place a glass sheet between the source and the detector. Do not change the distance between the detector and the source. Take an average reading over 30 s. Then determine and record the corrected number of counts per minute.

22. Place a second glass sheet between the source and the detector. Take an average reading over 30 s. Determine and record the corrected number of counts per minute.

23. Repeat Steps 21 and 22, using lead sheets rather than glass sheets.

24. Wash your hands thoroughly and check your hands with a radiation monitor before leaving the laboratory.

Questions for Part 1

1. Which shielding materials were effective in reducing the intensity of each type of radiation? Support your answers with data or observations.

2. How do the three types of radiation that you tested compare in terms of their penetrating ability?

3. Of the shielding materials tested, which do you conclude is the
   a. most effective in blocking radiation? Cite supporting evidence.
   b. least effective in blocking radiation? Cite supporting evidence.

4. Based on your observations, what properties of a material appear to affect its radiation-shielding ability?

Questions for Part 2

Analyze the graph that you prepared in Step 17:

5. By what factor did the intensity of radiation (measured in counts per minute) change when the initial distance was doubled?

6. Did this same factor apply when the distance was doubled again?

7. State the mathematical relationship between distance and intensity, using the factor identified in Question 5 and information in Figure 6.29.

> A *factor* is a number by which a value is multiplied to give a new value.

Questions for Part 3

8. How effective was doubling the shield thickness in blocking the radiation intensity by
   a. glass?   b. lead?

9. A patient receiving an X-ray is covered by a protective shield (Figure 6.30).
   a. What material would be a good choice for this apron?
   b. Why?

10. a. Which type of ionizing radiation from a source outside the body is likely to be most dangerous to living organisms?
    b. Why? Give evidence to support your answer.

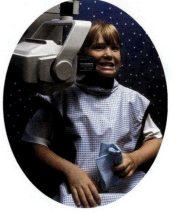

Figure 6.30 *What type of body shielding is used as this dental X-ray is obtained?*

B.5 NATURAL RADIOACTIVE DECAY

Shielding of Radiation

You know that alpha particles are positively charged. Why is this so?

An alpha particle is composed of two protons and two neutrons; it is identical in composition to the nucleus of a helium-4 atom. Since alpha particles have no electrons, they have a double-positive electrical charge and are often symbolized as a doubly charged helium-4 atom, $_2^4\text{He}^{2+}$. Alpha radiation (also called *alpha emission*) is released by many radioisotopes of elements with atomic numbers greater than 83.

Compared to a beta particle, an alpha particle has 5 to 50 times more energy and is over 7300 times more massive. However, the larger and slower alpha particles are easy to stop when they are outside the body. Once inside the body, however, the electrical charge and energy of alpha particles can cause great damage to tissues. This damage occurs over very short distances (about 0.025 mm).

> Because of their relatively large mass, slower velocities, and large (2+) electrical charge, alpha particles lose most of their energy within a short distance.

Because alpha particles are very powerful tissue-damaging agents once inside the body, alpha emitters in air, food, or water are particularly dangerous to human life. Fortunately, outside the body, alpha particles are easy to block. As you noted in Investigating Matter B.4 (page 504), alpha particles are even stopped within a few centimeters by air.

Figure 6.31 illustrates a radium-226 nucleus emitting an alpha particle. During this process, the radium nucleus loses two protons, so its atomic number drops from 88 to 86, and it becomes an isotope of a different element, radon. In addition to losing two protons, radium-226 loses two neutrons, so its mass number drops by 4. The net result is the formation of radon-222. The decay process can be represented by the following nuclear equation:

> Ionic charges, such as the 2+ of an alpha particle, are usually not included in nuclear equations.

$$\underset{\text{Radium-226}}{_{88}^{226}\text{Ra}} \longrightarrow \underset{\text{Alpha particle}}{_2^4\text{He}} + \underset{\text{Radon-222}}{_{86}^{222}\text{Rn}}$$

Atoms of two elements—helium and radon—have been formed from one atom of radium. (A radium compound is shown in Figure 6.32) Note that atoms are *not* necessarily conserved in nuclear reactions, as they are in chemical reactions. Atoms of different elements can appear on both sides of a nuclear equation. Total mass numbers and atomic numbers, however, *are* conserved. In the above equation for radium-226 decay, the reactant mass number equals the sum of mass numbers of products (226 = 4 + 222). Also, the atomic number of radium-226 equals the sum of atomic numbers of products (88 = 2 + 86). Both relations hold true for all nuclear reactions.

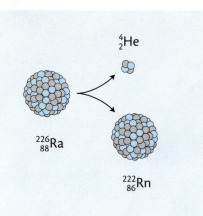

Figure 6.31 *Alpha-particle emission by radium-226. Consequently, its mass number decreases by four (2 p + 2 n) and its atomic number decreases by two (2 p).*

Figure 6.32 *Radium's name comes from the fact that small amounts of radium compounds, such as that on this antique watch face, emit enough radiation to glow in the dark. (See Figure 6.24, page 503.)*

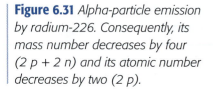

Beta particles are fast-moving electrons emitted from a nucleus. Because they are so much lighter than alpha particles and travel at very high velocities, beta particles have much greater penetrating ability than do alpha particles. On the other hand, beta particles are not as damaging to living tissue.

During **beta decay,** a neutron in a nucleus decays to a proton and an electron. The proton remains in the nucleus, but the electron is ejected. The electron emitted from the nucleus is a beta particle. The following equation describes the process:

$$\underset{\text{Neutron}}{{}_{0}^{1}n} \longrightarrow \underset{\text{Proton}}{{}_{1}^{1}p} + \underset{\substack{\text{Beta particle} \\ \text{(electron)}}}{{}_{-1}^{0}e}$$

Due to its negligible mass and negative electrical charge, a beta particle is assigned a mass number of 0 and an "atomic number" (nuclear charge) of -1. The net nuclear change due to beta emission is that a neutron is converted to a proton.

The following equation shows a nucleus of lead-210 undergoing beta decay: The nucleus loses one neutron and gains one proton. The mass number remains unchanged at 210, but the atomic number increases from 82 to 83. The new nucleus thus formed is that of bismuth-210:

$$\underset{\text{Lead-210}}{{}_{82}^{210}\text{Pb}} \longrightarrow \underset{\text{Bismuth-210}}{{}_{83}^{210}\text{Bi}} + \underset{\substack{\text{Beta particle} \\ \text{(electron)}}}{{}_{-1}^{0}e}$$

Once again, the sum of all mass numbers remains the same in this nuclear equation (210 on each side), and the sum of atomic numbers (the nuclear charge) remains unchanged (82 on each side).

Alpha and beta decay often leave nuclei in an energetically excited state. This type of excited state is described as *metastable* and is designated by the symbol *m.* For example, the symbol ${}^{99m}\text{Tc}$ represents a technetium isotope in a metastable or excited state. Energy from isotopes in such excited states is released as gamma rays—high-energy electromagnetic radiation that has as much or more energy than X-rays. Because gamma rays have neither mass nor charge, their release does not change the mass balance or charge balance in a nuclear equation. Table 6.8 summarizes the general changes involved in natural radioactive decay.

> Note that *n, p,* and *e* are symbols, respectively, for a neutron, a proton, and a beta particle. An emitted electron (beta particle) can also be symbolized by the Greek letter beta, ${}_{-1}^{0}\beta$

> Tc-99m is the most widely used radioisotope in medical diagnostic studies. Its half-life is 6 h.

| Table 6.8

CHANGES RESULTING FROM NUCLEAR DECAY

Decay Type	Symbol	Change in Atomic Number	Change in Total Neutrons	Change in Mass Number
Alpha	${}_{2}^{4}\text{He}^{2+}$, α	Decreases by 2	Decreases by 2	Decreases by 4
Beta	${}_{-1}^{0}e$, β	Increases by 1	Decreases by 1	No change
Gamma	γ	No change	No change	No change

Figure 6.33 *Relative penetrating ability of alpha (α) particles, beta (β) particles, and gamma (γ) rays. Gamma rays are the most penetrating; alpha particles the least.*

Gamma rays are the most penetrating of these three forms of ionizing radiation. Under most circumstances, however, they cause the least damage to tissue over comparable distances. Why is this so? Tissue damage is related to the extent of ionization created by the radiation, and is expressed as the number of ionizations within each unit of tissue. Alpha particles, which create the highest levels of ionization per tissue unit, cause considerable damage over a short range. However, protecting against them is easy. Beta and gamma radiation do less damage over a longer range, but it is more difficult to protect against them. The relative penetrating abilities of the three types of ionizing radiation are shown in Figure 6.33.

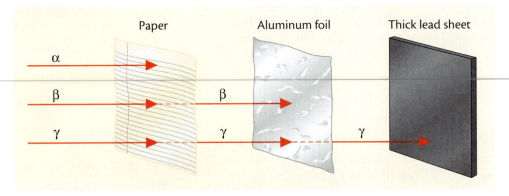

New isotopes produced by radioactive decay may also be radioactive and therefore undergo further nuclear decay. Uranium (U) (see Figure 6.34) and thorium (Th) are *parents* (reactants) in three natural **decay series** that start with U-238, U-235, and Th-232, respectively. Each decay series ends with formation of a stable isotope of lead (Pb). The decay series starting with uranium-238 contains 14 steps, as shown in Figure 6.35.

Figure 6.34 *Uranium ore contains U-238 and U-235 isotopes, each of which participates in a decay series, producing several generations of radioactive decay products.*

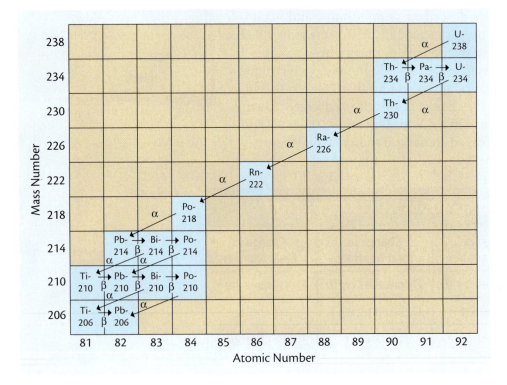

Figure 6.35 *The uranium-238 decay series. Diagonal lines show alpha decay; horizontal lines show beta decay. Here is how to interpret this chart: Locate radon-222 (Rn-222). The arrow pointing left shows that this isotope decays to polonium-218 by alpha (α) emission. This nuclear equation applies:*

$$^{222}_{86}Rn \longrightarrow {}^{218}_{84}Po + {}^{4}_{2}He.$$

B.6 NUCLEAR BALANCING ACT

The key to balancing nuclear equations is recognizing that both atomic numbers and mass numbers are conserved. Use the information in Table 6.8 (page 509) to complete the following exercise.

Sample Problem: *Cobalt-60 is one source of ionizing radiation for medical therapy. Complete this equation for the beta decay of cobalt-60:*

$$^{60}_{27}Co \longrightarrow ^{\ 0}_{-1}e + ?$$

Beta emission causes no change in mass number. Therefore, the new isotope will also have a mass number of 60. Thus the unknown product can be written as $^{60}?$. Because the atomic number increases by one during beta emission, the new isotope will have atomic number 28, one more than cobalt's atomic number. The periodic table indicates that atomic number 28 is nickel (Ni). The final equation is:

$$^{60}_{27}Co \longrightarrow ^{\ 0}_{-1}e + ^{60}_{28}Ni$$

Now answer these questions:

1. Write the appropriate symbol for the type of radiation given off in each reaction.

   a. The following decay process illustrates how archaeologists date the remains of ancient biological materials. Living organisms take in carbon-14 and maintain a relatively constant amount of it over their lifetimes. After death, no more carbon-14 is taken in, so the amount gradually decreases due to decay:

   $$^{14}_{6}C \longrightarrow ^{14}_{7}N + ?$$

   b. The following decay process takes place in some types of household smoke detectors:

   $$^{241}_{95}Am \longrightarrow ^{237}_{93}Np + ?$$

2. The two decay series beginning with Th-232 and U-238 are believed responsible for much of the thermal energy generated inside Earth. (Thermal contributions from the U-235 series are negligible; the natural abundance of U-235 is low.)
   Complete the following equations, which represent the first five steps in the Th-232 decay series, and identify the missing items *A, B, C, D,* and *E.* Each code letter represents a particular isotope or a type of radioactive emission. For example, in the first equation, Th-232 decays by emitting alpha radiation to form *A.* What is *A*?

   a. $^{232}_{90}Th \longrightarrow ^{4}_{2}\alpha + A$ d. $D \longrightarrow ^{4}_{2}He + ^{224}_{88}Ra$
   b. $^{228}_{88}Ra \longrightarrow ^{\ 0}_{-1}e + B$ e. $^{224}_{88}Ra \longrightarrow E + ^{220}_{86}Rn$
   c. $^{228}_{89}Ac \longrightarrow C + ^{228}_{90}Th$

> Thorium oxide (ThO_2) was used in gas-light mantles and more recently in outdoor camping lanterns, since it glows brilliant white as the fuel burns.

> An alpha particle can be symbolized as either $^{4}_{2}He$ or $^{4}_{2}\alpha$.

B.7 RADON

The gaseous element radon, which is the most massive of the noble gases, has always been a component of earth's atmosphere. It is a radioactive decay product of uranium. In the 1980s, unusually high concentrations of radioactive radon gas found in a relatively small number of U.S. homes became a public health concern.

Radon is produced as the radioisotope uranium-238 decays in the soil and in building materials. Once it is emitted, radon is transported throughout the environment in many ways. Some radon produced in the soil dissolves in groundwater. In other cases, radon gas seeps into houses through cracks in foundations and basement floors (see Figure 6.36).

In older houses, outdoor air entering through gaps in doors and windows can dilute the radon gas. However, to conserve energy, newer houses are built more airtight than are older houses. In a tightly sealed house, radon gas cannot mix freely with outdoor air or escape from the house. Consequently, radon gas concentrations may reach higher levels in newer homes. Remedies for high radon levels in houses include increasing ventilation and sealing cracks in floors (see Figure 6.37). Inexpensive radon test kits are available for home use.

You can locate this radioactive decay product as Rn-222 in Figure 6.35, page 510.

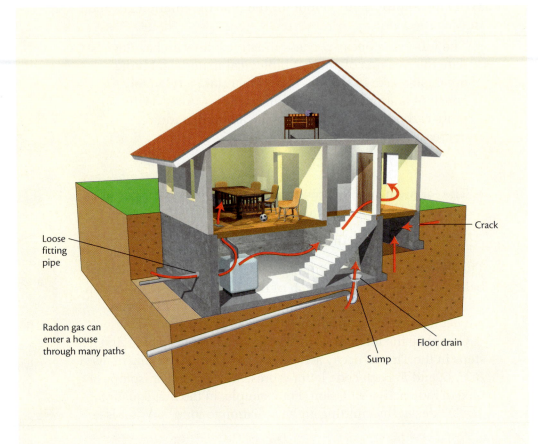

Figure 6.36 *Radon gas is naturally released from certain soils containing radioisotopes. Radon gas (red arrows) can enter a house by many paths.*

Figure 6.37 *Radon's high density and the fact that it enters a residence through foundation cracks means this radioactive gas can be removed by installing a venting system that withdraws air from the basement.*

The most serious danger of radon gas results from reactions that occur after it is inhaled. Radon decays to produce, in succession, radioactive isotopes of polonium (Po), bismuth (Bi), and lead (Pb). When radon gas is inhaled, it enters the body and is transformed, through radioactive decay, into these toxic heavy-metal ions, which cannot be exhaled as gases. These radioactive heavy-metal ions also emit potentially damaging alpha particles within the body.

Estimates indicate that about 6% of homes in the United States have radon levels higher than the exposure level recommended by the U.S. Environmental Protection Agency (EPA). See Figure 6.38. It is estimated that 10–15% of annual U.S. deaths from lung cancer are linked to the effects of indoor radon gas. These figures, although sobering, should be kept in perspective; about 80% of all U.S. lung-cancer deaths annually are attributed to cigarette smoking.

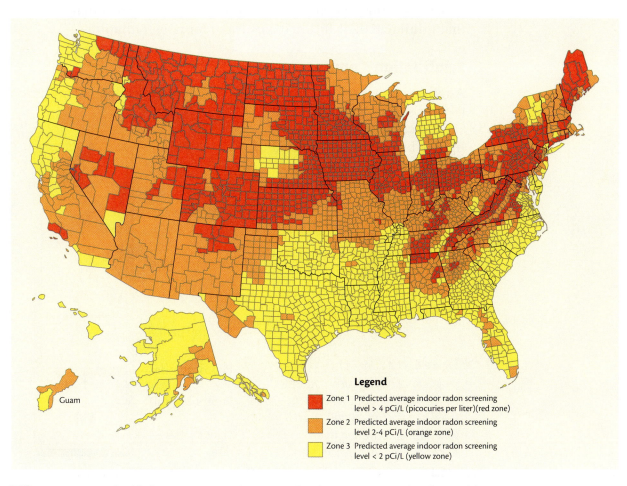

Legend

🟥 Zone 1 Predicted average indoor radon screening
level > 4 pCi/L (picocuries per liter)(red zone)

🟧 Zone 2 Predicted average indoor radon screening
level 2-4 pCi/L (orange zone)

🟨 Zone 3 Predicted average indoor radon screening
level < 2 pCi/L (yellow zone)

Figure 6.38 *Map highlighting U.S. areas by county that have experienced radon problems.*

B.8 NUCLEAR RADIATION DETECTORS

Detecting Radiation

The only way to detect radioactive decay is to observe the results of nuclear radiation interacting with matter. In each of the following detection methods, visible and/or electronically detectable changes in matter enable a technician to determine when such radiation is present.

Geiger-Mueller Tubes and Counters

The Geiger-Mueller tube contains argon gas that is ionized when nuclear radiation enters the tube. The ionized gas conducts an electric current; an electrical signal (counts per minute) is generated as the ions and components of radiation pass through the tube.

Film Badges

You learned earlier in this unit that emissions from radioactive materials will expose photographic film. Workers who handle radioisotopes wear film badges or other detection devices to monitor their exposure. The more ionizing radiation workers encounter, the greater the extent of exposure in their film badges. If workers receive doses in excess of the federal limits, they are temporarily reassigned to jobs that minimize their future exposure to ionizing radiation.

Scintillation Counters

Another detection approach involves devices called **scintillation counters** that contain solid substances whose atoms emit light when they are excited by ionizing radiation. In modern scintillation counters, the flashes of light are detected electronically. The scintillation counter probe illustrated in Figure 6.39 has a sodium iodide (NaI) detector that emits light when ionizing radiation strikes it.

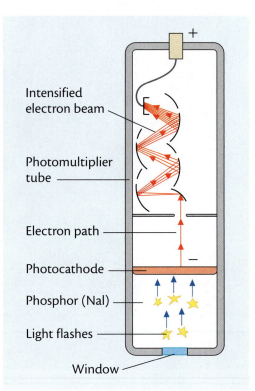

Figure 6.39 *A scintillation counter probe. Ionizing radiation causes flashes of light (scintillations) in the detector (NaI crystal). Each light flash is converted to an electron pulse that is increased many times as it moves through the photomultiplier tube.*

Intensified electron beam

Photomultiplier tube

Electron path

Photocathode

Phosphor (NaI)

Light flashes

Window

Solid-State Detectors

In another technique, **solid-state detectors** monitor changes in the movement of electrons through silicon-based semiconductors as they are exposed to ionizing radiation. These detectors are often used in research laboratories.

Cloud Chambers

Ionizing radiation can also be detected in a cloud chamber. You will learn about this detection method in the following investigation.

B.9 CLOUD CHAMBERS

Introduction

A **cloud chamber** is a container filled with air saturated with water or another vapor, similar to saturated air on a very humid day. If cooled, this air becomes supersaturated with vapor. This is an unstable condition (see page 55). When a radioactive source is placed near a cloud chamber filled with supersaturated air, the radiation ionizes gas molecules in the vapor as the radiation passes through the chamber. Vapor condenses to liquid on the ions that are formed, leaving a white trail along each passing radioactive emission, thus revealing the path of the particle or ray. Figure 6.40 is an image of particle tracks under such conditions.

 **Laboratory Video:
Cloud Chambers**

Cloud-chamber trails resemble the vapor trails from high-flying aircraft.

Figure 6.40 *A cloud chamber, showing many particle trails created by ionizing radiation. The radioactive source is located on the left side.*

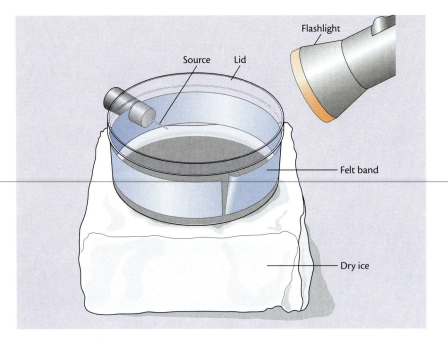

Figure 6.41 *A cloud chamber allows the path of ionizing radiation to be viewed with the naked eye or photographed.*

The cloud chamber you will use consists of a small plastic container and a felt band moistened with 2-propanol (isopropyl alcohol). The alcohol evaporates faster than does water, so it saturates the enclosed air more readily. You will chill the cloud chamber with dry ice to promote super-saturation and cloud formation (Figure 6.41).

Procedure

1. Fully moisten the felt band inside the cloud chamber with isopropyl alcohol. Also place dark paper on the chamber bottom and moisten it with a small quantity of alcohol.

2. Using gloves and forceps, quickly place a radioactive source in the chamber.

3. To cool the chamber, place it on a flat surface of crushed dry ice, ensuring that the chamber remains level.

4. Leave the chamber on the dry ice for 3–5 min.

5. Your teacher will adjust the room lighting. Focus a light source at an oblique angle (not straight down) through the container to illuminate the chamber base. If you do not observe any vapor trails, shine the light through the side of the chamber.

6. Observe the air inside the chamber near the radioactive source. Record your observations.

Questions

1. What differences, if any, did you observe among the tracks?

2. Which type of radiation do you think would make the most visible tracks? Why?

3. What is the purpose of the dry ice?

B.10 ENSURING PUBLIC SAFETY

Consider again the statements found in the CANT flyer (page 478) and also your planned presentation to the senior citizens of Riverwood. Within a small group of classmates, discuss the appropriateness of each of the following proposals for protecting the public from radiation hazards. For each proposal, identify statements in the CANT flyer that might encourage public groups to make such a proposal.

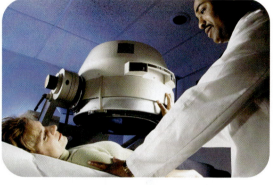

Figure 6.42 *Nuclear technology is widely used in medicine.*

1. Because all radiation is bad, government regulators should ban exposure to all forms of radiation.

2. Although there are several different types of ionizing radiation (alpha, beta, and gamma), government standards for protecting the public against possible dangers should treat all radiation exposures identically.

3. Because nuclear radiation can be harmful, physicians should be required by law to inform patients of any medical procedure that involves nuclear radiation. Patients can then reject the treatment if they have concerns about their exposure to radiation (Figure 6.42).

4. The government should provide a waste-disposal site for permanently storing all wastes that have ever been identified as radioactive (Figure 6.43).

In this section you studied the origins of ionizing radiation, sources of exposure to ionizing radiation, and some possible consequences of that exposure. Keep these ideas and concerns in mind as you explore uses of nuclear radiation in the next section.

Figure 6.43 *Should the government be responsible for permanently storing all wastes that have ever been identified as radioactive?*

SECTION B SUMMARY
Reviewing the Concepts

Radioisotopes in the environment create background radiation.

1. Define *background radiation*.
2. List five sources of background radiation.
3. Why does background radiation vary from one region to another?
4. Is it possible to eliminate background radiation? Explain.
5. What is the established U.S. background radiation limit for the general public expressed in
   a. mrem?
   b. Sv?
6. In what SI units is exposure to background radiation expressed?
7. What radiation dose would you receive
   a. during a five-hour jet aircraft flight?
   b. from a CT scan and a chest X-ray?
   c. if you lived in a brick house 25 mi (40 km) from a coal-fired power plant for one year?
8. How does radon gas get into houses?
9. Why do radon levels tend to be higher in energy-efficient houses?

Ionizing radiation has sufficient energy to break chemical bonds.

10. Why is the breaking of chemical bonds in living cells by ionizing radiation harmful?
11. What is a *mutation?*
12. How does ionizing radiation lead to an increase in mutation rates?
13. List the four factors that determine the extent of biological damage from radiation.
14. What types of human tissue are most susceptible to radiation damage?
15. At what radiation dose level would a person begin to experience nausea and hair loss? Express your answer in units of
    a. rem.   b. mrem.

Alpha, beta, and gamma radiation have different properties that determine their effects on living tissues.

16. Which type or types of ionizing radiation would
    a. be stopped by a glass window pane?
    b. penetrate a cardboard box?
    c. penetrate a thin sheet of plastic but not a thin sheet of lead?
17. Explain why materials that emit alpha radiation are more dangerous inside the body than outside the body.
18. What is the relationship between radiation intensity and distance from the radiation source?
19. Suppose a beta source gives a corrected radiation reading of 640 cpm at a distance of 3 cm. Predict the corrected reading at a distance of
    a. 6 cm.
    b. 12 cm.

The emission of nuclear radiation changes the composition of the nucleus.

20. a. What is the composition of an alpha particle?
    b. List two symbols used to represent an alpha particle.

21. a. What is a beta particle?
    b. List two symbols used to represent a beta particle.

22. How is a beta particle formed?

23. Why does the emission of an alpha or beta particle create an atom of a different element?

24. Copy and complete the following nuclear equations.

  a. $^{6}_{3}\text{Li} + ^{1}_{0}n \longrightarrow ^{4}_{2}\text{He} + \underline{\hspace{1cm}}$
  b. $^{42}_{19}\text{K} \longrightarrow ^{0}_{-1}e + \underline{\hspace{1cm}}$
  c. $^{235}_{92}\text{U} \longrightarrow \underline{\hspace{1cm}} + ^{231}_{90}\text{Th}$
  d. $^{1}_{0}n + \underline{\hspace{1cm}} \longrightarrow ^{142}_{56}\text{Ba} + ^{91}_{36}\text{Kr} + 3\ ^{1}_{0}n$

Ionizing radiation may be detected by its interaction with matter using a variety of methods.

25. Describe how each of the following devices can detect the presence of ionizing radiation:

  a. Geiger-Mueller tube and counter
  b. scintillation counter
  c. solid-state detector
  d. cloud chamber
  e. film badge

26. Would you expect alpha, beta, and gamma radiation to produce the same kinds of trails in a cloud chamber? Explain your answer.

Connecting the Concepts

27. High-level radioactive wastes are generally stored deep underground. Suggest two ways in which this method serves to keep people safe from excessive exposure to ionizing radiation.

28. In the 1940s and 1950s, some shoe stores invited customers to check the adequacy of shoe fit by X-raying their feet. Why did shoe stores discontinue that practice? (That shoe-store device was called a *fluoroscope.*)

29. Are the effects of shielding and distance the same for both ionizing and nonionizing radiation? Explain your answer.

30. Radon is an inert noble gas that is relatively harmless to living things. Explain why its presence in homes constitutes a health hazard for occupants.

31. Describe fundamental differences between nuclear and chemical reactions.

32. A student sets up a cloud chamber and sees no white trails. What are some possible explanations for this result?

33. Two students live next door to each other. One receives three times more annual radiation than the other. Explain how this could be possible.

34. Why do radiation detectors register counts even though no apparent source of radioactivity is in the vicinity?

35. A heavy apron is provided for a patient who receives a dental X-ray.

  a. What element is probably used in the apron?
  b. What is the purpose of the apron?
  c. Why does the dentist or hygienist leave the room while the X-ray device is turned on?

Extending the Concepts

36. In terms of radiation, how is a sunburn different from a suntan? How do sunscreens work to prevent both?

37. Identify the five fastest-growing metropolitan areas in the United States. Rank them according to the average level of ionizing radiation exposure that inhabitants in similar houses receive.

38. Investigate and explain whether or not beta particles can be distinguished from electrons.

39. Why does ionizing radiation break some chemical bonds but not others?

Using Radioactivity

Each radioisotope decays and emits ionizing radiation at its own special rate. Scientists have devised convenient ways to measure, analyze, and report how rapidly (or slowly) particular radioisotopes decay. In this section you will learn about radioactive decay rates, a characteristic that helps determine how useful or hazardous a radioisotope may be.

C.1 HALF-LIFE: A RADIOACTIVE CLOCK

How long does it take for a sample of radioactive material to decay? There is no simple answer to this question. However, by understanding how radioactive materials decay, scientists can predict how long a radioisotope used in a medical diagnostic test, for example, will remain active within the body; plan the long-term storage requirements for hazardous nuclear wastes; and estimate the ages of old organisms, civilizations, or rocks.

In 1904, Ernest Rutherford proposed the concept of *half-life* to describe the random process of radioactive decay. **Half-life** expresses the total time within which a radioactive atom has a 50-50 chance to undergo radioactive decay. Because one gram of any element contains well over 10^{21} atoms, the typical sample size involves a very large number of atoms. *Half-life* and *radioactive decay rate* have different meanings. Under normal conditions, the half-life for a particular radioisotope remains constant, while the radioactive decay rate represents the total number of atoms that have decayed within a certain period of time. The decay rate gradually decreases as the number of radioactive atoms decreases over time.

Statistics related to very large numbers provide two other useful ways to interpret half-life. The first—and perhaps most common—interpretation is that one half-life is the time required for one-half of the total radioactive atoms originally present in a sample to decay. The second interpretation is that at the end of one half-life, the rate of radioactive decay in a sample will decrease to one-half its original value. Table 6.9 lists the half-lives of several radioisotopes.

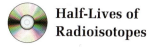
Half-Lives of Radioisotopes

DECAY EQUATIONS AND HALF-LIVES FOR FIVE RADIOACTIVE ISOTOPES

Radioisotope	Decay Equation	Half-Life
Hydrogen-3	$^{3}_{1}H \longrightarrow ^{3}_{2}He + ^{0}_{-1}e$	12.3 y
Carbon-14	$^{14}_{6}C \longrightarrow ^{14}_{7}N + ^{0}_{-1}e$	5.73×10^{3} y
Phosphorus-32	$^{32}_{15}P \longrightarrow ^{32}_{16}S + ^{0}_{-1}e$	14.3 d
Potassium-40	$^{40}_{19}K \longrightarrow ^{40}_{20}Ca + ^{0}_{-1}e$	1.28×10^{9} y
Radon-222	$^{222}_{86}Rn \longrightarrow ^{218}_{84}Po + ^{4}_{2}He$	3.82 d

| **Table 6.9**

Although it is not possible to predict when a particular radioactive atom will decay, each way of thinking about half-life is equally valid when working with the large numbers of atoms in typical chemical samples. The decay of carbon-14 is represented by the following equation:

$$^{14}_{6}\text{C} \longrightarrow \, ^{14}_{7}\text{N} + \, ^{0}_{-1}e$$

All living matter contains carbon and, therefore, a small amount of radioactive C-14. Thus, all living organisms emit a small but constant level of radioactivity. In living matter, decaying C-14 atoms are constantly replaced by new carbon atoms. After death, no new carbon atoms are taken in. The radioactive C-14 atoms continue to decay, so the longer an organism has been dead, the fewer C-14 atoms it contains.

Table 6.9 (page 521) indicates that C-14's half-life is 5730 years. If an organism contains 50.0 billion atoms of carbon-14 at the time of death, half of those atoms will have decayed after 5730 years pass, leaving 25.0 billion atoms of carbon-14. During the next 5730 years, another one-half of the atoms will decay, leaving 12.5 billion atoms of carbon-14. Therefore, if a sample from a previously living organism contains only one-fourth the number of C-14 atoms expected in a living organism, we can estimate that the sample is about 11 460 years old—that is, two half-lives must have passed since the organism died.

ChemQuandary 3

CARBON-14 DATING

It is not possible to determine the age of every artifact using carbon-14 dating. What kinds of materials might be good candidates for carbon-14 dating? What are some materials you could not date using carbon-14? Why does carbon-14 dating have a practical limit of about 50 000 years?

Every radioisotope has a specific half-life that is constant under normal circumstances. Half-lives of radioisotopes can be as short as a fraction of a second or as long as several billion years. For example, the half-life of polonium-212 is 3×10^{-7} seconds, while that of uranium-238 is 4.5 billion years. Thus, in one year all the atoms in a small sample of polonium-212 will probably have decayed, while well over 99% of the original uranium-238 atoms will still be present.

After 10 half-lives, only about 1/1000 or 0.1% of the original radioisotope atoms are still left to decay. (You can verify that statement with your own calculations.) This means that the rate of an isotope's radioactive decay has dropped to 0.1% of its initial level. This reduced level is often considered safe because it roughly approaches the level of normal background radiation.

Because there is no way to change the radioactive decay rate significantly for a particular isotope, radioactive waste disposal (or storage) can pose challenging problems, especially for radioisotopes with very long half-lives. You will examine that issue later in this unit.

In the following activity, you will model and explore the concept of half-life with heads-up and heads-down coins.

> In 2004, scientists reported that they could decrease the half-life of beryllium-7 by half a day (a 1% change) by trapping it in an electron-rich environment.

Modeling Matter

C.2 UNDERSTANDING HALF-LIFE

In this activity you will receive 80 pennies and a box. Place all pennies heads up to represent the starting sample of *headsium*. Assume each heads-up penny represents an atom of radioactive *headsium*. Its decay produces a tails-up penny—*tailsium*. Each shake of the closed box containing pennies represents one half-life. During this time a certain number of *headsium* nuclei will decay—flip over—to produce *tailsium*. You will investigate the relationship between the passage of time and the quantity of radioactive nuclei (heads-up pennies) that decay.

Complete these steps with your pennies:

1. Prepare a data table for recording the undecayed *headsium* and decayed *tailsium* atoms after each of four half-lives. Include initial values for passage of 0 half-lives.

2. Place the 80 pennies heads up in the box.

3. Close the box and shake it vigorously.

4. Open the box. Remove all atoms that decayed into *tailsium.* Record the number of undecayed (*headsium*) and decayed (*tailsium*) atoms after this first half-life.

5. Repeat Steps 3 and 4 three more times. You will now have simulated the passage of four half-lives. Record your results for each half-life.

6. Follow your teacher's instructions to obtain pooled class data for total undecayed *headsium* atoms remaining after each half-life.

7. Using your own data and the class-pooled data, prepare a graph by plotting the number of half-lives on the *x*-axis and the number of undecayed atoms remaining after each half-life on the *y*-axis. Plot and label two graph lines—one representing your own data and the other representing pooled class data.

Now answer these questions, based on your data:

1. a. Describe the appearance of your two graph lines. Are they straight or curved?
   b. Which set of data—yours or the pooled class data—provides a more convincing demonstration of half-life? Why?

2. About how many *headsium* nuclei would remain after three half-lives, if the initial sample had 600 *headsium* atoms?

3. If 190 *headsium* nuclei remain from an original sample of 3000 *headsium* nuclei, about how many half-lives must have passed?

4. Describe one similarity and one difference between your model based on pennies and actual radioactive decay. (*Hint:* Why did you pool the class data?)

5. How could you modify this model to demonstrate that different isotopes have different half-lives?

6. a. How many half-lives would be needed for one mole of a radioisotope to decay to 6.25% of the original number of atoms?
   b. Is it likely that any of the original radioactive atoms would still remain after
      i.   10 half-lives? Explain your answer.
      ii. 100 half-lives? Explain your answer.

7. a. In this simulation, can you predict when a particular *headsium* nucleus will "decay"?
   b. If you could follow the fate of an individual atom in a sample of radioactive material, could you predict when it would decay? Why or why not?

8. What other ideas could model the concept of half-life?

C.3 APPLICATIONS OF HALF-LIVES

Use what you learned about half-life to answer the following questions. Figure 6.44 may help.

Sample Problem: *The half-life of O-15 is 2.0 min; its radioactive decay produces N-15. How much O-15 will remain after 5.0 min, if the original sample contained 14.0 g O-15?*

First determine the total number of half-lives that 5.0 min represents:

$$5.0 \; \cancel{min} \times \frac{1 \; half\text{-}life}{2.0 \; \cancel{min}} = 2.5 \; half\text{-}lives$$

Because the total number of half-lives is not an integer, use Figure 6.44 to estimate the proportion of O-15 remaining. Locate 2.5 half-lives on the *x*-axis. From that point, move directly upward on the graph until you touch the curved line. Then move left until you intersect the *y*-axis. Read this point from the graph: about 0.18. This means that 18% of the original O-15 sample remains:

$$(14.0 \; g \; O\text{-}15) \times 0.18 = 2.5 \; g \; O\text{-}15 \; remain$$

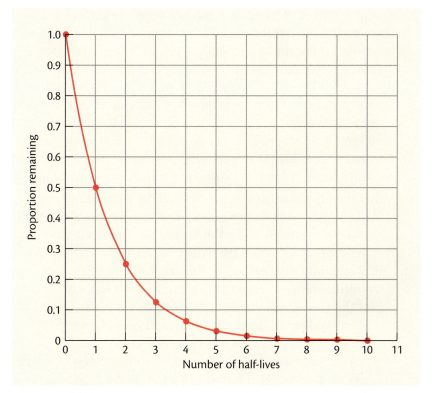

Figure 6.44 *This graph depicts the quantity of radioactive material remaining vs. total half-lives that have passed. What proportion of material would remain after four half-lives?*

You can check to see whether an answer of 2.5 g makes sense: After one half-life, 7.00 g O-15 would remain in the sample (the remaining mass represents N-15). After two and three half-lives, respectively, 3.50 g and 1.75 g O-15 would remain. The answer we found, 2.5 g, is between 3.50 g and 1.75 g, which is what we would expect.

1. Suppose you received $1000 (see Figure 6.45) and could spend one-half in the first year, one-half of the balance in the second year, and so on. (*Note:* One year corresponds to one half-life in this analogy.)

   a. If you spent the maximum allowed annually, at the end of which year would you have $31.25 left?

   b. How much money would be left after ten half-lives?

Figure 6.45 *If you initially had $1000 and spent half the money each year, how many years would it take until you had only $125?*

2. Potassium is a necessary nutrient for all living things and is the seventh most abundant element on earth's surface, composing about 1.5% of its crust. About 0.01% of natural potassium atoms are the radioisotope potassium-40. K-40 has a half-life of nearly 1.3 billion (1.3×10^9) years.

   a. Assuming earth is about 4.5 billion years old, how much of the K-40 at earth's formation remained after one half-life?

   b. Roughly how many times more K-40 was present when earth formed than is present now?

3. Strontium-90 is one of many radioisotopes generated by nuclear-weapon explosions. This isotope is especially dangerous if it enters the food supply. Strontium behaves chemically like calcium, because these elements are members of the same chemical family. Thus, rather than passing through the body, radioactive strontium-90 is incorporated into calcium-based material, such as bone. A nuclear test-ban treaty in 1963 ended most above ground weapons testing. But some Sr-90 released in previous testing is still present in the environment.

   a. Sr-90 has a half-life of 28.8 y. Track the decay of Sr-90 atoms that were present in the atmosphere in 1963 as follows:
      i. Using 1963 as year zero, when 100% of released Sr-90 was present, identify the years that represent the completion of one, two, three, four, and five half-lives.
      ii. Calculate the percent of the original 1963 Sr-90 radioisotope present at the end of each half-life.

   b. Plot the percent of the original 1963 Sr-90 radioactivity level on the y-axis and the years 1963 to 2110 on the x-axis. Connect the data points with a smooth curve.
      i. What percent of Sr-90 formed in 1963 is present now?
      ii. What percent will remain in 2100?

   c. Compare your graph to that shown in Figure 6.44 (page 525).
      i. How do the two graphs differ?
      ii. In what ways are they the same?

C.4 RADIOISOTOPES IN MEDICINE

For several decades following the discovery of X-rays and radioactivity, the public was fascinated by these scientific developments. In fact, some patent medicines advertised that they contained radium, claiming supposed benefits of that substance. Unfortunately, some older patent medicines actually contained radioactive materials and were very hazardous.

⌈The careful use of ionizing radiation and radioisotopes can be quite effective in medical diagnosis and treatment⌉ (see Figures 6.46 and 6.47). Such uses can be classified as either ⌈diagnostic or therapeutic.⌉ Diagnostic use helps doctors understand what is happening inside the body (see Figure 6.48), while therapeutic use involves treating a medical condition.

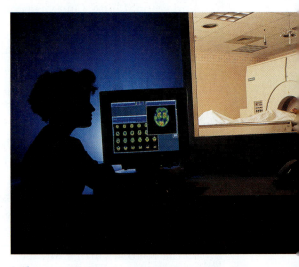

Figure 6.46 Injection of a radioisotope allows noninvasive imaging of internal organs.

A common diagnostic application uses *radioisotope-tracer studies,* based on detecting a radioisotope in particular parts of the body. Knowing that certain elements collect in specific parts of the body (for example, calcium in bones and teeth), physicians can investigate a given part of the body by using an appropriate radioisotope as a **tracer.** In a tracer study, the physician places radioisotopes with short half-lives in a patient's body. Such studies can identify cellular abnormalities and help physicians select appropriate therapies.

Tracers have properties that make them ideally suited to this task. First, radioisotopes behave the same as stable (nonradioactive) isotopes of the same element. Researchers can apply a solution of an appropriate tracer isotope to the body, or they can feed or inject the patient with a biologically active compound containing the radioactive tracer element. A detection system, as shown in Figure 6.46, then allows a medical technician to track the tracer's location throughout the body.

For example, iodine-123 is used to diagnose problems of the thyroid gland, which is located in the neck. A patient drinks a tracer solution containing sodium iodide (NaI), in which some of the iodide ions are the radioisotope I-123. The physician, using a radiation detection system, monitors the rate at which this tracer is taken up by the thyroid. A healthy thyroid will incorporate a known amount of iodine. An overactive or underactive thyroid will take up, respectively, more or less iodine. The physician compares the measured rate of I-123 uptake by the patient to the normal rate for an individual of the same age, gender, and weight, and then takes appropriate therapeutic action.

Technetium-99m (Tc-99m), a synthetic radioisotope, is the most widely used diagnostic radioisotope in medicine. It has replaced exploratory surgery as a way to locate tumors in the brain, thyroid, and kidneys. *Tumors* are areas of runaway cell growth; technetium concentrates where cell growth is fastest. A bank of radiation detectors around the patient's body can pinpoint the Tc-99m at the tumor's precise location.

Figure 6.47 *Radioisotopes are prepared and used with knowledge of their half-lives, thus ensuring appropriate levels of activity as tracers.*

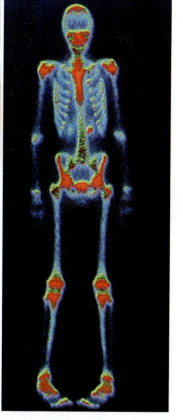

Figure 6.48 *Nuclear-medicine scan of bones of the entire human body.*

Physicians use *therapeutic radioisotopes* because they emit radiation that carries sufficient energy to destroy living tissue. In some cancer treatments, doctors kill cancerous cells with ionizing radiation. For thyroid cancer, a patient is administered a concentrated internal dose of radioiodine, which collects in and destroys the cancerous portion of the thyroid gland. In other cancer treatments, physicians may direct an external beam of ionizing radiation (from cobalt-60) at the cancerous spot. Physicians must administer treatments with great care; high-radiation doses also kill normal cells.

ChemQuandary 4

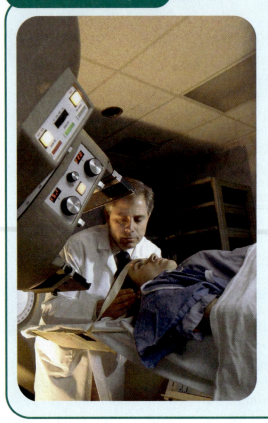

USING RADIOISOTOPES IN MEDICINE

One source of ionizing radiation is cobalt-60, which kills rapidly dividing cells. Consider two equally sized Co-60 samples shipped to two hospitals at the same time.

At one hospital, the Co-60 is used to treat dozens of individuals, while at the other hospital it is used only once or twice. Why would *both* hospitals dispose of their Co-60 samples after five years?

Radiosodium is usually administered as a sodium chloride solution.

Other medical applications include use of radiosodium (Na-24) to detect circulatory system abnormalities and radioxenon (Xe-133) to help locate lung embolisms (blood clots) and abnormalities. Table 6.10 summarizes medical uses for several radioisotopes.

You have learned that ionizing radiation, like that observed in a cloud chamber (page 515), is emitted from an unstable, radioactive isotope that eventually changes to a stable, nonradioactive atom. Do you think it would be possible to reverse that process, converting a stable isotope into an unstable, radioactive isotope? Think about it. This question will be addressed later in this section.

SELECTED MEDICAL RADIOISOTOPES

| Table 6.10

Radioisotope	Half-Life	Uses
Used as Tracers		
Technetium-99m	6.01 h	Measure cardiac output; locate strokes and brain and bone tumors
Gallium-67	78.3 h	Diagnosis of Hodgkin's disease
Iron-59	44.5 d	Determine the rate of red blood cell formation (these contain iron); anemia assessment
Chromium-51	27.7 d	Determine blood volume and life span of red blood cells
Hydrogen-3 (tritium)	12.3 y	Determine the volume of the body's water; assess vitamin D usage in body
Thallium-201	72.9 h	Cardiac assessment
Iodine-123	13.3 h	Thyroid function diagnosis
Used for Radiation Therapy		
Cesium-137	30.1 y	Treat shallow tumors (external source)
Phosphorus-32	14.3 d	Treat leukemia, a bone cancer affecting white blood cells (internal source)
Iodine-131	8.0 d	Treat thyroid cancer (external source)
Cobalt-60	5.3 y	Treat shallow tumors (external source)
Yttrium-90	64.1 h	Treat pituitary gland cancer internally with ceramic beads

C.5 NUCLEAR MEDICINE TECHNOLOGIES

Computers touch nearly all aspects of modern life. For example, health and medical science now employ two nuclear medicine technologies that rely heavily on computers to make sense of the large quantities of data obtained. These technologies are *positron emission tomography* (see Figure 6.49, page 530), which involves radioisotopes, and *magnetic resonance imaging.*

Positron emission tomography (PET) scans are based on a very unusual form of radioactive decay involving a few particular radioisotopes. Although most radioisotopes emit alpha, beta, and/or gamma radiation, a few radioisotopes emit radiation in the form of *positrons*. **Positrons** originate in the nucleus and have the same mass as beta particles (electrons). However, positrons differ from electrons in fundamental ways. Positrons have a positive electrical charge, while electrons are negatively charged. And positrons are not made of matter as we commonly understand it.

Scientific American Working Knowledge Illustration

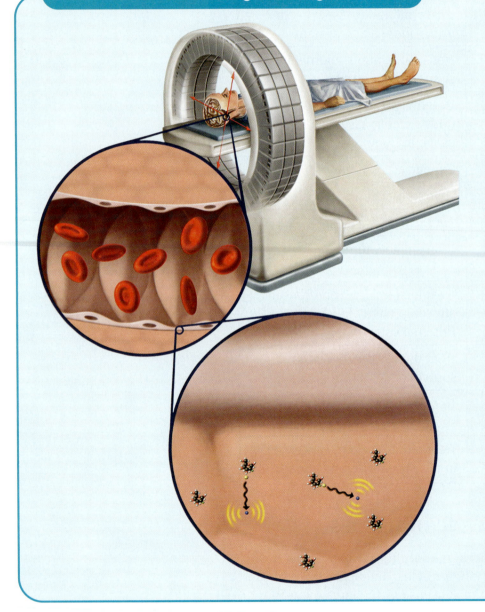

Figure 6.49 *Positron emission tomography (PET)*. *Cancerous tissue can be detected using positron emission tomography (PET) scans. In this illustration, a brain tumor is identified by detecting gamma radiation emitted when positrons collide with electrons. As depicted in the particulate-level representation (bottom image), sugar molecules are tagged with a radioisotope tracer that emits positrons (yellow spheres). When a positron collides with an electron (blue spheres) in the immediate vicinity, both particles are destroyed and produce two gamma rays (depicted as yellow waves). The gamma-ray pair is then detected.*

Positrons are composed of *antimatter*. When a positron encounters an electron, both particles are annihilated (destroyed) and produce two gamma rays that are emitted in opposite directions. PET detects these gamma-ray pairs and, with the help of computers, determines where they originated. By observing a large number of such events, a computer-generated image gradually emerges.

The radioisotope tracer that emits positrons in PET scans is attached to a sugar molecule. Physicians can accurately determine the movement of each tagged sugar molecule as it progresses through the body. Because cancers grow faster than normal tissues, cancerous tissue metabolizes more sugar in a given time than does normal tissue. The radioisotope tracer (sugar tag) eventually becomes more concentrated in regions of the body containing cancerous tissue. PET technology can thus detect and display metabolic activity. Doctors can use this information to investigate brain functioning without invasive surgery.

Magnetic resonance imaging (MRI) does not employ ionizing radiation; it relies on the properties of protons. MRI is an application of a laboratory process known as *nuclear magnetic resonance* (NMR), which was developed in the mid-twentieth century. NMR is a noninvasive technique that can identify atoms within a sample without altering the sample itself.

MRI imagery can produce useful images of soft tissues. A major benefit of MRI is that it does not rely on ionizing radiation. Unlike most other nuclear-medicine technologies, MRI uses radio waves of very low energies and involves no known health risks. Some patients were hesitant to undergo this procedure when it was called by its original name, nuclear magnetic resonance, due to fear evoked by the term *nuclear*. This unfounded fear prompted the name change to magnetic resonance imaging.

C.6 ARTIFICIAL RADIOACTIVITY

In 1919, Ernest Rutherford enclosed nitrogen gas in a glass tube and bombarded the sample with alpha particles. After analyzing the gas remaining in the tube, he found that some nitrogen atoms had been converted to an isotope of oxygen, according to the following equation:

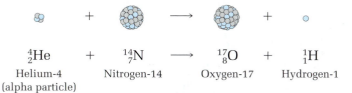

$$^4_2\text{He} \ + \ ^{14}_7\text{N} \ \longrightarrow \ ^{17}_8\text{O} \ + \ ^1_1\text{H}$$

Helium-4   Nitrogen-14   Oxygen-17   Hydrogen-1
(alpha particle)

Rutherford had produced the first synthetic or artificial **transmutation** of an element; that is, the first documented conversion of one element to another. He continued this work but was limited by the moderate energies of alpha particles then available. By 1930, scientists had developed particle accelerators that could produce highly energetic particles for bombardment reactions. Using these higher-energy particles, scientists then created many other synthetic atoms, some of which were radioactive.

The first **synthetic radioisotope** (one not occurring in nature) was produced in 1934 by French physicists Irène and Frédéric Joliot-Curie (the daughter and son-in-law of Marie and Pierre Curie; see Figure 6.50). They bombarded aluminum atoms with alpha particles, producing radioactive phosphorus-30 and neutrons:

$$\,^{27}_{13}\text{Al} + \,^{4}_{2}\text{He} \longrightarrow \,^{30}_{15}\text{P} + \,^{1}_{0}n$$

Since then, researchers have accomplished many transformations of one element to another and have synthesized new radioactive isotopes of various elements. Many of the diagnostic radioisotopes noted in Table 6.10 (page 529) are synthetic. Technetium-99m, for example, is both a synthetic element and a radioisotope.

Most synthetic radioisotopes are produced by bombarding elements with neutrons, which are captured by target nuclei. This process requires less energy than many other bombardment reactions because neutrons have no electrical charge and are not repelled by the positive charge of the nucleus. Such reactions produce radioactive nuclei that tend to emit beta particles, thus changing the atomic number and producing a different element.

The following examples show the formation of two synthetic radioisotopes often used as medical tracers, calcium-45 and iron-59.

$$\,^{44}_{20}\text{Ca} + \,^{1}_{0}n \longrightarrow \,^{45}_{20}\text{Ca}$$

$$\,^{58}_{26}\text{Fe} + \,^{1}_{0}n \longrightarrow \,^{59}_{26}\text{Fe}$$

Nuclear-bombardment reactions generally involve four particles:

▶ *Target nucleus:* the stable isotope that is bombarded.

▶ *Projectile particle* (bullet): the particle fired at the target nucleus.

▶ *Product nucleus:* the isotope produced in the reaction.

▶ *Ejected particle:* the lighter nucleus or particle emitted from the reaction.

For example, consider the Joliot-Curies' production of the first synthetic radioactive isotope, phosphorus-30. The four types of particles involved are identified as follows:

Reactions may release more than one ejected particle. For example, see the Cm-246 reaction on the next page.

$$\,^{27}_{13}\text{Al} \qquad + \qquad \,^{4}_{2}\text{He} \qquad \longrightarrow \qquad \,^{30}_{15}\text{P} \qquad + \qquad \,^{1}_{0}n$$

| Aluminum-27 | Alpha particle | Phosphorus-30 | Neutron |
| target nucleus | projectile particle | product nucleus | ejected particle |

C.7 NUCLEAR-BOMBARDMENT REACTIONS

Sample Problem: *Nobelium (No) can be produced by bombarding curium (Cm) atoms with nuclei of a low-mass element. What element serves as the projectile particle in this reaction?*

$$^{246}_{96}Cm + \ ? \longrightarrow \ ^{254}_{102}No + 4\,^{1}_{0}n$$

Because the sum of the product atomic numbers is 102, the projectile must have the atomic number 6, to make the sum of the reactant atomic numbers also 102. Therefore, the projectile must be carbon (atomic number = 6). The total mass numbers of products is 258 (254 + 4), indicating that the projectile must have been carbon-12 (258 − 246 = 12). The completed equation is as follows:

$$^{246}_{96}Cm \quad + \quad ^{12}_{6}C \quad \longrightarrow \quad ^{254}_{102}No \quad + \quad 4\,^{1}_{0}n$$

| Target nucleus | Projectile particle | Product nucleus | Ejected particles |

> As you learned earlier, completing nuclear equations involves balancing atomic numbers and mass numbers.

Complete the following equations by supplying the missing numbers or symbols. Name each particle. Then identify the target nucleus, projectile particle, product nucleus, and ejected particle.

1. $^{59}_{27}? + \,^{?}_{?}n \longrightarrow \,^{60}_{?}?$ (Scientists produce most medically useful isotopes by bombarding stable isotopes with neutrons. This process converts the original nuclei to radioactive forms of the same element.)

2. $^{96}_{42}? + \,^{?}_{?}H \longrightarrow \,^{97}_{43}? + \,^{1}_{0}?$ (Until it was synthesized in 1937, technetium was only an unfilled gap in the periodic table; all its isotopes are radioactive. Any technetium originally on earth has decayed. Technetium, the first element artificially produced, is now used in industry and medicine.)

3. $^{58}_{?}? + \,^{209}_{?}Bi \longrightarrow \,^{?}_{109}Mt + \,^{1}_{0}?$ (In 1992, a research group in Darmstadt, Germany, created element 109 by bombarding bismuth-209 nuclei. The name Meitnerium (Mt) honors Lise Meitner, the Austrian physicist who first proposed the idea of nuclear fission. See page 541.)

Not only does the ability to transform one element into another provide new and powerful technological capabilities, it also has changed our view of elements (see Figure 6.51).

Figure 6.51 *Collisions of nuclear particles can be studied with bubble chambers. Mixtures of liquefied noble gases (at low temperatures) boil and produce streams of tiny bubbles as they absorb energy from particles involved in these collision investigations.*

TRANSMUTATION OF ELEMENTS

Ancient alchemists searched in vain for ways to transform (transmute) lead or iron into gold. Has such transmutation now become a reality? From what you know about nuclear reactions, do you think that lead, iron, or mercury atoms could be changed to gold? If so, try to write equations for the possible reactions.

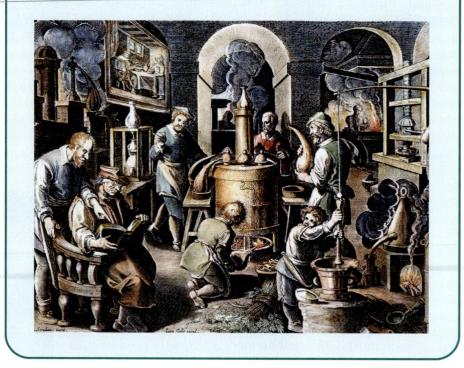

C.8 EXTENDING THE PERIODIC TABLE

Since 1940, nearly 20 **transuranium** elements—with atomic numbers greater than the atomic number of uranium (92)—have been added to the periodic table. These elements have been synthesized in nuclear reactions, usually conducted in accelerators known as *cyclotrons* (see Figure 6.52). From 1940 to 1961, Glenn Seaborg and coworkers at the University of California–Berkeley synthesized and identified ten new elements with atomic numbers 94 to 103, a prodigious feat. None of those new elements occurs naturally. All were made by high-energy bombardment of heavy nuclei with various particles. For example, alpha-particle bombardment of plutonium-239 produced curium-242:

$$\ce{^{239}_{94}Pu} + \ce{^{4}_{2}He} \longrightarrow \ce{^{242}_{96}Cm} + \ce{^{1}_{0}}n$$

> Seaborg and coworkers also identified over 100 new isotopes of various elements.

Figure 6.52 *Cyclotrons allow scientists to investigate high-energy bombardment of heavy nuclei with various particles. Cyclotron particle masses and collision energy increased dramatically over the past decades.*

Bombarding Pu-239 with neutrons yielded americium-241, a radioisotope now used in home smoke detectors:

$$^{239}_{94}\text{Pu} + 2\ ^{1}_{0}n \longrightarrow\ ^{241}_{95}\text{Am} +\ ^{0}_{-1}e$$

Seaborg received the 1951 Nobel Prize in Chemistry for his work. Albert Ghiorso, a colleague of Seaborg, led the way in producing several new elements beyond lawrencium (element 103). One is element 106, produced by bombarding a californium-249 target with a beam of oxygen-18 nuclei to produce an isotope of element 106. To honor Seaborg's pioneering work, element 106 was named *seaborgium* (Sg). Glenn Seaborg, shown in Figure 6.53, has been called the father of the modern periodic table.

Traditionally, the discoverer of an element selects its name. For example, when Marie Curie first discovered element 84, she named it *polonium* (Po) in honor of Poland, her home country.

Figure 6.53 *Glenn Seaborg and coworkers at the University of California-Berkeley synthesized and identified ten new elements beyond uranium. Can you identify any transuranium elements that they may have created?*

> Scientific research, like other human endeavors, can involve strong personalities, competition, and controversy.

Several scientific laboratories have claimed to have synthesized elements with atomic numbers greater than 92. For example, laboratories in both the former Soviet Union and the United States claimed the discovery of elements 104 and 105. USSR scientists proposed naming them *kurchatovium* (Ku) and *dubnium* (Db), while U.S. scientists proposed the names *rutherfordium* (Ru) and *hahnium* (Ha). The International Union of Pure and Applied Chemistry (IUPAC) examined the claims and supporting evidence for element discovery before recommending the official names. In 1997 the IUPAC approved the names *rutherfordium* and *dubnium*.

At present, the IUPAC recognizes official names and symbols for the first 111 elements. Although claims for the discovery of other elements have been filed, the IUPAC has not yet recommended their official names. Scientists temporarily identify such unnamed elements by Latin prefixes indicating their atomic numbers. For example, element 112 is temporarily named *ununbium* (un = 1, un = 1, bi = 2), which is symbolized as *Uub* until IUPAC officially recognizes the original discovery of that new element.

The production of transuranium elements has enriched our understanding of the atomic nucleus. With the synthesis and identification of elements beyond atomic number 92, the periodic table has expanded to fill the actinide series as well as nearly all of period 7.

The name proposed by discoveries of element 111—*roentgenium* (Rg)—was approved by IUPAC in 2004. It honors W. K. Roentgen, who discovered X-rays. Roentgen, you may recall, was the first scientist you learned about in this unit (see page 480).

Making Decisions

C.9 OPINIONS ABOUT RADIOACTIVITY

Some older people tend to associate nuclear technologies with the use of atomic weapons at the close of World War II and with the atomic-weapon threats of the Cold War in the 1950s and beyond. Some of these people are likely to be in the audience when you speak to the Riverwood senior citizens.

The following opinions might be expressed by such community members. Decide how you would respond to each opinion, using knowledge you have gained in this unit.

1. "I'm against having any isotopes in Riverwood. They're too dangerous."
2. "I don't want to live near anything that is radioactive."
3. "I don't know why scientists keep trying to make new elements; all the new ones are radioactive."

4. "I don't understand how cancer can be treated with radiation. I thought radiation *caused* cancer."

5. "We must outlaw radioactive material altogether to completely eliminate the possibility of a nation or terrorists using radioactive material to harm others."

Join with a classmate and share your responses to each opinion. How are your responses similar? How are they different?

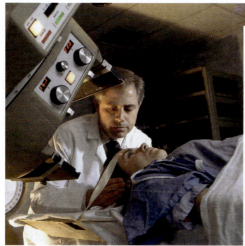

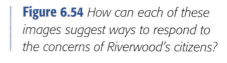

Figure 6.54 *How can each of these images suggest ways to respond to the concerns of Riverwood's citizens?*

SECTION C SUMMARY
Reviewing the Concepts

Half-life is used to characterize radioisotopes.

1. Define *half-life*.

2. How can the concept of half-life be used to determine a material's age?

3. The half-life of carbon-14 is 5730 years. Provide a rough estimate of the percent of the C-14 radioisotope that would be left after
   a. 24 h.
   b. 100 y.

4. The half-life of astatine-209 is 5.4 h. Estimate the percent of At-209 that would be left after
   a. 24 h.
   b. 100 y.

5. Phosphorus-32 has a half-life of 14.3 d. How long would it take for 1/8 of a sample of P-32 to remain unchanged?

6. Given a sample of 4.5 mol radon-222 (half-life = 3.82 d), how many moles of Rn-222 would remain undecayed after
   a. 3.82 d?
   b. 15.3 d?
   c. 28.0 d?

Radioisotopes can be used as tracers for diagnostic purposes.

7. How is metastable Tc-99m useful in medical diagnosis?

8. How is iron-59 used for diagnostic purposes?

9. Why do physicians use a radioisotope of iodine to detect thyroid problems?

10. Radioactive sodium chloride is appropriate for diagnosing circulatory problems, while radioactive xenon is helpful in searching for lung problems. Explain why each is used for its specific applications.

11. Medical personnel are selecting a radioisotope for diagnostic use. Why is each of the following considerations important in selecting the most suitable radioisotope?
    a. half-life
    b. mode of decay
    c. chemical properties of the element

Ionizing radiation emitted by some radioisotopes can be used for medical treatment.

12. How are cancer cells different from normal cells?

13. How do physicians use ionizing radiation to treat cancer?

14. What could happen if the source of ionizing radiation used to treat cancerous growth were
    a. too weak?
    b. too strong?

15. Name three radioisotopes that radiologists can use externally to treat cancer.

The conversion (transmutation) of one element to another can be accomplished by bombarding atomic nuclei with subatomic particles or other nuclei.

16. Define *transmutation.*

17. Why was Rutherford's ability to complete transmutations limited compared to the Curies' transmutation abilities?

18. Why are neutrons frequently used as projectiles in transmutation reactions?

19. Copy and complete each of the following transmutation equations:

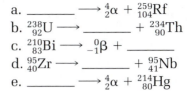

a. _____ $\longrightarrow$ $^{4}_{2}\alpha$ + $^{259}_{104}$Rf

b. $^{238}_{92}$U $\longrightarrow$ _____ + $^{234}_{90}$Th

c. $^{210}_{83}$Bi $\longrightarrow$ $^{0}_{-1}\beta$ + _____

d. $^{95}_{40}$Zr $\longrightarrow$ _____ + $^{95}_{41}$Nb

e. _____ $\longrightarrow$ $^{4}_{2}\alpha$ + $^{214}_{80}$Hg

Connecting the Concepts

20. A student wrote this statement in a homework assignment: "After one half-life, half of the mass of a material has disappeared." Do you agree or disagree? Explain.

21. Would carbon dating help scientists determine the age of dinosaur remains? Explain.

22. Explain why an externally applied alpha emitter is an ineffective treatment for tumors deep within the body.

23. Graph the data in the following table and determine the

   a. half-life of thorium-234.

   b. minimum number of days required for thorium-234 to decay to 20% of its original activity.

24. Scientists using carbon-14 dating generally do not use it to extend back further than seven half-lives.

   a. Explain why.

   b. What is the maximum number of years a substance can be dated using carbon-14 dating?

25. A radiologist injects a patient's bloodstream with a radioisotope tracer sample registering 10 000 cpm of radioactivity. Soon after, a technician draws 6.0 mL of blood, and the sample shows an activity of 10 cpm. What is the patient's total blood volume, assuming essentially no decrease in the activity of the radioisotope occurred during this clinical procedure?

Elapsed Time (days)	% Thorium-234 Activity Remaining
0	100
7	82
16	60
28	45
42	30
62	17
94	7

Extending the Concepts

26. Compare PET and MRI in terms of
    a. their method of operation.
    b. data they provide for diagnosis.
    c. radiation exposure for a patient.

27. Scientists originally were uncertain whether the oxygen gas produced during photosynthesis came from CO_2, H_2O, or both. How could radioisotope tracers be used to help settle that uncertainty?

28. A newly discovered element with an extremely short half-life is detected by analyzing its decay products. Explain how scientists can "work backward" to identify the original element.

29. Many gemstones are irradiated. Research and report on the reasons for this process.

30. Gold and other precious metals can be created from other metals by nuclear transformations.
    a. So why don't commercial firms do this as a source of profit?
    b. Is the synthetic precious metal distinguishable from the naturally occurring precious metal?

31. Research how the proton–neutron ratio is related to the stability of an atomic nucleus. Describe how you could use this ratio to predict the type of radioactive decay a particular atomic nucleus undergoes.

32. List foods that are currently irradiated with nuclear radiation before they are marketed for human consumption. Describe the process and evaluate its risks and benefits.

Nuclear Energy: Benefits and Burdens

In the 1930s, a bombardment reaction involving uranium unlocked a new energy source and led to the development of both nuclear power and nuclear weapons. This event marked the start of the nuclear age. How did scientists first unleash the enormous energy of the atom, and how have nuclear engineers harnessed atomic energy for both useful and destructive purposes?

D.1 UNLEASHING NUCLEAR FORCES

Shortly before the start of World War II, German scientists Otto Hahn and Fritz Strassman bombarded uranium with neutrons in the hope of creating a more massive nucleus and, thus, a new element. Much to their surprise, they found that one reaction product was atoms of barium, with only about half the atomic mass of the original target uranium atoms.

Figure 6.55 *Lise Meitner was first to suggest that nuclei might split due to neutron bombardment.*

The first to understand what had happened was the Austrian physicist Lise Meitner (Figure 6.55), then living in Sweden, who had previously worked with Strassman and Hahn. Meitner and her nephew Otto Frisch suggested that neutron bombardment had split the uranium atom into two parts of nearly equal mass. Other scientists quickly verified Meitner's explanation.

Lise Meitner fled to Sweden when Nazis assumed control of Germany and Austria.

Hahn and Strassman had actually triggered an array of related reactions. One of the reactions that produced barium is:

$$\underset{\text{Neutron}}{{}^{1}_{0}n} + \underset{\text{Uranium-235}}{{}^{235}_{92}U} \longrightarrow \underset{\text{Barium-140}}{{}^{140}_{56}Ba} + \underset{\text{Krypton-93}}{{}^{93}_{36}Kr} + \underset{\text{Neutrons}}{3\,{}^{1}_{0}n} + \text{Energy}$$

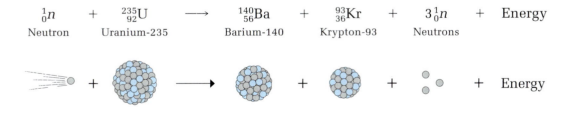

Splitting an atom into two smaller atoms is called **nuclear fission.** Scientists soon found that the uranium-235 nucleus can fission (split) into numerous pairs of smaller nuclei. The uranium usually did not split into two equal halves, but into one element accounting for about 60% of uranium's mass (such as barium) and another element equivalent to about 40% of uranium's mass (such as krypton). Here is another example of a nuclear fission reaction involving uranium-235:

$$\underset{\text{Neutron}}{_{0}^{1}n} + \underset{\text{Uranium-235}}{_{92}^{235}U} \longrightarrow \underset{\text{Xenon-143}}{_{54}^{143}Xe} + \underset{\text{Strontium-90}}{_{38}^{90}Sr} + \underset{\text{Neutrons}}{3_{0}^{1}n} + \text{Energy}$$

Fission of U-235 produces many other pairs of nuclei, such as Te-137 and Zr-97.

The nuclear fission of heavy atoms such as uranium releases a huge quantity of energy. Gram for gram, the released energy is at least a *million* times more than the energy of any chemical reaction. This is what makes nuclear explosions so devastating and nuclear energy so powerful.

Why does a nuclear reaction release much more energy than does a chemical reaction? Recall what you know about chemical reactions, such as burning petroleum. Chemical reactions involve breaking chemical bonds in reactants and making new chemical bonds in products. When bonds are stronger in products than in reactants, energy is released, often as thermal energy (heat). Thus, chemical energy is converted into thermal energy. There is no overall energy loss or gain. Similarly, mass is conserved in a chemical reaction. The nucleus of each atom, and thus its identity, remains intact in all chemical reactions. As a result, the number of atoms of each element remains unchanged; the atoms simply become rearranged. Balanced chemical equations illustrate this conservation of atoms and mass.

Not all nuclei are fissionable. U-235 is the only naturally occurring isotope that undergoes fission with lower-energy (thermal) neutrons. However, many synthetic nuclei (e.g., U-233, Pu-239, and Cf-262) also fission under neutron bombardment.

Nuclear reactions are also based on conserving energy and mass. However, during nuclear fission, very small quantities of mass are converted into appreciable energy. Where does this energy originate?

The origin of nuclear energy lies in the force that holds protons and neutrons together in the nucleus. This force, called the **strong force,** is fundamentally different from, and a thousand times stronger than, the electrical forces that hold atoms and ions together in chemical bonds. The strong force operates over very short distances, extending only across an atom's nucleus.

The forces holding nuclear particles together in the two atomic nuclei produced during U-235 fission are stronger than those in the nucleus of the uranium atom that was split. A small loss of mass results from forming two new nuclei and is converted into a large quantity of released energy.

If one kilogram of U-235 fissions, a mass of about one gram would be converted into energy.

How much mass and energy are involved? The mass loss is very small, often less than 0.1% of the total mass of the fissioning atom. Even so, the conversion of these small quantities of mass into energy accounts for the vast power of nuclear reactions.

Albert Einstein's famous equation relates mass and energy: $E = mc^2$. This equation indicates that the energy released (E) equals the mass lost (m) multiplied by the speed of light (a very large number) squared (c^2). If one gram of matter were fully converted to energy, the energy released would equal that produced by burning 700 000 gal of high-octane gasoline!

Such nuclear energy release has been harnessed by engineers to generate electricity (see Figure 6.56) and to create atomic weapons. However, the fission of one nucleus does not produce enough energy for practical use. How are fission reactions sustained to involve much larger quantities of nuclei?

Note from the equations on page 541–542 that another product of nuclear fission is the release of neutrons. These emitted neutrons can sustain the fission reaction by serving as reactants to split additional fissionable nuclei, which produce additional neutrons, which can split additional fissionable nuclei, and so on. The result is a **chain reaction** (see Figure 6.57).

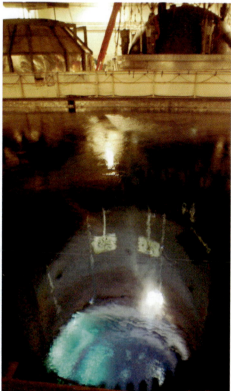

One gram of mass loss (1×10^{-3} kg) times the speed of light (3×10^8 m/s) squared equals 9×10^{13} J of energy.

Figure 6.56 *The core of a fission reactor, based on a nuclear chain reaction, emits visible light due to ionizing radiation released.*

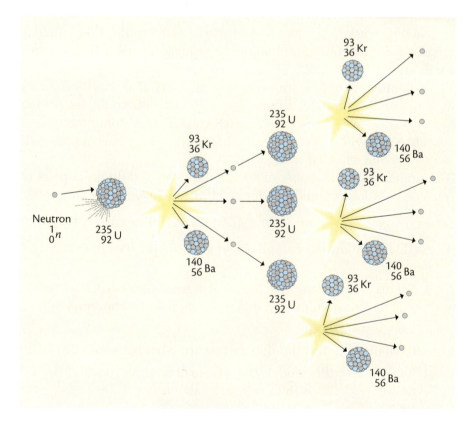

Figure 6.57 *A nuclear chain reaction. A neutron colliding with a uranium-235 nucleus initiates the reaction (left). The reaction continues and grows, as emitted neutrons encounter and split the nuclei of other fissionable atoms.*

Recall, however, that most of an atom is empty space. The probability that a neutron from a fission reaction will hit and split another fissionable nucleus depends on how much fissionable material is available. Unless a certain **critical mass** (minimum quantity) of fissionable material is present, the neutrons are not likely to encounter enough fissionable nuclei to sustain the reaction. However, if a critical mass of fissionable material is present, a chain reaction can occur, as depicted in Figure 6.57, page 543. Shortly after the first fission reactions were explained in 1939, scientists recognized that they could employ large-scale nuclear reactions in military weapons. Germany and the United States soon initiated projects to build atomic bombs during World War II. In 1945, U.S. planes dropped two such bombs onto Hiroshima and Nagasaki in Japan, which led rapidly to the end of the war.

More recently, nuclear engineers have used the energy produced by nuclear fission chain reactions to generate electricity. They carefully monitor and control the rate of fission for such uses. Nuclear power plants harness the enormous energy produced by nuclear fission reactions, while also minimizing the risks of an uncontrolled chain reaction. You will soon learn more about these design features.

Modeling Matter

D.2 THE TUMBLING DOMINO EFFECT

Chain reactions sustain nuclear fission reactions in applications such as electrical power generation and atomic weapons. In this activity, dominoes will model some aspects of a chain reaction.

Each domino that falls represents a nucleus that has been split through fission. Figure 6.58 shows one way that you could set up the dominoes so that making one domino fall causes other dominoes also to fall. Because you will model a specific fission reaction, your models will not match the one depicted in Figure 6.58.

The uranium-235 nucleus can fission into over 100 different pairs of nuclei. One way, for example, produces tellurium-137 and zirconium-97:

$$\,^{1}_{0}n + \,^{235}_{92}U \longrightarrow \,^{137}_{52}Te + \,^{97}_{40}Zr + 2\,^{1}_{0}n + Energy$$

1. As this equation shows, splitting *one* U-235 nucleus releases *two* neutrons.

   a. Set up all the dominoes you receive from your teacher so that each falling domino will make two more erect dominoes fall.

   b. Sketch your setup.

   c. Push over the first domino and record what happens.

   d. Explain how this models the release of neutrons during the fission of U-235 as in the equation above.

   e. What aspects of the U-235 fission reaction are not modeled by the behavior of your domino setup?

2. Now you will assemble a critical mass of atoms (dominos) to initiate a chain reaction.

   a. Set up the dominoes as in Step 1a.

   b. Assume that only dominoes (atoms) with seven total dots are fissionable.

   c. Remove all dominoes from your setup that do not have seven dots.

   d. Sketch your new setup.

   e. Push over the first domino and record what happens.

   f. In what way does this model help clarify the idea of *critical mass?*

   g. How can you ensure that you have a critical mass of "fissionable" dominoes in a particular setup?

3. Suppose you need to control the total neutrons emitted so that fission is just sustained.

   a. Set up the dominoes as in Step 1a.

   b. Devise a plan so that only *half* the dominoes will fall when you push over one domino. You should not remove any dominoes that are already set up.

   c. Describe and sketch your strategy.

   d. Try your plan and record what happens.

   e. What stopped the dominoes from falling?

   f. Use a domino model to describe how to control fission chain reactions.

4. Compare the domino arrangements in Questions 1 and 3.

   a. Which is a better model of an atomic-weapon explosion? Explain.

   b. Which is a better model of fission in a nuclear power plant? Explain.

5. Propose another way to model a nuclear chain reaction.

   a. Explain how your model illustrates features of a nuclear chain reaction.

   b. Explain some limitations of your model.

Figure 6.58 *Domino behavior can model key characteristics of a chain reaction.*

D.3 NUCLEAR POWER PLANTS

Figure 6.59 *A nuclear power plant. Notice the reactor containment building with its domed roof.*

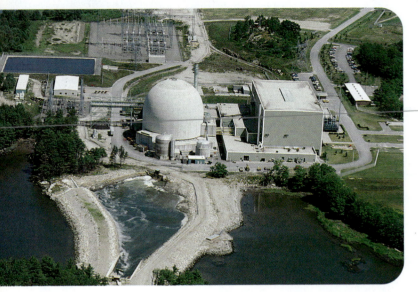

The first nuclear power reactors were designed and built during World War II. Since then, commercial companies have built many nuclear reactors to generate electricity, such as the nuclear power plant shown in Figure 6.59. In 2004, slightly more than 100 commercial nuclear reactors were generating electricity in the United States. Figure 6.60 shows where these reactors are located. On a global scale, an estimated 441 nuclear reactors in over 30 nations produce about one-sixth of the world's electricity.

Most conventional power plants generate electricity by burning fossil fuels to boil water and produce steam. A nuclear power plant operates in much the same way. However, instead of using fossil-fuel combustion to boil water, **nuclear power plants** use thermal energy released from nuclear-fission reactions to heat water and produce steam. The steam spins turbines of giant generators, producing electricity.

The essential parts of a nuclear power plant, diagrammed in Figure 6.61, include the fuel rods, control rods, moderator, generator, and cooling system.

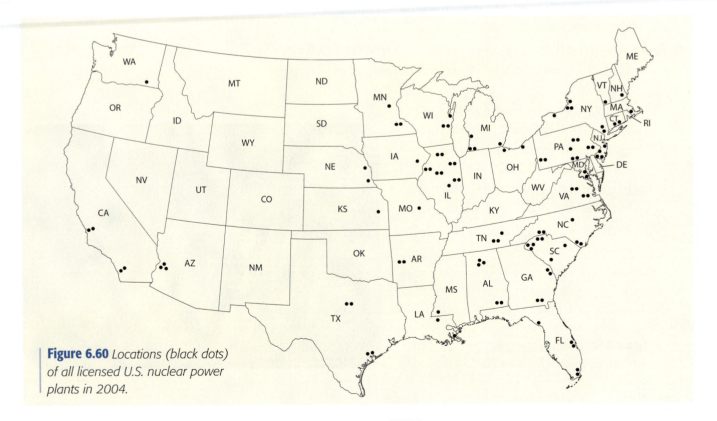

Figure 6.60 *Locations (black dots) of all licensed U.S. nuclear power plants in 2004.*

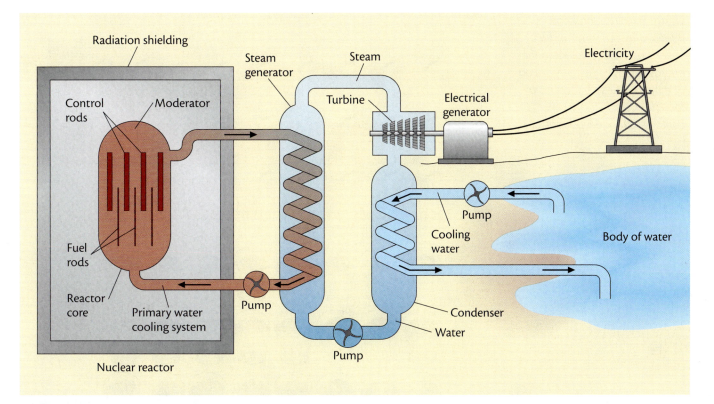

Figure 6.61 *Components of a nuclear power plant. Note the water flowing between the reactor and steam generator (red), and between the steam generator and outside body of water (blue). This water flow is important in controlling the power-plant operating temperature.*

A Nuclear Power Plant

Fuel Rods

Coal-fired power plants burn thousands of tons of coal daily. By contrast, nuclear reactor fuel occupies a small fraction of the volume needed for coal and is replenished only about once annually. Nuclear reactor fuel is small uranium dioxide (UO_2) pellets about the size and shape of short pieces of chalk. The energy in one uranium fuel pellet equals the energy in one ton of coal or 126 gal of gasoline (see Figure 6.62, page 548). One nuclear power plant uses as many as 10 million fuel pellets. These pellets are arranged inside long, narrow steel cylinders—the *fuel rods*. The fission chain reaction occurs inside these rods (see Figure 6.61).

Most uranium dioxide in fuel pellets is composed of the nonfissionable uranium-238 isotope. Only 0.7% of natural uranium is U-235, the fissionable isotope. In reactor fuel rods, U-235 composition has been enriched to about 3%, which is only a small fraction of total fuel-rod material. It is sufficient to sustain a chain reaction, but far less than enough to cause a nuclear explosion.

Weapons-grade uranium usually contains 90% or more of the fissionable U-235 isotope.

Figure 6.62 *The total energy released by one nuclear fuel pellet (top) is equivalent to the energy released by combustion of either of these quantities of fossil fuel. Each coal bucket contains about 65 pounds of coal.*

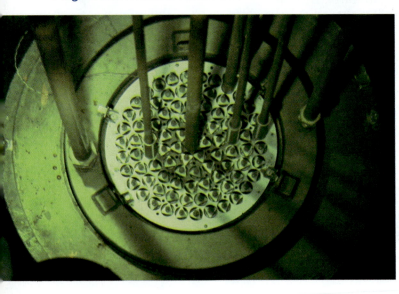

Figure 6.63 *A nuclear reactor core.*

Control Rods

The nuclei of some elements, such as boron or cadmium, can absorb neutrons very efficiently. Such materials are placed in *control rods*, which regulate the number of neutrons available for causing fission. Moving the control rods up or down (see Figure 6.63) between the fuel rods regulates the rate of the nuclear chain reaction. The reaction can be completely stopped by dropping the control rods all the way down between the fuel rods and absorbing nearly all neutrons released as U-235 fissions.

Moderator

In addition to fuel rods and control rods, the core of a nuclear reactor contains a *moderator*, which slows down high-speed neutrons. This allows the fuel rods to more efficiently absorb neutrons, enhancing the probability of fission.

Heavy water (with the formula D_2O, where D represents the hydrogen-2 isotope, deuterium), regular water (termed *light water* by nuclear engineers), and graphite (carbon) are the three most commonly used moderator materials.

Generator

In commercial nuclear reactors, the fuel rods and control rods are usually surrounded by a system of circulating water. In simpler reactors, the heat released by the fuel rods boils this water, and the resulting steam spins the turbines of the *electrical generator.* In another type of reactor, this water is superheated under pressure and does not boil. Instead, it circulates through a heat exchanger, where it boils the water contained in a second cooling loop. This type of reactor is illustrated in Figure 6.61 (page 547).

Cooling System

Steam moves past the turbines and travels through pipes where it is cooled by water drawn from a nearby body of water. The cooled steam condenses to liquid water and circulates inside the generator. So much thermal energy is generated that some steam must also be released into the air. The largest and most prominent feature of most nuclear power plants are tall, gracefully curved concrete cylinders, called *cooling towers,* where excess thermal energy is released. Some observers mistakenly assume that the cooling tower is the nuclear reactor.

The white plumes seen rising from such cooling towers are condensed steam, not smoke.

A nuclear reactor is designed to prevent the full escape of radioactive material if a malfunction causes the release of radioactive material, including cooling water, within the reactor itself. The core of a nuclear reactor is surrounded by concrete walls two to four meters thick. Further protection is provided by enclosing the reactor in a building with thick walls of steel-reinforced concrete designed to withstand a chemical explosion or an earthquake. The reactor building is also capped by a domed roof that can withstand high internal pressure.

The well-known nuclear accident in 1986 at one of the four reactors at the Ukraine's Chernobyl power plant occurred because too many control rods were withdrawn from the reactor and were not replaced fast enough. There was little control of the fission process, resulting in the buildup of considerable steam. The resulting explosion was not nuclear, but was due to the buildup of high-temperature steam and to the chemical reactions that it triggered. Unfortunately, the plant had been built without a surrounding concrete containment building, unlike currently operating U.S. reactors, so a large quantity of nuclear material from the reactor was also released directly into the environment.

In the Chernobyl accident, one chemical reaction that caused trouble was high-temperature steam reacting with carbon from the moderator, producing CO and H_2. When H_2 mixed with air, it became explosive.

Nuclear fission is not the only way to liberate nuclear energy. Soon you will learn about the kind of nuclear reaction that fuels the stars (and, indirectly, fuels all living matter).

Nuclear Power in Action: Submarines

With the possible exception of the space shuttle, nothing has as much technology crammed into as small a space as a nuclear-powered submarine. Nevertheless, you can be fully trained to operate the nuclear propulsion system on a submarine by the time you are 19—that is, if you pass Navy Lieutenant Mike Arnold's class. After six months of intensive training on land in Charleston, South Carolina, Lieutenant Arnold's students are ready for a challenging career under the oceans.

What Does a Submariner Do?

Before becoming a trainer at the Nuclear Power Training Unit, Navy Lieutenant Mike Arnold spent three years as a junior officer on the nuclear submarine *USS L. Mendel Rivers.* There he directed the operation of the nuclear propulsion plant, piloted through some of the most beautiful regions of the oceans, and, along with his crewmates, carried out defense missions.

Each submariner has a particular duty called a *watch*. Each watch is so distinct and essential to

the mission of the submarine that the entire crew on duty operates as a single entity. Individuals work in six-hour shifts, around the clock; no watch is ever left unattended.

Why Do Submarines Use Nuclear Power, While Most Ships, Land Vehicles, and Aircraft Have Combustion Engines?

Submarines must generate their own electricity and propel themselves underwater for months at a time. The key to the role of nuclear power, however, is the word *underwater*. Burning fuel (combustion) requires oxygen, but there is no way to provide enough usable oxygen underwater to fuel lengthy trips. Thus, controlled nuclear reactions are used to power submarines. Some of this energy is also used to generate electricity to operate the equipment and make life possible for the crew. For example, oxygen gas and freshwater are made from seawater using energy generated by nuclear reactions.

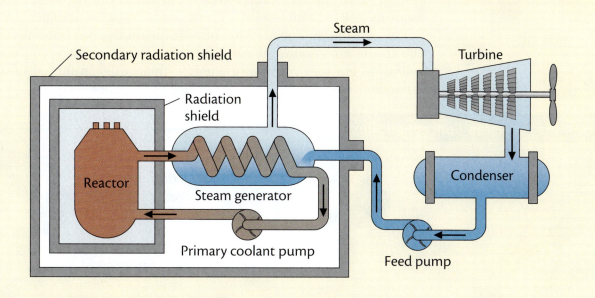

How Does the Nuclear Propulsion System Onboard a Submarine Work?

On a submarine, the turbines that generate electricity and the propellers are spun by steam. The energy used to make this steam comes from the controlled fission of uranium atoms in the core of the nuclear reactor. The design of the nuclear reactor is classified information, but the principles are not.

The heat generated by the fission reaction is transferred to water in a highly pressurized loop of piping called the *primary loop*. This water enters a steam generator, where it transfers heat to water running through a separate loop of piping, the *secondary loop*. The water in the primary loop goes back to the reactor. The water in the secondary loop is converted to steam and heads for the turbines, where the energy from the steam is converted to electricity and mechanical energy. The steam then goes through a condenser, where seawater is used to cool the steam to its liquid state, the condensed liquid heads back to the steam generator, and the cycle continues.

Questions to Explore

1. Investigate the dimensions of a submarine and find out how many crew share the living space during a typical voyage.

2. The world's first nuclear-powered submarine, the *USS Nautilus,* was launched in 1955. However, submarines were used in the Civil War, and even Alexander the Great's soldiers are said to have used them. Search the Web to learn how submarines were powered before the advent of nuclear propulsion.

3. Since the end of the Cold War era, some scientists have used Navy submarines for their underwater research. Find out what they have been studying at ocean depths by searching www.pbs.org/wgbh/nova/subsecrets/inside.html. While you are at this site, take a virtual tour of a submarine.

D.4 NUCLEAR FUSION

In addition to releasing energy by splitting massive nuclei (fission), large quantities of nuclear energy can be generated by *fusing*, or combining, small nuclei. **Nuclear fusion** involves forcing two relatively small nuclei to combine into a new, more massive nucleus. As with fission, the energy released by nuclear fusion can be enormous, again due to the conversion of mass into energy. Gram for gram, nuclear fusion liberates even more energy than does nuclear fission—between three and ten times more energy (Figure 6.64).

Scientific American Conceptual Illustration

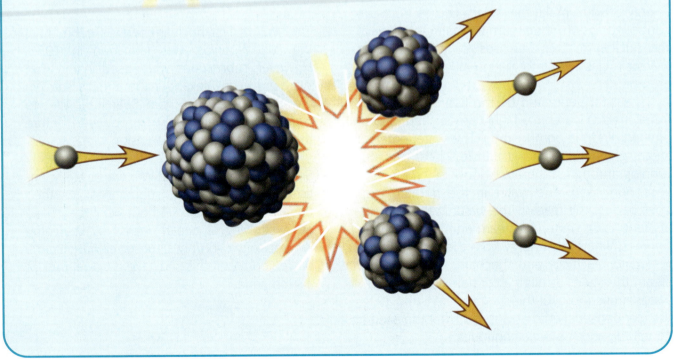

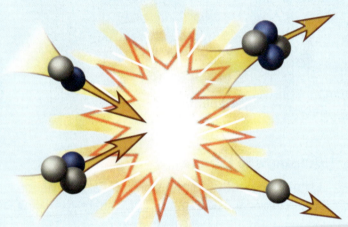

Figure 6.64 Nuclear fission and fusion. *Fusion (top) occurs when smaller nuclei (protons in blue, neutrons in grey) combine to form larger nuclei. Fission (bottom) occurs when a neutron of appropriate energy collides with certain large nuclei, creating two smaller nuclei. Both processes can release large quantities of energy, according to Einstein's famous equation, $E = mc^2$, where m is the mass converted into energy (E) and c is the speed of light. However, the total energy released decreases as the nuclei involved approach iron-56, the most stable nucleus.*

Nuclear fusion powers the Sun and other stars (Figure 6.65). Scientists believe that the Sun formed when a huge quantity of interstellar gas, mostly hydrogen, condensed under the force of gravity. As the volume of gas decreased, its temperature increased to about 15 million degrees Celsius, and hydrogen nuclei began fusing into helium. The nuclei that fused together were all positively charged and tended to repel one another. The high temperature gave each nucleus considerable kinetic energy, which helped overcome the repulsions.

Once fusion was started, the Sun began to shine, converting nuclear energy into radiant energy. Scientists estimate that the Sun, believed to be about 4.5 billion years old, is about halfway through its life cycle.

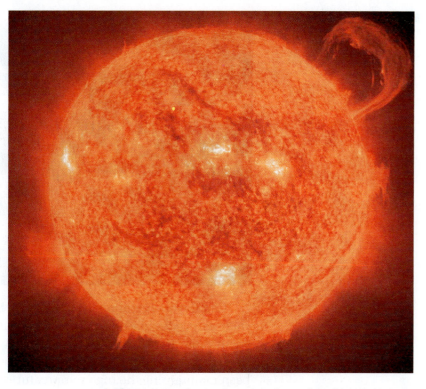

Figure 6.65 *The Sun's energy is produced by continuous nuclear-fusion reactions.*

The nuclear-fusion reactions occurring in the Sun are rather complicated, but the end result is the conversion of hydrogen nuclei into helium nuclei. The overall result can be summarized by this equation:

$$4\,^1_1\text{H} \longrightarrow\ ^4_2\text{He}\ +\ 2\,^0_{+1}e\ +\ \text{Energy}$$

Hydrogen-1   Helium-4   Positrons

How much energy does such a nuclear fusion reaction produce? Comparing the total mass of reactants to the total mass of products reveals that 0.006 900 5 g is lost when one gram of hydrogen-1 atoms fuse to produce one gram of helium-4 atoms. Using Einstein's equation, $E = mc^2$, the energy released through the fusion of one gram of hydrogen atoms (one mole H) is 6.2×10^8 kJ. Here is one way to put this very large quantity of energy into perspective: The nuclear energy released from the fusion of one gram of hydrogen-1 equals the thermal energy released by burning nearly 5000 gal of gasoline or 20 tons of coal.

Powerful military weapons incorporate nuclear fusion. The hydrogen bomb, also known as a *thermonuclear device,* is based on a fusion reaction that uses the thermal energy from the explosion of a small atomic (fission) bomb to initiate fusion.

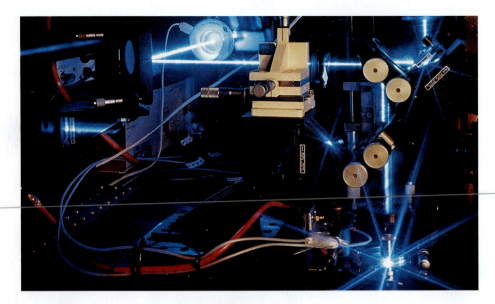

Figure 6.66 *A deuterium-tritium fuel pellet placed in the path of a laser beam, to initiate fusion.*

Can the energy of nuclear fusion be harnessed for beneficial purposes, such as producing electricity? This remains to be seen. Scientists have spent more than five decades pursuing this possibility. They have tried many schemes, but have not yet succeeded. The major difficulties have been maintaining the high temperatures needed for fusion while also containing the reactants and fused nuclei. Figure 6.66 shows one of these fusion experiments. The nuclear fusion facility at Princeton University has achieved temperatures sufficient to initiate fusion, but so far the total quantity of energy put into the process is more than the total energy released.

If scientists finally succeed in controlling nuclear fusion in the laboratory, there is still no guarantee that fusion reactions will become a practical source of energy. Low-mass isotopes needed to fuel such reactors are plentiful and inexpensive, but confinement of the reaction could be very costly. Furthermore, although the fusion reaction itself produces less radioactive waste than does nuclear fission, capturing the positrons and shielding the heat of the reaction could generate nearly as much radioactive waste as that produced now by fission-based power plants.

In splitting and fusing atoms, the nuclear energy that fuels the universe has been unleashed. Much good has arisen from it, but so have scientific, social, and ethical questions. Along with great benefits come great risks. How much risk is worth any potential benefits? In the next activity you will analyze risks and benefits associated with deciding how to travel, to see a friend.

> Electing *not to make a* decision is, in fact, also a decision—one with its own risks and benefits.

Making Decisions

D.5 THE SAFEST JOURNEY

Suppose you want to visit a friend who lives 500 mi (800 km) away, using your safest means of transportation. Insurance companies publish reliable statistics on the safety of different methods of travel. Using Table 6.11, answer the following questions:

RISK OF TRAVEL	
Mode of Travel	**Distance (Miles) Traveled at Which One Person in a Million Will Suffer Accidental Death***
Bicycle	10
Automobile	100
Train	120
Bus	500
Scheduled Airline	1900

*On average, chance of death is increased by 0.000 001.

Table 6.11 *The extent of risk associated with travel depends both on the risk of an accident and on the total time spent in that mode of travel.*

1. Assume there is a direct relationship between distance traveled and chance of accidental death (that is, assume that doubling the distance doubles your risk of accidental death).

   a. What is the risk factor (chance of accidental death) for traveling 500 mi by each mode of travel listed in the table? For example, Table 6.11 shows that the risk factor for biking increases by 0.000 001 for each 10 mi traveled. Therefore, the bike-riding risk factor in visiting your friend would be

   $$500 \text{ mi} \times \frac{0.000\ 001 \text{ risk factor}}{10 \text{ mi}} = 0.000\ 05 \text{ risk factor}$$

   b. What is the safest mode of travel (the one with the smallest risk factor)?
   c. What is the riskiest mode of travel (the one with the largest risk factor)?

2. a. List some benefits associated with each mode of transportation.
   b. List some risks associated with each mode of transportation.
   c. In your view, do benefits of riskier ways to travel outweigh their increased risks? Explain your reasoning.
   d. Identify some situations in which assumptions made in Question 1 would be invalid.

3. Do you think the statistics in Table 6.11 will apply 25 years from now? Explain.

4. What factors, beyond the risk to personal safety, would you include in a risk-benefit analysis before you decided on how to travel?

5. a. Which mode of travel would you choose? Why?
   b. Would another person's risk-benefit analysis always lead to your decision? Why or why not?

RISK-FREE TRAVEL?

Is there any way to travel to visit your friends or relatives that would be completely risk-free?

Would it actually be safer not to visit them at all? Why?

D.6 NUCLEAR WASTE: PANDORA'S BOX

Imagine that you live in a home that was once clean and comfortable, but now you have a major problem: You cannot throw away your garbage. The city forbids garbage removal because it has not decided what to do with the garbage. For nearly 50 years, your family has compacted, wrapped, and saved the garbage as efficiently as possible, but you are running out of room. Some bundles leak and are a health hazard. What can be done?

The U.S. nuclear power industry, the nuclear weapons industry, and medical and research facilities have a similar problem. Spent (used) nuclear fuel and radioactive waste products have been accumulating for nearly 50 years (see Figure 6.67). Some of these materials are still highly radioactive, while other materials—even initially—display low levels of radioactivity. It is uncertain where these materials, regardless of radioactivity levels, will be permanently stored or how soon.

Figure 6.67 *Although low-level radioactive waste does not require the same disposal methods as do spent fuel rods from nuclear reactors, serious hazards are posed by improper disposal of such wastes.*

Nuclear Wastes

Two broad categories of nuclear waste are *high-level* and *low-level*. **High-level nuclear wastes** are either (a) *products of nuclear fission,* such as those generated in a nuclear reactor; or (b) *transuranics,* products formed when the original uranium-235 fuel absorbs neutrons. For example, plutonium-239 is a transuranic material. **Low-level nuclear wastes** have much lower levels of radioactivity. These wastes include used nuclear laboratory protective clothing, diagnostic radioisotopes, and air filters from nuclear power plants. Figure 6.68 illustrates the composition of radioactive wastes in the United States.

Figure 6.68 emphasizes distinguishing *waste volume* from *radioactivity level* when considering nuclear wastes; high volume does not necessarily mean high radioactivity. For example, Figure 6.68 shows that military defense efforts produce the largest volume of radioactive wastes (57%), but this waste contributes slightly more than 4% of the total radioactivity in nuclear wastes. On the other hand, spent fuel from nuclear reactors occupies less than 1% of the volume of all radioactive wastes, yet it contributes 96% of the total radioactivity. Moreover, high-level nuclear wastes also can have extended half-lives, some as long as thousands of years.

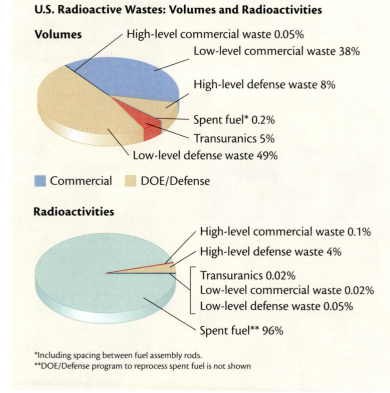

U.S. Radioactive Wastes: Volumes and Radioactivities

Volumes
- High-level commercial waste 0.05%
- Low-level commercial waste 38%
- High-level defense waste 8%
- Spent fuel* 0.2%
- Transuranics 5%
- Low-level defense waste 49%

■ Commercial   ■ DOE/Defense

Radioactivities
- High-level commercial waste 0.1%
- High-level defense waste 4%
- Transuranics 0.02%
- Low-level commercial waste 0.02%
- Low-level defense waste 0.05%
- Spent fuel** 96%

*Including spacing between fuel assembly rods.
**DOE/Defense program to reprocess spent fuel is not shown

Figure 6.68 *Volumes and radioactivities of radioactive waste.*

Current Disposal Methods

Because high-level and low-level nuclear wastes have different characteristics, people dispose of them differently. We can put low-level wastes into sealed containers and bury the containers in lined trenches 20 ft deep. Low-level military nuclear waste is disposed of at federal sites maintained by the Department of Energy. Since 1993, each state is responsible for the disposal of its own low-level commercial nuclear wastes. Groups of states in several regions have formed compacts in which all members use a disposal site in one state for low-level wastes from all compact members.

High-level nuclear waste disposal requires a very different approach. Spent nuclear fuel presents the greatest challenge. A commercial nuclear reactor (power plant) typically produces about 30 tons of spent fuel annually. That means that commercial nuclear power plants now operating in the United States generate roughly 3100 tons of waste annually. Approximately 22 000 tons of spent fuel are already stored in 34 states. Eventually, of course, the components of nuclear power-plant reactors will also become nuclear waste.

The U.S. nuclear weapons program contributed about 20 million liters of additional stored waste. The volume of military waste is much greater than the quantity of commercial waste. This waste is in the form of a sludge—the waste product of extracting plutonium from spent fuel rods in military reactors. (Nuclear weapons are created from the plutonium.) As radioisotopes in the waste gradually decay, they emit radiation and thermal energy. In fact, without external cooling, such waste can become hot enough to boil. This makes military waste containment a continual challenge.

Operators must replace up to one-third of a nuclear reactor's fuel rods annually. This is necessary because the uranium-235 fuel becomes depleted as fission occurs and because accumulated fission products interfere with the fission process. The spent fuel rods are highly radioactive, with some isotopes continuing to decay for many thousands of years. Table 6.12 lists the half-lives of three radioisotopes produced by nuclear fission. All of these and many more are contained in spent fuel rods.

By federal law, nuclear reactor waste must be stored on-site, usually in nuclear waste storage tanks, until a permanent repository is created. Available storage space on-site is limited, however. The federal government is responsible for the final storage of high-level radioactive waste, but it has not yet opened permanent long-term disposal sites for high-level wastes. Legal negotiations, congressional debates, and developmental work are in process to develop such a site at Yucca Mountain, Nevada. High-level nuclear wastes may eventually be stored there deep underground.

SOME RADIOISOTOPES FOUND IN SPENT FUEL RODS	
Isotope	**Half-Life**
Plutonium-239	24 110 y
Strontium-90	28.8 y
Barium-140	12.8 y

| **Table 6.12**

Investigating Long-Term Disposal Methods

The method of long-term radioactive waste disposal favored by the U.S. government (and by many other nations) is *mined geologic disposal*. The radioactive waste would be buried at least one kilometer below earth's surface and at least one kilometer from any water table in vaults that would presumably remain undisturbed.

To prepare radioactive waste for burial, spent fuel rods would first be allowed to cool for several decades in very large tanks of water. Over time, many of the radioisotopes would essentially decay, lowering the radioactivity level to a point where technicians could safely handle the materials. Such on-site nuclear waste storage would be only temporary; the tanks would require too much maintenance to remain safe for more than several decades.

Each nuclear reactor site already employs this cooling process for wastes.

Next, technicians would lock cooled radioactive wastes inside packages engineered to be leak-proof. The currently favored method is to transform the wastes by *vitrification* (see Figure 6.69) into a type of solid ceramic. Although the encased material would still be highly radioactive, the waste would be much less likely to leak or leach into the environment because of the glasslike envelope encasing it. Technicians could seal the vitrified radioactive wastes in containers made of

Vitrification is conversion of material into a glassy solid by application of heat.

glass, stainless steel, or concrete. The Savannah River vitrification plant in South Carolina is the world's largest facility of its kind. France has used vitrification for over a decade (nuclear reactors produce about 78% of France's electricity).

Figure 6.69 *One strategy to prevent radioactive waste from entering water supplies is to vitrify the waste—that is, convert it into a glassy solid.*

Nowhere in the world, however, has nuclear waste been permanently buried. The challenge is to find technically, politically, and socially acceptable sites. The Japanese government has even considered deep-ocean burial.

Some geologic sites formerly assumed to be stable enough for radioactive waste disposal were later discovered to be unsafe. For example, technicians buried some plutonium at Maxey Flats, Kentucky, in a rocky formation that geologists believed would remain stable for thousands of years. Within a decade, however, some buried plutonium had migrated dozens of meters away.

What are current plans to resolve our long-term nuclear waste disposal problems? By U.S. law, Congress has selected two sites for permanent radioactive waste disposal from options provided by the Department of Energy. These sites, located in regions of presumed geologic stability (see Figure 6.70), are the Waste Isolation Pilot Plant near Carlsbad, New Mexico, and Yucca Mountain, Nevada, an extinct volcanic ridge 100 miles northwest of Las Vegas.

Both proposals have generated considerable debate and controversy about site locations and the means of transporting radioactive wastes to them. Scientists, engineers, citizens, and politicians will continue to address legal, environmental, and engineering issues before making final decisions about these sites.

It will probably be at least 2010 before a suitable site is fully approved to receive high-level nuclear wastes for long-term storage.

Figure 6.70 *Storing nuclear waste requires an underground site that will minimize contact with local water resources and remain geologically stable over many centuries. This poses formidable challenges; the waste will remain dangerously radioactive for thousands of years.*

D.7 DISPOSING OF HIGH- AND LOW-LEVEL WASTE

Use Figure 6.68 (page 557) to answer the following questions:

1. Approximately what percent of nuclear waste is

   a. low-level waste?
   b. high-level (including transuranic) waste?

2. Which waste source accounts for the greatest volume of high-level nuclear waste?

3. Which two waste sources account for most of the low-level nuclear waste?

4. Should high-level or low-level nuclear wastes receive greater attention? Explain your answer.

SECTION D SUMMARY
Reviewing the Concepts

Some relatively large nuclei, when bombarded by neutrons, undergo nuclear fission.

1. What is *nuclear fission*?

2. Name three isotopes that can undergo nuclear fission.

3. Write a balanced nuclear equation for the fission of U-235 by a neutron, producing Br-87, La-146, and several neutrons.

4. Why does a nuclear reaction release more energy than does a chemical reaction?

5. State Einstein's mass-energy relationship and explain the meaning of each symbol.

6. Name the force that holds nuclear particles together and describe its characteristics.

7. Describe characteristics of a nuclear chain reaction.

8. Why is a critical mass of fissionable material needed to sustain a nuclear chain reaction?

The electricity produced by nuclear power plants originates from the energy released by fission in controlled chain reactions.

9. Describe how most conventional (non-nuclear) power plants generate electricity.

10. State the equivalent quantities of coal and petroleum needed to produce the total energy contained in one nuclear fuel pellet.

11. Why is U-235 used in nuclear power plants?

12. How does each of the following affect neutrons in a nuclear power plant?
    a. control rods
    b. moderator

13. Why is it impossible for the fuel in a nuclear power plant to cause a nuclear explosion?

14. List the three common moderators used in nuclear power plants.

15. The core of a nuclear reactor is surrounded by thick concrete walls. Give three reasons for these walls.

16. What is the composition of the white plumes often seen rising from nuclear power-plant towers?

Nuclear fusion is the combination of two relatively small nuclei into a new, more massive nucleus.

17. Why are high pressures and temperatures needed to initiate fusion reactions?

18. How much more energy can nuclear fusion produce than nuclear fission?

19. Why isn't nuclear fusion currently practical for generating electricity in power plants?

20. State the equivalent quantities of coal and gasoline needed to produce the energy released by the fusion of one gram of hydrogen-1.

21. What is a *thermonuclear weapon*?

22. Explain why both of these statements are true:
    a. Nuclear fusion has not been used as an energy source on earth.
    b. Nuclear fusion is earth's main energy source.

23. a. List and describe the two categories of nuclear wastes.
    b. What is the major difference between them?

24. In the United States, what are the two largest sources of radioactive waste? Refer to Figure 6.68, page 557.

25. Refer to Figure 6.68. Spent fuel makes up what percent of total U.S. radioactive wastes by
    a. volume?
    b. radioactivity?

26. Why do technicians have to regularly replace nuclear fuel pellets even if the pellets are still radioactive?

27. Compare current methods for disposing of high- and low-level nuclear wastes.

28. Why is vitrification a preferred method for handling and storing nuclear wastes?

29. a. Where are two potential U.S. sites for permanent radioactive waste disposal?
    b. Why were these particular sites selected?

Connecting the Concepts

30. Explain the difference between nuclear *fission* and nuclear *fusion*.

31. Sometimes burning is a good way to dispose of some types of extremely toxic material. Why would burning be an unacceptable plan for destroying nuclear waste?

32. Explain why the disposal of nuclear waste is a challenging issue in many nations. Use the concepts of radioactive decay, half-life, and radiation shielding in your answer.

33. How does nuclear fusion compare to a chemical reaction in which two hydrogen atoms combine to form a hydrogen molecule, H_2?

34. Construct a diagram showing energy transformations involved in producing electricity in a nuclear power plant. How would your diagram differ for a coal-fired power plant?

35. A simple way to minimize the need for long-term storage of radioactive waste might seem to be speeding up all radioactive decay rates involved. Why won't this plan work?

36. What are some factors complicating the cleanup of abandoned or improperly stored nuclear waste?

Extending the Concepts

37. Research how other countries dispose of their high-level nuclear waste and evaluate their methods in terms of risks and benefits.

38. When confronting problems surrounding high-level nuclear waste disposal, students sometimes propose loading the waste in a rocket and shooting it into the Sun. What are some risks and benefits of this plan?

39. Fusion reactions that produce iron and lighter elements can take place in the core of an ordinary star. However, elements with higher atomic numbers generally result from violent stellar explosions. Explain the difference.

40. What would be some advantages of nuclear fusion over nuclear fission for producing electricity?

41. In the late 20th century, a group of scientists mistakenly claimed they had accomplished "cold fusion." Research this event and its scientific impact.

42. Investigate conditions needed to sustain a critical-mass chain reaction in
    a. a nuclear reactor.
    b. a nuclear weapon.

COMMUNICATING SCIENTIFIC AND TECHNICAL INFORMATION

Throughout your chemistry studies up to this point, you have used scientific ideas to evaluate and draw your own conclusions regarding published claims in the CANT flyer. Your final task in this unit is to prepare and present what you have learned to help some Riverwood residents draw their own conclusions about nuclear concerns.

Unit 6 Overview

As you already know, your audience is composed of senior citizens at a local community center. They are concerned that regulations proposed by CANT will affect their ability to receive a full range of medical care. They are also alarmed by some CANT claims about nuclear waste disposal.

The senior citizens have invited your class to address the scientific aspects of statements made in the flyer from CANT so that they can evaluate the group's message more completely.

YOUR PRESENTATION

Each student group will be responsible for responding to one or more CANT statements. In preparing your presentation, coordinate with other groups working on statements that involve similar ideas.

For each statement, your presentation should address these points:

1. Decide whether the flyer statement is *true, false,* or *partially true.*

2. Explain the science or technology that the statement involves. For example, for statements about nuclear power plants, you should review how such plants work; for statements about radiation, you should explain different types of radiation.

3. Prepare a replacement statement, if necessary, that more accurately and evenhandedly addresses the same topic/issue.

Design your presentation to be understandable by adults with at least a high school education that most likely was completed many years ago.

When possible, include visual aids and everyday examples and applications. Also devise a way to evaluate the effect of your presentation. Try to assess whether or not the audience, following your presentation, is better able to make informed decisions regarding nuclear policies for Riverwood.

UNIT 7

Food: Matter and Energy for Life

HOW is food energy stored, transferred, and released?

Glucose

WHAT chemical roles do carbohydrates and fats play in human metabolism?

WHY are protein molecules essential to living organisms?

WHAT roles do vitamins, minerals, and additives play in foods we eat?

The Riverwood High School Parent-Teacher-Student Association (PTSA) requests your help on school vending machine policies. In this unit, you will analyze a typical student diet and make recommendations based on this analysis. Turn the page to find out more.

Ervin Kostecky, a parent and a member of Riverwood High School's Parent-Teacher-Student Association (PTSA), recently read information about quality of food sold in U.S. middle school and high school vending machines. Mr. Kostecky is concerned that Riverwood High School's food and beverage vending machines do not meet nutritional needs of the students who use them. He has requested that a committee be appointed to recommend school policies regarding these vending machines.

In this unit, you will learn about the chemistry of foods. You will learn how energy contained in food is stored and released and how substances in foods promote bodily growth and repair. You will also investigate the chemistry and nutritional roles of fats, carbohydrates, proteins, vitmins, and minerals. This information will help you make sound policy recommendations to be implemented next year for the school's vending machines.

As a member of the Vending Machine Policy Planning Committee, you will receive an inventory that documents a typical Riverwood High School student's diet over three days. Together with other committee members (your classmates), you will conduct several analyses of this three-day food inventory. By the end of the unit, your results will provide insight into items that would complement a typical Riverwood High School student's diet. You will then write a report summarizing your results and proposing recommended guidelines for selecting new items for the vending machines. This report will be used in the committee's discussions and, ultimately, in selecting future food and beverage items for school vending machines.

To aid your investigation, the PTSA has highlighted some characteristics of a healthful diet found in the *Dietary Guidelines for Americans 2005,* issued by the U.S. Department of Health and Human Services.

Keep the following dietary advice as well as chemistry principles you have learned in this course in mind throughout this unit:

▶ Consume a variety of nutrient-dense foods and beverages within and among the basic food groups.

▶ Balance calories from foods and beverages ingested with calories expended in physical activity.

▶ Choose foods that limit the intake of saturated and trans fats, cholesterol, added sugars, salt, and alcohol.

▶ Consume a sufficient quantity and variety of fruits and vegetables, while staying within energy needs.

Food as Energy

Where does the energy required for walking, running, playing sports, and even sleeping and studying come from? The answer is obvious: It comes from the food you eat. As you begin to analyze the three-day food inventory, you will learn the chemistry that will allow you to understand where that energy originates, how food energy is stored and used, and how decisions about eating can affect your well-being.

A.1 FOOD GROUPS

When considering foods, people often focus on particular categories, or *food groups*. A healthful diet includes foods from a variety of groups, such as those shown in Figure 7.1. Why? Figure 7.2 highlights the groups. What do you already know about the nutrient and energy contents of foods in each group?

Although it is healthful to consume foods from a variety of groups, the *Dietary Guidelines for Americans 2005* advise that people should increase their daily intake of fruits and vegetables, whole grains, and nonfat or low-fat milk and milk products. In addition, the *Dietary Guidelines* call for decreasing intake of foods rich in fats and sugars.

Some foods listed in the three-day inventory may be represented by categories in Figure 7.2, but others are not. Where would you classify foods not included in Figure 7.2? What range of food groups is represented in the diet you are analyzing? Find out in the next activity.

> The term *diet* does not necessarily imply a weight-loss plan; it refers to the pattern of food and drink one regularly consumes.

Figure 7.1 *An assortment of healthful foods.*

| Bread, cereal, grains, and pasta | Vegetables | Fruits | Fats, oils, sweets | Milk, yogurt, and cheese | Meat, poultry, fish, dry beans, eggs, and nuts |

Figure 7.2 *MyPyramid, a 2005 revision of the Food Pyramid*

A.2 DIET AND FOOD GROUPS

Your three-day food inventory analysis begins with the "big picture." That is, before analyzing the foods in terms of their characteristics—such as energy, fat, carbohydrate, protein, mineral, or vitamin content—you will determine what proportions of each food group the food inventory includes.

In particular, you will investigate whether a variety of food groups is represented within the inventory or if one or two food groups constitute most of the food. Your teacher may give you a prepared three-day food inventory. However, if you analyze a different food inventory (your own or one that you devise), these suggestions will help you:

▶ List each food item, snack, beverage, and dietary supplement consumed during each of three successive days.

▶ Express the quantity of each food item by estimating the number of servings, mass, or volume of the food or beverage. When possible, use food labels on each item to guide your quantity estimates.

For each item in the three-day inventory, indicate the food group to which it belongs. Also indicate if there are items that cannot be classified in any listed category (such as snacks or soft drinks). You may need to investigate food items about which you are unsure. Then complete the following steps:

1. Construct a data table so that the total quantity consumed within each food group can be easily recorded. Include four columns, one for each day, and one for the daily average over the three days.

2. Record the quantity consumed within each food group for each day. Then calculate and enter the average quantity for each group over the three days.

3. Using your data from Step 2, which food group or groups

   a. represent most of the foods within the diet?
   b. are minimally represented in the diet or not at all?

4. Overall, is there a reasonable balance of food groups in the inventory? If not, what dietary substitutions would lead to greater variety among food groups?

5. Complete the following steps for food items that were difficult to assign to a particular food group:

   a. List the item and the food group to which the item was assigned.
   b. Explain why it was difficult to assign this item to a food group.
   c. Explain how you decided to which group to assign each item.

6. Is it better to analyze food consumed over three consecutive days or only for one day? Explain.

7. Under what circumstances would basing your food analysis on a particular three-day average be misleading?

Nutrition Facts

Serving Size 1 oz (28g/about 20 chips)
Servings Per Container 5

Amount Per Serving		
Calories 150		Calories from Fat 80
		% Daily Value*
Total Fat 9g		14%
Saturated Fat 2g		10%
Trans Fat 0g		
Cholesterol 0mg		0%
Sodium 95mg		4%
Potassium 370mg		11%
Total Carbohydrate 14g		5%
Dietary Fiber 1g		4%
Sugars 0g		
Protein 2g		

Vitamin A 0%	•	Vitamin C 15%
Calcium 0%	•	Iron 2%

*Percent Daily Values are based on a 2,000 calorie diet.
Your daily values may be higher or lower depending on your calorie needs:

	Calories:	2,000	2,500
Total Fat	Less than	65g	80g
Sat Fat	Less than	20g	25g
Cholesterol	Less than	300mg	300mg
Sodium	Less than	2,400mg	2,400mg
Potassium	Less than	3,500mg	3,500mg
Total Carbohydrate		300g	375g
Dietary Fiber		25g	30g

Calories per gram:		
Fat 9	• Carbohydrate 4	• Protein 4

Figure 7.3 *Part of the label from a popular snack food. Note the mass and total Calories contained in one serving.*

A.3 SNACK-FOOD ENERGY

Introduction

Figure 7.3 shows part of a food label from a common snack food. Note the number of grams per serving; then note the Calories (150), which is the quantity of energy contained in one serving. How is this food-energy value determined? How much energy does 150 Cal actually represent?

The quantity of energy contained in a particular food can be determined by carefully burning a known mass of the item under controlled conditions and measuring how much thermal energy is released. You completed a similar investigation in Unit 3 when you determined a candle's heat of combustion (see Figure 7.4). This procedure is known as *calorimetry*, and the measuring device is called a *calorimeter*.

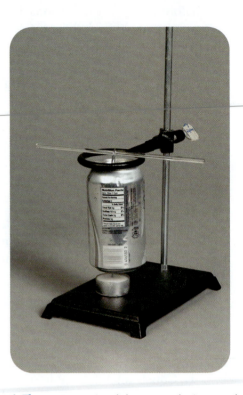

Figure 7.4 *Unit 3 laboratory device used to measure thermal energy released by a burning candle.*

Procedure

In this investigation, you will determine the energy contained in a sample of a snack food. Use the candle-burning procedure (pages 248–249) and this information to guide you as you design a procedure to determine the energy contained in this particular snack sample. (*Caution: Do not use any nut-based snack food for this investigation; some students may be allergic to such items.*) Before starting the investigation, complete the following steps:

**Lab Video:
Food Energy in
a Potato Chip**

▶ Write a complete procedure for determining the energy contained in this snack food.

▶ Sketch your laboratory setup.

▶ Construct a data table to record all necessary measurements and observations for two separate trials.

▶ Have your teacher check and approve your procedure and data table. After your teacher approves your procedure, you may start the investigation. Conduct two trials.

Data Analysis

If you know the mass of water and its temperature change, you can calculate the quantity of thermal energy that caused the temperature change. In this laboratory investigation, we assume that

Energy to heat the water = Energy released by the burning snack item

That assumes, of course, that no thermal energy is lost during this process.

You can calculate the quantity of thermal energy absorbed by the water sample from the specific heat capacity of water and the mass and temperature change of the water. Recall from Unit 3 that the specific heat capacity of liquid water is about 4.2 J/(g·°C). Thus, it takes about 4.2 J to raise the temperature of 1 g water by 1 °C. If you know the mass of water heated and its temperature change, you can calculate the thermal energy absorbed by the water. If necessary, refer to pages 249–250 to remind yourself how to complete this calculation.

Use data collected in this investigation to complete the following calculations:

1. Determine the mass (in grams) of the heated water.
2. Calculate the water's overall temperature change.
3. Calculate the total energy (in joules) required to heat the water.
4. The food **Calorie**, or *Cal* (written with an uppercase C), listed on food labels, is a much larger energy unit than the joule. One Calorie equals 4184 J, which can be conveniently rounded to 1 Cal = 4200 J.

   a. Calculate the total Calories used to heat the water.
   b. How many Calories were released by burning the item?

5. a. Calculate the energy released, expressed as Calories per gram (Cal/g) of snack item burned, for each trial.

   b. Calculate the average Calories per gram for the snack item.
   c. Use reported values from the package label to calculate the declared Calories per gram for the snack item.
   d. Calculate the percent difference between the declared value found on the label and your average experimental value:

$$\% \text{Difference} = \frac{|\text{Experimental value} - \text{Label value}|}{\text{Label value}} \times 100\%$$

> The precise specific heat capacity of water is 4.184 J/(g·°C).

> The *calorie* (cal), another common energy unit, is about one-fourth the size of a joule (1 cal = 4.184 J). One Calorie (1 Cal), which is a unit still used in the United States for food-energy values, equals one kilocalorie (1 kcal), or 1000 cal.

Questions

1. Which aspects of your laboratory setup and procedure might account for any difference between your Cal/g value and the corresponding label value?
2. How could your laboratory setup and procedure be improved to increase the accuracy of your results?
3. Consider the snack item shown in Figure 7.5. What procedural changes would be necessary to find the Calories per gram in the snack item shown?

Figure 7.5 *Describe how you would determine the quantity of energy contained in these potato chips.*

A.4 ENERGY FLOW: FROM THE SUN TO YOU

The snack item you burned released enough energy to raise the temperature of a sample of water by several degrees Celsius. Where did that energy come from?

The ultimate answer is easy: All food energy originates from sunlight. Through **photosynthesis,** green plants capture and use solar energy to make large molecules from smaller, simpler ones (see Figure 7.6). Recall from Unit 4 (page 348) that green plants, through photosynthesis, use solar energy to convert water and carbon dioxide into carbohydrates and oxygen gas. Although a variety of carbohydrates are produced, an equation for photosynthesis usually depicts the production of glucose:

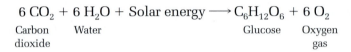

$$6\ CO_2 + 6\ H_2O + \text{Solar energy} \longrightarrow C_6H_{12}O_6 + 6\ O_2$$

Carbon   Water   Glucose   Oxygen
dioxide   gas

Figure 7.6 *Energy from the Sun was used to convert water and carbon dioxide into the molecules that constitute these onions. What type(s) of energy conversion may happen next?*

A diagram similar to that in Figure 7.7, showing the energy involved in the combustion of methane, appears on page 240.

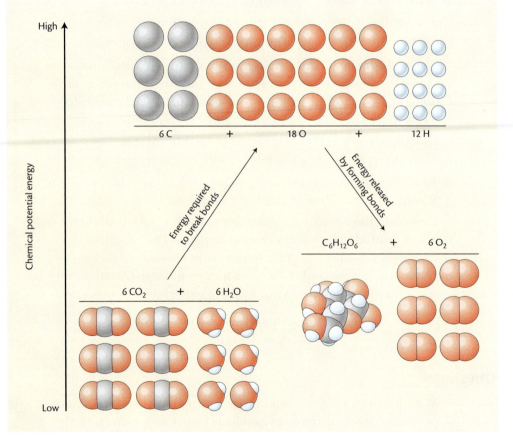

Figure 7.7 *Photosynthesis. Breaking bonds within carbon dioxide and water molecules requires energy. Energy is released when atoms combine to form glucose and oxygen molecules. The first step (bond breaking) requires more energy than the second step (bond forming) releases, so the overall process of photosynthesis is endothermic.*

For this reaction to occur, bonds between the carbon and oxygen atoms in carbon dioxide molecules and between the oxygen and hydrogen atoms in water molecules must be broken. The atoms must then recombine in a different arrangement to form glucose and oxygen molecules. Recall from Unit 3 that breaking bonds always requires energy, whereas bond formation releases energy. In photosynthesis, the bonds in carbon dioxide and water molecules require more energy to break than is released when chemical bonds in glucose and oxygen molecules form. The energy needed to drive this endothermic reaction, as the photosynthesis equation indicates, comes from the Sun. The potential energy diagram in Figure 7.7 shows the energy relationships in this process.

The Sun's radiant energy is converted to chemical energy stored within bonds of carbohydrate molecules. Living organisms release this chemical energy when they consume and metabolize carbohydrate molecules, converting them into lower-energy CO_2 and H_2O molecules. This chemical energy can be measured via calorimetry, as you did in Investigating Matter A.3. Material in A.5 addresses metabolism in more detail.

Energy originally delivered as sunlight continues to flow through ecosystems as carnivores (meat-eating animals) consume plant-eating animals. Eventually, plants and animals die and decay; organisms that aid in decomposition use the remaining stored-up energy. Thus, energy flows from the Sun to plants to herbivores and then to carnivores and decomposers.

The Sun's captured energy is dispersed and becomes less available as it is transferred from organism to organism. For example, as energy flows from one organism to another, some energy is dispersed into the environment as thermal energy. You may be surprised to learn that only a small fraction (about 10–15%) of food energy consumed by organisms is used for growth—for converting smaller molecules to larger molecules that become part of an animal's structure. Over half the energy contained in consumed food is used to digest food molecules. The supply of useful energy declines as energy continues to transfer away from its original source—the Sun. Figure 7.8 illustrates this decline in available energy.

> For a discussion of energy involved in chemical reactions, see pages 238–241.

> How would a potential energy diagram for the process of metabolism look?

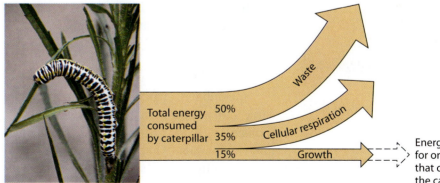

Figure 7.8 *The energy used by this caterpillar originated as solar energy. However, not all incident solar radiation is converted into usable energy. First, only a portion of the Sun's energy is actually stored as chemical energy in molecules within this plant. Next, only a fraction of that stored chemical energy is actually used by the caterpillar.*

ChemQuandary 1

HOW DOES YOUR GARDEN GROW?

From a chemical viewpoint, consider this question: As garden plants grow, what is the source of the material that causes their mass to increase?

A.5 ENERGY RELEASE AND STORAGE

Some energy in the food you eat is used soon after you consume it (Figure 7.9). The rest of the energy may be stored for later needs. These available reserves of bodily energy are mainly in fats and, to a much smaller extent, in carbohydrates. Figure 7.10 shows some foods rich in carbohydrates and fats.

Whether your body uses energy from food recently ingested or from stored fat, the release of energy from these molecules depends on a series of chemical reactions inside your cells. In *cellular respiration,* plants and animals use oxygen to break down complex organic molecules into carbon dioxide and water molecules. The energy required to break chemical bonds in reactant molecules is less than the energy released when chemical bonds in carbon dioxide and water form. Thus, energy is released in cellular respiration. You can think of this exothermic process as the reverse of the process shown in Figure 7.7 (page 572):

Figure 7.9 *A portion of the energy in food is expended soon after eating, particularly when engaging in energy-demanding activities such as bicycling. The remaining energy is then stored for later activity.*

Organic compounds + Oxygen ⟶ Carbon dioxide + Water + Energy

Specialized structures within each cell use the energy released to carry out a variety of tasks, such as energizing reactions, transporting molecules, disposing waste, storing genetic material, and synthesizing new molecules. That energy ultimately allows you to walk, talk, run, work, and think; it also provides energy to power your heart, lungs, brain, and other organs.

Carbohydrates, fats, and proteins can all be processed as fuel for your body. We will now examine cellular respiration through the oxidation of glucose, a carbohydrate. The overall equation, based on the oxidation of one mole of glucose is as follows:

$$C_6H_{12}O_6(aq) + 6\ O_2(g) \longrightarrow 6\ CO_2(g) + 6\ H_2O(l) + 686\ \text{Calories}$$

Figure 7.10 *Some foods that contain carbohydrates (top), and some foods rich in fats (bottom).*

Carbon dioxide and water are also produced from burning petroleum, coal, and other fossil fuels. See page 349.

This equation is the reverse of the equation for photosynthesis. See Figure 4.43, page 348.

Figure 7.11 *Products containing glucose.*

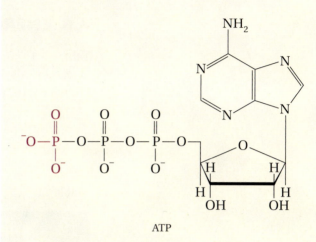

ATP

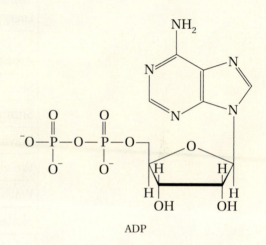

ADP

Figure 7.12 *Ionic forms of ATP and ADP. The colored atoms on ATP's left side are removed when ADP forms.*

This equation actually summarizes a sequence of more than 20 linked chemical reactions catalyzed by more than 20 different enzymes. Glucose (Figure 7.11) becomes oxidized molecule by molecule within cells throughout your body.

The energy needed to perform individual cellular functions is much less than the energy released by "burning" glucose molecules. Between its release from these energy-rich molecules and its later use in cells, energy is stored in biomolecules of adenosine triphosphate (ATP), shown in Figure 7.12.

As indicated by the following equation, more energy is required to break chemical bonds in adenosine diphosphate (ADP) and HPO_4^{2-} than is released when water and ATP form. The difference in these bond energies equals 7.3 Cal, which suggests that an ATP molecule can be regarded as an energy-storage site because energy is released when the reaction proceeds in the opposite direction.

$$7.3 \text{ Cal} + \text{ADP(aq)} + \underset{\substack{\text{Hydrogen} \\ \text{phosphate ion}}}{\text{HPO}_4^{2-}\text{(aq)}} \longrightarrow \text{H}_2\text{O(l)} + \text{ATP(aq)}$$

When this reaction is reversed, each mole of ATP releases 7.3 Cal. This conveniently small quantity of energy, compared to the 686 Cal produced as 1 mol of glucose is oxidized, is used to energize particular cellular reactions. Some of these reactions require less energy than that supplied by a single ATP molecule, whereas other steps require the total energy released by several ATP molecules.

Oxidation of one mole of glucose produces enough energy to add 38 mol of ATP to short-term cellular energy storage. Each day, your body stores and later releases energy from at least 6.02×10^{25} molecules (100 mol) of ATP.

Biomolecules such as ATP are compounds produced by chemical reactions associated with living systems.

A.6 ENERGY IN ACTION

Table 7.1 summarizes the average energy expended during various activities. Use the information provided in the table to answer questions appearing on the following page.

| Table 7.1

ENERGY EXPENDED IN VARIOUS ACTIVITIES				
Energy Expended (in Calories per Minute) for Individuals of Different Body Mass				
Activity	**46 kg (100 lb)**	**55 kg (120 lb)**	**68 kg (150 lb)**	**82 kg (180 lb)**
Sleeping	1	1	1	1
Sitting	1	1	1	2
Playing volleyball	2	3	4	4
Weight lifting	2	3	4	4
Walking (17 min/mi)	3	4	5	6
Skateboarding	4	5	6	7
Swimming	5	6	7	9
Wrestling	5	6	7	9
Bicycling (moderate)	6	7	8	10
Playing soccer	6	7	8	10
Doing step aerobics	6	7	8	10
Playing tennis	6	7	8	10
Playing basketball	6	8	10	12
Playing football	6	8	10	12
Playing ice hockey	6	8	10	12
Doing martial arts	8	10	12	14
Jogging (8 min/mi)	10	12	15	18

Sample Problem: *Table 7.1 indicates that a 55-kg (120-lb) person burns 5 Cal/min while skateboarding (Figure 7.13).*

> *a. This energy is supplied by ATP. If the oxidation of each mole of glucose ($C_6H_{12}O_6$) produces 38 mol ATP, how many moles of glucose will be needed to provide the energy for one hour of skateboarding?*
>
> *b. What mass of glucose does your answer to Question a represent? The mass of one mole of glucose is 180 g.*

To begin, find the moles of ATP needed for one hour of skateboarding:

$$\frac{5 \text{ Cal}}{1 \text{ min}} \times \frac{60 \text{ min}}{1 \text{ h}} = 300 \text{ Cal/h}$$

$$\frac{300 \text{ Cal}}{1 \text{ h}} \times \frac{1 \text{ mol ATP}}{7.3 \text{ Cal}} = 41 \text{ mol ATP/h}$$

Because 38 mol ATP can be obtained from each mole of glucose, the skateboarder's need for 41 mol ATP can be met by slightly more than one mole of glucose:

$$41 \text{ mol ATP} \times \frac{1 \text{ mol glucose}}{38 \text{ mol ATP}} = 1.1 \text{ mol glucose}$$

What mass of glucose does that represent? Because the skateboarder needs more than one mole of glucose, that mass must be greater than 180 g:

$$1.1 \text{ mol glucose} \times \frac{180 \text{ g glucose}}{1 \text{ mol glucose}} = 200 \text{ g glucose}$$

Now answer the following questions:

1. Assume your body produces approximately 100 mol ATP daily.
   a. How many moles of glucose are needed to produce that amount of ATP?
   b. What mass of glucose does your answer to Question 1a represent?

2. One minute of muscle activity requires about 0.0010 mol ATP for each gram of muscle mass. How many moles of glucose must be oxidized to energize 454 g (1 lb) of muscle to dribble a basketball for one minute?

3. a. How many hours do you typically sleep each night?
   b. How many Calories do you use during one night of sleep? (Refer to Table 7.1.)
   c. How many moles of ATP does this require?
   d. What kinds of activity require energy while you are sleeping?

Figure 7.13 *How much energy is expended in one hour of skateboarding?*

More than 200 g glucose would actually be needed because the energy-transfer steps are not 100% efficient.

A.7 ENERGY IN, ENERGY OUT

People follow various diets for different reasons. Some people may want to lose weight; others may want to gain weight. Still others put little thought into what they eat; instead, they merely eat for convenience or pleasure.

Some foods are needed to deliver important molecules to the body, regardless of their energy value. By contrast, other foods provide only energy. This latter category, which includes sugar-sweetened soft drinks, is sometimes described as furnishing "empty Calories."

If you want to lose weight, you must consume less energy than you expend. If your goal is to gain weight, you must do the opposite. If losing weight or gaining muscle mass is desired, then regular physical activity, such as that shown in Figure 7.14, is needed in addition to the particular diet you follow. Now you will explore the interplay between personal activity and diet.

> If you eat 100 Cal per day more than you burn up, you will gain about a pound each month.

Figure 7.14 *Technology is available for monitoring total energy expenditure.*

Making Decisions

A.8 GAIN SOME, LOSE SOME

Look again at Table 7.1 on page 576. The quantity of energy expended depends on the duration of a given activity and the person's body mass. For example, during one hour of soccer playing, a 55-kg (120-lb) person expends 420 Cal, whereas a 68-kg (150-lb) person expends 480 Cal in the same activity.

Use the information provided in Table 7.1 to answer the following questions.

Sample Problem: *A 46-kg student usually takes 10.0 min to bike home from school. Would this student burn more Calories by jogging home instead (at 8 min/mi for 15.0 min)?*

Bicycling requires 6 Cal/min × 10.0 min = 60 Cal

Jogging requires 10 Cal/min × 15.0 min = 150 Cal

Thus, jogging home burns more energy per trip—90 Cal more.

1. Consider eating an ice-cream sundae (Figure 7.15). Assume that two scoops of your favorite ice cream provide 250 Cal; the topping adds 125 Cal more.

   a. Assume that your regular diet (without that ice-cream sundae) just maintains your current body weight. If you eat the ice-cream sundae and wish to "burn off" those extra Calories,
      i.   for how many hours would you need to lift weights?
      ii. how far should you walk at 17 min/mi?
      iii. for how many hours would you need to swim?

   b. One pound of weight gain is equivalent to 3500 Cal. If you choose not to exercise, how much weight will you gain from eating the sundae?

   c. Now assume that you consumed a similar sundae once per week for 16 weeks. If you do not exercise to burn off the added Calories, how much weight will you gain?

2. Question 1 implies that eating an ice-cream sundae will cause weight gain unless you complete additional exercise.

   a. Can you think of a plan that would allow you to consume the sundae, do no additional exercise, and still *not* gain weight?

   b. Explain your answer.

   c. What concerns would you have about this plan?

Figure 7.15 *How does an extra helping of ice cream affect your daily balance between energy intake and expenditure?*

3. Suppose you drank six glasses (250 mL each) of ice water (0 °C) on a hot summer day.

   a. Assume your body temperature is 37 °C. How many joules of thermal energy would your body use in heating that ice water to body temperature? Recall that the specific heat capacity of water is about 4.2 J/(g·°C).

   b. How many Calories is this? Recall that 1 Cal equals about 4200 J.

   c. A serving of french fries contains 240 Cal. How many glasses of ice water would you need to drink to "burn off" the Calories consumed in one serving of french fries?

   d. Given your answer to Question 3c, does drinking large quantities of ice water seem like a reasonable strategy to lose weight? Explain.

4. Identify your favorite activity from Table 7.1. How many minutes would you need to engage in that activity to burn off the Calories in one serving of french fries (see Question 3c)?

A.9 ENERGY INTAKE AND EXPENDITURE

Energy Value of Food

Based on the chemical knowledge you have gained, you can find the total energy consumed in the three-day food inventory and estimate the energy expended based on typical activity levels.

Using appropriate resources suggested by your teacher, calculate and record the food energy (Calories) contained in each item in the three-day inventory.

1. Calculate the total Calories consumed each day.

2. Calculate the average Calories consumed per day.

3. Using your results from A.2 (page 569), how many Calories per day (on average) are supplied by each food group?

4. Using Table 7.1 (page 576) and other resources provided by your teacher, develop a list of typical daily activities for a high school student and record how much energy is expended during each activity listed.

Figure 7.16 *The balance between energy intake and expenditure affects one's body mass.*

5. Based on your list prepared in Question 4, calculate the total Calories expended over an average day. Calories are expended not only during the activities listed but also by normal, resting-state body processes. Be sure to include this expenditure in your calculation.

6. a. Based only on Calories consumed and expended, decide whether the three-day inventory you are evaluating would be appropriate for weight maintenance (see Figure 7.16). Explain your answer.

   b. Identify and describe any limitations in your analysis that may affect your decision in Question 6a.

7. Based on your analysis thus far, what types of items would you recommend for the Riverwood High School vending machines in terms of the energy they provide?

SECTION A SUMMARY
Reviewing the Concepts

Calorimetry can be used to determine the quantity of energy contained in a particular food sample.

1. List three common units used to express the energy content of food.

2. Sketch a simple calorimeter and label the purpose of each component.

3. Why is it important to know accurately the mass of water used in a calorimeter?

4. What is the mathematical relationship between a *Calorie* and a *calorie*?

5. Convert the following energy quantities:
   a. 4375 cal to Cal
   b. 76 932 J to Cal
   c. 289 Cal to J
   d. 12 226 000 cal to kJ

6. What is meant by the *specific heat capacity* of water?

7. a. How much thermal energy (in joules) is required to increase the temperature of a 115-g liquid water sample by 10.0 °C?
   b. Suppose that a 2.0-g food sample was burned in a calorimeter to provide the thermal energy involved in Question 7a. Assume all thermal energy released was used to heat the water.
      i. How many calories per gram were in the food sample?
      ii. How many Calories per gram were in the food sample?

All food energy originates from sunlight and is stored and released through a series of chemical reactions.

8. What process captures sunlight and transforms it into chemical energy?

9. Write a chemical equation for the production of glucose through photosynthesis.

10. Is photosynthesis an endothermic or exothermic chemical change? Explain.

11. Keeping in mind the law of conservation of energy, describe what happens to the food energy consumed by living creatures.

12. What is meant by an "energy-rich" molecule?

13. What is the relationship between photosynthesis and cellular respiration? Write chemical equations that support your answer.

14. What is the difference, in terms of molecular composition and stored energy, between ADP and ATP?

15. How do ADP and ATP permit the controlled use of energy from glucose?

The energy required for a physical activity depends on the particular activity, the total time involved, and the mass of the person engaged in the activity.

16. How much total energy (in Calories) is expended by a 68-kg (150-lb) person swimming for 35 min?

17. How much total energy (in joules) is expended by a 46-kg (100-lb) person walking for 56 min?

18. People expend energy even when they sit perfectly still. Why?

19. How many moles of ATP are required to provide energy for an 82-kg (180-lb) student to sit in class for 45 min?

20. Explain why two people performing exactly the same physical exercise may not burn the same total Calories.

21. Suppose a 55-kg (120-lb) person, whose regular diet and exercise just maintains body weight, adds one chocolate candy bar containing 354 Cal to their diet each day for 30 days without increasing the level of exercise. Predict how much weight the person will gain in 30 days.

Connecting the Concepts

22. Compare the process of cellular respiration to combustion.

23. Does calorimetry directly measure the quantity of thermal energy liberated by a food or fuel? Explain.

24. Describe problems that could arise in calorimetry by using
    a. a very small food sample.
    b. a very large food sample.

25. A student argues that eating a certain mass of chocolate or eating the same mass of apples results in the same gain in body weight. Explain why this is incorrect.

26. Figure 7.7 (page 572) shows an energy diagram for photosynthesis.
    a. Construct a similar energy diagram for cellular respiration.
    b. Compare the energy diagrams for photosynthesis and cellular respiration.

27. The term *respiration* is sometimes used to mean "breathing." What is the relationship, if any, between cellular respiration and breathing?

28. In reference to a local seafood restaurant, someone remarked, "When you eat a pound of fish, you're eating ten pounds of flies." Aside from its questionable value as a meal-promoting strategy, how accurate is this message? Why?

29. Explain why reduced-Calorie food products are sometimes described as *lite* or *light,* even though Calories are not a unit of mass.

30. You used similar calorimetry procedures in Unit 3 and in this unit to find the energy content of candles and food products, respectively. Does that imply that anything you can burn to heat water can also be used to fuel your body's metabolism? Explain.

Extending the Concepts

31. a. In what sense does the challenge of world hunger involve an "energy crisis"?
    b. In what sense is it a "resource crisis"?

32. Research and report on the difference between *under*nourishment and *mal*nourishment. Could either term ever apply to an overweight person?

33. From an energy standpoint, are there advantages to eating low on the food chain? For example, is it energetically more favorable to use 100 lb of grain or 100 lb of beef as a food source? Explain.

34. Why is it not possible to consume only pure ATP and eliminate some steps in metabolism?

35. Investigate the characteristics of a professionally designed chemistry laboratory calorimeter. Sketch its essential parts and explain its operation.

36. For hibernating animals, the storage of fat is critical to survival. Investigate a particular animal species to find out how it stores optimal quantities of fat for hibernation.

37. A student decides to lose some weight by not wearing a coat in cold winter weather. What knowledge of food energy might have inspired this idea? Does this plan have merit? Explain.

Energy Storage and Use

You are now ready to explore how the chemical energy contained in foods is stored, transferred, and used. Keep in mind the three-day food inventory you are evaluating and think about how this new knowledge may apply to your analysis.

B.1 CARBOHYDRATES: ONE WAY TO COMBINE C, H, AND O

All **carbohydrates** are composed of carbon, hydrogen, and oxygen. Glucose, which is the key energy-releasing carbohydrate in biological systems, has the molecular formula $C_6H_{12}O_6$. When such formulas were first established, chemists noted a 2:1 ratio of hydrogen atoms to oxygen atoms in carbohydrates, the same as in water. They were tempted to write the glucose formula as $C(H_2O)_6$, implying a chemical combination of carbon with six water molecules. Chemists even invented the term "carbohydrates" (water-containing carbon substances) for glucose and related compounds. Although chemists later determined that carbohydrates contained no water molecules, the name persisted. However, like water, carbohydrate molecules *do* contain O—H bonds in their structures.

Carbohydrate molecules may be simple sugars, such as glucose (see Figures 7.17 and 7.18), or chemical combinations of two or more simple sugar molecules. Simple sugars are called **monosaccharides,** molecules usually containing five or six carbon atoms. Glucose (like most other monosaccharides) exists principally in a ring form; however, glucose can also exist in a chain form, as shown in Figure 7.18. Do both forms have the same molecular formula?

> Sugars, starch, and cellulose are examples of carbohydrates.

Figure 7.17 *Many fruits and vegetables contain simple sugars such as glucose and fructose.*

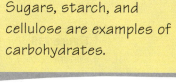

Ring form   Chain form

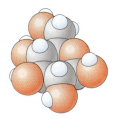

Figure 7.18 *Structural formulas for glucose. This carbohydrate, like most simple sugar and disaccharide molecules, is found primarily in ring form. How do ring and chain structures compare with one another?*

Condensation reactions were highlighted in Unit 3, pages 274–277, in the formation of esters and condensation polymers.

Sugar molecules composed of two monosaccharide units bonded together are called **disaccharides.** They are formed by a condensation reaction between two monosaccharides. Sucrose (table sugar, $C_{12}H_{22}O_{11}$) is a disaccharide composed of the ring forms of glucose and fructose, as illustrated in Figure 7.19.

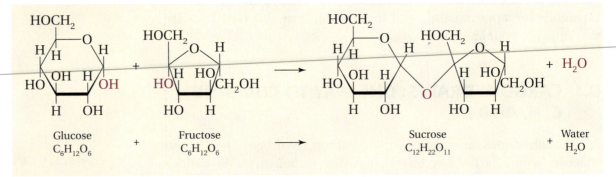

Glucose $C_6H_{12}O_6$ + Fructose $C_6H_{12}O_6$ → Sucrose $C_{12}H_{22}O_{11}$ + Water H_2O

Figure 7.19 *The formation of sucrose. Note that particular -OH groups react (shown in red), resulting in elimination of one H_2O molecule.*

The reaction that forms disaccharides—a condensation reaction—can also cause monosaccharides to form polymers. Such polymers, not surprisingly, are called **polysaccharides** (see Figure 7.20). Starch, which is a major component of grains and many vegetables, is a polysaccharide composed of glucose units. Cellulose, which is the fibrous or woody material of plants and trees, is another polysaccharide formed from glucose.

All polysaccharides are polymers of monosaccharide molecules.

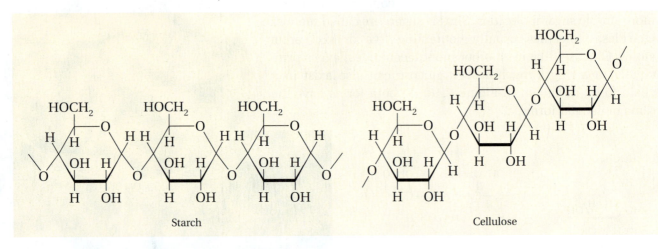

Starch

Cellulose

Figure 7.20 *Structural formulas for starch and cellulose, two polysaccharide molecules.*

Carbohydrates: Structure and Function

The major types of carbohydrates are summarized in Table 7.2. Compare the bonds in starch with those in cellulose. Due to structural differences, starch is easily digested by the body (see Figure 7.21), whereas cellulose is indigestible by humans. Such indigestible types of carbohydrates are commonly called *fiber.*

Carbohydrates and fats are the primary high-energy substances in the human diet. One gram of a carbohydrate provides 4 Cal of food energy. Nutritionists recommend that about 45–65% of dietary Calories come from high-fiber carbohydrates. Worldwide, most people obtain carbohydrates by eating grains such as rice, corn bread, wheat tortillas, bread, pasta, and beans. Other sources of carbohydrates include fruits, milk, and yogurt. Meats contain a small quantity of *glycogen,* the carbohydrate by which animals store glucose.

People in the United States tend to obtain more of their carbohydrates, on average, from wheat-based breads, potatoes, sugar-laden snacks and desserts, and soft drinks than is common in many other parts of the world. Currently, the average U.S. citizen consumes more than 65 kg (145 lb) of sugar annually, mainly from candy, desserts, and soft drinks.

Figure 7.21 *Breads and other foods contain starch, which is a digestible polysaccharide. Cellulose (found in wood, paper, and cotton) is a polysaccharide that humans cannot digest. What is a structural difference between digestible polysaccharides (such as starch) and cellulose?*

| Table 7.2

THE COMPOSITION OF COMMON CARBOHYDRATES

Classification and Examples	Composition	Formula	Common Name or Source
Monosaccharides		$C_6H_{12}O_6$	
Glucose	—		Blood sugar
Fructose	—		Fruit sugar
Galactose	—		—
Disaccharides	Monosaccharides	$C_{12}H_{22}O_{11}$	
Sucrose	Fructose + glucose		Cane sugar
Lactose	Galactose + glucose		Milk sugar
Maltose	Glucose + glucose		Germinating seeds
Polysaccharides	Glucose polymers	—	
Starch			Plants
Glycogen			Animals
Cellulose			Plant fibers

CHEMISTRY at Work

Food for Thought

The next time you chew gum, drink a canned or bottled beverage, or eat a bowl of your favorite breakfast cereal, think about corn. Gum, soda, cereal, ketchup, cookies, pastries, bread, and many other processed foods and drinks are sweetened or colored with products made from corn. Everything you eat and drink has connections to a farm, even if those ties are not readily apparent.

Sue Adams and her husband John work year-round to run an 1100-acre farm in Atlanta, Illinois. To protect the environment, they use integrated pest management and practice a technique called *no-till farming*—planting new crops directly in the residue of previously harvested crops. This method saves on labor and fuel. More importantly, this farming method reduces the loss of valuable topsoil and moisture, which dramatically reduces the need for materials such as fertilizers, pesticides, and herbicides.

In the spring, Sue and John plant half of their fields in corn and the other half in soybeans. As soon as the crops begin to grow, they scout the fields for evidence of damage by pests and, where necessary, apply special crop-protection products that decompose quickly in sunlight. Just like a chemist in a laboratory, Sue carefully monitors and measures how these products are used to ensure maximum protection for the crops and the environment.

Sue and John practice a precise, high-tech agriculture that is used on many of today's farms.

Ever Think of a Satellite as a Farming Tool?

Sue and John also follow a precise, high-tech agriculture system that is used on many of today's farms. One example is their use of global positioning system (GPS) technology. The GPS employs computers, monitoring units, and radio signals to send, receive, and process information from satellites to study objects on Earth's surface. On their farm, Sue uses GPS technology to plot with great precision which areas of her fields require additional fertilizers or other treatment. She even has a specially equipped all-terrain vehicle that can record and access information while she is out in the field.

Using GPS data, Sue and John apply fertilizer only to those areas of the farm that need it, minimizing the quantity of fertilizer they introduce to the environment. Such increased precision has also reduced the costs of fertilizing (materials, equipment, fuel, and labor) by as much as ten dollars an acre.

Earth and Its Food Supply: What's the Forecast?

In the 1700s, an English economist, T. Robert Malthus, published his "An Essay on the Principle of Population." The widely read pamphlet argued that unchecked population increases geometrically (1, 2, 4, 8, 16, . . .), while food supplies increase arithmetically (1, 2, 3, 4, 5, . . .). Malthus's theory suggested that one day the world might be unable to feed itself. Indeed, world population has grown significantly. In 1800, the world's population was about 1 billion; by 1999, it reached 6 billion.

Malthus s theory suggested that one day the world might be unable to feed itself.

In small groups, discuss answers to these questions.

1. Malthus' prediction has not become reality, at least for a majority of the world's inhabitants. Was any part of his thinking flawed, especially with regard to human population growth? Explain.

2. The world's population by the year 2020 will probably be almost 8 billion people. Why do you suppose there is renewed interest in Malthus's theory?

3. One of the issues with increased population is the total land required for human residence versus that needed for cropland. What new techniques might farmers, in collaboration with municipal planners, develop to allow for the space needed by both humans and agriculture?

4. No-till farming is said to have several advantages. Why do you suppose it has taken until now for farmers to use no-till techniques instead of the traditional technique of plowing fields?

5. In what ways do the Adamses' farming techniques reflect principles of Green Chemistry, that is, to prevent environmental problems before they occur?

6. a. What are direct advantages to the Adamses of using a greener, high-tech approach to farming?

   b. What are advantages to society of using these techniques?

B.2 FATS: ANOTHER WAY TO COMBINE C, H, AND O ATOMS

Unlike the terms *carbohydrate* and *protein, fat* has acquired its own general (and somewhat negative) meaning. From a chemical viewpoint, however, *fats* are just another major category of biomolecules with special characteristics and functions.

A significant part of a normal human diet, fats are present in meat, fish, poultry, oils, dairy products, nuts, and grains. When more food is consumed than is needed to satisfy energy requirements, much of the excess food energy is stored in the body as fat molecules. If food intake is not enough to meet the body's energy needs, then stored fat is "burned" to release energy to make up the difference.

Fats are composed of carbon, hydrogen, and oxygen atoms—the same three elements that compose carbohydrates. Fat molecules, however, contain fewer oxygen atoms and more carbon and hydrogen atoms. Thus, fats have a greater number of carbon-hydrogen bonds and fewer carbon-oxygen (and oxygen-hydrogen) bonds than do carbohydrate molecules. You can confirm this by comparing the structure of glyceryl tripalmitate, which is a typical fat molecule, shown in Figure 7.22, to that of glucose or starch molecules (Figures 7.18 and 7.20, respectively, pages 583 and 584).

When an organic substance is burned, it reacts with oxygen to form carbon dioxide, water, and thermal energy. Similarly, when fats, carbohydrates, and proteins are "burned" in the body, they are converted into carbon dioxide, water, and energy.

Fats: Structure and Function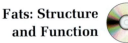

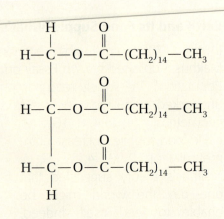

Figure 7.22 *The structural formula for glyceryl tripalmitate, a typical fat molecule.*

Figure 7.23 *Formation of a typical fat molecule, a triglyceride. The colored atoms interact to form the fat molecule plus water.*

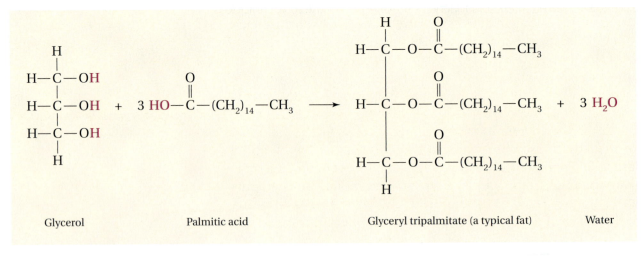

Glycerol   Palmitic acid   Glyceryl tripalmitate (a typical fat)   Water

Generally, fat molecules are nonpolar and only sparingly soluble in water. As you can see from Figure 7.22, fats have long hydrocarbon portions that prevent them from dissolving in polar solvents such as water. Their low water solubility and high energy-storing capacity give fat molecules, unlike carbohydrates, chemical properties similar to those of hydrocarbons.

The chemical properties of a fat molecule are due largely to the *fatty acid* groups in the molecule. **Fatty acids** are a class of organic compounds composed of long hydrocarbon chains with a carboxylic acid group (COOH) at one end. Two fatty acids are shown in Figure 7.24.

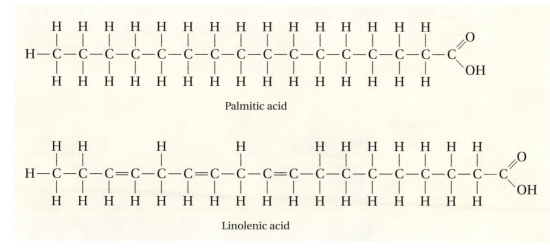

Palmitic acid

Linolenic acid

Figure 7.24 *Typical fatty acids.*

The typical fat molecule, a **triglyceride,** is a combination of a three-carbon alcohol molecule called *glycerol* and three fatty acid molecules.

1 Glycerol molecule + 3 Fatty acid molecules ⟶
1 Triglyceride (fat) molecule + 3 Water molecules

The formation of a triglyceride is shown in Figure 7.23. Each fatty acid molecule forms an ester linkage as it reacts with an —OH group of glycerol. This produces one water molecule per ester linkage. The main product of this condensation reaction is a triglyceride composed of three ester groups. Although the three fatty acids depicted in Figure 7.23 are identical, fats can contain two or three different fatty acids.

Because they have more carbon-hydrogen bonds than carbon-oxygen bonds, fats contain more stored energy per gram than do carbohydrates. In fact, one gram of fat contains over twice the energy stored in one gram of carbohydrate; 1 g fat is equivalent to 9 Cal, compared to 4 Cal for each gram of carbohydrate. Consequently, you must run more than twice as far or exercise twice as long to "work off" a given mass of fat compared to what you must do to "work off" the same mass of carbohydrate. For example, a typical glazed doughnut contains 11.6 g fat. Thus, the glazed doughnut has 11.6 g fat × 9 Cal/g fat = 104 Cal of energy from fat. It is not surprising that your body uses fat molecules to store excess food energy efficiently.

In general, the terms "fat" and "triglyceride" are used interchangeably.

The reaction that produces a fat molecule is similar to the reaction you observed in Unit 3 that produced the methyl salicylate ester (page 276).

B.3 SATURATED AND UNSATURATED FATS

Recall from Unit 3 that hydrocarbons can be saturated (containing only single carbon-carbon bonds) or unsaturated (containing double or triple carbon-carbon bonds). Likewise, hydrocarbon chains in fatty acids are either saturated or unsaturated. Look again at the fatty acids in Figure 7.24 (page 589). Can you identify each as either saturated or unsaturated?

Fats containing saturated fatty acids are called **saturated fats;** those containing some unsaturated fatty acids are known as **unsaturated fats.** A **monounsaturated fat** contains just one C=C double bond in its fatty acid components. A **polyunsaturated fat** contains two or more C=C double bonds in the fatty acid portion of a triglyceride molecule. Based on these definitions, is the fat depicted in Figure 7.22 (page 588) saturated, monounsaturated, or polyunsaturated?

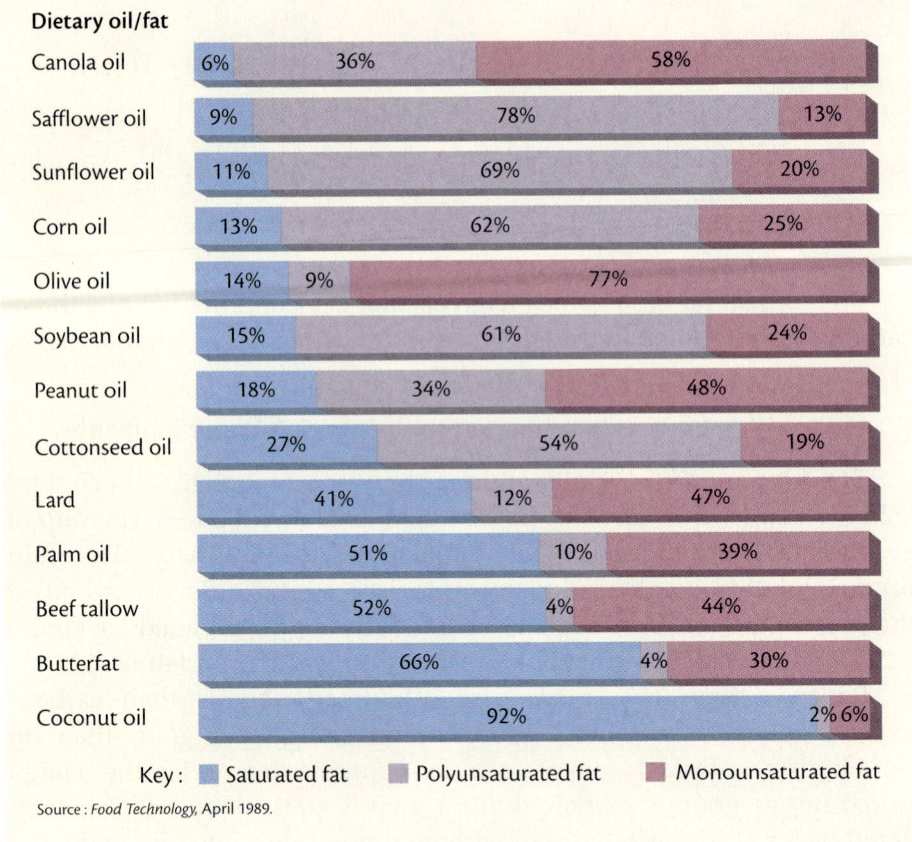

> Oils are fats that are liquids

Dietary oil/fat	Saturated fat	Polyunsaturated fat	Monounsaturated fat
Canola oil	6%	36%	58%
Safflower oil	9%	78%	13%
Sunflower oil	11%	69%	20%
Corn oil	13%	62%	25%
Olive oil	14%	9%	77%
Soybean oil	15%	61%	24%
Peanut oil	18%	34%	48%
Cottonseed oil	27%	54%	19%
Lard	41%	12%	47%
Palm oil	51%	10%	39%
Beef tallow	52%	4%	44%
Butterfat	66%	4%	30%
Coconut oil	92%	2%	6%

Key: Saturated fat   Polyunsaturated fat   Monounsaturated fat

Source: *Food Technology,* April 1989.

Figure 7.25 *The percent of polyunsaturated fat in fats from selected plant and animal sources.*

Triglycerides in animal fats are nearly all saturated and are solids at room temperature. However, fats from plant sources commonly are polyunsaturated or monounsaturated. In general, higher levels of unsaturation are associated with oils that have lower melting points. At room temperature these polyunsaturated fats are liquids. (See Figure 7.25).

Due to their C=C double bonds, unsaturated fats undergo addition reactions; saturated fats cannot undergo addition reactions. Thus, these two types of fats participate differently in body chemistry. Unsaturated fat molecules are much more chemically reactive. Increasing evidence suggests that saturated fats may contribute more to health problems than do some unsaturated fats. Saturated fats are associated with formation of *arterial plaque*, which are deposits of fatty material in blood vessel walls (see Figure 7.26). The result is a condition commonly known as "hardening of the arteries," or *atherosclerosis*, which is a particular threat to coronary (heart) arteries and arteries leading to the brain.

Figure 7.26 *An artery cross section, showing plaque (grey) on walls.*

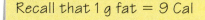
Developing Skills

B.4 CALORIES FROM FAT

A sample of butter (Fig 7.27) provides the following nutritional values for one serving, which is defined as 1 tablespoon (tbsp) or 14 g:

Total fat	10.9 g	Polyunsaturated fat	0.4 g
Saturated fat	7.2 g	Calories	100

Sample Problem: *Calculate the percent of polyunsaturated fat contained in the total fat in butter.*

To find the answer, divide the mass of polyunsaturated fat by the mass of total fat in butter and convert it to a percent value:

$$\frac{0.4 \text{ g}}{10.9 \text{ g}} \times 100\% = 4\%$$

Figure 7.27 *What type of fat is primarily found in butter?*

Thus, approximately 4% of butter's total fat is polyunsaturated.

1. Calculate the percent of saturated fat in the total fat in butter.

2. The saturated and polyunsaturated fat percent values don't total 100%.

   a. What does the "missing" percent value represent?
   b. How many grams of fat does this represent?

3. Calculate the total percent fat in one serving of butter.

4. The *Dietary Guidelines 2005* suggest that from 20–35% of total Calories in human diets should come from fats.
   A way to evaluate this is to compare total Calories from fat to total Calories delivered in that food serving.

   a. Determine the total Calories from fat in 1 tbsp butter.
   b. Calculate the percent of total Calories obtained from the fat in butter.

5. One serving (14 g) of margarine has 10 g fat and 90 Cal.

   a. Find the percent of Calories obtained from margarine fat.
   b. Compare the percent of total Calories from margarine fat to that value from the fat in butter.

Recall that 1 g fat = 9 Cal

Most fats, according to the *Dietary Guidelines*, should come from polyunsaturated and monounsaturated fatty acids, supplied by fish, nuts, and legumes.

B.5 HYDROGENATION

Partial **hydrogenation** adds hydrogen atoms to some C=C bonds in vegetable oil triglyceride molecules. The reaction with hydrogen converts a C=C double bond to a C—C single bond. Because this decreases the total unsaturated (C=C) sites, the partially hydrogenated product becomes more saturated, and the original liquid oil becomes semisolid. The ability to change an oil from a liquid to a solid, depending on the extent of hydrogenation, allows food manufacturers to control the consistency and softness of their food products. Such partially hydrogenated fats are formed in margarine, vegetable shortening, deep-fried foods, and many snack foods.

The reaction between hydrogen and C=C double bonds in a polyunsaturated fat is typical of how C=C bonds in most alkenes react. Generally, C=C bonds in alkenes can be converted to C—C single bonds when they react with a variety of substances, including hydrogen.

You learned in Unit 3 (page 268) that a double bond in an alkene prevents adjacent carbon atoms in the double bond from rotating around the bond axis. The double-bonded carbon atoms align as shown in Figure 7.28. This inflexible arrangement between carbon atoms creates the possibility of *cis–trans isomerism*.

In **cis–trans isomerism**, two identical functional groups can be in one of two different molecular positions. Both groups can be on the same side of the double bond (the cis isomer), or they can be across the double bond from each other (the trans isomer), as illustrated in Figure 7.29.

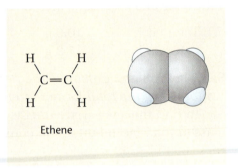

Ethene

Figure 7.28 *The structural formula for ethene.*

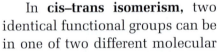

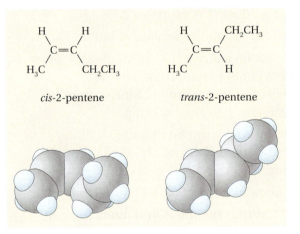

cis-2-pentene   *trans*-2-pentene

Figure 7.29 *A comparison of* cis-2-pentene *and* trans-2-pentene. *Note positions of the functional groups.*

In the example of 2-pentene, each arrangement creates a different compound, even though the compounds have identical molecular formulas (C_5H_{10}). In *cis*-2-pentene, the molecule has the two hydrogen atoms on the same side (cis) of the double-bond plane. (See Figure 7.29.) In *trans*-2-pentene, the two hydrogen atoms are positioned on opposite sides (trans) of the double-bond plane.

Cis–trans isomerism, which was just described for an alkene, is also possible in unsaturated fats because fatty acid chains often contain C=C bonds. Unsaturated fatty acids in foods that have not been hydrogenated typically have their double bonds in the cis arrangement, as shown in Figure 7.30. During hydrogenation, some cis double bonds break and reform in the trans arrangement. Although questions have arisen about the nutritional safety of *trans*-fatty acids, studies addressing these concerns have not produced conclusive answers. However, the *2005 Dietary Guidelines* advise that people should "keep *trans*-fatty acid consumption as low as possible."

Currently, most U.S. citizens obtain about 33% of their total food Calories from fats, near the upper limit of 35% recommended by the *Dietary Guidelines*. Additionally, the *Guidelines* suggest that less than 10% of total food Calories should come from saturated fats.

High fat consumption is a factor in several modern health problems, including obesity and atherosclerosis. Most dietary fat consumed in the United States comes from processed meat, poultry, fish, and dairy products. Fast foods and deep-fried foods—such as hamburgers, french fries, fried chicken, and many snack items—add even more dietary fat. In addition, if your intake of food energy is higher than what you expend in physical activity, your body converts excess proteins and carbohydrates into fat for storage.

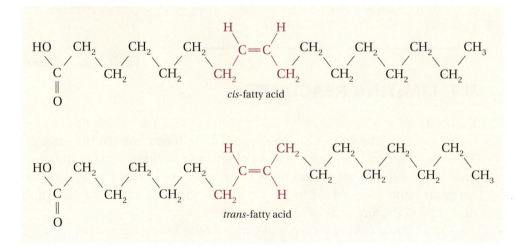

| **Figure 7.30** *Cis- and trans- fatty acids.*

B.6 FATS IN THE DIET

Your favorite ice-cream flavor is available as frozen yogurt. The advertising promotes frozen yogurt as a "reduced-fat alternative" to ice cream. Here are the nutritional data:

	Serving	Calories	Saturated Fat	Unsaturated Fat
Ice cream	115 g	310	11 g	8 g
Frozen yogurt	112 g	200	2.5 g	3.5 g

1. Calculate and compare the total percent fat (by mass) contained in each dessert.

2. Another way to compare these two desserts is to examine the percent of Calories from fat in each.

   a. Does either ice cream or frozen yogurt meet the guideline of 35% or less Calories from fat?

   b. Does either dessert meet the guideline of 10% or less Calories from saturated fat?

> Recall that 1 g of fat contains 9 Cal of food energy.

ChemQuandary 2

FAT-FREE FOOD?

You compared ice cream to frozen yogurt in terms of their fat content. You might consider *fat-free* ice cream as another frozen dessert. Consider this ice-cream label carefully. Is fat-free ice cream actually fat free? How do you know? What will happen to the food energy provided by fat-free ice cream if it is not immediately used by the body? Do you think it is wise to base a diet entirely on fat-free foods? Why or why not?

Vanilla Chocolate Swirl
CONTAINS 100 CALORIES COMPARED TO 150 CALORIES IN OUR REGULAR ICE CREAM, A 33% REDUCTION.

Nutrition Facts
Serving Size: 1/2 Cup (71g)
Servings Per Container: 14

from excessive consumption of these ingredients.

Amount Per Serving

Calories 100   Calories from Fat 0

	% Daily Value*
Total Fat 0g	1%
Saturated Fat 0g	0%
Cholesterol 0mg	0%
Sodium 50mg	2%
Total Carbohydrate 20g	7%
Dietary Fiber 6g	24%
Sugars 4g	
Sugar Alcohol 3g	
Protein 4g	

B.7 LIMITING REACTANTS

Biochemical reactions in your body convert fats, carbohydrates, and—under extreme dietary conditions—proteins in foods you eat into energy. These biochemical reactions, like all chemical reactions, require the presence of a complete set of reactants to produce the desired product. Furthermore, the amount of product produced by a chemical reaction depends on the amounts of reactants present.

Think about baking a cake and consider the following recipe:

2 cups flour	1 tablespoon baking powder
2 eggs	1 cup water
1 cup sugar	1/3 cup oil

The proper combination of these quantities (and some thermal energy) will produce one cake. What if you have 14 cups flour, 4 eggs, 9 cups sugar, 10 tablespoons baking powder, 10 cups water, and $3\frac{1}{3}$ cups oil? How many complete cakes can you bake?

Well, 14 cups of flour is enough for 7 cakes (2 cups flour per cake). And there is enough sugar for 9 cakes (1 cup sugar per cake). The supplies of baking powder, water, and oil are sufficient for 10 cakes. However, it is not possible to make 10, 9, or even 7 cakes with the available ingredients.

Why? Because only 4 eggs are available, which is enough for 2 cakes. The supply of eggs limits the number of cakes you can make. Excess quantities of other ingredients (flour, sugar, baking powder, water, and oil) remain unused. If you want more than 2 cakes, you will need more eggs.

In chemical terminology, the eggs in this cake-making analogy represent the *limiting reactant* (also called the *limiting reagent*). The **limiting reactant** is the starting material (reactant) that is used up entirely when a particular chemical reaction occurs. This starting material *limits* how much product can form.

In chemical reactions, just as in baking, materials react in certain fixed ratios. These ratios are referred to as the **stoichiometry** of the reaction. The ratios are indicated in chemical equations by the coefficients given for each substance. Consider the equation for the cellular respiration ("burning") of glucose:

$$C_6H_{12}O_6 \ + \ 6\,O_2 \ \longrightarrow \ 6\,CO_2 \ + \ 6\,H_2O \ + \ \text{Energy}$$

| 1 Glucose molecule | 6 Oxygen molecules | 6 Carbon dioxide molecules | 6 Water molecules | |

Sample Problem: *Suppose you have 5 glucose molecules available to react with 60 oxygen molecules. Which substance will become the limiting reactant in this reaction?*

From the equation, you can see that 1 glucose molecule reacts with 6 oxygen molecules. That means that 5 glucose molecules would require 30 oxygen molecules to react completely. On the other hand, to use up all 60 oxygen molecules, 10 glucose molecules are required. Which of these two scenarios is actually possible in this reaction? In other words, which reactant—oxygen or glucose—will be used up completely in this reaction?

Because 60 oxygen molecules are available, all the glucose can react, with some oxygen left unreacted. Alternatively, to react completely, the 60 oxygen molecules would require 10 glucose molecules; however, only 5 glucose molecules are available. Thus, glucose is the limiting reactant. Because the reaction stops after the 5 glucose molecules react with 30 oxygen molecules, 30 oxygen molecules will remain unreacted ('excess') at the end of the reaction.

> Recall from Unit 4 that stoichiometry involves quantitative relationships among reactants and products in a reaction described by its chemical equation.

> Limiting reactants were introduced in Unit 4, page 353.

The idea of limiting reactants applies equally well to living systems. The shortage of a key nutrient or reactant can severely affect the growth or health of plants and animals. In many biochemical processes, a product from one reaction becomes a reactant for other reactions. If a reaction stops because one substance is completely consumed (the limiting reactant), all reactions that follow it will also stop.

Fortunately, in some cases, alternate reaction pathways are available. If the body's glucose supply is depleted, for example, glucose metabolism cannot occur. One backup system oxidizes stored body fat in place of glucose. More drastically, under starvation conditions, structural proteins are broken down and used for energy. Producing glucose from protein is much less energy efficient than producing glucose from carbohydrates; if dietary glucose becomes available later, glucose metabolism reactions start up again.

Alternate reaction pathways are not a permanent solution. If the intake of a vital nutrient is consistently inadequate, that nutrient may become a limiting reactant in biochemical processes and affect personal health.

Modeling Matter

B.8 LIMITING-REACTANT ANALOGIES

In Unit 4, you considered a container filled with super-bounce balls as an analogy for the kinetic molecular behavior of gases (see page 325). Another analogy, a cake recipe, introduced you to the idea of limiting reactants. Think about this and other limiting-reactant analogies (see Figure 7.31) as you answer the following questions.

1. Consider the cake-making analogy (pages 594–595). Assume that you have 26 eggs and the quantities of all the other ingredients, as previously specified.

   a. Which ingredient now limits the total number of cakes you can make?

   b. How many total cakes can you make under the conditions specified?

   c. When the limiting reactant is fully consumed, how much of each other ingredient will be left over?

2. A restaurant prepares carry-out lunch boxes. Each box consists of 1 sandwich, 3 cookies, 2 paper napkins, 1 milk carton, and 1 container. The current inventory is 60 sandwiches, 102 cookies, 38 napkins, 41 milk cartons, and 66 containers.

   a. As carry-out lunch boxes are prepared, which item will be used up first?

   b. Which item is the limiting reactant?

   c. How many complete carry-out lunch boxes can be assembled from this inventory?

Figure 7.31 *What are key ingredients needed to make this triple-stacker s'more? How could the "limiting reactant" concept apply here?*

3. Add 2.5 mL of starch suspension to each tube.

4. Add 2.5 mL of 0.5% amylase solution to each tube.

5. Insert a stopper into each tube. Hold the stopper in place with your thumb or finger and shake each tube well for several seconds.

6. Leave the room-temperature test tubes in the laboratory overnight as directed by your teacher.

7. Give your teacher the test tubes that are to be refrigerated.

8. Wash your hands thoroughly before leaving the laboratory.

Figure 7.46 *Remember to label all test tubes, so that you can identify them later in the investigation.*

Day 2: Evaluating the Results

9. Prepare a hot-water bath by adding about 100 mL of tap water to a 250-mL beaker. Add a boiling chip. Warm the beaker on a hot plate. (Heat to just below boiling—hot, but not boiling.)

10. Add 5 mL of Benedict's reagent to each tube. Replace each stopper, being careful not to mix the stoppers. Hold the stopper in place with a thumb or finger and shake each tube well for several seconds.

11. Ensure that all tubes are still clearly labeled. Remove the stoppers and place the test tubes into the hot-water bath.

12. Heat the test tubes in the hot-water bath until the solution in at least one tube has turned yellow or orange. Then continue heating for two to three more minutes.

13. Use tongs to remove the test tubes from the hot-water bath. Arrange them in a test-tube rack in order of increasing pH.

14. Observe and record the color of the contents of each tube.

15. Share your data with your classmates, as directed by your teacher.

16. Wash your hands thoroughly before leaving the laboratory.

> Benedict's reagent also provides a test for the presence of glucose in urine, a symptom of diabetes.

Questions

1. a. At which general temperature range (cooled, room temperature, or heated) did the enzyme perform most effectively?

   b. At which pH value did the enzyme perform most effectively?

2. Write a summary statement about the effects of temperature and pH on this enzyme-catalyzed reaction.

Your cells synthesize many enzymes and other kinds of proteins to keep you alive. The amino acids used to synthesize these proteins are best obtained through a diet that provides appropriate amounts of protein. You will now decide whether a particular diet meets those protein needs.

C.8 PROTEIN CONTENT

You have analyzed the three-day food inventory in terms of energy that the diet provides and fat and carbohydrate molecules that it delivers. Now consider whether the food provided supplies the recommended amounts of a key building block of living material—protein.

Use the food inventory to answer the following questions. Refer to Table 7.4 (page 609), if necessary.

1. What is the average total mass of protein (in grams) consumed daily?

2. What other information would you need to know regarding the person who follows this food-intake pattern to evaluate the appropriateness of the protein supplied by these foods?

3. a. Does the person consume a good balance of essential amino acids? Explain your answer.

   b. What types of snack foods are high in protein (especially types that might be included in the vending machines at Riverwood High School)?

SECTION C SUMMARY
Reviewing the Concepts

Proteins, major structural components of living creatures, fulfill many cellular roles.

1. Name three types of tissue in your body for which protein is the main structural component.

2. The name *protein* comes from a Greek word that means "of prime importance." Why is this name appropriate?

3. List five cellular functions in which proteins are particularly important.

4. Name three food items composed primarily of protein.

5. Why are proteins considered polymers?

6. What chemical elements do proteins contain?

7. How many Calories would be provided by the metabolism of 3.6 g protein? (Recall that 1 g protein = 4 Cal.)

8. What is the chemical composition of an enzyme?

Amino acids are the chemical subunits that make up proteins.

9. How does the relatively small number of different amino acids account for the vast variety of proteins found in nature?

10. What is a *peptide bond?* Use structural formulas to illustrate your answer.

11. How does the total number of amino acids vary within protein molecules?

12. What is the protein DRI value for

    a. a 4-month-old infant?
    b. a 36-year-old male?

13. Write structural formulas for the following molecules:

    a. a dipeptide of glycine and cysteine
    b. a tripeptide abbreviated Asp–Ala–Cys

14. Explain the meaning of the following terms:

    a. complete protein
    b. essential amino acid
    c. complementary proteins

15. On which two functional groups is the name "amino acid" based?

Some protein molecules function as enzymes, that is, biological catalysts that speed up cellular reactions.

16. a. How are enzymes similar to other catalysts?
    b. How are enzymes different from other catalysts?

17. What would be the effect if all enzyme activity in the human body suddenly stopped? Explain.

18. Describe how high temperatures affect the ability of most enzymes to function.

19. Explain why enzymes can speed up only certain chemical reactions.

20. a. Explain the interaction of an active site and a substrate in terms of a lock-and-key analogy.
    b. Describe at least one limitation to the lock-and-key analogy for enzyme activity.

Connecting the Concepts

21. a. In what ways are proteins similar to carbo-hydrates and fats?
    b. In what ways are proteins different?

22. If a steak is left on a barbecue grill too long, it turns black. What does that observation suggest about the chemical composition of protein in the meat?

23. Although many plants contain high levels of protein, vegetarians must be more concerned than nonvegetarians about including adequate protein in their diets.

    a. How can vegetarians ensure that they obtain all the chemical building blocks needed to build required proteins?
    b. Explain your answer.

24. Using simple diagrams, including structural formulas, sketch how an enzyme might help form a peptide bond between two amino acids.

25. If a person followed a daily diet of 55 g fat, 75 g protein, and 85 g carbohydrate, how much energy would be provided through the metabolism of each? Show your calculations.

26. In Unit 1, you learned that fish species need a water pH range within which they can live. Given what you know about enzymes from the amylase tests in this section, explain why this is so.

27. Explain how the protein that you eat—whether from beef, turkey, beans, nuts, or tofu—is transformed into human-body proteins.

Extending the Concepts

28. The genetic code in DNA carries blueprints for making proteins in the human body. Explain how DNA helps determine the body's physical development and functioning.

29. The phrase "form follows function" is particularly applicable to enzymes. Explain how form and function are closely related for enzyme molecules.

30. Obtain information on the condition known as *ketosis*. What are its causes and effects? How can a high-protein diet lead to ketosis?

31. Explain why hydrogen peroxide is an effective antiseptic. How does your answer relate to enzyme action?

32. Research and report on chemical and physical properties of one or more amino acids used by the human body in building protein molecules.

33. What is a *zwitterion?* Under what conditions do amino acids become zwitterions?

Other Substances in Foods

The focus of this unit thus far has been on macronutrients in foods: proteins, carbohydrates, and fats. However, there are other substances in food, often in trace amounts: *vitamins, minerals,* and sometimes *additives.* Dietary guidelines imply that vitamins and minerals play vital roles within your body. These micronutrients occur naturally in many foods and may also be added to foods, such as cereal or bread, to improve their quality as sources of various micronutrients. Food manufacturers use food additives for different reasons, as you will soon learn. What do vitamins and minerals do in your body? What purposes do food additives serve?

D.1 VITAMINS

Vitamins are biomolecules necessary for growth, reproduction, health, and life. Each vitamin is required in only a tiny amount. The total quantity of *all* vitamins required daily by an adult is only about 0.2 g; "a little goes a long way" with vitamins (see Figures 7.47 and 7.48). How much is "enough"? The actual quantity depends on age and gender, as is suggested by Table 7.5 (page 620).

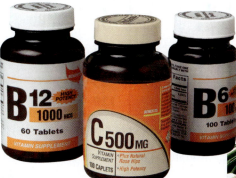

Figure 7.47 *Vitamin supplements are useful whenever sufficient vitamins are not provided in one's diet.*

Vitamins perform very specialized tasks. Vitamin D, for example, helps move calcium ions from your intestines into the bloodstream. Without vitamin D, your body would not use much of the calcium you ingest. Some vitamins function as **coenzymes,** which are organic molecules that interact with enzymes and enhance their activity. For example, the B-vitamins act as coenzymes in releasing energy from food molecules. Figure 7.49 illustrates how a coenzyme functions.

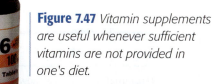

Figure 7.48 *Many foods such as leafy greens, liver, milk, eggs, and whole grains are rich in vitamins.*

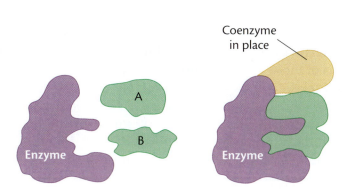

Figure 7.49 *A vitamin serving as a coenzyme.*

Age or Condition	Vit A (μg/d)	Vit D (μg/d)	Vit E (mg/d)	Vit K (μg/d)	Vit C (mg/d)	Thiamin (B₁) (mg/d)	Riboflavin (B₂) (mg/d)	Niacin (mg/d)	Vit B₆ (mg/d)	Vit B₁₂ (μg/d)	Folate (μg/d)
Males											
9–13 yrs	**600**	5*	**11**	60*	**45**	**0.9**	**0.9**	**12**	**1.0**	**1.8**	**300**
14–18 yrs	**900**	5*	**15**	75*	**75**	**1.2**	**1.3**	**16**	**1.3**	**2.4**	**400**
19–30 yrs	**900**	5*	**15**	120*	**90**	**1.2**	**1.3**	**16**	**1.3**	**2.4**	**400**
31–50 yrs	**900**	5*	**15**	120*	**90**	**1.2**	**1.3**	**16**	**1.3**	**2.4**	**400**
50–70 yrs	**900**	10*	**15**	120*	**90**	**1.2**	**1.3**	**16**	**1.7**	**2.4**	**400**
>70 yrs	**900**	15*	**15**	120*	**90**	**1.2**	**1.3**	**16**	**1.7**	**2.4**	**400**
Females											
9–13 yrs	**600**	5*	**11**	60*	**45**	**0.9**	**0.9**	**12**	**1.0**	**1.8**	**300**
14–18 yrs	**700**	5*	**15**	75*	**65**	**1.0**	**1.0**	**14**	**1.2**	**2.4**	**400**
19–30 yrs	**700**	5*	**15**	90*	**75**	**1.1**	**1.1**	**14**	**1.3**	**2.4**	**400**
31–50 yrs	**700**	5*	**15**	90*	**75**	**1.1**	**1.1**	**14**	**1.3**	**2.4**	**400**
50–70 yrs	**700**	10*	**15**	90*	**75**	**1.1**	**1.1**	**14**	**1.5**	**2.4**	**400**
>70 yrs	**700**	15*	**15**	90*	**75**	**1.1**	**1.1**	**14**	**1.5**	**2.4**	**400**
Pregnant	**770**	5*	**15**	90*	**85**	**1.4**	**1.4**	**18**	**1.9**	**2.6**	**600**
Nursing	**1300**	5*	**19**	90*	**120**	**1.6**	**1.6**	**17**	**2.0**	**2.8**	**500**

DIETARY REFERENCE INTAKES (DRIs) FOR SELECTED VITAMINS

Note: Recommended Dietary Allowances (RDAs) are in **bold type.** RDAs are established to meet the needs of almost all (97 to 98%) individuals in a group. Adequate Intakes (AIs) are followed by an asterisk (*). AIs are believed to cover the needs of all individuals in the group, but a lack of data prevent being able to specify with confidence the percent of individuals covered by this intake.

Source: Food and Nutrition Board, National Academy of Sciences: National Research Council, *Dietary Reference Intakes 2004.*

| Table 7.5

Vitamins and Solubility

Recall that "like dissolves like." See page 72.

Long before the term *vitamin* was introduced early in the last century, people discovered that small quantities of certain substances were necessary to maintain health. One example of vitamin deficiency is *scurvy,* which once was common among sailors; this condition is characterized by swollen joints, bleeding gums, and tender skin. Although early seafarers (in the 1700s and 1800s) did not know what caused scurvy, they commonly loaded citrus fruit onboard, which they ate during long voyages to prevent scurvy. Scurvy is now known to be caused by vitamin C deficiency, a vitamin supplied by citrus fruit. In addition to vitamin C, about a dozen different vitamins have been identified over the past century, each critical to reactions occurring within the human body. Table 7.6 documents how some of those vitamins support human life.

Vitamins are classified as either *fat-soluble* or *water-soluble* (see Table 7.6). Water-soluble vitamins with polar functional groups pass directly into the bloodstream. They are not stored in the body; they must be ingested daily. Some water-soluble vitamins, including the B vitamins and vitamin C, are also destroyed by heat in cooking.

VITAMINS BY CATEGORY, SHOWING SOURCES AND DEFICIENCY CONDITIONS

Vitamin	Main Sources	Deficiency Condition
Water-soluble		
B₁ (thiamin)	Liver, milk, pasta, bread, wheat germ, lima beans, nuts	Beriberi: nausea, severe exhaustion, paralysis
B₂ (riboflavin)	Red meat, milk, eggs, pasta, bread, beans, dark green vegetables, peas, mushrooms	Severe skin problems
Niacin	Red meat, poultry, enriched or whole grains, beans, peas	Pellagra: weak muscles, loss of appetite, diarrhea, skin blotches
B₆ (pyridoxine)	Muscle meats, liver, poultry, fish, whole grains	Depression, nausea, vomiting
B₁₂ (cobalamin)	Red meat, liver, kidneys, fish, eggs, milk	Pernicious anemia, exhaustion
Folate (folic acid)	Kidneys, liver, leafy green vegetables, wheat germ, peas, beans	Anemia
Pantothenic acid	Plants, animals	Anemia
Biotin	Kidneys, liver, egg yolk, yeast, nuts	Dermatitis
C (ascorbic acid)	Citrus fruits, melon, tomatoes, green peppers, strawberries	Scurvy: tender skin; weak, bleeding gums; swollen joints
Fat-soluble		
A (retinol)	Liver, eggs, butter, cheese, dark green and deep orange vegetables	Inflamed eye membranes, night blindness, scaling of skin, faulty teeth and bones
D (calciferol)	Fish-liver oils, fortified milk	Rickets: soft bones
E (tocopherol)	Liver, wheat germ, whole-grain cereals, margarine, vegetable oil, leafy green vegetables	Breakage of red blood cells in premature infants, oxidation of membranes
K (menaquinone)	Liver, cabbage, potatoes, peas, leafy green vegetables	Hemorrhage in newborns; anemia

| **Table 7.6**

Your body absorbs fat-soluble vitamins into the blood from the intestine with assistance from fats in the food you eat. Because the nonpolar structures of fat-soluble vitamins allow them to be stored in body fat, it is not necessary to consume fat-soluble vitamins daily. In fact, because fat-soluble vitamins accumulate within the body, they can build up to toxic levels if ingested in excessively large quantities (megadoses).

D.2 VITAMINS IN THE DIET

1. Carefully planned vegetarian diets are nutritionally balanced. Individuals who follow a *vegan* diet do not consume any animal products, including eggs and milk. Because of their dietary limitations, vegans must ensure that they obtain the recommended daily allowances of two particular vitamins.

   a. Use Table 7.6 (page 621) to identify these two vitamins and briefly describe the effect of their absence in the diet.

   b. How might individuals following a vegan diet avoid this problem?

2. Complete the following table about yourself, using data from Tables 7.5 (page 620) and 7.7.

Figure 7.50 *How does the vitamin content of raw and steamed broccoli compare?*

Vegetable (one-cup serving)	Your RDA		Total Servings to Supply Your RDA	
	B₁	C	B₁	C
Green peas				
Broccoli				

   a. Would any of your entries change if you were of the opposite gender? If so, which entry or entries?

   b. Based on your completed table, why do you think variety is essential in a person's diet?

   c. Why might vitamin deficiencies pose problems even if people received adequate supplies of food Calories?

3. Nutritionists recommend eating fresh fruit rather than canned fruit and raw or steamed vegetables instead of canned or boiled vegetables (see Figures 7.50 and 7.51).

   a. What does food freshness have to do with vitamins?

   b. To what type of vitamins might nutritionists be referring?

   c. Is such advice sound? Explain.

Figure 7.51 *Why do dietary guidelines usually favor raw fruit over canned fruit?*

VITAMIN B₁ AND C CONTENT OF SOME VEGETABLES

Vegetable (1-cup serving)	Vitamin (in mg)	
	B₁ (thiamin)	C (ascorbic acid)
Green peas	0.387	58.4
Lima beans	0.238	17
Broccoli	0.058	82
Potatoes	0.15	30

| **Table 7.7**

D.3 VITAMIN C

Introduction

Vitamin C, also called *ascorbic acid,* is a water-soluble vitamin. It is among the least stable vitamins because it reacts readily with oxygen gas, and exposure to light or heat can decompose it. In this investigation, you will find out how much vitamin C is contained in some popular beverages, including fruit juices, milk, and soft drinks.

This investigation is based on a chemical reaction of ascorbic acid (vitamin C) with iodine (I_2). A colored solution of iodine oxidizes ascorbic acid, forming the colorless products dehydroascorbic acid, hydrogen ions, and iodide ions:

$$I_2 \quad + \quad C_6H_8O_6 \quad \longrightarrow \quad C_6H_6O_6 \quad + \quad 2\,H^+ \quad + \quad 2\,I^-$$

| Iodine | Ascorbic acid (vitamin C) | Dehydroascorbic acid | Hydrogen ion | Iodide ion |

Figure 7.52 shows the structures of ascorbic and dehydroascorbic acid.

You will complete a **titration** (Figure 7.53), a common laboratory procedure used to determine concentrations of substances in solution. This investigation involves adding a known amount of one reactant (iodine solution) slowly from a Beral pipet to a second reactant (ascorbic acid in the beverage) in a wellplate until just enough has been added for a complete reaction. The completion of the reaction, the **endpoint,** is signaled by a color change. Knowing the chemical equation for this reaction, you can then calculate the unknown amount of the second reactant (ascorbic acid) from the measured volume and known concentration of the the the iodine-solution *titrant.*

The titration endpoint in the beverage-containing well of the wellplate is the point where a dark blue–black color appears and does not disappear with additional stirring. This color is due to the reaction of excess iodine with starch. First, you add a starch suspension to the beverage sample to be tested. Next, an iodine solution of known concentration is added drop by drop from a Beral pipet.

As long as ascorbic acid is present, the added iodine is quickly converted to colorless iodide ions; you will observe no bluish black iodine-starch product. When all the available ascorbic acid has been oxidized to colorless dehydroascorbic acid, the next drop of added iodine solution reacts with the starch, producing the bluish black color that signals the endpoint.

Lab Video: Vitamin C

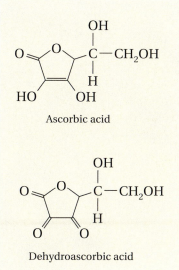

Ascorbic acid

Dehydroascorbic acid

Figure 7.52 *Molecular structures of ascorbic and dehydroascorbic acid.*

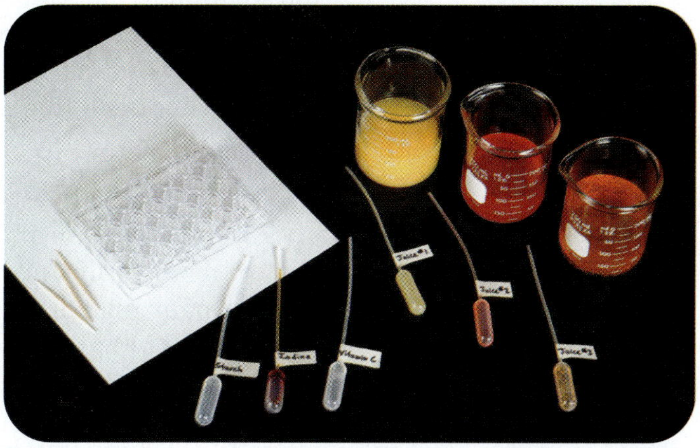

Figure 7.53 *Material needed for fruit-juice titration.*

You will begin by completing a titration involving a solution of known vitamin C concentration. The data will allow you to find the mass of ascorbic acid that reacts with one drop of iodine solution. You can then calculate the mass (in milligrams) of vitamin C present in a 25-drop sample of each beverage. This information will allow you to rank the tested beverages in terms of the mass of vitamin C that each beverage contains.

Before starting this investigation, prepare a suitable data table. Leave room in the data table to record the total drops of vitamin C solution in 1.0 mL (Step 1) and the total drops of iodine needed to reach the endpoint (Step 5). Provide a horizontal row for each beverage that you investigate.

Procedure

Part 1: Investigating the Iodine Solution

1. Fill a Beral pipet with vitamin C solution. Then determine how many drops of vitamin C solution delivered by that pipet represent a volume of 1.0 mL.

2. Fill a second Beral pipet with iodine solution. Determine the total drops delivered by that pipet that equal 1.0 mL, just as in Step 1.

3. Add 25 drops vitamin C solution into a well of a clean 24-well wellplate. This vitamin C solution has a known concentration of 1.0 mg vitamin C per milliliter of solution.

4. Add 1 drop of starch suspension to the same well.

5. Place a sheet of white paper underneath the wellplate; it will help you detect the appearance of color. Add iodine solution one drop at a time, carefully counting drops, to the well that contains the starch and vitamin C mixture. After adding each drop of iodine solution, use a toothpick to gently stir the resulting mixture. Add and count the iodine solution drop by drop, with stirring, until the solution in the well remains bluish black for 20 s. If the color fades before 20 s have elapsed, add another drop of iodine solution. Record the total drops of iodine solution needed to reach the endpoint (the appearance of the first permanent bluish-black color).

6. The concentration of the vitamin C solution used is 1.0 mg/mL.

   a. What total volume of vitamin C solution did you test?
   b. How many milligrams of vitamin C did that volume contain?
   c. How many total drops of iodine did you use?

   Use that information to calculate how many milligrams of vitamin C react with 1 drop of iodine solution:

$$25 \text{ drops vitamin C} \times \frac{1 \text{ mL vitamin C}}{?? \text{ drops vitamin C}} \times \frac{1 \text{ mg vitamin C}}{1 \text{ mL vitamin C}} \times \frac{1}{?? \text{ drops I}_2} = \underline{\quad} \text{ mg vitamin C per drop I}_2 \text{ solution}$$

$$\text{(Step 1)} \qquad\qquad\qquad\qquad \text{(Step 5)}$$

Suppose there were 30 drops of vitamin C solution in 1.0 mL (Step 1) and that it took 22 drops of iodine solution to reach the endpoint (Step 5). These data lead to the result that 0.038 mg vitamin C reacts with 1 drop iodine solution. Complete calculations to confirm this result.

Part 2: Analyzing Beverages for Vitamin C

7. Using a Beral pipet, add 25 drops of the beverage to be tested to a clean well of a wellplate.

8. Add one drop of starch suspension to the same well.

9. With a Beral pipet, slowly add iodine solution one drop at a time to the well containing the beverage and starch mixture (see Figure 7.54). After each addition, stir the resulting mixture with a toothpick. Count the total drops of iodine solution that are added to reach the endpoint. (*Note:* Colored beverages may not produce a true bluish-black endpoint color. For example, red beverages may make the endpoint appear purple.)

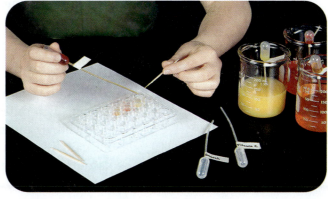

10. Continue adding iodine solution drop by drop, with stirring, until the solution in the well remains bluish black for 20 s. If the color fades before 20 s have elapsed, add another drop of iodine solution. Record the total drops of iodine solution needed to reach the endpoint.

| **Figure 7.54** *Proper technique for a wellplate titration.*

11. Repeat Steps 7–10 for each beverage you have been assigned to investigate.

12. Use your data from Step 10 and the results from Step 6 to calculate the milligrams of vitamin C contained in 25 drops of each assigned beverage.

13. Rank the tested beverages in terms of how much vitamin C each contains, from the highest quantity to the lowest.

14. Wash your hands thoroughly with soap and water before leaving the laboratory.

Questions

1. Among the beverages tested, in your opinion, were any vitamin C levels

   a. unexpectedly low? If so, explain.
   b. unexpectedly high? If so, explain.

2. Imagine that you added too many drops of iodine solution during the titration and missed the true endpoint. Will this procedural error increase or decrease your calculated value of milligrams of vitamin C in the sample? Explain.

D.4 MINERALS: ESSENTIAL WITHIN ALL DIETS

Minerals in ionic form are essential for life support. Some are quite common; others are likely to be found in large quantities only on research-laboratory shelves.

Some minerals become part of the body's structural material, such as bones and teeth. Others help enzymes do their jobs. Still others help maintain the health of the heart and other organs. The thyroid gland, for example, uses a miniscule quantity of iodine (only millionths of a gram daily) to produce the vital hormone *thyroxine* (Figure 7.55). The rapidly growing field of *bioinorganic chemistry* explores how minerals function within living systems.

Of the more than 100 known elements, only 32 are believed essential to support human life. For convenience, essential minerals are divided into **major minerals** (also called **macrominerals**), and **trace minerals** (also called **microminerals**).

Your body contains rather large quantities, at least 100 mg per kilogram of body mass, of each of the seven macrominerals. Each trace mineral is present in relatively small quantities, less than 100 mg per kilogram of body mass in an average adult. However, trace minerals are just as essential in a human diet as are macrominerals. Any essential mineral, whether a macromineral or a trace mineral, can become a limiting reactant if it is not present in sufficient quantity.

Figure 7.55 *The thyroid gland does not function properly without proper levels of iodine in one's body. Iodine deficiency causes goiter (depicted here).*

The essential minerals and their dietary sources, functions, and deficiency conditions are listed in Table 7.8. Several other minerals, including arsenic (As), cadmium (Cd), and tin (Sn), are known to be needed by laboratory test animals. These minerals and perhaps other trace minerals may also be essential to *human* life. You may be surprised to learn that the widely known poison arsenic might be an essential mineral. In fact, many substances beneficial in low doses become toxic in higher doses. Table 7.9 (page 628) summarizes the recommended daily doses for several macrominerals and trace minerals.

ESSENTIAL DIETARY MINERALS

Mineral	Typical Food Sources	Deficiency Conditions
Macrominerals		
Calcium (Ca)	Milk, dairy products, canned fish	Rickets in children; osteomalacia and osteoporosis in adults
Chlorine (Cl)	Table salt, meat, salt-processed foods	—
Magnesium (Mg)	Seafood, cereal grains, nuts, dark green vegetables, cocoa	Heart spasms, anxiety, disorientation
Phosphorus (P)	Meat, dairy products, nuts, seeds, beans	Blood cell disorders, gastrointestinal tract and renal dysfunction
Potassium (K)	Orange juice, bananas, dried fruits, potatoes	Poor nerve function, irregular heartbeat, sudden death during fasting
Sodium (Na)	Table salt, meat, salt-processed foods	Headache, weakness, thirst, poor memory, appetite loss
Sulfur (S)	Protein (e. g. meat, eggs, legumes)	Conditions related to deficiencies in sulfur-containing essential amino acids
Trace minerals		
Chromium (Cr)	Animal and plant tissue, liver	Loss of insulin efficiency with age
Cobalt (Co)	Animal protein, liver	Conditions related to deficiencies in cobalt-containing vitamin B_{12}
Copper (Cu)	Egg yolk, whole grains, liver, kidney	Anemia in malnourished children
Fluorine (F)	Seafood, fluoridated water	Dental decay
Iodine (I)	Seafood, iodized salt	Goiter
Iron (Fe)	Meats, green leafy vegetables, whole grains, liver	Anemia, tiredness, apathy
Manganese (Mn)	Whole grains, legumes, nuts, tea, leafy vegetables, liver	—
Molybdenum (Mo)	Whole grains, legumes, leafy vegetables, liver, kidney	Weight loss, dermatitis, headache, nausea, disorientation
Selenium (Se)	Meat, liver, organ meats, grains, vegetables	Muscle weakness, Keshan disease (heart-muscle disease)
Zinc (Zn)	Shellfish, meats, wheat germ, legumes, liver	Anemia, growth retardation

| Table 7.8

DIETARY REFERENCE INTAKES (DRIs) FOR SELECTED MINERALS

Age or Condition	Calcium (mg/d)	Phosphorus (mg/d)	Magnesium (mg/d)	Iron (mg/d)	Zinc (mg/d)	Iodine (μg/d)
Males						
9–13 years	1300*	1250	240	8	8	120
14–18 years	1300*	1250	410	11	11	150
19–30 years	1000*	700	400	8	11	150
31–50 years	1000*	700	420	8	11	150
50–70 years	1200*	700	420	8	11	150
> 70 years	1200*	700	420	8	11	150
Females						
9–13 years	1300*	1250	240	8	8	120
14–18 years	1300*	1250	360	15	9	150
19–30 years	1000*	700	310	18	8	150
31–50 years	1000*	700	320	18	8	150
50–70 years	1200*	700	320	8	8	150
> 70 years	1200*	700	320	8	8	150
Pregnant	1000*	700	350	27	11	220
Nursing	1000*	700	310	9	12	290

Recommended Dietary Allowances (RDAs) are in **bold type.** RDAs are established to meet the needs of almost all (97 to 98%) individuals in a group. Adequate Intakes (AIs) are followed by an asterisk (*). AIs are believed to cover the needs of all individuals in the group, but a lack of data prevent being able to specify with confidence the percent of individuals covered by this intake.

Source: Food and Nutrition Board, National Academy of Sciences: National Research Council, *Dietary Reference Intakes 2004.*

| Table 7.9

Developing Skills

D.5 MINERALS IN THE DIET

Use the values provided in Table 7.9 to answer the following questions.

Sample Problem: *How many cups of broccoli would a 16-year-old female need to eat daily to reach her daily iron allowance? (1 cup broccoli = 1.1 mg iron) Assume that broccoli is her only dietary source of iron.*

The iron DRI for a 16-year-old female is 15 mg per day:

$$15 \text{ mg iron} \times \frac{1 \text{ cup broccoli}}{1.1 \text{ mg iron}} = 14 \text{ cups broccoli}$$

1. One slice of whole-wheat bread contains 0.8 mg iron.

    a. How many slices of whole-wheat bread would supply your daily iron allowance? (Assume that this is your only dietary source of iron.)

    b. Predict health consequences of not consuming an adequate amount of iron.

2. One cup of whole milk contains 288 mg calcium. How much milk would you need to drink daily to meet your daily allowance for that mineral if milk were your only dietary source of calcium?

3. One medium pancake (Figure 7.56) contains about 27 mg calcium and 0.4 mg iron.

a. Does one pancake provide a greater percent of your DRI for calcium or for iron?

b. Explain your answer.

4. a. What total mass of each of these minerals do you need to consume daily?
  i. calcium
  ii. phosphorus

b. Why are the values in your answer to Question 4a higher than DRI values for other listed essential minerals? (*Hint:* Consider how calcium and phosphorus are used in the body.)

c. List several good dietary sources of
  i. calcium.
  ii. phosphorus.

d. Predict the health consequences due to a deficiency of
  i. calcium.
  ii. phosphorus.

e. i. Would a particular age group or gender especially be affected by those consequences?
  ii. If so, which age group or gender and why?

5. Most table salt, sodium chloride (NaCl), includes a small amount of added potassium iodide (KI).

a. Why do you think KI is added to table salt?

b. If you do not use iodized salt, what other kinds of food could you use as sources of iodine?

Figure 7.56 *Which foods are good sources of minerals such as calcium?*

"Iodized" salt refers to such products.

D.6 FOOD ADDITIVES

Small amounts of vitamins and minerals occur naturally in foods. Some foods, especially processed foods such as packaged snacks or frozen entrees, also contain small amounts of **food additives.** Manufacturers add these substances during processing to increase the nutritive value of foods or to enhance their storage life, visual appeal (see Figure 7.57), or ease of production. A food label might provide the ingredient information shown below. Can you guess the identity of the food product from only this label?

> Sugar, bleached flour (enriched with niacin, iron, thiamin, and riboflavin), semisweet chocolate, animal and/or vegetable shortening, dextrose, wheat starch, monocalcium phosphate, baking soda, egg white, modified corn starch, salt, nonfat milk, cellulose gum, soy lecithin, xanthan gum, mono- and diglycerides, BHA, BHT.

This list shows quite a collection of ingredients. You probably recognize the major ingredients, such as sugar, flour, shortening, and baking soda; and some of the additives, such as vitamins (niacin, thiamin, and riboflavin) and minerals (iron and monocalcium phosphate). However, you probably do not recognize the food additives xanthan gum (an *emulsifier* that helps produce uniform, nonseparating water–oil mixtures—see Figure 7.59, page 632) and BHA and BHT (butylated hydroxy-anisole and butylated hydroxy-toluene, which are antioxidants).

Food additives have been used since ancient times. For example, salt has been used for centuries to preserve foods, and spices disguised the flavor of food that was no longer fresh. To make foods easier and less expensive to distribute and store, most manufacturers, especially manufacturers of processed foods, rely on food preservatives. Table 7.10 summarizes major categories of food additives. The structural formulas of several common additives are shown in Figure 7.58.

Color and taste additives often enhance the commercial appeal of food products. In the following investigation, you will analyze several commonly used food-coloring additives (dyes).

Figure 7.57 *Food-coloring additives are often used to make food items more attractive and appetizing.*

FOOD ADDITIVE TYPES, PURPOSES, AND EXAMPLES | Table 7.10

Additive Type	Purpose	Examples
Anticaking agents	Keep foods free-flowing	Sodium ferrocyanide
Antioxidants	Prevent fat rancidity	BHA and BHT
Bleaches	Whiten foods (flour, cheese); hasten cheese maturation	Sulfur dioxide (SO_2)
Coloring agents	Increase visual appeal	Carotene (natural yellow color); synthetic dyes
Emulsifiers	Improve texture, smoothness; stabilize oil–water mixtures	Cellulose gums, dextrins
Flavoring agents	Add or enhance flavor	Salt, monosodium glutamate (MSG), spices
Humectants	Retain moisture	Glycerin
Leavening agents	Give foods light texture	Baking powder, baking soda
Nutrients	Improve nutritive value	Vitamins, minerals
Preservatives and antimycotic agents (growth inhibitors)	Prevent spoilage, microbial growth	Propionic acid, sorbic acid, benzoic acid, salt
Sweeteners	Impart sweet taste	Sugar (sucrose), dextrin, fructose, aspartame, sorbitol, mannitol

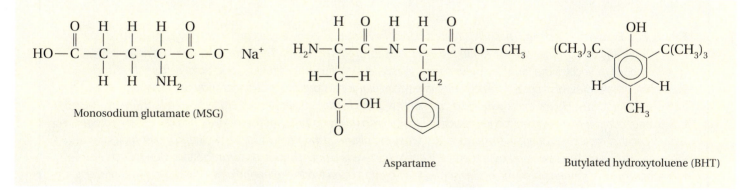

| **Figure 7.58** *The molecular structures of MSG, aspartame, and BHT, a preservative.*

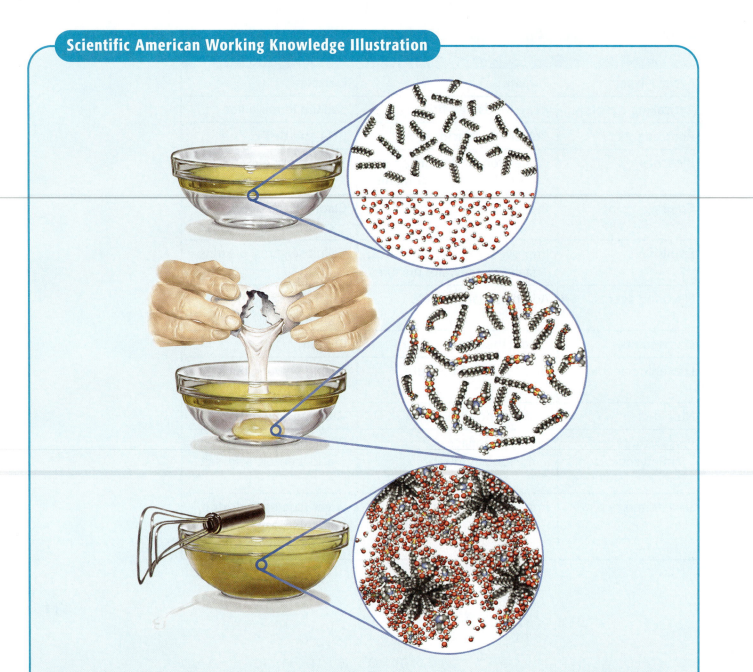

Figure 7.59 Emulsifier action. *These illustrations depict how lecithin, a molecule contained in egg yolk, emulsifies (mixes together) an oil-water system. In the top view, oil is layered over water. The particulate-level enlargement reveals that oil molecules are long, nonpolar carbon chains quite different in structure and polarity from the small, polar water molecules, also shown. Water-water intermolecular forces are much stronger than are oil-water intermolecular forces. Hence these two liquids do not mix. Each lecithin molecule, the emulsifying agent (middle view), has a long, nonpolar carbon chain and a smaller, polar region involving phosphorus, oxygen, nitrogen, and hydrogen atoms. Lecithin molecules orient themselves so that they simultaneously interact with both oil and water molecules. Once an egg is cracked open and added and the entire mixture is whisked, micelles form (bottom view). Each micelle is a spherical arrangement of oil molecules surrounded by and attracted to the nonpolar region of lecithin molecules. The polar regions of lecithin molecules align at the outside of a micelle, causing the entire sphere to intermingle among polar water molecules. This stable arrangement permits oil molecules to remain evenly dispersed through water. Emulsification is used to make numerous food products, including mayonnaise and ice cream.*

D.7 ANALYZING FOOD-COLORING ADDITIVES

Introduction

Many candies contain artificial coloring agents to enhance their visual appeal. Colorless candies would be quite dull! In this investigation, you will analyze the food dyes in two commercial candies and compare them with the dyes in food-coloring products.

Lab Video: Food Coloring Analysis

You will separate and identify the food dyes by **paper chromatography.** This technique uses a solvent (called the *mobile phase*) and paper (the *stationary phase*). Paper chromatography is based on relative differences in attraction between (a) dye molecules and solvent vs. (b) dye molecules and paper. As the solvent and dye mixture moves up the paper, dye molecules that are more strongly attracted to the paper will more readily leave the solvent and separate out onto the paper, leaving a colored spot on the paper. Dye molecules less attracted to paper (and more attracted to solvent) will travel further up the paper before they adhere to the paper later. Thus, characteristic areas representing different dye molecules will appear on the chromatography paper.

To analyze the results of this chromatography investigation, you will calculate the R_f value for each spot. The R_f value, as illustrated in Figure 7.60, is a ratio involving the distance traveled by each dye compared to the distance the solvent has moved up the paper. The actual distance that the dye and solvent travel might vary from trial to trial, but the R_f value will remain constant for a particular dye. Thus chemists use the R_f value, not the actual distance traveled, to identify dye molecules.

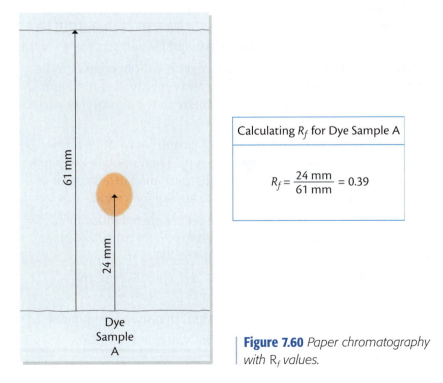

Calculating R_f for Dye Sample A

$$R_f = \frac{24 \text{ mm}}{61 \text{ mm}} = 0.39$$

Figure 7.60 *Paper chromatography with R$_f$ values.*

Procedure

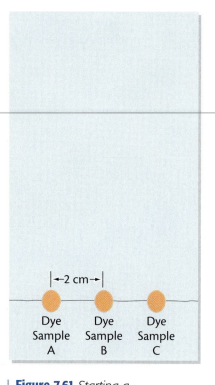

Figure 7.61 *Starting a chromatography investigation.*

Dye Sample A   Dye Sample B   Dye Sample C

|←2 cm→|

Figure 7.62 *Spotting a sample on chromatography paper.*

1. Obtain one piece each of two different commercial candies from your teacher, as well as a food-coloring sample. The food coloring and candies should all be the same color.

2. Put each candy into a separate well of a wellplate. Note which candy is in each well. Add 5–10 drops of water to each well. Stir the mixture in each well with a separate toothpick until the colored coating completely dissolves from the candy. Add 3 to 4 drops of the food coloring sample to a third well. Observe and record the initial colors of each sample.

3. Obtain a strip of chromatography paper, handling it only by its edges. With a pencil (do not use a pen), draw a horizontal line 2 cm from the bottom of the paper strip and another horizontal line 3 cm from the top. Label the strip as shown in Figure 7.61.

4. Next, place a spot of each dye solution on the bottom line; the spots should not be large. To do this, use separate toothpicks for each of the three samples. Place a drop of the first candy's colored solution with a toothpick, as shown in Figure 7.62.

5. Allow the drop of solution to sit until the spot it makes stops spreading out on the paper. Then apply a second drop of the same sample on top of the spot. Repeat this procedure for the second candy's colored solution and also for the food-coloring sample. You will then have three spots along the bottom pencil line on the chromatography paper (Figure 7.61). If the chromatography paper provided is narrow, you may need to use a second piece of paper.

6. Obtain a chromatography chamber. Place a mark on the outside of the chamber approximately 1 cm up from the bottom. Fill the chromatography chamber with the solvent (water) to a depth of 1 cm.

7. Lower the spotted chromatography paper's bottom edge into the chromatography chamber until edge rests evenly in the solvent (water). Be sure the colored spots remain above the solvent surface. Cover the chromatography chamber.

8. Allow the solvent to move up the paper until solvent reaches the top penciled line (3 cm from the top). Then remove the paper from the chamber and, using a pencil, mark the farthest point of the solvent's path. Allow the paper to air-dry overnight.

9. After the paper has dried, record the colors observed for the dye sample and candy solutions.

10. Measure the distance (in cm) from the initial pencil line where you placed the spots to the *center* of each dye spot. Record these distances in your data table.

11. Measure and record the distance (in cm) that the solvent moved.

12. Calculate the R_f value of each dye spot you observed. Record these values.

Questions

1. Why was it important to use a pencil rather than a pen to mark lines on the chromatography paper in Step 3?

2. Which sample solution, if any, created a single spot rather than several spots?

3. Which dye in the sample solutions had the greatest attraction for the

   a. paper?
   b. solvent?

4. a. Based on your data, do any of the three samples contain any of the same dyes?

   b. What evidence did you use to answer Question 4a?

5. The candy and food-coloring packages list each dye that they contain. Compare this information with your experimental results.

   a. What similarities did you find?
   b. What differences did you find?
   c. If you found differences, what are some possible reasons for them?

D.8 REGULATING ADDITIVES

Both processed and unprocessed foods may contain contaminants that were not deliberately added, such as pesticides, mold, antibiotics used to treat animals, insect fragments, food-packaging materials, or dirt. We presume that food purchased in grocery stores and restaurants is safe to eat. In most cases, this is true. Nonetheless, in the past, some food additives and contaminants were identified as or suspected of posing hazards to human health.

Safe food in the United States is required by law—the Federal Food, Drug, and Cosmetic Act of 1938. This act authorized the Food and Drug Administration (FDA) to monitor food's safety, purity, and wholesomeness. This act has been amended to address concerns about pesticide residues; artificial colors and food dyes; potential cancer-causing agents (*carcinogens*); and **mutagens,** which are agents that cause mutations or changes in DNA. Food manufacturers must complete a battery of tests and provide extensive evidence about the safety of any proposed food product or additive. Any new food product must earn FDA approval before it can be marketed.

According to the amended Federal Food, Drug, and Cosmetic Act, ingredients that were known not to be hazardous and were in use for a long time prior to the act were exempted from testing. These substances, rather than legally defined as additives, constitute the "generally recognized as safe" (GRAS) list. The GRAS list, periodically reviewed in light of new findings, includes items such as salt, sugar, vinegar, vitamins (for example, vitamin C and riboflavin), and some minerals.

In accord with the Delaney Clause, which was added to the act in the 1950s, every proposed new food additive must be tested on laboratory animals (usually mice). The Delaney Clause specifies that "no additive shall be deemed to be safe if it is found to induce cancer when it is ingested by man or animal." Thus, approval of a proposed additive is denied if it causes cancer in test animals.

Since the 1950s, great advances have occurred in science and technology. Improvements in chemical-analysis techniques permit scientists to detect even smaller amounts of potentially harmful substances. Consequently, scientists can now detect and study food contaminants that may have always been present but previously were undetected.

This new information has resulted in a greater understanding of potential risks associated with human exposure to particular food additives. People now recognize that amounts of additives comparable to those causing cancer in test animals are often vastly greater than would ever be encountered in a human diet. In light of this, Congress passed the Food Quality and Protection Act in 1996. This legislation, which replaced the Delaney Clause, states that manufacturers may use food additives that present "negligible risk."

Many concerns about food additives still remain. Sodium nitrite ($NaNO_2$), for instance, is a color stabilizer and spoilage inhibitor used in many cured meats, such as hot dogs and lunch meats. Nitrites are particularly effective in inhibiting the growth of the bacterium *Clostridium botulinum,* which produces botulin toxin. This toxin is the cause of botulism, an often fatal disease. Sodium nitrite, however, may be a carcinogen. In the stomach, nitrites are converted to nitrous acid:

$$NaNO_2(aq) + HCl(aq) \longrightarrow HNO_2(aq) + NaCl(aq)$$

| Sodium nitrite | Hydrochloric acid | Nitrous acid | Sodium chloride |

Nitrous acid can then react with compounds formed during protein digestion, producing *nitrosoamines,* known as potent carcinogens. An example of this reaction is shown below. The concentration of carcinogenic compounds produced, however, is generally below the toxic threshold.

$$HNO_2 + R\text{---}N\text{---}H \longrightarrow R\text{---}N\text{---}N{=}O + H_2O$$
$$\qquad\quad |\qquad\qquad\qquad |$$
$$\qquad\quad R\qquad\qquad\qquad R$$

NITRITE ADDITIVES

Consider the following benefits and risks of using nitrites to preserve meats:

NITRITE RISK TABLE		
	Using Nitrites	**Eliminating Nitrites**
Benefit	Decreases risk of forming botulin toxin	Decreases risk of forming possible carcinogens
Risk	Increases risk of forming possible carcinogens	Increases risk of forming botulin toxin

Based on this information, do you think food manufacturers should use nitrites to preserve meats?

D.9 ARTIFICIAL SWEETENERS

Because many people believe they should reduce the amount of sugar in their diets, many use low-Calorie sweeteners. *Saccharin* (found in Sweet'N Low) was the first sugar substitute used extensively in the United States. There has been some controversy over saccharin use because early investigations conducted with rats suggested that massive amounts of saccharin may cause cancer. Further investigation indicated that the link between saccharin and cancer was very weak. Currently, it is generally believed that saccharin is safe for human consumption. Today, saccharin is used in many candies, baked goods, jellies, and jams.

The sugar-substitute *aspartame* (NutraSweet and Equal) is an ingredient in many diet beverages and in thousands of other food products. Aspartame (Figure 7.63) is a chemical combination of two natural amino acids, aspartic acid and phenylalanine, neither of which, by itself, tastes sweet. One gram of aspartame contains roughly the same food energy as one gram of table sugar (4 Cal), but aspartame tastes 200 times sweeter than sugar. Because we need smaller quantities of aspartame to sweeten a product, it is a low-Calorie alternative to sugar—one that is also safe for people with diabetes. Annually, thousands of tons of aspartame are used in the United States to sweeten diet drinks and foods. Aspartame decomposes at cooking temperatures. However, Splenda, an artificial sweetener based on *sucralose* (a chlorinated carbohydrate; see Figure 7.63), can be used directly in cooking.

 Molecular Design of an Artificial Sweetener

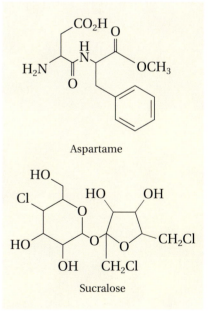

Aspartame

Sucralose

| **Figure 7.63** *Molecular structures of aspartame and sucralose.*

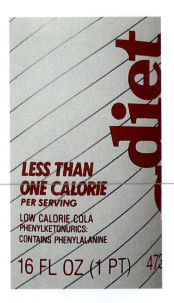

Figure 7.64 *The warning to phenylketonurics found on a can of diet drink containing aspartame.*

Although no serious warning about aspartame has been issued for the general population, aspartame does pose a health hazard to *phenylketonurics*—individuals who cannot properly metabolize phenylalanine. An FDA-required warning, "Phenylketonurics: Contains Phenylalanine," on the label of foods with aspartame highlights the potential risk (see Figure 7.64).

Individuals with specific medical conditions may need to avoid certain foods and food additives. For example, people with diabetes (see Figure 7.65) must regulate their intake of carbohydrates, including sugars. People with high blood pressure (hypertension) must avoid excess sodium. Some people have food allergies. If such restrictions apply to you, you must always read food labels. Sometimes a new ingredient will be used in a food product that you have used safely in the past; this new ingredient may put you at risk.

Figure 7.65 *A diabetic individual checks her blood glucose level.*

Developing Skills

D.10 FOOD ADDITIVE SURVEY

Being aware of what you eat is a wise habit to develop. In this activity, you will investigate food labels and consider additives.

1. Collect the labels from five or six packaged foods in your home. Select no more than two samples of the same type of food. For example, do not select labels from more than two breakfast cereals or two canned soups. Bring your collected labels to class.

2. From the ingredient listings on the labels, select three additives.

3. Complete a summary table with the following format:

   a. List the three additives in a vertical column along the left side.
   b. Make four vertical columns to the right of the listed additives with these headings: "Food Product Where Found," "Purpose of Additive (if known)," "Chemical Formula," and "Other Information."

4. Use Table 7.10 (page 631) to review the purposes of food additives. Then answer the following question for each food additive in your summary table: Why do you think this additive is included in this food?

5. What alternatives to food additives can you propose to prevent food spoilage?

D.11 ANALYZING VITAMINS AND MINERALS

So far, you have analyzed the three-day food inventory in terms of fats, carbohydrates, and proteins. Now decide whether the inventoried foods provide adequate quantities of vitamins and minerals.

1. Your teacher will specify two vitamins and two minerals for you to analyze.

2. Using references provided by your teacher, analyze each food item in your three-day inventory for the presence of these two vitamins and two minerals. Record these values.

3. Determine the average intake of these two vitamins and two minerals over the three days covered in the three-day food inventory. Record these data.

4. a. Does your analyzed food inventory provide enough of each specified vitamin and mineral?
   b. If not, what types of food can be added to increase the intake of these vitamins and minerals? Identify the food groups to which these food types belong.

5. Some people believe that eating a well-balanced diet provides all the vitamins a person needs to remain healthy. Others suggest that vitamin supplements can enhance personal health. Proponents of both opinions have engaged in considerable advertising to promote their viewpoints.

   a. After considering your data, which viewpoint do your results support?
   b. What could you do to help people make more informed decisions about this issue?

Keeping Food Supplies Safe for Consumers

NEWS BRIEFS

SUFFOLK
19TH HORSE DIES: A horse from Suffolk has died and is suspected to be the 19th known victim of "moldy corn poisoning" in Virginia. The horse's owners, who were not identified, could not recall where they purchased the corn,

State inspectors widen area for corn feed testing

Associated Press
RICHMOND — Inspectors from the Virginia Department of Agriculture expanded their search Thursday for a toxin in corn feed that has killed 18 horses.

Inspectors began taking samples of from manufacturers and dealers who mix their own feed in the area south of Interstate 64 and east of Interstate 95. The samples will be tested for Fumonisin B1, a mycotoxin that causes leukoence-phalomalacia, also called moldy corn poisoning.

These actual newspaper headlines illustrate the deadly effects that certain microscopic toxins can have. The consumers of the food discussed in these articles were horses, but similar toxins can endanger human consumers as well.

Mary Trucksess is a chemist working to understand and eradicate food toxins. She is a research chemist and chief of the Bioanalytical Chemistry Branch in the Division of Natural Food Products, Office of Plant and Dairy Food and Beverages at the U.S. Food and Drug Administration (FDA). Mary and her fellow chemists at the FDA help ensure the quality of foods and drugs (prescription and nonprescription) to protect people from harmful contaminants. Mary is currently investigating *fumonisins*—toxins found mainly in corn that are caused by a common soil fungus. Fumonisins can cause horses to develop a fatal disease of the brain. The FDA is investigating these toxins to make certain that they do not pose a threat to humans.

Mary's office analyzes other foods, such as peanuts and peanut butter, for aflatoxins. Aflatoxins form when certain molds (*Aspergillus flavus* or *Aspergillus para-*

Currently Mary is investigating *fumonisins*— toxins found mainly in corn.

siticus) begin to grow on peanuts. If not controlled, these and other toxins could threaten the health of consumers.

Mary also manages and directs other chemists as they conduct research. The chemists in her branch often discuss with her any problems they encounter in their research. She reviews progress reports from the chemists and recommends different approaches as needed.

For several years, Mary, who holds undergraduate and doctoral degrees in chemistry, has supervised the work of high school honor students who have been selected to intern at the FDA. Mary develops specific projects for the students and trains them in laboratory procedures and data interpretation. Mary has also been a coordinator for Project SEED. This summer program for economically disadvantaged high school students offers them the opportunity to work with a mentor in a chemistry laboratory. For information about this program, visit http://www.acs.org/education/student/projectseed.html

For Research and Practice

1. When preparing food, consumers can unknowingly create conditions in which bacteria thrive.
   a. What should you do to keep your food safe?
   b. What temperature range is the danger zone for food, and why is it called that?
2. According to a CNN report, "A survey by the Centers for Disease Control and Prevention shows that men between the ages of 18 and 29 violate food safety practices more than any other group." Members of this group are also more likely to experience food poisoning than those of any other group. Design a food safety ad campaign that will target men between 18 and 29.

SECTION D SUMMARY
Reviewing the Concepts

Vitamins are organic molecules that are necessary for basic life functions.

1. Vitamins are not considered foods, yet they are vital to healthful diets. Explain.

2. What is a coenzyme?

3. Give an example of a function of a vitamin.

4. List some typical symptoms of dietary deficiency of
   a. vitamin B_{12}.
   b. vitamin A.
   c. vitamin D.

5. Vitamins are regarded as micronutrients. What does this mean regarding the quantity of vitamins required daily?

6. What molecular properties determine whether a vitamin is fat-soluble or water-soluble?

7. Why should people take water-soluble vitamins more regularly than fat-soluble ones?

8. Give two examples of how DRIs for vitamins and minerals vary, based on age and gender.

A titration is a common laboratory procedure to determine the amount of a solute in a particular solution.

9. Describe key steps in a vitamin C titration and explain how this process can determine the quantity of vitamin C in a beverage.

10. What is meant by the *endpoint* of a titration?

11. Vitamin C is colorless in solution, yet you measured its concentration in Investigating Matter D.3 (see page 625). What reactions allowed you to detect its presence in a solution?

12. Suppose you find in a vitamin C titration that 0.035 mg vitamin C reacts with 1 drop iodine solution. How much vitamin C is in a sample that turns bluish black after adding
    a. 1 drop of iodine solution?
    b. 12 drops of iodine solution?

13. For what juices would the vitamin C titration endpoint be difficult to observe? Give examples supporting your answer.

Minerals are elements that are essential for human life.

14. Name three essential minerals.

15. Of the more than 100 known elements, how many are believed essential for human life?

16. What is the difference between a *macromineral* and a *trace mineral*?

17. What roles do minerals play in the body?

18. List a macromineral and a trace mineral and describe a deficiency condition for each.

19. What are some dietary sources of
    a. magnesium?   c. iron?
    b. potassium?   d. molybdenum?

20. Regarding minerals in the diet, some people believe that "if a little is good, more must be better." Evaluate this idea, particularly considering such minerals as arsenic (As) and cadmium (Cd).

21. What determines if a substance is considered an additive or a basic component of a food?

22. List two typical food additives and their functions.

23. Give an example of a food additive used to
    a. increase nutritive value.
    b. improve storage life.
    c. enhance visual appeal.
    d. ease production.

24. a. What is a carcinogen?
    b. What is a mutagen?

25. How does the Food Quality and Protection Act differ from the Delaney Clause?

26. Aspartame's energy value is the same as that of sugar, 4 Cal/g. Why, then, is aspartame useful as a low-Calorie artificial sweetener?

27. Explain why aspartame can't be used in baking.

28. a. What is an R_f value?
    b. How is it determined?

29. In a paper-chromatography investigation, the solvent moves 5.8 cm, and a food-dye component moves 3.9 cm. What is the R_f for the food-dye? Show calculations to support your answer.

30. What properties allow chromatography to separate particular components of a solution?

31. A student tests two samples of dye solution with paper chromatography. In A, the solvent moves 6.2 cm, and the dye moves 4.1 cm. In B, the solvent moves 5.3 cm, and the dye moves 4.1 cm. Do these data support that A and B contain the same dye? Explain.

Connecting the Concepts

32. Although high cooking temperatures can destroy some vitamins in foods, they rarely affect the quality of minerals in foods. Explain.

33. Why with certain food additives, do food standards use the term "generally recognized as safe" (GRAS), rather than "always safe" or "100% safe"?

34. Would paper chromatography provide any useful information if one or more sample components were colorless? Explain.

35. How could you modify Investigating Matter D.3 (pages 623–625) to find out how much vitamin C is lost during cooking a food item?

36. One cup of spinach supplies 51 mg calcium and 1.7 mg iron. Does spinach provide a greater percent of your DRI for calcium or for iron? Show calculations to support your answer.

37. Suppose a government banned all food additives. Overall, do you think the ban would produce positive or negative results? Explain.

38. Suppose one component in a chromatography sample was quite volatile, tending to evaporate when placed on paper. What effect might this property have on the experimental results?

39. All substances described in this unit consist of atoms, molecules, or ions. With that in mind, why do you think some people describe food additives as "chemicals," but do not describe carbohydrates and fats also as chemicals?

40. In a paper chromatography investigation, a sample fails to move up the paper when water is used as a solvent. How could the investigation be modified to address this problem?

Extending the Concepts

41. Investigate some additional types of chromatography, such as gas chromatography and column chromatography. Explain how each type works.

42. Modern food additives are generally safe and healthful. This was not always the case in the past. Research the history of food additives and the unexpected impact of some early additives on human health.

43. When administered in large doses, a prospective food additive is found to cause cancer in laboratory rats. Should this evidence be used to withhold the approval for its use in human foods? Explain your answer.

44. Research some food-preparation methods that tend to preserve vitamin content. Discuss the chemical concepts that can account for the effectiveness of such methods.

45. Liver is a rich source of many trace minerals. Explain this based on liver physiology.

46. Research *goiter*. When and where has it been most prevalent? Why? What specific product makes it much less common today?

47. Some historians claim that the most significant contribution to the success of the British Navy in the 1700s was adding sauerkraut to shipboard food supplies. Investigate and explain the possible connection.

SCHOOL VENDING CHOICES: GUIDING PTSA'S DECISIONS

Now that you have completed a series of analyses of the three-day food inventory, it is time to meet with other members of the Vending Machine Policy Planning Committee to consider policies for Riverwood High School's vending machines and to convey your recommendations to the PTSA. Review the results of your analyses and think of points you would like to raise about students' eating habits. Use the guidelines for a healthful diet provided by the PTSA (page 567) and what you have learned in this unit to decide on the content and organization of your report.

As a group, the team (your class) should make recommendations about the types of items that should be supplied in the food and beverage vending machines. Focus on generalized recommendations accompanied by specific examples, such as: "The vending machines should offer food options high in calcium, including [examples]." In your discussion, be sure to support proposed recommendations with evidence and data from your three-day food inventory analyses.

After your team has prepared its list of recommendations, each team member should write a report to submit to the PTSA. The following material provides some guidelines for writing your report.

CONSIDER YOUR AUDIENCE

As you studied this unit, you learned much about food chemistry; thus you may decide to include some technical terminology in your report. If you do, make sure you explain these terms so that PTSA members can understand your findings and recommendations. Some of the members may have studied chemistry (they may know about bonding and other fundamental chemistry ideas) but may be unfamiliar with food chemistry.

WRITTEN REPORT

Your written report should include the following sections:

1. *Introduction.* The introduction should be a brief summary that highlights the characteristics of the three-day food inventory. In one or two paragraphs, present an overview of what you found in your food-inventory analyses. In addition, list your top three policy recommendations for the food and beverage vending machines. Do not include detailed explanations in this section.

2. *Background Information.* The background provides information that the PTSA members will need to understand your report. The following questions should be answered in this background section:
   ▶ What are fats, carbohydrates, and proteins? You may include information about each substance's molecular structure, energy content, and function in the human body.
   ▶ Why do humans need fats, carbohydrates, and proteins in their diets?
   ▶ What roles do vitamins and minerals play in the diet?
   ▶ You may include structural formulas and other diagrams; however, remember that you need to guide your readers regarding what to look for in any diagrams.

3. *Data Analysis.* The body of your report should be based on your analysis of the data. In this section, you bring together detailed data that you gathered concerning energy, fats, carbohydrates, proteins, vitamins, and minerals. Organize and present the information so that it will make sense to your readers. If helpful, include graphs and tables.

   In this data analysis section, you may wish to address questions about the food inventory such as the following:

   ▶ What is the average daily Calorie intake of Riverwood High School students? How reasonable is this value? Does the average daily Calorie intake accurately represent each day in the inventory or does it vary widely from day to day?
   ▶ What percent of total Calories are supplied by carbohydrates, fats, and proteins? Are these values consistent with recommendations in the *Dietary Guidelines for Americans 2005*?
   ▶ What percent of fats are saturated and unsaturated? Do these levels meet *Dietary Guidelines*?
   ▶ Which vitamins and minerals are present at adequate levels? Which are lacking?

4. *Conclusions.* Here is where you will provide further details about recommendations that you highlighted at the beginning of your report. Discuss each recommendation based on your findings and the guidelines for a healthful diet summarized at the beginning of this unit. List representative foods and beverages for vending that either (a) meet one or more of your proposed recommendations or (b) do *not* meet your recommendations. Explain how you decided on those particular items. Keep in mind that nutritional value is not the only consideration; taste, appearance, and whether students would actually eat or drink the proposed vending-machine items are also important.

LOOKING BACK

This unit focused on food, a mixture of chemical substances that you encounter daily. You can now explain how food you eat provides both energy you need for daily living and structural components for growth.

You can attach deeper chemical meaning to terms such as *carbohydrate* and can better evaluate consequences associated with deciding to consume or not consume certain foods.

The next time you hear someone remark, "You are what you eat," smile and reply that you know some chemistry that is behind the true meaning of those words!

THE SCIENTIFIC METHOD
vs. SCIENTIFIC METHODS

Scientists deepen their knowledge and understanding of the natural world by observing and manipulating their environment. The inquiry approach used by scientists to solve problems and seek knowledge has led to vast increases in understanding how nature works. Many efforts have been made to formalize and list the steps that scientists use to generate and test new knowledge. You may have been asked to learn the "steps" of the Scientific Method, such as Make Observations, Define the Problem, and so on. But no one comes to an understanding of how scientific inquiry works by simply learning a list of steps or definitions of words.

Although you may not go on to become a research scientist, it is important that all students acquire the ability to conduct scientific inquiry. Why? Everyone is confronted daily with endless streams of facts and claims. What should be accepted as true? What should be discarded? Having well-developed ways to evaluate and test claims is essential in deciding between valid and deceptive information.

What abilities are needed to conduct scientific inquiry? According to the National Science Education Standards (NSES), they include the abilities to

- identify questions and concepts that guide scientific investigations.
- design and conduct scientific investigations.
- use technology and mathematics to improve investigations and communications.
- formulate and revise scientific explanations and models using logic and evidence.
- recognize and analyze alternative explanations and models.
- communicate and defend a scientific argument.

These skills are necessary for both doing and learning science. They are also important skills to evaluate information in daily living. You can only acquire these skills through practice—by doing exercises and investigations such as those contained in this textbook.

Even with all the abilities listed above, doing science is a complex behavior. The NSES outlines the ideas that all students should know and understand about the practices of science:

- Scientists usually inquire about how physical, living, or designed systems function.
- Scientists conduct investigations for a wide variety of reasons.
- Scientists rely on technology to enhance the gathering and manipulation of data.
- Mathematics is essential in scientific inquiry.
- Scientific explanations must adhere to criteria such as: a proposed explanation must be logically consistent; it must abide by the rules of evidence; it must be open to questions and possible modification; and it must be based on historical and current scientific knowledge.
- Results of scientific inquiry—new knowledge and methods— emerge from different types of investigations and public communication among scientists.

The last statement above acknowledges that—despite generalizations that can be listed—there are many paths to gaining new scientific knowledge. That is what makes studying scientific processes so important. Learning how to acquire the abilities to do and understand scientific inquiry may be the most important and useful thing you learn in this course.

NUMBERS IN CHEMISTRY

Chemistry is a quantitative science. Most chemistry investigations involve not only measuring, but also a search for the meaning among the measurements. Chemists learn how to interpret as well as perform calculations using these measurements.

SCIENTIFIC NOTATION

Chemists often deal with very small and very large numbers. Instead of using many zeros to express very large or very small numbers, they often use scientific notation. In scientific notation, a number can be rewritten as the product of a number between 1 and 10 and an exponential term—10^n, where n is a whole number. The exponential term is the number of times 10 would have to be multiplied or divided by itself to yield the appropriate number of digits in the number. For instance, 10^3 is $10 \times 10 \times 10$, or 1000; $3.5 \times 10^3 = 3500$.

> **Sample Problem 1:** *Express the distance between New York City and San Francisco, 4 741 000 meters, using scientific notation.*
>
> 4 741 000 m = $(4.741 \times 1\ 000\ 000)$ m or **4.741×10^6 m**
>
> **Sample Problem 2:** *Express the amount of ranitidine hydrochloride in a Zantac tablet, 0.000 479 mol, using scientific notation.*
>
> 0.000 479 mol = 4.79×0.0001 mol or **4.79×10^{-4} mol**

It is easier to assess magnitude and to perform operations with numbers written in scientific notation than with numbers fully written out. As you will see, it is also easier to communicate the precision of the measurements involved.

Rules for adding *and* subtracting *using scientific notation:*

Step 1.   Convert the numbers to the same power of 10.

Step 2.   Add (subtract) the non-exponential portion of the numbers. *The power of ten remains the same.*

Example

Add $(1.00 \times 10^4) + (2.30 \times 10^5)$.

Step 1.   A good rule to follow is to express all numbers in the problem to the highest power of ten. Convert 1.00×10^4 to 0.100×10^5.

Step 2. $(0.100 \times 10^5) + (2.30 \times 10^5) = 2.40 \times 10^5$

Rules for multiplying *using scientific notation:*

Step 1. Multiply the nonexponential numbers.

Step 2. Add the exponents.

Step 3. Convert the answer to scientific notation.

Sample Problem: *Multiply (4.24 × 10²) by (5.78 × 10⁴).*

Steps 1 and 2. $(4.24 \times 5.78) \times (10^{2+4}) = 24.5 \times 10^6$

Step 3. Convert to scientific notation $= 2.45 \times 10^7$

Rules for dividing *using scientific notation:*

Step 1. Divide the nonexponential numbers.

Step 2. Subtract the denominator exponent from the numerator exponent.

Step 3. Express the answer in scientific notation.

Sample Problem: *Divide (3.78 × 10⁵) by (6.2 × 10⁸).*

Steps 1 and 2. $\dfrac{3.78}{6.2} \times (10^{5-8}) = 0.61 \times 10^{-3}$

Step 3. Convert to scientific notation $= 6.1 \times 10^{-4}$

Practice Problems*

1. Convert the following numbers to scientific notation.
   a. 0.000 036 9
   b. 0.0452
   c. 4 520 000
   d. 365 000

2. Carry out the following operations:
   a. $(1.62 \times 10^3) + (3.4 \times 10^2)$
   b. $(1.75 \times 10^{-1}) - (4.6 \times 10^{-2})$
   c. $\dfrac{6.02 \times 10^{23}}{12.0}$
   d. $\dfrac{(6.63 \times 10^{-34}\ \text{J·s}) \times (3.00 \times 10^8\ \text{m s}^{-1})}{4.6 \times 10^{-9}\ \text{m}}$

*Answers to odd-numbered problems can be found on page A-17.

DIMENSIONAL ANALYSIS

Dimensional analysis, also called the *factor-label method,* is used by scientists to keep track of units in calculations and to help guide their work in solving problems. You already used this method to convert from one type of metric unit to another. The method is helpful in setting up problems and also in checking work.

Dimensional analysis consists of three basic steps:

Step 1. Identify equivalence relationships in order to create suitable conversion factors.

Step 2. Identify the given unit(s) and the new unit(s) desired.

Step 3. Arrange each conversion factor so that each unit to be converted can be divided by itself (and thus cancelled).

Sample Problem 1: *In an exercise to determine the volume of a rectangular object, your laboratory partner measured the object's length as 12.2 in (inches). However, measurements of the object's width and height were recorded in centimeters. In order to calculate the object's volume, convert the object's measured length to centimeters.*

Step 1. Find the equivalence relating centimeters and inches.

$$2.54 \text{ cm} = 1 \text{ in}$$

Step 2. Identify the given unit and the new unit.

Given unit = in   new unit = cm

Step 3. Create a fraction so that the "given" unit (in) can be divided and thus cancelled.

$$12.2 \text{ in} \times \frac{2.54 \text{ cm}}{1 \text{ in}} = 31.0 \text{ cm}$$

Sample Problem 2: *How many seconds are in 24 hours?*

Step 1. Identify the equivalence:

1 hr = 60 min   1 min = 60 s

Step 2. Given unit: hr   new unit: s

Step 3. Arrange for the given unit to cancel and progress to the desired unit:

$$24 \text{ hr} \left(\frac{60 \text{ min}}{1 \text{ hr}} \right) \times \left(\frac{60 \text{ s}}{1 \text{ min}} \right) = 86\ 400 \text{ s}$$

3. The distance between two European cities is 4 741 000 m. That may sound impressive, but to put all those digits on a car odometer is slightly inconvenient. Kilometers are a better choice for measuring distance in this case. Convert the distance to kilometers.

4. The density of aluminum is 2.70 g/cm³. What is the mass of 234 cm³ of aluminum?

SIGNIFICANT FIGURES

Significant figures are all the digits in a measured or calculated value that are known with certainty plus the first uncertain digit. Numerical measurements have some inherent uncertainty. This uncertainty comes from the measurement device as well as from the human making the measurement. No measurement is exact. When you use a measuring device in the laboratory, read and record each measurement to one digit beyond the smallest marking interval on the scale.

Guidelines for Determining Significant Figures

Step 1. All digits recorded from a laboratory measurement are called significant figures.

The measurement of 4.75 cm has three significant figures.

Note: If you use a measuring device that has a digital readout, such as a balance, you should record the measurement just as it appears on the display.

Measurement	Number of Significant Figures
123 g	3
46.54 mL	4
0.33 cm	2
3 300 000 nm	2
0.033 g	2

Step 2. All non-zero digits are considered significant.

---

*Answers to odd-numbered problems can be found on page A-17.

A-6   Appendix B: Numbers in Chemistry

Step 3. There are special rules for zeros. Zeros in a measurement or calculation fall into three types: middle zeros, leading zeros, and trailing zeros.

Middle zeros are always significant.

303 mm — a middle zero—always significant. This measurement has three significant figures.

A leading zero is never significant. It is only a placeholder, not a part of the actual measurement.

0.0123 kg — two leading zeros—never significant. This measurement has three significant figures.

A trailing zero is significant when it is to the right of a decimal point. This is not a placeholder. It is a part of the actual measurement.

23.20 mL — a trailing zero—significant to the right of a decimal point. This measurement has four significant figures.

The most common errors concerning significant figures are (1) reporting all digits found on a calculator readout, (2) failing to include significant trailing zeros (14.15<u>0</u> g), and (3) considering leading zeros to be significant—0.002 g has only one significant figure, not three.

Practice Problem*

5. How many significant figures are in each of the following?
   a. 451 000 m
   b. 4056 V
   c. 6.626×10^{-34} J·s
   d. 0.0065 g
   e. 0.0540 mL

Using Significant Figures in Calculations

Addition and subtraction: The number of *decimal places* in the answer should be the same as in the measured quantity with the smallest number of *decimal places*.

Sample Problem: *Add the following measured values and express the answer to the correct number of significant figures.*

$$
\begin{array}{r}
1259.1 \ \ \ \text{g} \\
2.365 \ \text{g} \\
+ \ 15.34 \ \ \text{g} \\
\hline
1276.805 \ \text{g}
\end{array}
\quad = \mathbf{1276.8 \ g}
$$

---

*Answers to odd-numbered problems can be found on page A-17.

Multiplication and division: The number of *significant figures* in the answer should be the same as in the measured quantity with the smallest number of *significant figures*.

> **Sample Problem:** *Divide the following measured values and express the answer to the correct number of significant figures.*

$$\frac{13.356 \text{ g}}{10.42 \text{ mL}} = 1.2817658 \text{ g/mL} = \textbf{1.282 g/mL}$$

Practice Problem*

6. Report the answer to each of these using the correct number of significant figures.

    a. 16.27 g + 0.463 g + 32.1 g

    b. 42.04 mL − 3.5 mL

    c. 15.1 km × 0.032 km

    d. $\dfrac{13.36 \text{ cm}^3}{0.0468 \text{ cm}^3}$

Note: Only measurements resulting from scale readings or digital readouts carry a limited number of significant figures. However, values arising from direct counting (such as 25 students in a classroom) or from definitions (such as 100 cm = 1 m, or 1 dozen = 12 things) carry unlimited significant figures. Such "counted" or "defined" values are regarded as exact.

*Answers to odd-numbered problems can be found on page A-17.

CONSTRUCTING GRAPHS

Graphs are of four basic types: pie charts, bar graphs, line graphs, and x-y plots. The type chosen depends on the characteristics of the data displayed.

Pie charts show the relationship of the parts to the whole. This presentation helps a reader visualize the magnitude of difference among various parts. They are made by taking a 360° circle and dividing it into wedges according to the percent of the whole represented by each part.

Bar graphs and **line graphs** compare values within a category or among categories. The horizontal axis (x-axis) is used for the quantity that can be controlled or adjusted. This quantity is the **independent variable.** The vertical axis (y-axis) is used for the quantity that is influenced by the changes in the quantity on the x-axis. This quantity is the **dependent variable.** For example, a bar graph could present a visual comparison of the fat content (dependent variable on the y-axis) of types of cheese (independent variable on the x-axis). Such a graph would make it easy to choose a cheese snack with a low-fat content. Bar graphs can also be useful in studying trends over time.

Graphs involving **x-y plots** are commonly used in scientific work. Sometimes it is difficult to decide if a graph is a line graph or an x-y plot. In an x-y plot, it is possible to determine a mathematical relationship between the variables. Sometimes the relationship is the equation for a straight line ($y = mx + b$), but other times it is more complex and may require transformation of the data to produce a simpler graphical relationship. The first example below refers to a straight-line or direct relationship.

Sample Problem 1: *A group of entrepreneurs was considering investing in a mine that was said to contain gold. To verify this claim, they gave several small, irregular mined particles to a chemist, who was told to use nondestructive methods to analyze the samples.*

The chemist decided to determine the density of the small samples. The chemist found the volume of each particle and determined its mass. The data collected are shown in the table below.

Particle	Volume (mL)	Mass (g)
1	0.006	0.116
2	0.012	0.251
3	0.015	0.290
4	0.018	0.347
5	0.021	0.386

The chemist then used these data to construct an *x-y* plot.

Gold Particle Data

Mass/Volume

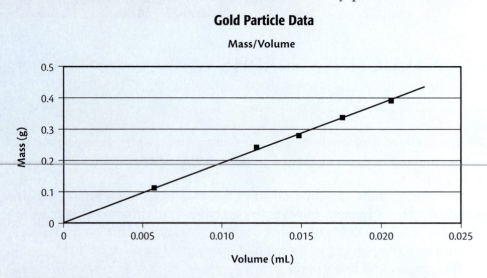

Since the *x-y* plot is linear, the sample materials are likely the same. If point (0, 0) is included (a sample of zero volume has zero mass), the slope is about 19 g/mL, close to gold's density (19.3 g/mL), so these are likely gold particles.

> **Sample Problem 2:** *The graph below is a plot of data gathered at constant temperature involving the volume of 2.00 mol of ammonia (NH_3) gas measured at various pressures. Determine whether the measurements are related by a simple mathematical relationship.*

Volume of 2.00 mol NH_3 at Different Pressures

Pressure/Volume Relationship

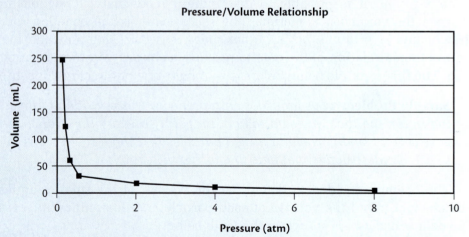

When the best smooth graph line is not a straight line (or direct relationship), the data can be manipulated to see if any other simple mathematical relationship is possible. Several of the most common types of mathematical relationships are inverse, exponential, and logarithmic. Each relationship has unique characteristics that can often be identified from the graphical presentation. Knowing the mathematical relationship allows scientists to interpret the data. In this case, it appears that as the pressure increases the volume decreases, which is a characteristic of an inverse relationship. In testing this theory, the value of 1/V (the inverse of the volume) can be calculated, recorded in another column of the table, and plotted versus the pressure.

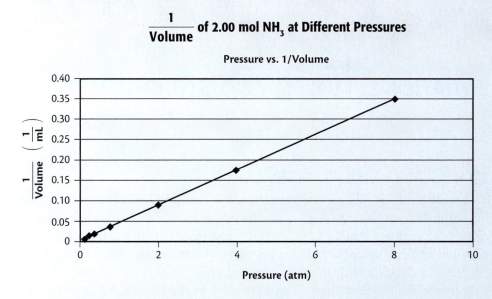

$\dfrac{1}{\text{Volume}}$ of 2.00 mol NH$_3$ at Different Pressures

This graph exhibits a straight line showing that pressure is directly related to the inverse of the volume. This leads to the mathematical result that pressure and volume are inversely related. If this mathematical manipulation had not resulted in a straight line, some other reasonable relationships might have been considered and tested.

EQUATIONS, MOLES, AND STOICHIOMETRY

MOLES AND MOLAR MASS

The **mole** is regarded as a counting number—a number used to specify a certain number of objects. **Pair** and **dozen** are other examples of counting numbers. A mole equals 6.02×10^{23} objects. Most often, things that are counted in units of moles are very small—atoms, molecules, or electrons.

The modern definition of a mole specifies that one mole is equal to the number of atoms contained in exactly 12 g of the carbon-12 isotope. This number is named after Amedeo Avogadro, who proposed the idea, but never determined the number. At least four different types of experiments have accurately determined the value of Avogadro's number—the number of units in one mole. Avogadro's number is known to eight significant figures, but three will be enough for most of your calculations—6.02×10^{23}.

The modern atomic weight scale is also based on C-12. Compared to C-12 atoms with a defined atomic mass of exactly 12, hydrogen atoms have a relative mass of 1.008. Therefore, one mole of hydrogen atoms has a mass of 1.008 g. One mole of oxygen atoms equals 15.9994 g—also called the molar mass of oxygen. The total mass of one mole of a compound is found by adding the atomic weights of all of the atoms in the formula and expressing the sum in units of grams.

Sample Problem: *Calculate the mass of one mole of water.*

$$2 \text{ mol H} \times \frac{1.008 \text{ g H}}{1 \text{ mol H}} = 2.016 \text{ g H}$$

$$1 \text{ mol O} \times \frac{16.00 \text{ g O}}{1 \text{ mol O}} = 16.00 \text{ g O}$$

molar mass of water = 2.016 g + 16.00 g = 18.02 g

Practice Problems*

Find the molar mass (in units of g/mol) for each of the following:

1. Acetic acid, CH_3COOH
2. Formaldehyde, $HCHO$
3. Glucose, $C_6H_{12}O_6$
4. 2-Dodecanol, $CH_3(CH_2)_9CH(OH)CH_3$

*Answers to odd-numbered problems can be found on page A-17.

GRAM-MOLE CONVERSIONS

Conversions between grams and moles can be readily accomplished by using the technique of dimensional analysis (see Appendix B).

Sample Problem 1: *What mass (in grams) of water contains 0.25 mol H_2O?* The mass of one mole of water (18.02 g) is found as illustrated earlier. Two factors can be written, based on that relationship:

$$\frac{1 \text{ mol } H_2O}{18.02 \text{ g } H_2O} \text{ and } \frac{18.02 \text{ g } H_2O}{1 \text{ mol } H_2O}$$

The second conversion factor is chosen so that each unit to be converted is divided by itself and thus cancelled. Units for the answer are g H_2O, as expected.

$$0.25 \text{ mol } H_2O \times \frac{18.02 \text{ g } H_2O}{1 \text{ mol } H_2O} = 4.5 \text{ g } H_2O$$

Sample Problem 2: *How many moles of water molecules are present in a 1.00-kg sample of water?*

$$1.00 \text{ kg } H_2O \times \frac{1000 \text{ g } H_2O}{1 \text{ kg } H_2O} \times \frac{1 \text{ mol } H_2O}{18.02 \text{ g } H_2O} = 55.5 \text{ mol } H_2O$$

Practice Problems*

5. Acetic acid, CH_3COOH, and salicylic acid, $C_7H_6O_3$, can be chemically combined to form aspirin. If a chemist uses 5.00 g salicylic acid and 10.53 g acetic acid, how many moles of each compound are involved?

6. Calcium chloride hexahydrate, $CaCl_2 \cdot 6H_2O$, can be sprinkled on sidewalks to melt ice and snow. How many moles of that compound are in a 5.0-kg sack of that substance?

A QUANTITATIVE UNDERSTANDING OF CHEMICAL FORMULAS

Calculating percent composition from a formula

The percent by mass of each component found in a sample of material is called its **percent composition.** To find the percent of an element in a particular compound, first calculate the molar mass of the compound. Then find the total mass of the element contained in one mole of the compound. Then divide the mass of the element by the molar mass of the compound and multiply the result by 100%.

Sample Problem 1: *Calculate the molar mass of sucrose, $C_{12}H_{22}O_{11}$.*

12 mol C (12.01 g/mol) = 144.1 g
22 mol H (1.008 g/mol) = 22.18 g
11 mol O (16.00 g/mol) = 176.0 g
Molar mass of $C_{12}H_{22}O_{11}$ = 342.3 g

*Answers to odd-numbered problems can be found on page A-17.

Sample Problem 2: *Find the mass percent of each element in sucrose (rounded to significant figures).*

$$\% \ C = \frac{mass \ C}{mass \ C_{12}H_{22}O_{11}} \times 100\% = \frac{144.1 \ g \ C}{342.3 \ g \ C_{12}H_{22}O_{11}} \times 100\% = 42.10\% \ C$$

$$\% \ H = \frac{mass \ H}{mass \ C_{12}H_{22}O_{11}} \times 100\% = \frac{22.18 \ g \ H}{342.3 \ g \ C_{12}H_{22}O_{11}} \times 100\% = 6.48\% \ H$$

$$\% \ O = \frac{mass \ O}{mass \ C_{12}H_{22}O_{11}} \times 100\% = \frac{176.0 \ g \ O}{342.3 \ g \ C_{12}H_{22}O_{11}} \times 100\% = 51.42\% \ O$$

Practice Problems*

7. Barium sulfate, $BaSO_4$, is commonly used to detect gastrointestinal tract abnormalities; it is administered as a water suspension by mouth prior to X-ray imaging. Find the mass percent of each element in barium sulfate.

8. Sodium acetate is a common component in commercial thermal packs. Find the mass percent of each element in sodium acetate, $NaCH_3COO$.

Deriving formulas from percent composition data

An **empirical formula** gives the relative numbers of each element in a substance, using the smallest whole numbers for subscripts. The empirical formula of a compound can be calculated from percent composition data. A formula requires the relative numbers of moles of each element, so percent values must be converted to grams and grams to moles. It is easiest to assume 100 g of compound. Then each percent value is equal to the total grams of that element.

Sample Problem 1: *A hydrocarbon (the only elements in the compound are hydrogen and carbon) consists of 85.7% carbon and 14.3% hydrogen by mass. What is its empirical formula?*

Step 1. Assume 100 g of compound—thus 85.7 g are carbon and 14.3 g are hydrogen.

Step 2. Use dimensional analysis to find the total moles of each element in the compound.

$$85.7 \ \cancel{g \ C} \times \frac{1 \ mol \ C}{12.01 \ \cancel{g \ C}} = 7.14 \ mol \ C$$

$$14.3 \ \cancel{g \ H} \times \frac{1 \ mol \ H}{1.008 \ \cancel{g \ H}} = 14.2 \ mol \ H$$

Step 3. Determine the smallest whole-number ratio of moles of elements by dividing all mole values by the smallest value.

$$\frac{7.14 \ mol \ C}{7.14} = 1 \ mol \ C$$

$$\frac{14.2 \ mol \ H}{7.14} = 1.99 \ mol \ H$$

The ratio of moles C to moles H = 1:2, so the empirical formula for the compound must be CH_2.

*Answers to odd-numbered problems can be found on page A-16.

Sample Problem 2: *The percent composition of one of the oxides of nitrogen is 74.07% oxygen and 25.93% nitrogen. What is the empirical formula of that compound?*

Step 1. 100 g of compound consists of 74.07 g oxygen and 25.93 g nitrogen.

Step 2. $25.93 \text{ g N} \times \dfrac{1 \text{ mol N}}{14.01 \text{ g N}} = 1.851 \text{ mol N}$

$74.07 \text{ g O} \times \dfrac{1 \text{ mol O}}{16.00 \text{ g O}} = 4.629 \text{ mol O}$

Step 3. $\dfrac{1.851 \text{ mol N}}{1.851} = 1 \text{ mol N}$

$\dfrac{4.629 \text{ mol O}}{1.851} = 2.50 \text{ mol O}$

Step 4. This ratio (1:2.5) does not consist entirely of whole numbers. Thus all the numbers in the ratio must be multiplied by a number that converts the decimal to a whole number. In this case the number is 2, and the empirical formula becomes N_2O_5.

Practice Problems*

9. The percent composition by mass of an industrially important substance is 2.04% H, 32.72% S, and 65.24% O. What is the formula of this compound?

10. Determine the empirical formula of a compound that contains (by mass) 38.71% C, 9.71% H, and 51.58% O.

MASS RELATIONSHIPS IN CHEMICAL REACTIONS

Since the total number of atoms is conserved in a chemical reaction, their masses must also be conserved, as expected from the law of conservation of mass. In the equation for the formation of water from the elements hydrogen and oxygen, $2H_2(g) + O_2(g) \rightarrow 2H_2O(l)$, 2 molecules of hydrogen gas and 1 molecule of oxygen gas combine to form 2 molecules of water. One could also interpret the equation this way: 2 mol hydrogen gas react with 1 mol oxygen gas to form 2 mol water. Using the molar mass of each substance, the mass relationships in the table below can be determined. The ratio of moles of hydrogen gas to moles of oxygen gas in forming water will be 2:1. If 10 mol hydrogen gas are available, 5 mol oxygen gas are required.

$2 H_2(g)$ +	$O_2(g)$ →	$2 H_2O(l)$
2 molecules	1 molecule	2 molecules
2 mol	1 mol	2 mol
2 mol × (2.02 g/mol)	1 mol × (32.00 g/mol)	2 mol × (18.02 g/mol)
4.04 g	32.00 g	36.04 g

*Answers to odd-numbered problems can be found on page A-17.

Solving problems involving the masses of products and/or reactants is conveniently accomplished by dimensional analysis. And remember, all numerical problems involving chemical reactions involve a correctly balanced equation.

Sample Problem: *Find the mass of water formed when 10.0 g hydrogen gas completely reacts with oxygen gas.*

$$2 \; H_2(g) + O_2(g) \rightarrow 2 \; H_2O(l)$$

Step 1. Find the moles of hydrogen gas represented by 10.0 g, using the molar mass of H_2.

$$10.0 \; \cancel{g\,H_2} \times \frac{1 \; mol \; H_2}{2.02 \; \cancel{g\,H_2}} = 4.95 \; mol \; H_2$$

Step 2. Find the total moles of H_2O produced by 4.95 mol H_2. From the balanced equation, you know that for every 2 mol H_2, 2 mol H_2O are produced.

$$4.95 \; \cancel{mol \; H_2} \times \frac{2 \; mol \; H_2O}{2 \; \cancel{mol \; H_2}} = 4.95 \; mol \; H_2O$$

Step 3. Find the mass of H_2O that contains 4.95 mol H_2O by using the molar mass of water.

$$4.95 \; \cancel{mol \; H_2O} \times \frac{18.02 \; g \; H_2O}{1 \; \cancel{mol \; H_2O}} = 89.2 \; g \; H_2O$$

Most chemistry students find it is more convenient to set up all three steps in one extended calculation. Assure yourself that all units divide and cancel except for grams of water (g H_2O)—an appropriate way to express the answer sought in the problem.

$$10.0 \; \cancel{g\,H_2} \times \underbrace{\frac{1 \; mol \; H_2}{2.02 \; \cancel{g\,H_2}}}_{\substack{\text{Molar mass} \\ \text{of } H_2}} \times \underbrace{\frac{2 \; \cancel{mol \; H_2O}}{2 \; \cancel{mol \; H_2}}}_{\substack{\text{Coefficients} \\ \text{in equation}}} \times \underbrace{\frac{18.02 \; g \; H_2O}{1 \; \cancel{mol \; H_2O}}}_{\substack{\text{Molar mass} \\ \text{of } H_2O}} = 89.2 \; g \; H_2O$$

Practice Problems*

11. Find the mass of copper(II) oxide formed if 2.0 g copper metal completely reacts with oxygen gas.

$$2 \; Cu(s) + O_2(g) \rightarrow 2 \; CuO(s)$$

12. What mass of water is produced when 25.0 g methane gas reacts completely with oxygen gas?

$$CH_4(g) + 2 \; O_2(g) \rightarrow 2 \; H_2O(g) + CO_2(g)$$

*Answers to odd-numbered problems can be found on page A-17.

ANSWERS TO PRACTICE PROBLEMS

Appendix B

1. **a.** 3.69×10^{-5}
   b. 4.52×10^{-2}
   c. 4.52×10^{6}
   d. 3.65×10^{5}
3. 4741 km
5. **a.** 3
   b. 4
   c. 4
   d. 2
   e. 3

Appendix D

1. 60.05 g/mol
3. 180.16 g/mol
5. 0.1754 mol acetic acid, 0.0362 mol salicylic acid
7. 58.54% Ba, 13.74% S, 27.42% O
9. H_2SO_4
11. 2.5 g CuO

A

absolute temperature scale
See kelvin temperature scale

absolute zero
the lowest temperature theoretically obtainable, which is 0 K (zero kelvins), or $-273\ °C$

acid–base reaction
See neutralization

acid rain
fog, sleet, snow, or rain with a pH lower than about 5.6 due to dissolved gases such as SO_2, SO_3, and NO_2

acidic solution
an aqueous solution with a pH < 7; it turns litmus from blue to red

acids
ions or compounds that produce hydrogen ions, H^+ (or hydronium ions, H_3O^+), when dissolved in water

activation energy
the minimum energy required for the successful collision of reactant particles in a chemical reaction

active site
the location on an enzyme where a substrate molecule becomes positioned for a reaction

activity series
the ranking of elements in order of chemical reactivity

addition polymer
a polymer formed by repeated addition reactions at double or triple bonds within monomer units

addition reaction
a reaction at the double or triple bond within an organic molecule

adsorb
to take up or hold molecules or particles to the surface of a material

aeration
the mixing of air into a liquid, as in water sprayed into air, or as in water flowing over a dam

alcohol
a nonaromatic organic compound containing one or more –OH groups

alkali metal family
the group of elements consisting of lithium, sodium, potassium, rubidium, cesium, and francium

alkaline
a basic solution containing an excess of hydroxide ions (OH^-)

alkane
a hydrocarbon containing only single covalent bonds

alkene
a hydrocarbon containing one or more double covalent bonds

alkyne
a hydrocarbon containing one or more triple covalent bonds

allotropes
two or more forms of an element in the same state that have distinctly different physical and/or chemical properties

alloy
a solid solution consisting of atoms of two or more metals

alpha decay
the radioactive decay of a nucleus that is accompanied by the emission of an alpha particle

alpha particles (α)
high-speed, positively charged particles emitted during the decay of some radioactive elements; consist of a helium nucleus, $^4_2He^{2+}$

amine group
a nitrogen atom (with attached hydrogen atoms) connected to a carbon chain

amino acid
an organic molecule containing a carboxylic acid group and an amine group; serves as a protein building block

anion
a negatively charged ion

anode
an electrode in an electrochemical cell at which oxidation occurs

aqueous solution
a solution in which water is the solvent

aquifer
a structure of porous rock, sand, or gravel that holds water beneath Earth's surface

aromatic compound
a ringlike compound, such as benzene, that can be represented as having alternating double and single bonds between carbon atoms

atmosphere
(a) a unit of gas pressure (atm); (b) the gaseous envelope surrounding Earth, composed of four layers: troposphere, stratosphere, mesosphere, and thermosphere

atom
the smallest particle possessing the properties of an element

atomic mass
the mass of a particular atom of an element

atomic number
the number of protons in an atom; this value distinguishes atoms of different elements

atomic weight
the average mass of an atom of an element as found in nature

average
See mean

Avogadro's law
equal volumes of all gases, measured at the same temperature and pressure, contain the same number of gas molecules

Avogadro's number
6.02×10^{23} particles, which represents one mole of particles

B

background radiation
the relatively constant level of natural radioactivity that is always present

balanced chemical equation
See chemical equation

barometer
a device that measures atmospheric pressure

base unit
an SI unit that expresses a fundamental physical quantity (such as temperature, length, or mass)

bases
ions or compounds that produce OH^- ions when dissolved in water

basic solution
an aqueous solution with a pH > 7; the solution turns litmus from red to blue

batch process
a series of discrete steps or operations involved in producing a substance or other product; *see also* continuous process

battery
a device composed of one or more connected voltaic cells that supplies electrical current

beta decay
the radioactive decay of a nucleus accompanied by the emission of a beta particle

beta particles (β)
negatively charged particles emitted during the decay of some radioactive elements; composed of high-speed electrons

biodiesel
an alternative fuel or fuel additive for diesel engines made from various materials such as new or recycled vegetable oils and animal fats

biomolecules
large organic molecules found in living systems

boiling point
the temperature at which a liquid boils at normal atmospheric pressure

bond, chemical
See chemical bond

Boyle's law
the pressure and volume of a gas sample at constant temperature are inversely proportional; $PV = k$

branched-chain alkane
an alkane in which at least one carbon atom is bonded to three or four other carbon atoms

branched polymer
a polymer formed by reactions that create numerous side chains rather than linear chains

brittle
a property of a material that causes it to shatter under pressure

buffer
a substance or combination of dissolved substances capable of resisting changes in pH when limited quantities of either acid or base are added

C

Calorie (Cal)
a unit of energy; thermal energy required to raise the temperature of one kilogram of water by one degree Celsius; commonly used to express quantity of food energy; informally called *food calorie*; 1 Cal = 1000 cal; 1 kJ = 4.184 Cal

calorie (cal)
a unit of energy; thermal energy required to raise the temperature of one gram of water by one degree Celsius; 1 J = 4.184 cal

carbohydrates
substances such as sugars and starches that are composed of carbon, hydrogen, and oxygen atoms; a main source of energy in foods

carbon chain
carbon atoms chemically linked to one another, forming a chainlike molecular structure

carbon cycle
the movement of carbon atoms within Earth's ecosystems, from carbon storage as plant and animal matter, through release as carbon dioxide due to cellular respiration, combustion, and decay, to reacquisition by plants

carboxylic acid
an organic compound containing the —COOH group

carcinogens
substances known to cause cancer

catalyst
a substance that speeds up a chemical reaction but is itself unchanged

catalytic converter
the reaction chamber in an auto exhaust system designed to accelerate the conversion of potentially harmful exhaust gases to nitrogen gas, carbon dioxide, and water vapor

cathode
an electrode in an electrochemical cell at which reduction occurs

cathode ray
a beam of electrons emitted from a cathode when electricity is passed through an evacuated tube

cation
a positively charged ion

Celsius temperature scale (°C)
SI temperature scale; originally based on the freezing point of water (0 °C) and its normal boiling point (100 °C)

CFCs
See chlorofluorocarbons (CFCs)

chain reaction
in nuclear fission, a reaction that is sustained because it produces enough neutrons to collide with and split additional fissionable nuclei

Charles' law
the volume of a gas sample at constant pressure is directly proportional to its kelvin temperature; $V = kT$

chemical bond
the attractive force that holds atoms or ions together; *see also* covalent bonds; ionic bonds

chemical change
an interaction of matter that results in the formation of one or more new substances

chemical energy
energy stored in the chemical bonds of substances

chemical equation
a symbolic expression summarizing a chemical reaction, such as $2 H_2(g) + O_2(g) \longrightarrow 2 H_2O(g)$

chemical formula
a symbolic expression representing the elements contained in a substance, together with subscripts that indicate the relative numbers of atoms of each element

chemical properties
properties only observed or measured by changing the chemical identity of a sample of matter

chemical reaction
the process of forming new substances from reactants that involves the breaking and forming of chemical bonds

chemical species
See species

chemical symbol
an abbreviation of an element's name, such as N for nitrogen or Fe for iron

chemistry
a branch of natural science dealing with the composition of matter and the structures, properties, and reactions of substances

chlorination
adding chlorine (as gas or hypochlorite) in water treatment to kill harmful microorganisms

chlorofluorocarbons (CFCs)
synthetic substances previously used as aerosol propellants, cooling fluids, and cleaning solvents; CFCs lead to stratospheric ozone destruction through production of chlorine radicals

cis–trans isomerism
isomers based on arrangement about a double bond in a molecule; the *cis-* isomer has attachments on the same side of the double bond, while the *trans-* isomer has attachments on opposite sides of double bond

cloud chamber
a container filled with supersaturated air that, when cooled and exposed to ionizing radiation, produces visible trails of condensation, tracing paths taken by radioactive emissions

CNG
See compressed natural gas (CNG)

coating
a treatment in which a material is physically or chemically attached to a product's surface, typically for protection

coefficients
the numbers in a chemical equation that indicate the relative numbers of units of each reactant or product involved in the reaction

coenzyme
an organic molecule that interacts with an enzyme to facilitate or enhance its activity

collision theory
for a reaction to occur, reactant molecules must collide in proper orientation with sufficient kinetic energy

colloid
a mixture containing solid particles small enough to remain suspended and not settle out

color standard
colored solutions of known concentration that allow estimating an unknown solution concentration based on color-intensity comparisons to the color standards

colorimeter
an instrument for determining the concentration of a colored solution based on color-intensity measurements

colorimetry
a chemical analysis method that uses color intensity to determine solution concentration

combustion
a chemical reaction with oxygen gas that produces thermal energy and light; burning

complementary proteins
multiple protein sources that provide adequate amounts of all essential amino acids when consumed together

complete protein
a protein source for humans containing adequate amounts of all essential amino acids

complex ion
a single central atom or ion, usually a metal ion, to which other atoms, molecules, or ions are attached

compound
a substance composed of two or more elements bonded together in fixed proportions; a compound cannot be broken down into simpler substances by physical means

compressed natural gas (CNG)
natural gas condensed under high pressures (160–240 atm) and stored in metal cylinders; CNG can serve as a substitute for gasoline or diesel fuel

concentration
See solution concentration

condensation
converting a substance from its gaseous state to liquid or solid state

condensation polymer
a polymer formed by repeated condensation reactions of one or more monomers

condensation reaction
the chemical combination of two organic molecules, accompanied by the loss of water or other small molecule

condensed formula
a chemical formula that provides additional information about bonding; for example, the condensed formula for propane, C_3H_8, is $CH_3-CH_2-CH_3$ or $CH_3CH_2CH_3$

conductor
a material that allows electricity (or thermal energy) to flow through it

confirming test
a laboratory test giving a positive result if a particular chemical species is present

continuous process
an uninterrupted sequence or flow of operations involved in producing a substance or other product; *see also* batch process

control
in an experiment, a trial that duplicates all conditions except for the variable under investigation

conversion factor
a factor by which a quantity expressed in one set of units is multiplied to convert it to another set of units

corrosion
a chemical action causing the gradual destruction of a metal's surface through oxidation–reduction processes

covalent bond
a linkage between two atoms involving the sharing of one pair (single bond), two pairs (double bond), or three pairs (triple bond) of electrons

cracking
the process by which hydrocarbon molecules from petroleum are converted to smaller molecules, using thermal energy and a catalyst

critical mass
the minimum mass of fissionable material needed to sustain a nuclear chain reaction

cross-linking
polymer chains interconnected by chemical bonds; causes polymer rigidity

crude oil
unrefined liquid petroleum as it is pumped from the ground by oil wells

curie (Ci)
a formerly used expression of a sample's radioactive intensity, equal to radiation emitted by one gram of radium-226, or 37 billion atoms undergoing radioactive decay per second; the corresponding SI unit is the becquerel (Bq), where $1\ Bq = 2.703 \times 10^{-11}\ Ci$

cycloalkane
a saturated hydrocarbon containing carbon atoms joined in a ring

D

data
objective pieces of information, such as information gathered in a laboratory investigation

decant
to pour slowly from the top of a liquid, leaving sediment or other solid material behind in the container

decay
See radioactive decay

decay product
isotopes formed from radioactive decay

decay series
the decay of a particular radioisotope, which yields a different radioisotope that, in turn, decays; this sequential decay process may continue through several more radioisotopes until a stable nucleus is produced

degrees Celsius (°C)
See Celsius temperature scale

density
the mass per unit volume of a given material that is often expressed as g/cm^3

derived unit
an SI unit formed by mathematically combining two or more base units

detergent
a compound or mixture of compounds to assist cleaning that does not form insoluble precipitates with Mg^{2+} or Ca^{2+} as soaps do

diatomic molecule
a molecule made up of two atoms, such as chlorine gas, Cl_2, or carbon monoxide, CO

dimer
a molecule composed of two monomers

dipeptide
a molecule consisting of two amino acids bonded together

direct water use
water consumed by an end user

disaccharide
a sugar molecule (such as sucrose) composed of two monosaccharide units bonded through a condensation reaction

dissolve
the process of a solute interacting with a solvent to form a solution

distillate
the condensed products of distillation

distillation
a process that separates liquid substances based on differences in their boiling points; *see also* fractional distillation

doping
adding impurities to a semiconductor to enhance its electrical conductivity

dot structure
See electron-dot structure

double covalent bond
a bond in which four electrons are shared between two adjacent atoms

ductile (ductility)
a property of a material that permits it to be stretched into a wire without breaking

dynamic equilibrium
see equilibrium

E

electric current
the flow of electrons, as through a wire connecting electrodes in a voltaic cell

electrical conductivity
the ability to transmit an electric current

electrical potential
the tendency for electrical charge to move through an electrochemical cell (based on an element's relative tendency to lose electrons when in contact with a solution of its ions); it is measured in volts (V)

electrochemistry
the study of chemical changes that produce or are caused by electrical energy

electrode
a strip of metal or other conductor serving as a contact between an ionic solution and the external circuit in an electrochemical cell

electrolysis
the process by which a chemical reaction is caused by passing an electrical current through an ionic solution

electrolyte
an ionic solution; dissolved ions allow the passage of an electrical charge through a solution

electromagnetic radiation
radiation ranging from low-energy radio waves to high-energy X-rays and gamma rays; includes visible light

electromagnetic spectrum
comprising the full range of electromagnetic radiation frequencies; *see also* electromagnetic radiation

electrometallurgy
the use of electrical energy to add electrons to metal ions, thus reducing them

electron
a particle possessing a negative electrical charge; electrons surround the nuclei of atoms

electron shell
See shells (electron)

electron-dot formula
See Lewis dot structure

electron-dot structure
See Lewis dot structure

electronegativity
an expression of the tendency of an atom to attract shared electrons within a chemical bond

electroplating
the deposition of a thin layer of metal on a surface by an electrical process involving oxidation-reduction reactions

electrostatic precipitation
a pollution-control method in which combustion waste products are electrically charged and then collected on plates of opposite electric charge

elements
the fundamental chemical substances from which all other substances are made

endothermic
a process that requires the addition of energy

endpoint
point where a titration is stopped, usually because an appropriate indicator just changes color

engineered ceramics
ceramic-based materials with properties designed for particular applications, such as high-temperature stability, wear resistance, electrical insulation, corrosion resistance, high strength and low weight, and thermal insulation

enzymes
biological catalysts present in all cells

equilibrium
the point in a reversible reaction where the rate of products forming from reactants is equal to the rate of reactants forming from products; also called dynamic equilibrium

essential amino acids
amino acids not synthesized in adequate quantities by the human body and that must be obtained from protein in the diet

ester
an organic compound containing the -COOR group, where R represents any stable arrangement of bonded carbon and hydrogen atoms

evaporation
changing from the liquid state to gaseous state

exothermic
a process that involves the release of energy

extrapolation
the process of estimating a value beyond a known range of data points

F

family (periodic table)
See group

fats
energy-storage molecules composed of carbon, hydrogen, and oxygen; they have fewer oxygen atoms and more carbon and hydrogen atoms than do carbohydrates

fatty acids
organic compounds made up of long hydrocarbon chains with carboxylic acid groups at one end

filtrate
the liquid collected after filtration

filtration
the process of separating solid particles from a liquid by passing the mixture through a material that retains the solid particles

fission, nuclear
See nuclear fission

flocculation
a process in which an insoluble material is suspended in or precipitated from a solution; used in water purification

fluorescence
the emission of visible light when exposed to radiant energy (usually ultraviolet radiation)

fluoridation
adding small quantities of fluoride ion (F^-) to treated water supplies with the intended purpose of promoting dental health

food additive
a substance added during food processing that is intended to increase nutritive value or enhance storage life, visual appeal, or ease of production

force
a push or pull exerted on an object; expressed by the newton (N), an SI unit

formula unit
a group of atoms or ions represented by a compound's chemical formula; simplest unit of an ionic compound

fossil fuel
a fuel (such as coal, petroleum, or natural gas) believed formed from plant or animal remains buried under Earth's surface for millions of years

fraction
(a) a mixture of petroleum-based substances with similar boiling points and other properties; (b) one of the distillate portions collected during distillation

fractional distillation
a process of separating a mixture into its components by boiling and condensing the components; *see also* distillation

free radicals
atoms or molecules with unstable arrangements and numbers of electrons

freezing point
the temperature at which a liquid turns into a solid; it is the same value as its melting point

frequency (ν)
the number of waves that pass a given point each second; in other words, the rate of oscillation; for electromagnetic radiation, the product of frequency and wavelength equals the speed of light

fuel cell
a device for directly converting chemical energy into electrical energy by chemically combining a fuel (such as hydrogen gas) with oxygen gas; does not involve combustion

functional group
an atom or a group of atoms that imparts characteristic properties to an organic compound

fusion, nuclear
See nuclear fusion

G

gamma decay
the radioactive decay of a nucleus accompanied by the emission of high-energy gamma rays

gamma radiation
high-energy electromagnetic radiation (gamma rays) emitted during the decay of some radioactive elements

gamma ray (γ)
high-energy electromagnetic radiation emitted during the decay of some radioactive elements

gas constant (R)
the constant used in the ideal gas law, $PV = nRT$; $R = 0.0821$ L·atm/(mol·K)

gaseous state
the state of matter having no fixed volume or shape

geothermal energy
energy extracted from hot water or steam from Earth's crust

gray (Gy)
the unit expressing the quantity of ionizing radiation delivered to tissue; it equals one joule absorbed per kilogram of body tissue

Green Chemistry
the design of chemical products and processes that require fewer resources and less energy and that reduces or eliminates reliance on and generation of hazardous substances

greenhouse effect
the trapping and returning of infrared radiation to Earth's surface by atmospheric substances such as water and carbon dioxide

greenhouse gases
atmospheric substances that absorb infrared radiation, such as CO_2, N_2O, and CH_4

groundwater
water from an aquifer or other underground source

group (periodic table)
a vertical column of elements in the periodic table; also called a family; group members share similar properties

H

Haber-Bosch process
the industrial synthesis of ammonia from hydrogen and nitrogen gases, which involves high pressure and temperature accompanied by a suitable catalyst

half-cell
a metal (or other electrode material) in contact with a solution of ions; forms one-half of a voltaic cell

half-life
the time for a radioactive substance to lose half of its radioactivity from decay; at the end of one half-life, 50% of the original radioisotope remains

half-reaction
a chemical equation explicitly showing either a loss of electrons (oxidation) or a gain of electrons (reduction); any oxidation–reduction reaction can be expressed as the sum of two half-reactions

halogen family
the group of elements consisting of fluorine, chlorine, bromine, iodine, and astatine

hard water
water containing high concentrations of calcium (Ca^{2+}), magnesium (Mg^{2+}), and/or iron(III) (Fe^{3+}) ions

heat
See thermal energy

heat capacity
See specific heat capacity

heat of combustion
the thermal energy released when a specific quantity of a material burns

heavy metals
metallic elements with high atomic weights, such as lead, mercury, and cadmium

heavy-metal ions
ions of heavy metals; these tend to be harmful to living matter

heterogeneous mixture
a mixture that is not uniform throughout

high-level nuclear waste
radioactive waste products associated with nuclear fission or products formed when neutrons are absorbed by nuclear fuel

histogram
a graph indicating the frequency or number of instances of particular values (or value ranges) within a set of related data

homogeneous mixture
a mixture that is uniform throughout; in other words, a solution

hybrid vehicle
a vehicle that combines two or more power sources; the combination of gasoline and electric power is the most common design

hydrocarbon
a molecule composed only of carbon and hydrogen atoms

hydrogenation
a chemical reaction that adds hydrogen atoms to an organic molecule

hydrologic cycle
See water cycle

hydrometallurgy
the industrial extraction of metals from ores at ordinary temperatures by water-based processes

hydrosphere
all parts of Earth where water is found, including oceans, clouds, ice caps, glaciers, lakes, rivers, and underground water supplies

I

ideal gas
a gas sample that behaves under all conditions as described by the kinetic molecular theory or by the ideal gas law

ideal gas law
the mathematical relationship that describes the behavior of an ideal gas sample, $PV = nRT$, where P = gas pressure; V = gas volume; n = moles of gas; R = gas constant, 0.0821 L·atm/(mol·K); and T = kelvin temperature

incomplete combustion
the partial burning of a fuel; the incomplete combustion of carbon, for example, produces carbon monoxide rather than carbon dioxide

indirect water use
water consumed in the preparation, production, or delivery of goods and services

infrared (IR) radiation
electromagnetic radiation just beyond the red (low-energy) end of the visible spectrum

insoluble
cannot dissolve in a particular solvent to any appreciable or directly observable extent

insulator
a material that does not conduct heat or electricity

intermolecular forces
forces of attraction among molecules

International System of Units (SI)
the modernized metric system

ions
electrically charged atoms or groups of atoms; negative ions are called anions, and positive ions are called cations

ion exchange
a process for softening water in which hard-water ions (such as Ca^{2+} and Mg^{2+}) are exchanged for other ions (such as Na^+)

ionic bond
a linkage among anions and cations involving attractions between opposite electrical charges; bonding among anions and cations produces an ionic solid

ionic compound
a substance composed of positive and negative ions

ionize
to become converted to ions

ionizing radiation
nuclear radiation and high-energy electromagnetic radiation with sufficient energy to produce ions by ejecting electrons from atoms and molecules

IR
See infrared (IR) radiation

irradiation
exposure to radiation of any kind, particularly to ionizing radiation

isomer
a molecule that has the same formula as another molecule and differs from it only by the arrangement of atoms or bonds

isomerization
a chemical change involving the rearrangement of atoms or bonds within a molecule without changing its molecular formula

isotopes
atoms of the same element with differing numbers of neutrons

IUPAC
the International Union of Pure and Applied Chemistry, a nongovernmental organization of worldwide chemistry societies; for example, IUPAC is the recognized authority for establishing standards and procedures for naming elements and other substances

K

kelvin temperature scale (K)
the absolute temperature scale, where zero kelvins (0 K) represents the theoretical lowest possible temperature; 0 °C = 273 K

kelvins (K)
See kelvin temperature scale (K)

kilopascal (kPa)
the SI pressure unit equal to 10^3 pascals; *see also* pascal (Pa)

kinetic energy
energy associated with the motion of an object

kinetic molecular theory (KMT)
observed gas behavior and gas laws explained by rapidly moving particles (gas molecules) that are relatively far apart and change direction only through collisions with each other or the container walls

kinetics
the study of reaction rates

L

law of conservation of energy
energy can change form but cannot be created or destroyed in any chemical reaction or physical change

law of conservation of matter
matter is neither created nor destroyed in any chemical reaction or physical change

Le Châtelier's principle
the predicted shift in the equilibrium position that partially counteracts the imposed change in conditions

Lewis dot structure
the representation of atoms, ions, and molecules where valence-electron dots surround each atom's symbol; it is useful for indicating covalent bonding; also called electron-dot structure

limiting reactant
the starting substance that is used up first in a chemical reaction; sometimes called the limiting reagent

liquid state
the state of matter with a fixed volume but no fixed shape

lithosphere
the solid outer layer of Earth, which also includes land areas under oceans and other bodies of water

low-level nuclear waste
radioactive waste products with relatively low levels of radioactivity, such as discarded protective clothing from nuclear laboratories

luster
the reflection of light from the surface of a material

M

macrominerals
minerals essential to human life and occurring in relatively large quantities (at least 5 g) in the body; also called major minerals

macroscopic
large enough to be seen by an unaided eye

magnetic resonance imaging (MRI)
a non-invasive computerized method of imaging soft human tissues by use of a powerful magnet and radio waves

major minerals
See macrominerals

malleable
the property of a material that permits it to be flattened without shattering

mass
the quantity of matter in a object; the kilogram (kg) is the SI base unit of mass

mass number
the sum of the number of protons and neutrons in the nucleus of an atom of a particular isotope

matter
anything that has mass and occupies space

mean
an expression of central tendency obtained by dividing the sum of a set of values by the number of values in the set; also known as the average value

mechanical filtering
a pollution-control method in which the combustion of waste products passes into filters that trap particles too large to pass through

median
within an ascending or descending set of values, the number that represents the middle value with an equal number of values above and below it

melting point
the temperature at which a solid turns into a liquid; it is the same value as its freezing point

mesosphere
the layer of the atmosphere directly above the stratosphere

metal
a material possessing properties such as luster, ductility, conductivity, and malleability

metalloid
a material with properties intermediate between those of metals and nonmetals

meter (m)
the SI base unit of length

microminerals
minerals essential to human life and occurring in relatively small quantities (less than 5 g) in the body; also called trace minerals

microwaves
electromagnetic radiation located between ultrahigh frequency radio waves and infrared radiation

mineral
(a) a naturally occurring solid substance commonly removed from ores to obtain a particular element of interest or value; (b) an inorganic substance needed by the body to maintain good health

mixture
a combination of materials in which each material retains its separate identity

models
tools to understand and interpret natural phenomena that can be physical, mathematical, or conceptual and represent, explain, or predict observed behavior; models are particularly helpful in chemistry to account for molecular-level interactions

molar concentration (M)
See molarity

molar heat of combustion
the quantity of thermal energy released from burning one mole of a substance

molar mass
the mass (usually in grams) of one mole of a substance

molar volume
the volume occupied by one mole of a substance; at 0 °C and 1 atm, the molar volume of any ideal gas is 22.4 L

molarity (M)
the concentration determined by dividing the total moles of solute by the solution volume (expressed in liters); also known as molar concentration

mole (mol)
the SI unit for amount of substance, equal to 6.02×10^{23} units, where the unit may be any specified entity; it is the chemist's "counting" unit

molecule
the smallest particle of a substance retaining all the properties of that substance

molecular formula
a chemical expression indicating the total atoms of each element contained in one molecule of a particular substance

molecular substance
a substance composed of molecules, such as H_2O and CH_4

monomer
a compound whose molecules can react to form the repeating units of a polymer

monosaccharide
a simple sugar (such as glucose or fructose) that cannot be hydrolyzed to produce other sugars

monounsaturated fat
a fat molecule containing one carbon–carbon double bond

mutagen
a material that causes mutations in DNA

mutation
changes in the structure of DNA that may result in production of altered protein material

N

natural water
untreated water gathered from natural sources such as rivers, ponds, or wells

negative oxidation state
the negative number assigned to an atom in a compound when that atom has greater control of bonding electrons than the control exerted by one or more atoms to which it is bonded

neutral solution
a water solution in which H^+ (H_3O^+) and OH^- concentrations are equal; pH = 7 at 25 °C

neutralization
combining an acid and a base in amounts that result in the elimination of all excess acid or base

neutron
a particle without electrical charge; found in the nucleus of an atom

newton (N)
the SI unit of force that is roughly equal to the force exerted by a mass of 100 g at Earth's surface

nitrogen cycle
the movement of atmospheric nitrogen atoms through Earth's ecosystems via collection by bacteria, conversion into ammonia or ammonium ions, conversion into nitrate ions, uptake by plants, passage through the food chain, release as ammonia or ammonium ions, and conversion back into atmospheric nitrogen

nitrogen fixation
chemically converting atmospheric nitrogen gas into nitrogen compounds; this can be done by lightning, by the action of microorganisms, or by industrial means (*see* Haber-Bosch process)

noble gas family
an unreactive element belonging to the last (right-most) group on the periodic table

nonconductor
a material that does not allow electrical current (or thermal energy) to flow through it

nonideal gas
a gas sample that fails to behave as described by the kinetic molecular theory, or a gas sample whose behavior does not follow the ideal gas law; nonideal gas behavior is mainly observed under conditions of low gas temperature and/or high gas pressure

nonionizing radiation
electromagnetic radiation in the visible and lower-energy regions of the electromagnetic spectrum with insufficient energy to form ions when it transfers energy to matter

nonmetal
a material possessing properties such as brittleness, lack of luster, and nonconductivity; nonmetals are often insulators

nonpolar molecule
a molecule that has an even distribution of electrical charge with no regions of partial positive and negative charge

nonrenewable resource
a resource in limited supply that cannot be replenished by natural processes over the time frame of human experience

nuclear fission
splitting a nucleus into two smaller nuclei

nuclear fusion
the combination of two nuclei to form a new, more massive nucleus

nuclear power plant
a facility where thermal energy generated by the controlled fission of nuclear fuel drives a steam turbine, producing electrical power; in other words, a facility that converts nuclear energy into electricity

nuclear radiation
a form of ionizing radiation that results from changes in the nuclei of atoms

nucleus, atomic
the dense, positively charged central region of an atom that contains protons and neutrons

O

octane rating
a measure of the combustion quality of gasoline compared to the combustion quality of isooctane; the higher the number, the higher the octane rating; also called octane number

OIL RIG
a mnemonic device for remembering the processes of *oxidation* and *reduction*: Oxidation Is Loss, Reduction Is Gain (of electrons)

oil shale
sedimentary rock containing a material (kerogen) that can be converted to crude oil

ore
a rock or other solid material from which it is profitable to recover a mineral containing a metal or other useful substances

organic chemistry
the branch of chemistry dealing with hydrocarbons and their derivatives

osmosis
the passage of a solvent from a dilute solution to a more concentrated one through a semipermeable membrane that allows the passage of solvent molecules (such as water) but not solute particles (ions or molecules)

oxidation
any process in which one or more electrons can be considered as lost by a chemical species

oxidation state
the apparent state of oxidation of an atom; also called oxidation number

oxidation–reduction (redox) reaction
a chemical reaction in which oxidation and reduction simultaneously occur

oxidized
see oxidation

oxidizing agent
a species that causes another atom, molecule, or ion to become oxidized; the oxidizing agent becomes reduced in this process

oxygenated fuel
a fuel with oxygen-containing additives, such as methanol, that increase the octane rating and reduce harmful emissions

ozone hole
a region of significant thinning of the stratospheric ozone layer; it occurs seasonally, most prominently over the South Pole

ozone shield
the protective layer of stratospheric ozone that absorbs intense solar ultraviolet radiation that is harmful to living organisms

P

paper chromatography
a method for separating substances that relies on solution components having different attractions to the solvent (mobile phase) and paper (stationary phase)

particle
entities such as alpha particles, atoms, ions, or molecules; *see also* particulates

particulate level
the realm of unseen atoms, molecules, and ions in contrast to observable macroscopic entities

particulate pollution
air contaminants made up of small particles suspended in the atmosphere

particulates
solid particles such as dust, mist, fumes, soot, and smoke that do not settle, but remain in the air; generated by human activities or natural processes

parts per billion (ppb)
an expression of concentration; the number of units of solute found in one billion units of solution

parts per million (ppm)
an expression of concentration; the number of units of solute found in one million units of solution

pascal (Pa)
the SI pressure unit; equal to one newton of force applied per square meter, $1 \text{ Pa} = 1 \text{ N/m}^2$

peptide bond
the chemical bond that links amino acids together in peptides and proteins

percent composition
the percent by mass of each component in a material; or, specifically, the percent by mass of each element within a compound

percent recovery
the proportion of sought material recovered in a process

period (periodic table)
a horizontal row of elements in the periodic table

periodic relationship
regular patterns among chemical and physical properties of elements arrayed in the periodic table

periodic table of the elements
an arrangement of elements in order of increasing atomic number, such that elements with similar properties are located in the same vertical column (group)

petrochemical
any organic compound produced from petroleum or natural gas

pH
an expression of the acidic or basic character of a solution; at 25 °C, solutions with pH values lower than 7 are acidic, solutions above 7 are basic, and neutral solutions equal 7; pH is based on a solution's hydrogen ion (H^+ or H_3O^+) concentration

photochemical smog
a potentially hazardous mixture of secondary pollutants formed by solar irradiation of certain primary pollutants in the presence of oxygen

photon
an energy bundle of electromagnetic radiation that travels at the speed of light

photosynthesis
the process by which green plants and some microorganisms use solar energy to convert water and carbon dioxide to carbohydrates (stored chemical energy)

physical change
a change in matter in which the identity of the material involved does not change

physical property
a property that can be observed or measured without changing the identity of the sample of matter

polar bond
a covalent bond where electrons are shared unequally due to an electronegativity difference in the bonded atoms

polar molecule
a molecule with regions of partial positive and negative charge resulting from the uneven distribution of electrical charge

pollutant
an undesirable contaminant that adversely affects the chemical, physical, or biological characteristics of the environment

polyatomic ion
an ion composed of two or more atoms, such as the ammonium cation, NH_4^+, or acetate anion, $C_2H_3O_2^-$

polymer
a molecule composed of very large numbers of identical repeating units

polysaccharide
a polymer composed of many monosaccharide units

polyunsaturated fat
a fat molecule containing two or more carbon–carbon double bonds

positive oxidation state
the positive number assigned to an atom in a compound when that atom has less control of its electrons than it has as a free element

positron
a positively charged subatomic particle with the same mass as an electron; the antimatter counterpart to the electron

positron emission tomography (PET)
a technique for examining metabolic activity in tissues (particularly in the brain) by measuring blood flow containing tracers that emit positrons

postchlorination
adding a low level of chlorine to treated water to kill and prevent later bacterial infestation

potable water
water that is safe to drink

potential energy
energy associated with position

prechlorination
a process in which chlorine is added early in water treatment to kill bacterial infestation

precipitate
an insoluble solid substance that has separated from a solution

precipitation
(a) forming a chemical precipitate; (b) moisture falling from the atmosphere as rain, snow, sleet, or hail

pressure
equals force applied per unit area; in SI, pressure is expressed in pascals (Pa)

primary air pollutant
a contaminant that directly enters the atmosphere; it is not initially formed by reactions of airborne substances

products
substances formed in a chemical reaction

protein
a major structural component of living tissue made from many linked amino acids

proton
a particle possessing a positive electrical charge that is found in the nuclei of all atoms; the total protons in an element's atom equals its atomic number

pure substance
See substance

pyrometallurgy
the use of high temperatures (thermal energy) to process metals or their ores

Q

qualitative test
a chemical test indicating the presence or absence of an element, ion, or compound in a sample

quantitative test
a chemical test indicating the amount or concentration of an element, ion, or compound in a sample

R

rad
a unit that expresses the quantity of ionizing radiation absorbed by tissue; 1 rad = 0.01 Gy

radiation
energy emitted in the form of electromagnetic waves or high-speed particles; refers to both ionizing radiation and nonionizing radiation

radioactive decay
a change in an atom's nucleus due to the spontaneous emission of alpha, beta, or gamma radiation

radioactivity
the spontaneous emission of nuclear radiation

radioisotopes
radioactive isotopes

range
the difference between the highest and lowest values in a data set

reactants
starting materials in a chemical reaction

reaction rate
an expression of how fast a particular chemical change occurs

redox reaction
See oxidation–reduction reaction

reduced
See reduction

reducing agent
a species that causes another atom, molecule, or ion to become reduced; the reducing agent, in turn, becomes oxidized in this process

reduction
any process in which one or more electrons can be considered as gained by a chemical species

reference solution
a solution of known composition used as a comparison in chemical tests

reflectivity
the proportion of radiation that a material reflects rather than absorbs

rem
a unit that expresses the ability of radiation to cause ionization in human tissue; 1 rem = 0.01 Sv

renewable resource
a resource that can be replenished by natural processes over the time frame of human experience

reverse osmosis
a water-treatment process involving the separation of dissolved ions by forcing water through a semipermeable membrane under high pressure

reversible reaction
a chemical reaction in which products form reactants at the same time that reactants form products

R_f value
the ratio of the distance a given substance moves to the distance the solvent moves in paper chromatography

S

salt bridge
a connection that allows a voltaic cell's two half-cells to be in electrical contact without mixing; specifically, a tube containing an electrolyte (such as potassium chloride solution) that completes the internal circuit of a voltaic cell

sand filtration
separating solid particles from a liquid by passing the mixture through sand

saturated fat
a fat molecule containing only single carbon–carbon bonds within its fatty acid components

saturated hydrocarbon
a hydrocarbon consisting of molecules in which each carbon atom is bonded to four other atoms

saturated solution
a solution in which the solvent has dissolved as much solute as it can retain stably at a specified temperature

scintillation counter
a detector of ionizing radiation that measures light emitted by atoms that have been excited by ionizing radiation

scrubbing, wet
a pollution-control method involving an aqueous solution that removes particles and sulfur oxides from industrial combustion processes

secondary air pollutant
a contaminant generated in the atmosphere by chemical reactions between primary air pollutants and natural components of air

semiconductor
a solid substance, such as silicon, with electrical conductivity between that of conductors and nonconductors at normal temperatures

shells (electron)
energy levels surrounding an atom's nucleus within which one or more electrons reside; outer-shell electrons are commonly called valence electrons

SI
See International System of Units (SI)

sievert (Sv)
an SI unit that expresses the dose equivalent of absorbed radiation that causes same biological effects as one gray of gamma rays

single covalent bond
a bond in which two electrons are shared by the two bonded atoms

smog
a potentially hazardous combination of smoke and fog; *see also* photochemical smog

soft water
water that is not hard; it contains relatively low concentrations of calcium (Ca^{2+}), magnesium (Mg^{2+}), and iron(III) (Fe^{3+}) ions

solid state
the state of matter having a fixed volume and fixed shape

solid-state detector
a device used to monitor changes in the movement of electrons through semiconductors as they are exposed to ionizing radiation

solubility
the quantity of a substance that will dissolve in a given quantity of solvent to form a saturated solution at a particular temperature

solubility curve
a graph indicating the solubility of a particular solute at different temperatures

solute
the dissolved species in a solution; the solute is usually the smaller component in a solution.

solution
a homogeneous mixture of two or more substances

solution concentration
the quantity of solute dissolved in a specific quantity of solvent or solution

solvent
the dissolving agent in a solution; the solvent is usually the larger component in a solution.

species
a general name used in chemistry for atoms, molecules, ions, free radicals, or other well-defined entities

specific heat capacity
the quantity of thermal energy needed to raise the temperature of 1 g of a material by 1 °C; the expression commonly has units of J/(g °C)

standard temperature and pressure (STP)
conditions of 0 °C and 1 atm

states
see gaseous state, liquid state, and solid state

stoichiometry
the relationships by which quantities of substances involved in a chemical reaction are linked and calculated

STP
See standard temperature and pressure

straight-chain alkane
an alkane consisting of molecules in which each carbon atom is linked to no more than two other carbon atoms

stratosphere
the layer of the atmosphere just above the troposphere

strong acid
an acid that fully ionizes in solution to liberate H^+ (H_3O^+); no molecular form of the acid remains

strong base
a base that fully liberates OH^- in solution

strong force
the force that holds protons and neutrons together in an atom's nucleus

structural formula
a chemical formula showing the arrangement of atoms and covalent bonds in a molecule, in which each electron pair in a covalent bond is represented by a line between the symbols of two atoms

structural isomers
substances involving rearrangement of atoms or bonds within a molecule but sharing a common molecular formula

subatomic particles
particles smaller than an atom; commonly regarded as electrons, protons, and neutrons

subscript
the number printed below the line of type indicating the total atoms of a given element in a chemical formula; in H_2O, for example, the subscript 2 specifies total H atoms

substance
an element or a compound; that is, a material with a uniform, definite composition and distinct properties

substituted alkenes
alkene molecules that contain, in addition to carbon and hydrogen atoms, one or more atoms of other elements, such as oxygen, nitrogen, chlorine, or sulfur

substrate
a molecule that interacts with an enzyme and undergoes a reaction

superconductivity
the ability of a material to conduct an electrical current with zero electrical resistance; with present technology, operating superconductors must be extremely cold

supersaturated solution
a solution containing a higher concentration of solute than a saturated solution at a specified temperature

surface water
water found on Earth's surface, such as oceans, rivers, and lakes

suspension
a mixture containing large, dispersed solid particles that can settle out or be separated by filtration

synergistic interaction
an interaction where the combined effect of several factors is greater than the sum of their separate effects

synthetic radioisotope
a radioactive isotope produced solely by human activity

synthetic substance
a substance produced solely by human activity

T

tap water
water delivered through plumbing lines

temperature
extent of "hotness" or "coldness" of a sample and related to the average kinetic energy of its particles; the SI base unit is the kelvin (K) or degree Celsius (°C)

temperature inversion
an atmospheric condition where a cool air mass is trapped beneath a less-dense warm air mass; it most frequently occurs in a valley or over a city

tetrahedron
a regular triangular pyramid; the four bonds of each carbon atom in an alkane point to the corners of a tetrahedron

thermal energy
the energy a material possesses due to its temperature; also informally called heat

thermoplastics
plastics that soften or melt when heated, and return to their rigid form when cooled, such as nylon, polyvinyl chloride, and polyethylene

thermosphere
the highest layer of the atmosphere

titration
a laboratory procedure for determining the concentrations of dissolved substances

trace minerals
See microminerals

tracer
a readily-identifiable material, such as a radioisotope, used to diagnose disease or to determine how the body is responding to treatment

transmutation
the conversion of one element to another either naturally or artificially

transuranium element
any element with an atomic number greater than 92 (uranium)

triglyceride
a fat molecule composed of a simple three-carbon alcohol (glycerol) and three fatty acid molecules

tripeptide
a molecule composed of three amino acids bonded together

troposphere
the layer of the atmosphere closest to Earth's surface where most clouds and weather are located

turbidity
cloudy conditions caused by suspended solids in a liquid

Tyndall effect
the scattering of a beam of light caused by reflection from suspended particles

U

ultraviolet (UV) radiation
electromagnetic radiation beyond the violet end of the visible spectrum; overexposure to this radiation can cause skin damage

unsaturated fat
a fat molecule containing one or more carbon–carbon double bonds; it is monounsaturated if each fat molecule has a single double bond and polyunsaturated if it has two or more double bonds

unsaturated hydrocarbon
a hydrocarbon molecule containing one or more double or triple bonds

unsaturated solution
a solution containing a lower concentration of solute than a saturated solution contains at a specified temperature

V

valence electrons
electrons in the outermost shell of an atom; these relatively loosely held electrons often participate in bonding with other atoms or molecules

viscous
pertaining to "thick" liquid with resistance to flow; high viscosity

visible radiation
electromagnetic radiation that the human eye can detect; visible radiation has wavelengths from about 400 nm (violet) to 700 nm (red)

vitamin

a biomolecule necessary for growth, reproduction, health, and life

vitrification

the conversion of material into a glassy solid by application of high temperatures

volatile organic compounds (VOCs)

reactive carbon-containing substances that readily evaporate into air, such as components of gasoline and organic solvents

voltaic cell

an electrochemical cell in which a spontaneous chemical reaction produces electricity

W

water cycle

repetitive processes of rainfall (or other precipitation), run-off, evaporation, and condensation that circulate water within Earth's crust and atmosphere; also called the hydrologic cycle

water softener

a device containing an ion-exchange resin that is used to treat water by removing ions causing water hardness

water treatment

refers to cleaning and purification processes applied to water

wavelength (λ)

the distance between corresponding points of two consecutive waves; for electromagnetic radiation, the product of frequency and wavelength equals the speed of light

weak acid

an acid that does not fully ionize in solution to liberate H^+ (H_3O^+) but remains primarily in molecular form

weak base

a base that does not fully liberate OH^- in solution but remains primarily in molecular form

wellplate

a clear, plastic laboratory device containing multiple indentations (wells), where small samples of solutions can be held and tested or chemical reactions studied

wet scrubbing

See scrubbing, wet

X

X-ray

high-energy electromagnetic radiation that cannot penetrate dense materials such as bone or lead but can penetrate less dense materials

Z

zero oxidation state

the value of the oxidation state of an element's atoms when not chemically combined with any other element

ILLUSTRATION AND PHOTO

CHEMCOM PROJECT CREDITS

ChemCom is the product of teamwork involving individuals from all over the United States over more than twenty years. The American Chemical Society is pleased to recognize all who contributed to *ChemCom*.

The team responsible for the fifth edition of *ChemCom* is listed on the copyright page. Individuals who contributed to the initial development of *ChemCom*, to the first edition in 1988, the second edition in 1993, the third edition in 1998, and the fourth edition 2002 are listed below.

Principal Investigator: W. T. Lippincott

Project Manager: Sylvia Ware

Chief Editor: Henry Heikkinen & Conrad L. Stanitski

Contributing Editor: Mary Castellion

Assistant to Contributing Editor: Arnold Diamond

Editor of Teacher's Guide: Thomas O'Brien & Patricia J. Smith

Revision Team: Diane Bunce, Gregory Crosby, David Holzman, Thomas O'Brien, Joan Senyk, Thomas Wysocki

Editorial Advisory Board: Joseph Breen, Glenn Crosby, James DeRose, I. Dwaine Eubanks, Lucy Pryde Eubanks, Regis Goode, Henry Heikkinen (chair), Mary Kochansky, Ivan Legg, W. T. Lippincott (ex officio), Steven Long, Nina McClelland, Lucy McCorkle, Carlo Parravano, Robert Patrizi, Max Rodel, K. Michael Shea, Patricia Smith, Susan Snyder, Conrad Stanitski, Jeanne Vaughn, Sylvia Ware (ex officio)

Writing Team: Rosa Balaco, James Banks, Joan Beardsley, William Bleam, Kenneth Brody, Ronald Brown, Diane Bunce, Becky Chambers, Alan DeGennaro, Patricia Eckfeldt, Dwaine Eubanks (dir.), Henry Heikkinen (dir.), Bruce Jarvis (dir.), Dan Kallus, Jerry Kent, Grace McGuffie, David Newton (dir.), Thomas O'Brien, Andrew Pogan, David Robson, Amado Sandoval, Joseph Schmuckler (dir.), Richard Shelly, Patricia Smith, Tamar Susskind, Joseph Tarello, Thomas Warren, Robert Wistort, Thomas Wysocki

Steering Committee: Alan Cairncross, William Cook, Derek Davenport, James DeRose, Anna Harrison (ch.), W. T. Lippincott (ex officio), Lucy McCorkle, Donald McCurdy, William Mooney, Moses Passer, Martha Sager, Glenn Seaborg, John Truxall, Jeanne Vaughn

Consultants: Alan Cairncross, Michael Doyle, Donald Fenton, Conrad Fernelius, Victor Fratalli, Peter Girardot, Glen Gordon, Dudley Herron, John Hill, Chester Holmlund, John Holman, Kenneth Kolb, E. N. Kresge, David Lavallee, Charles Lewis, Wayne Marchant, Joseph Moore, Richard Millis, Kenneth Mossman, Herschel Porter, Glenn Seaborg, Victor Viola, William West, John Whitaker

Synthesis Committee: Diane Bunce, Dwaine Eubanks, Anna Harrison, Henry Heikkinen, John Hill, Stanley Kirschner, W. T. Lippincott (ex officio), Lucy McCorkle, Thomas O'Brien, Ronald Perkins, Sylvia Ware (ex officio), Thomas Wysocki

Evaluation Team: Ronald Anderson, Matthew Bruce, Frank Sutman (dir.)

Field Test Coordinator: Sylvia Ware

Field Test Workshops: Dwaine Eubanks

Field Test Directors: Keith Berry, Fitzgerald Bramwell, Mamie Moy, William Nevill, Michael Pavelich, Lucy Pryde, Conrad Stanitski

Pilot Test Teachers: Howard Baldwin, Donald Belanger, Navarro Bharat, Ellen Byrne, Eugene Cashour, Karen Cotter, Joseph Deangelis, Virginia Denney, Diane Doepken, Donald Fritz, Andrew Gettes, Mary Gromko, Robert Haigler, Anna Helms, Allen Hummel, Charlotte Hutton, Elaine Kilbourne, Joseph Linker, Larry Lipton, Grace McGuffie, Nancy Miller, Gloria Mumford, Beverly Nelson, Kathy Nirei, Elliott Nires, Polly Parent, Mary Parker, Dicie Petree, Ellen Pitts, Ruth Rand, Kathy Ravano, Steven Rischling, Charles Ross, Jr., David Roudebush, Joseph Rozaik, Susan Rutherland, George Smeller, Cheryl Snyder, Jade Snyder, Samuel Taylor, Ronald Tempest, Thomas Van Egeren, Gabrielle Vereecke, Howard White, Thomas Wysocki, Joseph Zisk

Field Test Teachers: Vincent Bono, Allison Booth, Naomi Brodsky, Mary D. Brower, Lydia Brown, George Bulovsky, Kay Burrough, Gene Cashour, Frank Cox, Bobbie Craven, Pat Criswell, Jim Davis, Nancy Dickman, Dave W. Gammon, Frank Gibson, Grace Giglio, Theodis Gorre, Margaret Guess, Yvette Hayes, Lu Hensen, Kenn Heydrick, Gary Hurst, Don Holderread, Michael Ironsmith, Lucy Jache, Larry Jerdal, Ed Johnson, Grant Johnson, Robert Kennedy, Anne Kenney, Joyce Knox, Leanne Kogler, Dave Kolquist, Sherman Kopelson, Jon Malmin, Douglas Mandt, Jay Maness, Patricia Martin, Mary Monroe, Mike Morris, Phyllis Murray, Silas Nelson, Larry Nelson, Bill Rademaker, Willie Reed, Jay Rubin, Bill Rudd, David Ruscus, Richard Scheele, Paul Shank, Dawn Smith, John Southworth, Mitzi Swift, Steve Ufer, Bob Van Zant, Daniel Vandercar, Bob Volzer, Terri Wahlberg, Tammy Weatherly, Lee Weaver, Joyce Willis, Belinda Wolfe

Field Test Schools: California: Chula Vista High, Chula Vista; Gompers Secondary School, San Diego; Montgomery High, San Diego; Point Loma High, San Diego; Serra Junior-Senior High, San Diego; Southwest High, San Diego. Colorado: Bear Creek Senior High, Lakewood; Evergreen Senior High, Evergreen; Green Mountain Senior High, Lakewood; Golden Senior High, Golden; Lakewood Senior High, Lakewood; Wheat Ridge Senior High, Wheat Ridge. Hawaii: University of Hawaii Laboratory School, Honolulu. Illinois: Project Individual Education High, Oak Lawn. Iowa: Linn-Mar High, Marion. Louisiana: Booker T. Washington High, Shreveport; Byrd High, Shreveport; Caddo Magnet High, Shreveport; Captain Shreve High, Shreveport; Fair Park High, Shreveport; Green Oaks High, Shreveport; Huntington High, Shreveport;

CHART OF THE ELEMENTS

Element	Symbol	Atomic Number	Atomic Weight	Element	Symbol	Atomic Number	Atomic Weight
Actinium	Ac	89	[227]	Mendelevium	Md	101	[258]
Aluminum	Al	13	26.98	Mercury	Hg	80	200.59
Americium	Am	95	[243]	Molybdenum	Mo	42	95.94
Antimony	Sb	51	121.76	Neodymium	Nd	60	144.24
Argon	Ar	18	39.95	Neon	Ne	10	20.18
Arsenic	As	33	74.92	Neptunium	Np	93	[237]
Astatine	At	85	[210]	Nickel	Ni	28	58.69
Barium	Ba	56	137.33	Niobium	Nb	41	92.91
Berkelium	Bk	97	[247]	Nitrogen	N	7	14.01
Beryllium	Be	4	9.012	Nobelium	No	102	[259]
Bismuth	Bi	83	208.98	Osmium	Os	76	190.23
Bohrium	Bh	107	[264]	Oxygen	O	8	16.00
Boron	B	5	10.81	Palladium	Pd	46	106.42
Bromine	Br	35	79.90	Phosphorus	P	15	30.97
Cadmium	Cd	48	112.41	Platinum	Pt	78	195.08
Calcium	Ca	20	40.08	Plutonium	Pu	94	[244]
Californium	Cf	98	[251]	Polonium	Po	84	[209]
Carbon	C	6	12.01	Potassium	K	19	39.10
Cerium	Ce	58	140.12	Praseodymium	Pr	59	140.91
Cesium	Cs	55	132.91	Promethium	Pm	61	[145]
Chlorine	Cl	17	35.45	Protactinium	Pa	91	231.04
Chromium	Cr	24	52.00	Radium	Ra	88	[226]
Cobalt	Co	27	58.93	Radon	Rn	86	[222]
Copper	Cu	29	63.55	Rhenium	Re	75	186.21
Curium	Cm	96	[247]	Rhodium	Rh	45	102.91
Darmstadtium	Ds	110	[281]	Roentgenium	Rg	111	[272]
Dubnium	Db	105	[262]	Rubidium	Rb	37	85.47
Dysprosium	Dy	66	162.50	Ruthenium	Ru	44	101.07
Einsteinium	Es	99	[252]	Rutherfordium	Rf	104	[261]
Erbium	Er	68	167.26	Samarium	Sm	62	150.36
Europium	Eu	63	151.96	Scandium	Sc	21	44.96
Fermium	Fm	100	[257]	Seaborgium	Sg	106	[266]
Fluorine	F	9	19.00	Selenium	Se	34	78.96
Francium	Fr	87	[223]	Silicon	Si	14	28.09
Gadolinium	Gd	64	157.25	Silver	Ag	47	107.87
Gallium	Ga	31	69.72	Sodium	Na	11	22.99
Germanium	Ge	32	72.64	Strontium	Sr	38	87.62
Gold	Au	79	196.97	Sulfur	S	16	32.07
Hafnium	Hf	72	178.49	Tantalum	Ta	73	180.95
Hassium	Hs	108	[277]	Technetium	Tc	43	[98]
Helium	He	2	4.003	Tellurium	Te	52	127.60
Holmium	Ho	67	164.93	Terbium	Tb	65	158.93
Hydrogen	H	1	1.008	Thallium	Tl	81	204.38
Indium	In	49	114.82	Thorium	Th	90	232.04
Iodine	I	53	126.90	Thulium	Tm	69	168.93
Iridium	Ir	77	192.22	Tin	Sn	50	118.71
Iron	Fe	26	55.85	Titanium	Ti	22	47.87
Krypton	Kr	36	83.80	Tungsten	W	74	183.84
Lanthanum	La	57	138.91	Uranium	U	92	238.03
Lawrencium	Lr	103	[262]	Vanadium	V	23	50.94
Lead	Pb	82	207.2	Xenon	Xe	54	131.29
Lithium	Li	3	6.941	Ytterbium	Yb	70	173.04
Lutetium	Lu	71	174.97	Yttrium	Y	39	88.91
Magnesium	Mg	12	24.31	Zinc	Zn	30	65.41
Manganese	Mn	25	54.94	Zirconium	Zr	40	91.22
Meitnerium	Mt	109	[268]				

Note: A value in square brackets is the mass number of the isotope with the longest half-life.